Springer Series in Optical Sciences   Volume 57

Edited by Theodor Tamir

Springer Series in Optical Sciences

Editorial Board: D.L.MacAdam A.L.Schawlow K.Shimoda A.E.Siegman T.Tamir

Managing Editor: H.K.V. Lotsch

Volumes 1–41 are listed on the back inside cover

E.-G. Neumann

Single-Mode Fibers

Fundamentals

With 105 Figures

Springer-Verlag Berlin Heidelberg GmbH

Professor Dr. ERNST-GEORG NEUMANN

Bergische Universität Gesamthochschule Wuppertal, Fachbereich Elektrotechnik,
Fuhlrottstrasse 10, D-5600 Wuppertal, Fed. Rep. of Germany

ISBN 978-3-662-13699-7   ISBN 978-3-540-48173-7 (eBook)
DOI 10.1007/978-3-540-48173-7

to Maike

Preface

Single-mode fibers are the most advanced means of transmitting information, since they provide extremely low attenuation and very high bandwidths. At present, long distance communication by single-mode fibers is cheaper than by conventional copper cables, and in the future single-mode fibers will also be used in the subscriber loop. Since single-mode fibers have many applications, a variety of people need to understand this modern transmission medium. However, waveguiding in single-mode fibers is much more difficult to understand than waveguiding in copper lines.

A single-mode fiber is a dielectric waveguide operated at optical wavelengths. Since 1961, I have been involved in experimental and theoretical research on dielectric rod waveguides in the microwave region. From the experiments, I learned much about the properties of a wave guided by a dielectric rod or a glass fiber, especially about its behavior at waveguide discontinuities like bends, gaps, or the waveguide end. Since 1972, my co-workers and I have also been investigating dielectric waveguides at optical frequencies, and since 1973 I have lectured on "Optical Communications". These activities have shown that there is a need for a tutorial introduction to the new technical field of single-mode fibers. In this book the physical fundamentals are emphasized and the mathematics is limited to the absolutely necessary subjects. Besides presenting a physical explanation of waveguiding in single-mode fibers, it is also the aim of this book to give an overview of the knowledge accumulated in this field. Many references are given in the text to the original papers in technical journals or conference proceedings.

E.-G. Neumann

Acknowledgements

In writing a book, the author needs the help and advice of many people. I gratefully acknowledge the help of colleagues both in university and in industry, of assistants, of co-workers, and of many students. Particular thanks are extended to Prof. Dr.-Ing. H. Chaloupka, Dr. rer. nat. F. Krahn, Dr.-Ing. D. Rittich, and Dipl.-Ing. J. Streckert.

Contents

1. Introduction

1.1 Historical Note

Single-mode fibers are dielectric waveguides for optical waves. Although dielectric waveguides have a history that goes back as far as 1910 [Hondros and Debye 1910], they have never been used for long distance communication at microwave frequencies because of the losses in the dielectric materials available.

Then, in 1966, Kao and Hockham [1966] proposed to use glass fibers as dielectric waveguides at optical frequencies for long distance communications. At that time, the attenuation of the best optical glasses available was of the order of 1000 dB/km. However, mainly by purifying the materials, it has been possible to reduce the losses by several orders of magnitude: In 1972, fibers with an attenuation below 20 dB/km were reported [Kapron et al. 1970a,b]. At present, fibers with an attenuation as low as 0.2 dB/km are commercially available, and a large number of lightwave communication systems are in use.

Because of the availability of semiconductor light sources and silicon photodetectors and the difficulties of jointing single-mode fibers, the first generation of optical communication systems used multimode graded-index fibers at wavelengths of about $0.85\,\mu\mathrm{m}$. For repeater distances of typically 10 km, they could transmit a binary signal with a maximum bit rate of about 100 Mb/s.

Since the signal attenuation decreases with wavelength, it is advantageous to use longer wavelengths [Kimura and Daikoku 1977]. Therefore, in the second generation of fiber optic systems, the wavelength was shifted to $1.3\,\mu\mathrm{m}$, but multimode graded index fibers were still being used. Because of the smaller attenuation of about 0.4 dB/km, link lengths of up to 50 km were possible; however, because of the spread in time delay for the several hundred modes propagating in multimode fibers, the maximum bit rate was of the order of 100 Mb/s.

The third generation of fiber optic systems offered higher transmission capacities by using single-mode fibers, while continuing to operate in the $1.3\,\mu\mathrm{m}$ band. Third generation systems operating at data rates of 565 Mb/s are presently being installed.

The loss minimum for silica fibers is at a wavelength of $1.55\,\mu\mathrm{m}$. The lowest attenutation ever reported is 0.154 dB/km at $1.55\,\mu\mathrm{m}$ [Kanamori et al. 1986]. Practical cables can have losses of 0.2 dB/km. Because of this low loss, the fourth generation, which is still in its research and development stage, will use single-mode fibers at this long wavelength, thus allowing repeater spacings of more than 100 km and bit rates of more than 1 Gb/s [Lilly and Walker 1984].

Additional future generations of optical communication systems will use coherent optical carriers and heterodyne receivers, or fibers made of new low-loss materials like heavy-metal fluoride glasses instead of silica glass.

In 1983, the first single-mode system has been introduced into commercial use and it is to be expected that, in the future, single-mode fibers will dominate in long distance communication. Presumably, single-mode fibers will be used also for shorter links, e.g. in local area networks (LAN's) [Rocher 1985; Cochrane et al. 1986] or subscriber loops [Kaiser 1985; Krumpholz 1985a; Krumpholz 1985b; Kaiser 1986]. First, these systems will operate with cheap LED-emitters [de Bortoli and Moncalvo 1986], but later, the transmission capacity can easily be upgraded by replacing the LED's with semiconductor lasers. Far in the future, further upgrading may be accomplished by introducing heterodyne receivers and frequency division multiplexing [Kaiser 1985; Khoe and Dieleman 1985; 1986].

The history of fiber-optic communications, in general, has been reviewed several times [Miller S.E. and Tillotson 1966; Kapany 1967; Miller et al. 1973a: Miller et al. 1973b; Clarricoats 1976; Börner 1980; Li 1983; Suematsu 1983; Kapron 1984a; Niizeki 1984; Kapron 1985; Henry 1985].

In contrast to this, the history of single-mode fibers has been described only a few times [Hooper et al. 1985; Seikai et al. 1985; Hooper 1986]. Optical dielectric-waveguide modes in glass fibers were first analyzed [Snitzer and Hicks 1959; Snitzer 1961] and observed experimentally [Osterberg et al. 1959; Snitzer and Osterberg 1961] in 1959. When research on optical fiber communication started in 1966, the main interest was in single-mode fibers [Kao and Hockham 1966; Krumpholz 1970; Kao et al. 1970; Börner 1971; Krumpholz 1971]. However, because of the difficulties of launching and jointing, at the beginning of the 1970's, the interest shifted away from single-mode fibers to multimode fibers. However, at the end of the 1970's, it became obvious that the bandwidth of long graded-index multimode fibers is very limited, and single-mode fibers became very interesting again.

1.2 Multimode and Single-Mode Fibers

The relatively small bandwidth of multimode fibers is the main reason for the strong trend to introduce single-mode fibers with bandwidths that are wider by a factor of at least 100. With a view to the future, only single-mode fibers can be used when one wants to replace the present direct photo-receivers, which resemble the first crystal radio receivers of the 1920's, by more sensitive and frequency-selective heterodyne receivers. Moreover, most integrated optical components use single-mode dielectric channel waveguides, so that only single-mode fibers can be coupled effectively to integrated optical repeaters.

Other advantages of single-mode fibers over multimode fibers are that they have lower fiber attenuation, lower splice and connector loss, larger production tolerances, lower cost, they preserve coherence and the degree of polarization, and exhibit useful nonlinear effects [Kapron 1984a]. In multimode fibers, attenuation and pulse broadening depend on the launch conditions, which makes these quantities difficult to define, to measure, and to calculate in advance. In contrast, in single-mode fibers, the propagation characteristics are independent of the launch conditions. The properties of single-mode fibers for long distance telecommunications have been compared with those of multimode fibers by Gambling and Matsumara [1979].

It has been argued in the past that it would be difficult to couple light into the tiny core of a single-mode fiber and that the losses at connectors and splices would be very high. However, in the factory one can nowadays couple a semiconductor laser to a short section of single-mode fiber, a so called pigtail, with a loss of about 3 dB. The pigtail can easily be fusion-spliced to the system fiber with an insertion loss of the order of only 0.1 dB. Today, single-mode fiber connectors with losses of less than 1 dB are also commercially available.

Because of the advantages of single-mode fibers, many optical and electrical engineers and technicians, as well as managers, in industry and in telecommunication administration, will have to study the principles and applications of single-mode fibers. Even those engineers who know multimode fibers will find that some concepts cannot be transfered to the new technology. For instance, it is easy to understand how light waves are guided by a multimode fiber: in step-index fibers by multiple total reflection of the rays at the core-cladding interface, and in graded index fibers by continuous ray bending. To understand waveguiding by single-mode fibers is far more difficult.

Because of the tiny core of single-mode fibers, the methods of geometrical optics [Cornbleet 1983] fail to describe the wave properties adequately. One has to use the more accurate and more complicated methods of wave optics for analyzing single-mode fibers. The concept of light rays of "zero" diameter cannot be applied. One has to consider light beams of finite diameter, which have a natural tendency to increase their width. Thus, diffraction effects have to be taken into account. Therefore, the ray path approach used with multimode fibers is no longer applicable, and electromagnetic field theory must be used.

Of course, most of the many textbooks on optical communications [Kapany 1967; Kapany and Burke 1972; Marcuse 1974; Arnaud 1976; Unger 1977a; Miller S.E. and Chynoweth 1979; CSELT 1980; Adams 1981; Barnoski 1981, Grau 1981; Marcuse 1981a; Sharma A.B. et al. 1981; Marcuse 1982a; Okoshi 1982; Kersten 1983; Snyder and Love 1983; Cancellieri and Ravalioli 1984; Unger 1984, Unger 1985; Geckeler 1986a] also report on the wave theory of fibers, from which one can deduce information on wave propagation in single-mode fibers. However, the complexity of the theory even for the simple step-index profile makes it difficult to extract general rules for the properties of single-mode fibers. Special properties of single-mode fibers are described in more than 4000 original publications in the literature. Since these papers are scattered over about 20 technical journals, it may be difficult to locate a paper covering a special problem.

3

Of course, many short reviews of single-mode fiber technology have been published in the technical journals [Kapany and Burke 1961; Snitzer 1961; Krumpholz 1971; Ramsay et al. 1975; Börner and Maslowski 1976; Kimura 1979; Yeh 1979; Kimura 1980; Li 1980; Iwahashi 1981; Midwinter 1981; Murata and Inagaki 1981; Garrett and Todd 1982; Kimura and Yamada 1982; Lilly 1982; Hooper et al. 1983; Keck 1983b; Midwinter 1983a; Midwinter 1983b; Pocholle 1983; Jeunhomme 1984; Nakagawa 1984; Shinohara 1984; Hooper et al. 1985; Monerie 1985; Zeidler 1985; Wismeyer 1985; Yamanouchi 1985; Hooper and Smith 1986; Jeunhomme 1986]. There is one book solely devoted to single-mode fibers [Jeunhomme 1983] and one book containing reprints of original journal papers on single-mode fibers [Thomson-CSF 1985].

1.3 Aim and Organization of the Book

Since neither a tutorial introduction nor a comprehensive overview of the original literature is available, this book intends to give an introduction to the new technical field of single-mode fibers, on the one hand, and to provide an overview of the literature on the other hand. Since the sources and receivers used for single-mode transmission systems are similar to those known for multimode fiber-systems, this book confines itself to a description of the properties of the single-mode fiber as a transmission medium.

The actual makers and users of single-mode fibers, such as component and system designers, manufacturing and sales engineers, technical and marketing mangagers, instructors, technicians, etc., usually may not share a great interest in mathematics and electromagnetic field theory. Because this book is written mostly for those engineers and not primarily for the few scientists working in optical waveguide theory, the material is presented from a physical point of view rather than from a mathematical one, preferring an intuitive, descriptive approach. In general, the formulas are given without derivation. Instead, the physical quantities are explained in detail, typical values for the quantities are given and it is discussed when and why it is necessary to know the actual value of a quantity. Of course, references are given, so that the more interested reader can find the missing derivations.

As can be estimated from the size of the list of references at the end of this book, a large body of knowledge on single-mode fibers has been compiled by many scientists and engineers in the past. Though the field of single-mode fibers is a relatively new one, there is already an embarrassing variety of definitions, symbols, formulas, methods of analysis, measuring methods, etc. Of course, the reason for this is that single-mode fibers are a very promising but still rapidly changing technology.

During the preparation of this book, I tried to collect as many of the publications related to the subject as possible. After having collected the papers, there was the problem of how to select the material to be presented in this book. Of course, the information must be important, correct, and not obsolete.

However, even in a very special limited field, e.g. methods for measuring the refractive-index profile of fibers, it is difficult to assess different techniques: "A comparison of accuracies expected from each method is exceedingly hard to give. It is hard enough to get a feeling for the accuracy of the method which one is familiar with, but it is far more difficult to assess some other method which one only knows theoretically" [Marcuse and Presby 1980].

Facing this difficulty, a book author can either subjectively select a few interesting topics and treat them in depth, or he can try to give an overview of the total knowledge. In the first method, there is the danger of omitting important material, and in the second method, of presenting irrelevant or incorrect information.

Since the field of single-mode fibers is in a state of rapid evolution and since standard procedures have not been agreed upon for most problems, I preferred the second method. I tried to evaluate all collected papers with the exception of those that I subjectively felt to be not very important, incorrect, or obsolete. This explains the large number of references made in the text and the length of the list of references at the end of this book.

In each section, those items that seem to be of fundamental importance or seem to achieve general acceptance will be described first and more completely than the rest. Of course, my selection of items is subjective and I invite the reader to propose amendments.

When new concepts and novel terminology are being invented so fast, it is hard for everyone to keep up with and agree on definitions. Standardization of terms, definitions, and measuring methods for single-mode fibers is in progress. With respect to the terminology, I tried to adhere to the recommendations of CCITT [1983] and IEEE [1984]. When there are differences between the terms recommended by CCITT and IEEE (e.g. fibre and fiber, or cut-off wavelength and cutoff wavelength), I chose the recommendation of IEEE.

Because of the bulk of information, it has been impossible, to state all assumptions made in an analysis, or to name all devices and their properties necessary for a measurement. This book only intends to give a survey of the most important effects, quantities, definitions, properties, methods, components, etc. Therefore, information about the details will sometimes be sparse. Before actually applying e.g. a measurement method described, or using a value for a quantity given in this book, the reader is cautioned to consult the original papers cited in the text. Since complete references will be given for each problem, a reader wanting more detail on a particular subject should have no difficulties locating suitable references in which he can find the missing details. I recommend consultation primarily of the cited books, reviews, and recent papers and then, secondly, of older full papers. Finally, in order to learn about almost everything that has been published about a certain subject, the reader may study the older letters and conference papers.

It is important to have simplified theories for the teaching of optical waveguide sciences [Hung-chia 1983b]. Therefore, it is necessary to explain waveguiding by single-mode fibers in a simple manner.

There are many possible methods for explaining single-mode fibers. The usual way is to start from Maxwell's equations, the wave equation, and the boundary conditions, to obtain rather complicated formulas for the field components and the wavelength of the wave guided by the fiber. One can also start with a dielectric slab waveguide, which represents a mathematically simple model of the circular fiber. However, in order to understand single-mode fibers, I think it best to use neither of these two strategies nor to start from coaxial cables, metallic hollow waveguides [Hung-chia 1983], or from multimode fibers.

As we shall see later, there is a strong similarity between a laser beam propagating in air and the wave guided by a single-mode fiber. Therefore, Chap. 2 of this book starts by considering a laser beam in air. Then, in physical terms and without using formulas, it is explained how an electromagnetic wave can be guided both by a straight fiber and by a curved fiber. Chapter 2 is meant to provide enhanced physical insight into waveguiding by single-mode fibers. Reading this chapter will help the reader to gain an intuitive feeling for light propagation in single-mode fibers.

It has already been stated, that the methods of geometrical optics cannot be applied to understand single-mode fibers and that wave-optical methods must be used for their description. Therefore, in order to gain a deeper understanding of single-mode fibers, it is necessary to summarize some methods of describing electromagnetic fields (Chap. 3).

Knowing the behaviour of a laser beam in air is important for understanding wave propagation in single-mode fibers, the coupling between a laser and a single-mode fiber, the coupling between two fiber ends, and the radiation from the end of a fiber. Therefore, Chap. 4 deals with the most important properties of the so-called Gaussian beams propagating in a homogeneous medium.

The properties of the wave propagating in a single-mode fiber depend on the special form of the refractive index profile. In practice, there is a large variety of different index profiles. In general, the engineer does not know the exact form of the index profile. Moreover, it is difficult to measure the index profile to a high degree of accuracy. Even if one does know the index profile exactly, the wave properties can in general, only be calculated by a numerical analysis.

In spite of these difficulties, there are many properties of single-mode fibers that can be discussed independently of the actual refractive index profile. These fundamental properties are compiled in Chap. 5. Of course, in that chapter we do not discuss methods for calculating the properties of the electromagnetic wave from the dielectric properties of the waveguide. In this volume, provisionally, one can regard the wave properties as quantities that can in principle be measured.

In order to obtain simple formulas, the radial field distribution of the fundamental mode is often approximated by a simple Gaussian function. Use of this Gaussian approximation is made throughout this book. Since the step-index fiber is a well-known reference fiber, some important formulas will additionally be given for this special type of fiber.

For any single-mode fiber, the wavelength must not become smaller than a certain value, because other waveforms, so called higher-order modes begin to propagate at smaller wavelengths. Higher-order waves are undesirable, since they reduce the fiber bandwidth and introduce signal distortions and noise. Some properties of these unwanted higher order modes are described in Chap. 6.

The following three chapters discuss the fundamentals of launching guided waves at the fiber input end, of the radiation from the fiber output end, and of the jointing of two fiber ends.

A very important property of a single-mode fiber is the transverse extent of the guided wave beam. In Chap. 10, several definitions for this so called "spot size" are summarized and the relations between the spot size and the width of the exit radiation pattern are discussed.

The most important application of single-mode fibers is signal transmission. Chapter 11 addresses the questions of fiber bandwidth, pulse dispersion and impairments in signal quality.

In the following chapter, information is given about passive components for single-mode transmission systems, such as splices, connectors, or directional couplers. Active components such as optical sources and detectors will not be considered in this book.

Since it is difficult, in general, to calculate the waveguide properties in advance, it is very important to know methods and equipment for measuring the quantities that are necessary to design a single-mode transmission system. Therefore, the last chapter summarizes single-mode fiber measuring techniques, most of which are different from those used for measuring multimode fibers.

At the end of the book, there appears a list of references, arranged alphabetically by the first author's name and chronologically for each author. In one particular year, the list begins with the papers which the author published alone, followed by papers with one coauthor, and finally by papers with more than one coauthor.

In the text, reference to a paper is made by stating the name of the author, the names of the first two authors, or − for more than two authors − the name of the first author followed by "et al.". Sometimes, for identification, it is necessary to add the initials of the first authors. Usually, the references in the text end with the year of publication. In those cases, where this information is not sufficient to identify a paper, a lower-case letter (a, b, c,...) is added to the number of the year both in the text and in the list of references.

Because of the multitude of original papers and because of the limited time for preparing the manuscript for this book, it was not possible to verify each result taken from an original paper. The reader is therefore kindly asked to write to me indicating errors and possible improvements. Furthermore it was not always possible to search systematically for the first author of a particular theory or method in order to give due credit to him.

2. Physical Explanation of Waveguiding by Single-Mode Fibers

Because the core diameter of a single-mode fiber is not very large as compared to the wavelength, one is not allowed to use the mental image of light rays of "zero" diameter, which in geometric optics are introduced as the geometrical trajectories along which light travels from one point to another [Felsen 1985]. Waveguiding in single-mode fibers cannot be explained as in multimode fibers by total internal reflection of rays at the core boundary (for step-index fibers) or by continuous ray bending in graded-index fibers.

A single-mode fiber does not guide a single ray, but a single physical light beam with a finite diameter. Such light beams can also propagate freely in air, the red beam of a helium-neon laser is a well-known example. One can couple a free beam into the front end of a fiber. In the fiber, it will be propagated as a guided wave, i.e. it will remain in the fiber and also follow slight bends in the fiber. At the end of the fiber, the wave will be re-radiated into the air as a free beam.

In this chapter, we make use of the similarities between a free wave beam in a homogeneous medium and the wave guided by a single-mode fiber to explain in physical terms and without using formulas, why a single-mode fiber can guide a light beam.

2.1 Free Wave Beam in a Homogeneous Medium

Imagine the red light beam emitted by a helium-neon laser, oscillating in the fundamental transverse mode. In order to study the properties of this beam, we imagine using a very small semiconductor photodiode to probe the light intensity distribution. Because the number of electrons generated by the photoelectric effect is proportional to the number of photons impinging on the sensitive surface of the photodiode, the measured photocurrent is proportional to the power of the incident light wave. Since the area of the photodiode is constant, we have a sensor for measuring the light power relative to unit cross-sectional area, i.e. the light intensity.

In order to measure the intensity distribution in the beam cross section, we scan our small probe transversely through the beam and observe the intensity as a function of the transverse coordinates. When the laser oscillates in its fundamental mode, the beam is circularly symmetric, and a two-dimensional scan is not necessary. It is sufficient to measure the intensity along one radial straight line passing normally through the beam axis. We will find a bell-shaped intensity distribution with maximum intensity on the beam axis. Since there are no sharp beam boundaries defined by nature, we use the common definition of the beam boundary: in a given cross section, it is that circle, on which the intensity is 13.5 % (i.e. $1/e^2$) of the maximum value of the intensity on the beam axis. The radius of the beam boundary is called the spot size of the beam; the beam diameter equals two spot sizes. One can calculate, that 86 % of the total power of the light beam are transmitted within the beam boundary and that only 14 % are transmitted outside the beam boundary.

For the beam emitted by the plane, partially transparent exit mirror of a typical helium-neon laser, the beam diameter measured will be of the order of one millimeter. If we measure the beam diameter in different cross sections along the laser beam, we will find that near the laser's output window, the beam diameter is approximately constant. Near the laser, we have a parallel beam. However, some meters from the laser, the beam diameter begins to increase. The location of the minimum beam diameter, which in this case is in the output window, is called the beam waist.

We can generalize our finding by stating that it is not possible in an homogeneous medium to produce a wave beam with a beam diameter which remains constant all along the beam. The physical reason for this is the fact that a wave beam with a finite transverse diameter can be thought to be composed of a spectrum of uniform plane waves with infinite transverse field extent propagating in slightly different directions. Therefore, an initially parallel light beam of finite diameter will somewhere begin to diverge.

The same physical effect can also be observed at the light beam transmitted through a small hole in a diaphragm: at the hole, a light beam is produced which has a finite diameter. In terms of geometrical optics, one would expect behind the hole a light beam of constant diameter and a sharp boundary between the light region and the shadow region. But in reality, for hole diameters of the order of a wavelength, the beam diameter is known to increase with the distance from the hole. The effect that light also penetrates into the shadow region is called diffraction; therefore the increase of the beam diameter in an homogeneous medium can be called diffraction spreading.

Up to now, we have only discussed the intensity distribution. For a deeper understanding of wave beams, it is also necessary to consider the wavefronts and the direction of energy propagation. As with water waves on the surface of a lake, the wavefronts are simply defined as those surfaces where at a fixed point in time the electromagnetic vibration has the same phase, e.g. a positive maximum. It can be shown [Neumann and Rudolph 1974; Simon et al. 1986] that the energy propagates (approximately) normal to the wavefronts. From this fact, one can conclude that in the beam waist, the wavefront is a plane normal to the beam axis. In the region of increasing beam width, energy has to flow away from the axis; therefore the wavefronts have to be curved.

In Fig. 2.1, the wave beam is represented by two solid curves which present the intersection between the beam boundary and a longitudinal plane containing the beam axis. The dashed curves represent some wavefronts in the beam.

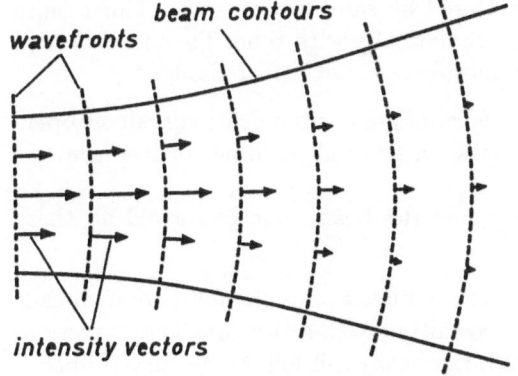

Fig. 2.1. Wavefronts, beam contours, and intensity vectors of a free wave beam

The pointers normal to the dashed wavefronts represent intensity vectors, which point in the direction of local energy transport and which have a length proportional to the intensity at that location. This picture gives an intuitive idea of the intensity distribution and the energy flow in a free wave beam.

Considered in time, the wavefronts propagate (approximately) in the directions given by the intensity vectors. After one period of the optical vibration, the wavefront arrives at the original position of the neighboring wavefront. Therefore, the normal distance between two wavefronts is proportional to the velocity of the wavefronts, the so called phase velocity. Since the wavefronts, which at the beam waist are planes, become curved, the phase velocity cannot be constant; its value around the beam axis must be slightly larger than at points far from the axis [Gloge 1964].

Vice versa, the increased value of the local phase velocity can be used to explain the effect of diffraction spreading in simple terms: the higher local phase velocity around the beam axis causes the wavefronts to bend, the intensity vector develops a component pointing away from the beam axis. Thus, power also flows in the radial direction, causing the beam diameter to increase. Since the total power transmitted by the wave does not change along the beam axis, the power transmitted per unit area decreases with increasing beam diameter. Therefore, the vectors representing the intensity become shorter with increasing spot size of the beam (Fig. 2.1).

Of course, the free light beam described above can be used for optical communications through the atmosphere. But this method has several drawbacks:

1. In order that the receiving optical antenna, a lens or a parabolic mirror, intercepts as much optical power as possible, the beam diameter at the location of the receiver must be comparable to the diameter of the lens or mirror. Then it is difficult to point the narrow pencil beam of light in the direction of the receiver.
2. Since the light beam propagates along a straight path, no obstacle must be within a volume with a diameter of some beam diameters around the beam axis. In a homogeneous medium it is not possible to guide the free wave beam along a curved path.

3.   In the atmosphere, light is scattered by rain, fog, or snow. This results
      in signal losses which can vary dramatically with time. Therefore, optical
      communication through the atmosphere is not very reliable.

For these reasons, there are only few optical communication systems operating with free beams, for instance between two tall buildings or between two
satellites in space.

By using single-mode fibers to guide the beam, one can avoid all these
disadvantages of free beams:

1.   As we will see shortly, the core of the fiber avoids the diffraction spread-
      ing, so that the beam can be transmitted with a very small cross section.
2.   If the axis of the fiber is curved, the beam will follow the curved fiber.
3.   Since the light beam propagates in a highly transparent glass, the sig-
      nal transmitted will not be influenced by the changing properties of the
      atmosphere.

2.2 Wave Beam Guided by a Straight
Single-Mode Fiber

As we have seen, the unwanted effect of diffraction spreading of free wave beams
with a diameter which is not much larger than the wavelength, can be attributed
to the fact that the phase velocity is slightly higher near the beam axis than at
points far from the beam axis. Therefore, in order to avoid diffraction spreading,
one has to look for methods for reducing the phase velocity near the beam axis.

For a plane wave with infinite transverse field extent, a so called uniform
plane wave (Sect. 4.5), there is a very simple method for reducing the phase
velocity: simply increase the refractive index of the material. It seems natural
to assume that this method also works locally. Thus, in order to decrease the
local phase velocity around the beam axis, one has to increase the refractive
index in a region around the axis. By this argument, one is naturally led to
the construction of a single-mode fiber consisting of a cylindrical core with a
refractive index which is slightly higher than that of the surrounding medium,
the cladding.

To explain waveguiding by a straight single-mode fiber, Fig. 2.2a once more
represents a free wave beam in a homogeneous medium with refractive index
n_2 by the plane wavefront in the beam waist, another curved wavefront, and
some intensity vectors. The beam spreads because of diffraction.

In Fig. 2.2b, it is assumed that a uniform plane wave is incident on the
beginning of a single-mode fiber with core and cladding indices n_1 and n_2,
respectively. The core index n_1 must be larger than the cladding index n_2. Note
the constant length of the intensity vectors! In the uniform plane wave, there
is no tendency for diffraction spreading. Hence, one can study the influence of
the fiber core separately. The local value of the phase velocity is now reduced
in the core region. The wavefronts in the core are retarded relative to those

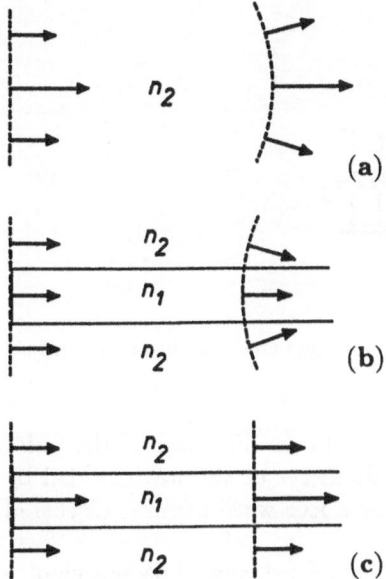

Fig. 2.2. Waveguiding by a straight single-mode fiber is explained by the effect of compensating the diffraction spreading by the reduced local phase velocity in the fiber core, see text

in the cladding. Since the electromagnetic field is a continuous function of the spatial coordinates, the phase is also continuous, and the wavefront is not torn to pieces at the core-cladding boundary. During propagation, the wavefronts bend in the manner shown in Fig. 2.2b. Since the intensity vector is normal to the wavefront, it gets a component pointing toward the fiber axis, meaning that power flows towards the axis. This increases the intensity near the axis. The fiber core provides continuous focusing of the light. This effect caused by the fiber core can be called the focusing effect.

Finally, in Fig. 2.2c, a wave beam with a finite diameter is incident on the beginning of a straight single-mode fiber. The waist of the beam is assumed to be located at the fiber input face, i.e. there we have a plane wavefront. Now, we will observe both effects at the same time: on the one hand, the beam diameter tends to increase because of diffraction, and, on the other hand, it tends to decrease because of the continuous focusing provided by the fiber core. For small values of the beam diameter, diffraction will prevail (as in Fig. 2.2a) whereas for large values of the beam diameter, focusing will prevail (as in Fig. 2.2b). There will be a certain intermediate value of the beam diameter for which the two tendencies compensate: if a free beam with this matched beam diameter is incident on the beginning of the single-mode fiber, it will be transmitted with wavefronts which remain plane and with a constant beam diameter.

This consideration leads to the most important method to excite the wave beam guided by the fiber at the fiber input end (Fig. 2.3). A converging free wave beam (Chap. 4) illuminates the fiber core with diameter $2a$. The beam axis must be aligned with the fiber axis, the waist of the incident free wave

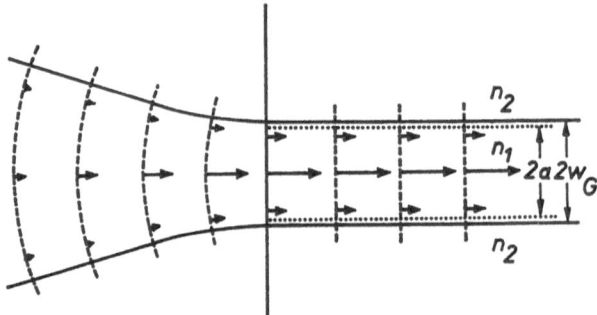

Fig. 2.3. Launching the fundamental fiber mode by a matched Gaussian beam incident on the fiber front face

must be located on the fiber input face, and the beam diameter at the waist has to be matched to the characteristic field diameter of the wave guided by the fiber. The excitation of the guided mode by a free wave beam is discussed in more detail in Sect. 7.2.2.

Since the transverse field distribution of the guided wave does not change along the fiber, it is called an eigenwave or mode of the fiber. In general, a fiber can guide more than one mode (Chap. 6). The guided mode described before is called the fundamental mode of the fiber. Other names for this wave are HE_{11} wave or LP_{01} mode. The spot size of the matched incident beam measured at the beam waist, i.e. at the front face of the fiber, is called the spot size w_G of the fundamental mode of the fiber (Sect. 10.1.2).

In summarizing the result of our physical discussion, one can state that a *straight* single-mode fiber is able to guide a wave beam since its core with higher refractive index, compensates for the diffraction spreading.

A similar description of the physics of waveguiding by a straight single-mode fiber has been given by Snyder [1981]: waveguiding is based on "balance between the spread in ray directions due to diffraction on one hand and their containment by the refractive index gradient on the other hand".

In a similar way, optical waves can be guided in a medium in which the refractive index decreases quadratically with distance from an axis [Kogelnik and Li 1966]. The main difference between such an untrunctated parabolic index medium and a single-mode fiber is that the first can guide many different types of waves (modes), whereas the latter can guide only one mode.

2.3 Wave Beam Propagating in a Curved Single-Mode Fiber

After having explained waveguiding by a straight fiber, we still have to explain why a wave beam, which in a homogeneous medium propagates along a straight path, can also follow a *curved* single-mode fiber [Neumann 1982a].

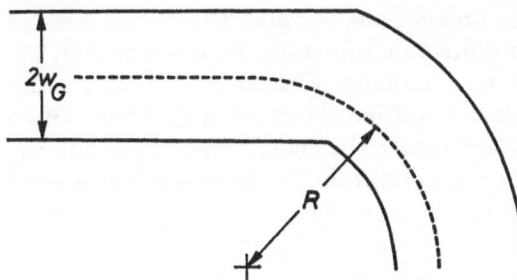

Fig. 2.4. Waveguiding by a curved single-mode fiber is explained by the field shift, which causes the local phase velocity to be reduced at the inner side of the wave beam

Consider a straight section of single-mode fiber connected to a bent section with constant radius of curvature R. The dashed line in Fig. 2.4 represents the axes of the straight and bent fiber sections. Assume that the fundamental mode of the fiber is incident along the straight waveguide section. The guided beam is represented by its boundaries (solid lines). When the wave beam arrives at the transition to the curved section, the local direction of power flow is parallel to the axis of the straight fiber section and it seems feasible to assume that the beam will initially continue to follow its former straight path. This causes the beam axis to be displaced relative to the fiber axis. The wave beam is shifted away from the axis of curvature to the outside of the bent fiber [Marcuse 1976b; Miyagi and Yip 1976; Gambling et al. 1977a]. The field boundaries which at the straight fiber were symmetrical about the fiber axis, are shifted to the outside.

If we now divide the beam cross section into an inner and an outer half with reference to the axis of curvature, more than half of the fiber core cross section will be found in the inner half of the beam cross section. This means, that the average refractive index in the inner half of the beam cross section is higher than that in the outer half. Therefore, the wavefronts in the inner half propagate more slowly than those in the outer half, i.e. the wavefronts turn. Since the power propagates normal to the wavefronts, the beam follows a curved path. Nature chose the field offset in such a way that the beam follows the curved fiber.

Summarizing, waveguiding in a bent fiber can be explained by the field shift. But note however, that part of the transmitted power is radiated from the bend, resulting in a loss increase. The attenuation of the guided wave caused by radiation from bends is discussed in more detail in Sect. 5.9.7.

By the effects described, continuous focusing and beam bending due to field offset, the beam power remains confined to the fiber. Neglecting radiation from bends or other discontinuities, the wave beam can leave the fiber only at the end of the fiber. The fiber acts as a kind of light pipe.

2.4 Free Beam Radiated from the End of a Single-Mode Fiber

When the fundamental mode guided by a single-mode fiber arrives at the free end of the fiber, a reflected guided wave is produced at the glass-air interface causing a small power loss of at most 4 % (Sect. 5.9.5). Most of the power enters the air. Since the core is no longer present, the focusing effect vanishes and there remains only the tendency for diffraction spreading discussed in Sect. 2.1. Therefore the end of a single-mode fiber radiates a free diverging wave beam into the air. Its beam waist with a plane wavefront is located in the fiber endface and its spot size measured in the waist equals the characteristic spot size w_G of the fundamental fiber mode. The radiation from the fiber end is described in more detail in Chap. 8.

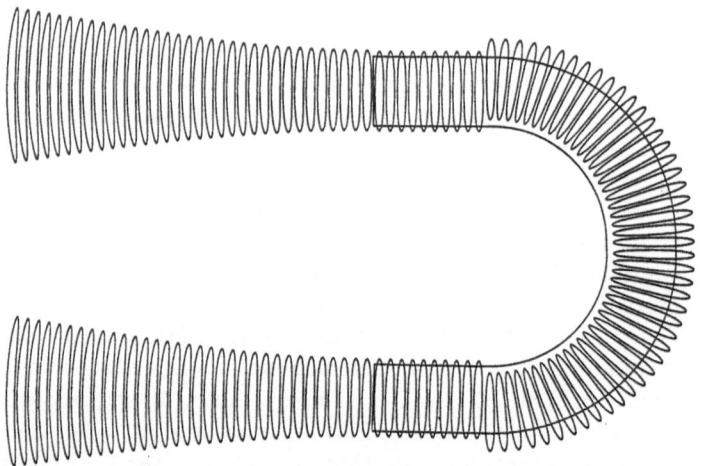

Fig. 2.5. A free wave beam launches the fundamental fiber mode, which is guided by straight and curved fiber sections and radiated from the fiber end

The launching of the fundamental fiber mode by a converging free wave beam, the guiding of the fundamental mode by straight and bent fiber sections, and the radiation of a diverging free beam from the fiber end are represented schematically in Fig. 2.5. Instead of using beam boundaries, wavefronts, and intensity vectors to describe the beams, they are represented in this figure by electric field lines (Sect. 3.5.1). The field of the wave propagating in the curved section will be explained in more detail in Sect. 5.9.7.

3. Electromagnetic Fields

A light beam propagating freely in air and the fundamental mode guided by a single-mode fiber are electromagnetic waves, i.e. electric and magnetic fields which vary in time and depend on the location of the point of observation. In order to understand single-mode fibers, it is necessary to know how to describe these electromagnetic fields.

There are many forms of electromagnetic waves. Physically, these waves are caused by two effects: firstly, a time varying electric field causes a magnetic field, and, secondly, a time varying magnetic field causes an electric field by induction. Thus, electric and magnetic fields are linked twice; this is the main physical reason for the existence of propagating electromagnetic waves. Quantitatively, the first effect is described by the first of Maxwell's equations and the second by Maxwell's second equation.

These equations form a set of two coupled partial differential equations of first order for determining the electric and the magnetic field. In order to solve this set of coupled differential equations, one first has to decouple them. As is shown in many text books on electrodynamics [e.g. Jackson D.J. 1975], by taking the curl of Maxwell's second equation, one gets a partial differential equation of second order for the electric field only, the so called wave equation. Having solved the wave equation for the electric field, one can calculate the corresponding magnetic field from Maxwell's second equation. When calculating the electromagnetic field of waves guided by a fiber, one must take into account that the fields must be finite on the fiber axis, that they decrease to zero for points far from the axis, and that the tangential field components must be continuous at interfaces between two different transparent media. These conditions lead to the so called characteristic equation, which can be solved for characteristic parameters of the guided light waves such as wavelength, group velocity, or spot size.

In this book, we will not derive the solutions of Maxwell's equations, i.e. the formulas for the electromagnetic fields of wave beams in free space or of the modes of a fiber. The reason for this is simply that, compared to the theory of coaxial cables or hollow metallic waveguides, the theory of optical dielectric waveguides is rather complicated. Even in the simplest case of a fiber with constant refractive index in the core, a so called step-index fiber, it is not possible to derive an exact closed form expression for the wavelength of the fundamental mode, for example. One has to resort to numerical methods for solving the characteristic equation and to graphical representations of the results of the computations.

Therefore, in this volume, we will be satisfied with stating the most important results of the theory without derivation. Of course, for the more interested reader, references will be given where the derivation of the results can be found. In the projected second volume, a review of methods for analyzing single-mode fibers will be presented.

In order to understand the equations describing electromagnetic fields, we will recapitulate in this chapter on some fundamentals.

3.1 Coordinate Systems

In optical communications, it is usual to describe the location of points in space either in cartesian coordinates, in cylindrical polar coordinates, or in spherical polar coordinates (Fig. 3.1).

In cartesian coordinates, the point of observation P with the coordinates x, y, and z is simply the point of intersection of three orthogonal planes with distances x, y and z from the origin.

Similarly, in cylindrical polar coordinates, the point with the coordinates ϱ, ϕ, and z is the intersection of a circular cylinder with radius ϱ around the z-axis, a half-plane terminated by the z-axis with azimuth angle ϕ measured against the x-axis, and a plane normal to the z-axis with a distance z from the origin.

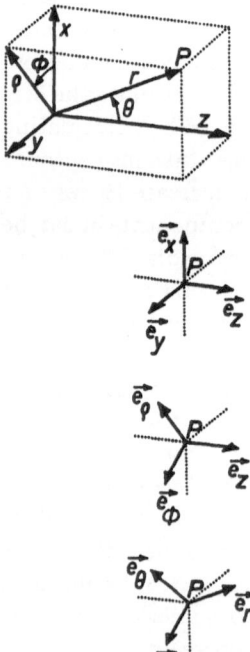

Fig. 3.1. Cartesian, cylindrical polar, and spherical polar coordinates used in this book, and unit vectors in these coordinate systems attached to the point of observation

Finally, in spherical polar coordinates, the point with the coordinates r, θ, and ϕ is the intersection point of a sphere with radius r, a circular cone with the polar angle θ measured against the z-axis, and the same half-plane as above characterized by the azimuth angle ϕ.

3.2 Vectors

In this book, vectors are denoted by boldface letters. For characterizing vectors, it is necessary to introduce orthogonal unit vectors attached to the point of observation.

In cartesian coordinates, the unit vectors e_x, e_y,, and e_z point along the directions of the coordinate axes x, y, or z respectively, (Fig.3.1). Likewise, in cylindrical polar coordinates, the unit vectors e_ϱ, e_ϕ, and e_z point in the directions of increasing radius ϱ, increasing azimuth ϕ, and increasing axial distance z, respectively. For spherical polar coordinates, the unit vectors e_r, e_θ, and e_ϕ are normal to the sphere, the cone, and the half-plane respectively.

Note, that the unit vectors must be thought of as being attached to the point of observation P. For cartesian coordinates, the unit vectors have constant directions whereas in cylindrical or spherical polar coordinate systems, the unit vectors may change their directions when the point of observation changes.

3.3 Vector Fields

Having defined the unit vectors, an arbitrary vector, e.g. the vector of the electric field strength E can be described by the vector sum of three vector components:

$$E = E_x e_x + E_y e_y + E_z e_z \quad , \tag{3.1}$$

where the quantities E_x, E_y, and E_z are the scalar components of the vector E in cartesian coordinates. In an analogous manner, a vector can be represented by its components in cylindrical or spherical polar coordinates.

Since a field vector depends both on the time t and the location of the point of observation, the independent variables, e.g ϱ, ϕ, z, and t are added in parenthesis to the symbols of the vector or its components: For instance, $E_x(\varrho, \phi, z, t)$ denotes the instanteneous value of the x-component of the electric field vector at a point P characterized by the cylindrical polar coordinates ϱ, ϕ, and z. Be careful not to confuse information on the field component (the subscript) and the information on the location (the argument).

The formulas describing the field distribution both for free beams and for the modes of single-mode fibers become especially simple if the field vectors are decomposed into cartesian components, whereas cylindrical polar coordinates are used to describe the location of the point P.

3.4 Complex Notation for Sinusoidal Quantities

When a signal is transmitted via an optical wave, the time variation of the field components is very complicated. But an arbitrary function of time can be always described by a Fourier series or a Fourier integral, i.e. it can be decomposed into sinusoidal functions of time. Therefore, to begin with, it suffices to discuss optical waves with field components which vary sinusoidally in time, so-called monochromatic waves.

If the transmission properties of a certain (linear and time invariant) optical system are known for monochromatic waves, one can predict its properties for fields with an arbitrary time dependence: just decompose the field at the input of the optical system into harmonic time components, calculate each component at the output end, and superimpose all components there.

Consider the case of an electric field component, e.g. E_x, varying sinusoidally in time:

$$E_x(\varrho, \phi, z, t) = E_x(\varrho, \phi, z) \cos\left[\omega t + \Phi_x(\varrho, \phi, z)\right] \quad . \tag{3.2}$$

In this equation, the real functions $E_x(\varrho, \phi, z)$ and $\Phi_x(\varrho, \phi, z)$ denote the amplitude and the zero phase angle (i.e. the phase angle for $t = 0$), respectively, of the sinusoidal oscillation of the x-component of the electric field vector at the location with the coordinates ϱ, ϕ, z. Note, that both the amplitude and the zero phase angle depend on the location but not on time!

For sinusoidally varying voltages or currents it is very useful to introduce a complex notation, which can also be extended to sinusodially varying electromagnetic fields. First, we combine the information about the amplitude and the zero phase angle in a new complex function by the definition

$$\underline{E}_x(\varrho, \phi, z) = E_x(\varrho, \phi, z) \exp[\mathrm{j}\Phi_x(\varrho, \phi, z)] \quad . \tag{3.3}$$

This function is called the phasor of the x-component of the electric field vector. By definition, its magnitude equals the field amplitude, and its phase angle equals the zero phase angle. In this book, we use underlining in order to represent complex quantities. Please note also, that the phasor is defined by using the field *amplitude* and not by using the root mean square (RMS) value of the field component.

Now, with the help of the phasor, the instantaneous value of the field component can be obtained by multiplying the phasor (3.3) by the time function $\exp(\mathrm{j}\omega t)$ and by taking the real part of the product:

$$E_x(\varrho, \phi, z, t) = \mathrm{Re}\{\underline{E}_x(\varrho, \phi, z) \exp(\mathrm{j}\omega t)\} \quad . \tag{3.4}$$

Just use Euler's equation to see that the right hand side of (3.4) equals the right hand side of (3.2). Usually, only the expression for the complex phasor will be given in this book. But remember, that for discussing the physical significance of an equation for a field quantity, it is necessary to return to the instantaneous value of the quantity by using (3.4).

The complex scalar phasors of the electric field components can be combined to form a complex vector:

$$\underline{\boldsymbol{E}} = \underline{E}_x \boldsymbol{e}_x + \underline{E}_y \boldsymbol{e}_y + \underline{E}_z \boldsymbol{e}_z \quad . \tag{3.5}$$

Complex vectors are represented by underlined boldface letters.

3.5 Graphical Representation of Electromagnetic Fields

Since it is not easy to understand electromagnetic fields represented by equations only, it is useful to illustrate the formulas pictorially. In this book, we will use several kinds of graphical representation of electromagnetic fields, these are described below.

3.5.1 Field Lines

First, there are the well-known field lines. These are tangential to the local direction of the vector field to be represented and have a density proportional to the local amplitude of the field. Full lines will be used to represent electric field lines and dashed lines to represent magnetic field lines. We will plot either the field lines projected onto a cross section or a longitudinal section or represent them stereoscopically, see for example Fig. 4.10.

3.5.2 Field Vectors

For representing the field distribution of a mode in the cross section of a fiber, another method can be used: calculate the electric field vector at a fixed moment in time for several points in the cross section and draw these vectors into a picture of the cross section. Using this method, one obtains graphical information on the polarization properties of the mode, see e.g. Fig. 4.14.

3.5.3 Wavefronts

In Chap. 2, which explained waveguiding by single-mode fibers, we have already used the concept of wavefronts (Fig. 2.1) which also are called equiphase surfaces [Barlow H.M. and Cullen 1953]. In order to learn more about these wavefronts, we discuss the movement of a wave crest in a general wave field. The concepts of a local wavelength and a local phase velocity will be introduced for later use.

Since the cosine function has its maximum value of unity when its argument is an integer multiple of 2π, it follows from (3.2) that the points of maximum field strength E_x are situated on surfaces defined by the equation

$$\omega t + \Phi_x(\varrho, \phi, z) = n2\pi \quad , \qquad n = \ldots -2, \ -1, \ 0, \ 1, \ 2, \ \ldots \quad . \tag{3.6}$$

With increasing time, these wavefronts move in the direction of decreasing values of the zero phase angle Φ_x.

We now restrict ourself to a certain wavefront, i.e. one characterized by a fixed value of the integer n. At the time t_0 it will be located at a surface in three-dimensional space defined by the equation

$$\Phi_x(\varrho, \phi, z) = n2\pi - \omega t_0 \quad . \tag{3.7}$$

A little later, at time $t_0 + \Delta t$, the wavefront in question will be shifted to a neighboring surface defined by

$$\Phi_x(\varrho, \phi, z) = n2\pi - \omega t_0 - \omega \Delta t \quad , \tag{3.8}$$

i.e. to points in space where the value of the zero phase angle Φ_x is smaller by the amount $\omega \Delta t$.

The local normal distance between the wavefront at t_0 and at $t_0 + \Delta t$ is denoted Δs. From vector analysis, it is known that the change of the scalar function Φ_x normal to the wavefront per unit length is equal to the magnitude of the vector $\operatorname{grad} \Phi_x$. Since $|\operatorname{grad} \Phi_x|$ determines the rate of phase change normal to the wavefront, it is called the phase constant or phase-change coefficient and is given a new symbol β:

$$\beta = |\operatorname{grad} \Phi_x| \quad . \tag{3.9}$$

In the literature, the quantity β is sometimes called the propagation constant. Since there is also a the complex propagation constant (Sect. 11.1)

$$\gamma = \alpha' + j\beta \quad , \tag{3.10}$$

which combines the information about the attenuation coefficient α' (Sect. 5.9.1) and the phase constant β, the name phase constant will be used for β.

Since β is the change of the scalar function Φ_x per unit length normal to the wavefront, the change $\Delta \Phi_x$ in going from the first to the second location of the wavefront is $\beta \Delta s$, where Δs is the normal distance between these surfaces. On the other hand, from (3.7,8), the change of the function $\Phi_x(\varrho, \phi, z)$ is seen to be $\omega \Delta t$. Equating the two expressions for $\Delta \Phi_x$ gives

$$\beta \Delta s = \omega \Delta t \quad , \quad \text{or} \tag{3.11}$$

$$\frac{\Delta s}{\Delta t} = \frac{\omega}{\beta} \quad . \tag{3.12}$$

Therefore, the local value of the velocity of the wavefront, the so called phase velocity v_p, is given by:

$$v_{\mathrm{p}} = \frac{\Delta s}{\Delta t} = \frac{\omega}{\beta} \quad . \tag{3.13}$$

By plotting several wavefronts with different values of the integer n at a fixed moment in time one gets an idea of the local direction of wave propagation, which is normal to the wavefronts. We use dashed lines to represent the wavefronts. The distance between neighboring wavefronts equals the local value of the wavelength and is denoted by λ_l. Since the wavelength of a uniform plane wave in a vacuum will be denoted λ without any index, the subscript "l" meaning "local" has been introduced.

Since the wavelength λ_l is the distance Δs moved by the wavefront within a period $\Delta t = T = 1/f$ of the optical oscillation, one obtains from (3.13) the following relation between the wavelength and the phase constant:

$$\lambda_l = \frac{2\pi}{\beta} \quad . \tag{3.14}$$

3.5.4 Lines of Constant Amplitude

One can also represent the field distribution graphically by surfaces of constant amplitude. A surface of constant amplitude is defined by the equation:

$$E_x(\varrho, \phi, z) = \text{constant} \quad . \tag{3.15}$$

By plotting equi-amplitude surfaces [Barlow H.M. and Cullen 1953] for several different values of the constant, one gets a very vivid impression of the amplitude distribution. If the values of the constant differ e.g. by a factor of $10^{1/2} = 3.16$, the field level changes by $5\,\mathrm{dB}$ between neighboring surfaces of constant field strength.

The wavefronts and the surfaces of constant amplitude are surfaces in three-dimensional space. Usually, we only plot the intersections of these surfaces with a transverse or longitudinal plane, i.e. lines of constant phase or constant amplitude, see e.g. Fig. 4.6.

Let us mention that it is quite easy to measure both the wavefronts and the equi-amplitude contours experimentally at microwave frequencies using scaled models of single-mode fibers, a small probe antenna, and an amplitude/phase receiver [Neumann 1975; Albrecht and Neumann 1979].

If one of the three components of the electric field is dominant, the equi-amplitude contours can also be considered as iso-intensity plots.

3.5.5 Beam Boundary

For wave beams with a simple form, it is not necessary to represent the amplitude distribution by many lines of constant amplitude. As already done in Chap. 2, it is sufficient to draw the boundary of the beam, i.e. the surface where the light intensity is 13.5 % of the maximum value in the same cross section, see e.g. Fig. 2.1.

3.5.6 Intensity Vector

In order to get an impression of the energy transport by an electromagnetic wave, one can calculate at several points in a longitudinal plane the local value of the intensity vector, i.e. the time averaged Poynting's vector (3.16), and draw these vectors; their direction gives the direction of power flow and their length is proportional to the local value of the intensity, see e.g. Fig. 2.1.

3.5.7 Regions of High Intensity

The transverse intensity distribution in a free beam or in a mode guided by a fiber can be represented by plotting several lines of constant intensity in a cross section (iso-intensity plot). In a simplified version, only one of these lines is plotted, on which the intensity is a certain percentage (e.g. 13.5 %) of the maximum value of the intensity. The area surrounded by this line is hatched. A similar picture would be obtained by microphotographing the endface of the fiber using a high-contrast film.

Instead of hatching the regions of high intensity, one can also draw only the limits of the regions of high intensity, see. e.g. Fig. 4.14. These curves give a rough idea of the transverse distribution of the power flow.

3.5.8 Field Profiles

For completeness, reference must be made to the method of plotting one of the field components as a function of one of the coordinates of the point of observation. A very important case of these field profiles is the radial dependence of the dominant component of the electric field, e.g. $E_x(\varrho)$ (Fig.5.6).

In conclusion, there are several methods for graphical representation of electromagnetic fields. In this book, we will use all the methods mentioned, sometimes also in combination, for instance by plotting in one picture both field vectors and areas of high intensity (e.g. Fig. 4.14), or both wavefronts and lines of constant amplitude (Fig. 4.6).

3.6 The Poynting Vector and the Transmitted Power

The Poynting vector is another name for the intensity vector which we explained in the introductory Chap. 2. This vector points in the local direction of power flow and its magnitude equals the time average of the power density, i.e. the power transmitted per unit cross-sectional area. When using the complex notation, it is easy to calculate the time averaged Poynting's vector S:

$$S = \tfrac{1}{2}\mathrm{Re}\{\underline{E} \times \underline{H}^*\} \quad , \tag{3.16}$$

which means that one has to form the vector product of the complex vector $\underline{E}$ of the electric field and the complex conjugate $\underline{H}^*$ of the magnetic field vector, and then to take half of the real part of the product. The factor of one half is the average of the cosine function squared and it appears in (3.16) because we defined the phasor using the amplitude and not the RMS-value (3.3). Since the magnetic field is essentially proportional to the electric field, one can remember that the intensity vector is proportional to the square of the field amplitude. Because of the definition of the vector product, Poynting's vector is normal to both the electric and the magnetic field vector.

Since the Poynting vector gives the power transmitted per unit cross-sectional area, the total power P transmitted by a wave beam is obtained by integrating the Poynting vector over the total cross section A of the wave beam:

$$P = \int_A \boldsymbol{S} \cdot d\boldsymbol{A} \ , \tag{3.17}$$

where the differential vector $d\boldsymbol{A}$ is normal to the cross-sectional area and has a magnitude which equals the area of the element of the cross section.

Whereas in optics we have no meters for measuring the strength of the electric and magnetic field vectors directly, the transmitted optical power can be measured easily using an optical power meter (see Sect. 13.2). If the theoretical relation between the field strength on the axis of the wave beam and the transmitted power calculated from (3.17) is known, it can be inverted to determine the unknown optical field strength from the measured power.

3.7 Coherence

With single-mode fibers, it is very important to know whether the radiation is coherent or incoherent, since e.g. the launching efficiency (Chap. 7), the pulse broadening (Chap. 11), and the total intensity, all depend on the coherence properties of the radiation. Therefore, it is necessary to introduce some concepts from the theory of partial coherence [Born and Wolf 1975; Perina 1985].

3.7.1 Interference

Since coherence is intimately linked with the phenomenon of interference, we first consider the superposition of two sinusoidal waves with the same frequency and with the same linear polarization derived from the same source. The electric fields of the two waves at a fixed point in space are denoted by $\boldsymbol{E}_1$ and $\boldsymbol{E}_2$. The fields superimpose and the resulting electric field $\boldsymbol{E}$ is the sum of $\boldsymbol{E}_1$ and $\boldsymbol{E}_2$. For fixed amplitudes of both fields, the resulting electric field will depend on the phase difference between the two oscillations: when the partial fields oscillate in phase, the total field will be maximum, and for a phase difference of π, it will be a minimum.

Because most optical detectors today produce an electrical signal proportional to the optical power incident on the sensitive surface, we consider the intensity of the resulting field. Since the intensity is essentially the square of the total electric field, the total intensity S becomes

$$S = S_1 + S_2 + 2\sqrt{S_1 S_2} \cos \Delta\phi \quad , \tag{3.18}$$

where S_1 and S_2 are the intensities of the two partial waves and where $\Delta\phi$ is the phase difference between the two fields at the point of observation.

From (3.18), it can be seen that the total intensity S does not equal the sum $S_1 + S_2$ of the intensities of the partial waves. A third term $2\sqrt{S_1 S_2} \cos \Delta\phi$ has to be added, which depending on the phase difference $\Delta\phi$ can be positive or negative. If the partial waves are in phase, the maximum of intensity exceeds the sum of the intensities of the superimposed waves. If they are in antiphase, the resulting intensity is smaller than the sum of the partial intensities. Depending on the phase difference $\Delta\phi$, the waves superimpose constructively or destructively. This effect is called "interference". Therefore, the third term in (3.18) can be called the "interference" term.

3.7.2 Coherent and Incoherent Waves

Now, in real optical waves, the electric field never depends strictly sinusoidally on time (Sect. 3.7.6). The amplitude of the oscillation is approximately constant, but the phase fluctuates randomly in time. Because of the random phase fluctuations of the two wave fields E_1 and E_2 considered above, the phase difference $\Delta\phi$ will also fluctuate in time, i.e. the interference term in (3.18) also changes with time. For fluctuations of the phase difference which are very small as compared to 2π, the interference term will fluctuate only slightly around the value determined by the average phase difference, i.e. interference effects still occur.

But if the magnitude of the fluctuation of the phase difference is large as compared to 2π, the interference term will fluctuate randomly between its extreme values of $\pm 2(S_1 S_2)^{1/2}$. The time average of the interference term will be zero, and the total time averaged intensity is simply the sum of the intensities of the two individual waves:

$$S = S_1 + S_2 \quad . \tag{3.19}$$

Therefore, as for random noise voltages, the powers (intensities) add up instead of the amplitudes. Two waves with a phase difference that fluctuates by much more than 2π, are called incoherent, whereas sinusoidal waves with a fixed phase difference are called coherent. Thus coherence relates to the time stability of the phase difference.

To sum up: coherent waves have a relatively stable phase difference, the principle of superposition holds for the electric and magnetic field strengths, but not for the intensities. The total intensity depends on the phase difference between the waves. On the other hand, for incoherent waves, the phase difference fluctuates randomly by amounts larger than 2π, interference effects vanish, and the principle of superposition holds for the time averaged intensities and not for the fields. The total intensity is then independent of the average phase difference between the waves.

Between the extreme cases of perfectly coherent waves and completely incoherent waves discussed so far, there is of course a gradual transition. Waves which are neither perfectly coherent nor totally incoherent are called "partially coherent."

3.7.3 Degree of Coherence and Visibility of Interference Fringes

At this point, it is useful to define a quantitative measure for the coherence of two wave fields, namely the "complex degree of coherence" γ_{12} [Born and Wolf 1975]. Essentially, it is the normalized time average of the analytical signal (the time varying phasor) of the first field multiplied by the complex conjugate of the analytic signal of the second field. The user of single-mode fibers usually is interested only in the modulus $|\gamma_{12}|$ of the complex degree of coherence, which is called for short the "degree of coherence."

When two partially coherent waves with intensities S_1 and S_2 and the same linear polarization are superimposed, the resulting total intensity S is given by [Born and Wolf 1975; Shibata N. et al. 1984]

$$S = S_1 + S_2 + 2|\gamma_{12}|(S_1 S_2)^{1/2} \cos \Delta\phi \quad , \tag{3.20}$$

where $|\gamma_{12}|$ is the degree of coherence and $\Delta\phi$ the average phase difference between the fields. For sinusoidal fields, the degree of coherence is unity,

$$|\gamma_{12}| = 1 \quad , \tag{3.21}$$

and (3.20) reduces to (3.18). For incoherent fields, the degree of coherence is zero

$$|\gamma_{12}| = 0 \quad , \tag{3.22}$$

and (3.20) reduces to (3.19). Therefore, the degree of coherence is a measure for the capability of two waves to interfere.

The degree of coherence $|\gamma_{12}|$ is related to the so called "visibility of interference fringes" V, which can easily be measured by an interference experiment: superimpose the two fields, change the average phase difference $\Delta\phi$ between the fields (e.g. by shifting the point of observation through several wavelengths or by slightly changing the optical frequency), and observe the resulting intensity: it will change between a maximum value $S_{\max}$ and a minimum value $S_{\min}$. The fringe visibility V then is defined by

$$V = (S_{\max} - S_{\min})/(S_{\max} + S_{\min}). \tag{3.23}$$

In the special case that the two waves have *equal* intensities

$$S_1 = S_2 \quad , \tag{3.24}$$

we have from (3.20)

$$S_{\max} = 2S_1(1 + |\underline{\gamma}_{12}|) \tag{3.25}$$

and

$$S_{\min} = 2S_1(1 - |\underline{\gamma}_{12}|) \quad , \tag{3.26}$$

and therefore,

$$V = |\underline{\gamma}_{12}| \quad . \tag{3.27}$$

Thus, the degree of coherence $|\underline{\gamma}_{12}|$ can easily be determined from the fringe visibility for equal wave intensities [Born and Wolf 1975].

For perfectly coherent waves, because of the assumption of equal field strengths, the minimum intensity will be zero and the visibility $V = 1$. On the other hand, for incoherent waves, the average total intensity is independent of the relative average phase angle and, therefore, $S_{\max} = S_{\min}$ and $V = 0$.

Practically, two waves are considered to be highly coherent when the degree of coherence exceeds 0.88, partially coherent for values less than 0.88, and incoherent for values significantly less than 0.88 [CCITT 1983].

Following this discussion of the coherence between two waves, we now consider the question of the temporal and spatial coherence of a single optical wave which can be traced back to the coherence between two waves.

3.7.4 Coherence Time and Coherence Length

For defining the temporal coherence of a single wave, the wave is divided (e.g. by a half mirror) into two waves of equal amplitude. One wave is delayed relative to the other by a variable time t_d. Then the visibility V of the fringes produced by superimposing the delayed and the undelayed waves is measured as a function of the delay t_d. For zero time delay, the two waves are correlated, interference effects are observed and the visibility is unity. With increasing time delay, because of the phase fluctuations in real waves, interference effects will become smaller, i.e. the visibility will decrease gradually to zero.

The visibility function $V(t_d)$ is essentially the autocorrelation function of the field strength vs time function. Its width equals that time delay up to which the two parts of the wave can be considered coherent.

At this point, it is necessary to quantitatively define the width of the visibility curve. Later we will have to define the widths of other curves, e.g. the width of a spectral line (Sect. 3.7.6), or the duration of a pulse (Sect. 11.5).

A universal measure of the width of a function $f(x)$ is the so-called effective or RMS width. In order to define the RMS width, one introduces so-called moments M_n of the function $f(x)$ by [Grau 1981]:

$$M_n = \int\limits_{-\infty}^{+\infty} x^n f(x) dx \quad . \tag{3.28}$$

In the integrand the function $f(x)$ is multiplied by the nth power of the variable x.

The moment of zero order M_0 is the area under the curve representing the function $f(x)$. The first order moment M_1 divided by the zero order moment M_0 defines the center of gravity x_c of the function:

$$x_c = M_1/M_0 = \left[\int\limits_{-\infty}^{+\infty} x f(x) dx \right] \left[\int\limits_{-\infty}^{+\infty} f(x) dx \right]^{-1} \quad , \tag{3.29}$$

as can be recognized by assuming a very narrow rectangular function $f(x)$.

As a universal measure of the width of the function $f(x)$, the RMS width σ_x is defined as the square root of the mean squared weighted deviation of the argument x from the central argument x_c:

$$\sigma_x^2 = \frac{1}{M_0} \int\limits_{-\infty}^{+\infty} (x - x_c)^2 f(x) dx \quad . \tag{3.30}$$

By expanding the square of the difference in the integrand, the RMS width can be expressed in terms of the moments M_0, M_1, and M_2:

$$\sigma_x^2 = \frac{M_2}{M_0} - \left(\frac{M_1}{M_0} \right)^2. \tag{3.31}$$

Thus, in order to determine the RMS width, one has to measure the function $f(x)$ and to calculate the three moments M_0, M_1, and M_2. In order to simplify the analysis, it is often assumed that the actual function $f(x)$ can be approximated by a Gaussian function:

$$f(x) = \frac{1}{\sqrt{2\pi}\sigma_x} \exp[-0.5(x - x_c)^2/\sigma_x^2] \quad . \tag{3.32}$$

The factor in front of the Gaussian function is chosen to make the zero order moment $M_0 = 1$. In the argument of the Gaussian function, the width is expressed in terms of the RMS width σ_x.

When the argument x deviates by one effective width σ_x from the center of gravity x_c, i.e. for $|x - x_c| = \sigma_x$, the Gaussian function is $\exp(-0.5) = 0.607$, i.e. for a Gaussian function, the RMS half width is the width measured between the location of the maximum and the point where the function has decreased by a factor of 0.607.

In practice, other quantities are used to characterize the width of the $f(x)$-curve. Assuming a Gaussian form of the function $f(x)$, the following relations hold between the RMS width and other practical measures for the width.

The 50% halfwidth Δx_{50} is obtained by setting the Gaussian function equal to 0.5 in (3.32):

$$\Delta x_{50} = (2\ln 2)^{1/2}\sigma_x = 1.177\sigma_x. \tag{3.33}$$

The $1/e$ halfwidth Δx_{37} is obtained by setting the argument of the Gaussian function equal to 1:

$$\Delta x_{37} = 2^{1/2}\sigma_x = 1.414\sigma_x \quad . \tag{3.34}$$

Note, that some authors use the full widths which are twice the halfwidths defined above.

Having discussed how to characterize the width of a general function $f(x)$, we return to the question of how to define a measure for the width of the visibility versus time delay curve. The "coherence time" t_c is defined as the RMS width of the square of the visibility function $V(t_d)$ [Born and Wolf 1975]. Thus, for the function $f(x)$ in (3.30) one has to take $V^2(t_d)$. If the visibility function can be assumed to be Gaussian [Unger 1977a], it can be written

$$V(t_d) = \exp\left(-\frac{t_d^2}{4t_c^2}\right), \tag{3.35}$$

i.e. for a time delay t_d of one coherence time t_c, the visibility is 0.78.

Since the time delay between the two waves is usually caused by a difference in path length, it is also useful to define a so-called "coherence length" L_c as the product of the coherence time and the velocity of light in free space:

$$L_c = t_c c \quad . \tag{3.36}$$

If the path difference is small as compared to the coherence length, then the time delay is small compared to the coherence time and the partial waves will superimpose coherently, i.e. interference will be observed. On the other hand, for a path difference much larger than the coherence length, the partial waves will add incoherently, i.e. there will be no interference.

3.7.5 Coherence Area

For defining the spatial coherence of a single wave, the fields at two different points are considered at the same moment of time. If there is a fixed phase difference, the wave is said to be spatially coherent, and if the phase difference fluctuates randomly by more than 2π, the wave is called spatially incoherent.

The spatial coherence of a wave can be characterized by the so-called "coherence area", which is the area in a plane perpendicular to the direction of propagation over which light can be considered highly coherent, i.e. where the degree of coherence exceeds 0.88 [CCITT 1983].

When a coherent beam is injected into a multimode fiber, the phase relationship is destroyed, because each mode travels with a slightly different phase velocity. When the light emerges after traversing a length of fiber, it superimposes constructively and destructively in a very complex fashion, depending upon the precise phase of each mode and its spatial distribution [Piazzolla and de Marchis 1980]. In the fiber endface, one observes a complicated "speckle" pattern, which, due to wavelength variations, fluctuates randomly in time.

In contrast, when only one mode is supported in a fiber, coherence is retained. The field of the fundamental fiber mode is spatially completely coherent [Machida et al. 1979; Shibata N. et al. 1981c]. This property of single-mode fibers is very important, e.g., when using heterodyne detection for increasing the receiver sensitivity.

3.7.6 Relation Between Linewidth and Coherence Time

The coherence time of a quasi monochromatic wave is related to the width of the spectrum or the so called linewidth. In this section, we will discuss the causes of the finite width of the spectrum of an optical wave, introduce measures for the linewidth, and discuss the relation between the coherence time and the linewidth.

Even the best lasers do not produce an optical field strength which varies exactly sinusoidally in time. The reason for this is that the laser output consists of photons radiated both by stimulated and by spontaneous emission. The "stimulated" photons produce a sinusoidal carrier whereas the "spontaneous" photons add randomly and cause a random phase modulation [Siegman 1971; Unger 1977a].

There is another useful picture for the light emission from a real optical source: the source can be considered to emit many strictly sinusoidal wave trains of finite duration in a random manner. The average duration of a single wave train is the coherence time t_c of the emitted light [Born and Wolf 1975]. There is no fixed phase relationship between successive wave trains. The superposition of the wave trains results in wave field with random fluctuations of the phase angle. There are also amplitude fluctuations, but for laser sources they are negligibly small [Grau 1978].

Expressed in electrical engineering terms, a real optical source emits a sinusoidal wave which is phase modulated by a noise voltage. From Fourier's theorem, it follows that the spectrum of this wave has a bell-shaped form with a finite width (we simplify this aspect by excluding lasers oscillating in several longitudinal cavity modes). Since the optical wave is not yet modulated by the signal to be transmitted, the width of the spectral line is called the "natural linewidth".

There are several measures for the width of the spectrum: the spectrum can either be considered a function of wavelength or a function of frequency; one can define its width by the $\frac{1}{2}$ or $\frac{1}{e}$-points; one can use either the half width or the full-width; and finally, one can consider either the amplitude or the power spectrum.

31

In theoretical papers, the linewidth is most often characterized by the root-mean-square linewidth σ_λ (RMS linewidth) of the power spectrum in the wavelength domain. The spectral power density

$$P_\lambda(\lambda) = \frac{dP}{d\lambda} \tag{3.37}$$

is defined as the optical power dP within a wavelength interval divided by this interval and measured as a function of optical wavelength λ.

The RMS linewidth σ_λ is defined by (3.30), where the variable x has to be replaced by λ and where the function $f(x)$ has to be replaced by the power spectrum $P_\lambda(\lambda)$ [Unger 1977a]

$$\sigma_\lambda^2 = \frac{1}{P} \int\limits_0^\infty (\lambda - \lambda_c)^2 P_\lambda(\lambda) d\lambda = \frac{M_2}{M_0} - \left(\frac{M_1}{M_0}\right)^2 \quad . \tag{3.38}$$

The moment of zero order M_0 is simply the spectral power density integrated over all wavelengths, i.e. the optical power transmitted by the wave

$$M_0 = \int\limits_0^\infty P_\lambda(\lambda) d\lambda = P \quad . \tag{3.39}$$

The ratio of the first order moment to the zero order moment

$$\frac{M_1}{M_0} = \frac{1}{P} \int\limits_0^\infty \lambda P_\lambda(\lambda) d\lambda = \lambda_c \quad , \tag{3.40}$$

defines the center or carrier wavelength λ_c as can be seen by assuming a very narrow rectangular spectrum.

The definition (3.38) for the RMS linewidth holds for any form of the power spectrum, but in order to determine the RMS linewidth, it is necessary to measure this form with a spectrometer and to calculate the three moments M_0, M_1, and M_2 by evaluating the integrals (3.28).

In order to simplify the characterization of the linewidth, the practitioner often assumes a Gaussian form of the spectrum

$$P_\lambda(\lambda) = P_\lambda(\lambda_c) \exp\left(\frac{-0.5(\lambda - \lambda_c)^2}{\sigma_\lambda^2}\right). \tag{3.41}$$

In (3.41), the width of the power spectrum is characterized by the RMS linewidth σ_λ.

For a Gaussian form of the spectrum, the relations between the more common 50 % halfwidth $\Delta\lambda_{50}$, the $1/e$ halfwidth $\Delta\lambda_{37}$ and the RMS linewidth σ_λ are (3.33; 34):

$$\sigma_\lambda = 0.849 \Delta\lambda_{50} = 0.707 \Delta\lambda_{37} \quad . \tag{3.42}$$

The full width at half maximum (FWHM), which is used most often in practice, is

$$\Delta\lambda_{\text{FWHM}} = 2.36\sigma_\lambda \quad . \tag{3.43}$$

For sources with very small linewidths, the spectral width is often expressed as a frequency interval instead of a wavelength interval. By differentiating the relation

$$f = \frac{c}{\lambda} \quad , \tag{3.44}$$

between vacuum wavelength λ, light velocity c in vacuum, and optical frequency f with respect to wavelength and taking into account that the relative width of the spectrum is narrow, one obtains

$$\frac{\sigma_f}{f} = \frac{\sigma_\omega}{\omega} = \frac{\sigma_\lambda}{\lambda} \quad , \tag{3.45}$$

where σ_f and σ_ω are the RMS widths in the frequency and angular frequency domain, respectively. Equation (3.45) states that the relative spectral width is the same in the frequency and the wavelength domain.

In classical optics, frequencies are often expressed in terms of the spectroscopic wave number

$$\overline{\nu} = \frac{1}{\lambda} \quad , \tag{3.46}$$

which is the number of wavelengths per centimeter. Therefore, linewidths are sometimes given in the unit cm^{-1}. The relation between the spectroscopic wave number $\overline{\nu}$ and the optical frequency f is:

$$\overline{\nu} = \frac{f}{c} \quad . \tag{3.47}$$

Large linewidths are usually expressed in nanometers, while small linewidths are expressed in megahertz (MHz) or gigahertz (GHz). For an operating wavelength of 1.3 μm, a linewidth of $\sigma_\lambda = 1$ nm corresponds to a bandwidth of $\sigma_f = 177$ GHz.

Since the optical wave emitted can be thought of as consisting of sinusoidal wave trains of duration t_c, the width of the spectrum can be expected to be of the order of $1/t_c$. Quantitatively, there exists the following relation between the coherence time t_c defined as the RMS width of the visibility curve squared, and the RMS width σ_ω of the spectrum [Born and Wolf 1975]:

$$t_c \geq \frac{1}{2\sigma_\omega} \quad . \tag{3.48}$$

The equality sign only holds when the visibility function is Gaussian. For the spectra encountered in optics, the inequality sign in (3.48) may be replaced by the order of magnitude sign.

The coherence length defined by (3.36) can be obtained from the full-width-half-maximum spectral linewidth $\Delta\lambda_{\text{FWHM}}$ using

$$L_c = \frac{0.76\lambda^2}{\Delta\lambda_{\text{FWHM}}} \quad . \tag{3.49}$$

In the recommendation of CCITT [1983], the factor 0.76 is missing in the formula relating the coherence length and the linewidth, but neither the coherence length nor the measure used for the linewidth are defined exactly.

The ideal optical source produces an electric field that is strictly sinusoidal in time. The spectrum of this so-called "monochromatic" wave consists of a single line of zero width. Because of the inverse relationship between the linewidth and the coherence length, the coherence length is infinite.

For a wavelength of $1.3\,\mu$m, typical values of the FWHM linewidths, coherence times, and coherence lengths are listed in Table 3.1 for LED's, multimode lasers, and single frequency lasers.

Table 3.1. Typical values for the linewidth, coherence time, and coherence length ($\lambda = 1.3\,\mu$m)

Source	$\Delta\lambda_{\text{FWHM}}$	t_c	L_c
LED	100 nm	40 fs	10 μm
Multimode LD	5 nm	1 ps	0.2 mm
Single-frequency LD	5.6×10^{-4} nm (100 MHz)	8 ns	2 m

The temporal coherence characteristics of a single-frequency semiconductor laser have been measured using a Michelson interferometer [Yamabayashi and Saruwatari 1983].

4. Gaussian Beams

As has been explained in Sect. 2.2, the wave propagating in a single-mode fiber is very similar to a free wave beam in air; it can easily be launched by an incident free beam, and produces a free beam radiated from the fiber end. For these reasons, it is worth while studying the basic properties of free wave beams propagating through a homogeneous medium. We have already described some qualitative properties of wave beams in a homogeneous medium in Sect. 2.1. Relying on physical intuition, we introduced the concepts of the transverse intensity distribution, the spot size, the beam waist, the wavefronts, and the intensity vector. We also learned that a beam has a natural tendency to diffraction spreading. In this chapter, we will discuss the properties of these so-called Gaussian beams more quantitatively.

Before turning to the mathematical details, it is interesting to note that although wave beams are natural phenomenona, they have only been studied in detail since the invention of the laser [Goubau and Schwering 1961; Boyd G.D. and Gordon 1961]. Since that time, we have learned that a Gaussian wave beam is a simple fundamental physical phenomenon. Nowadays, one even decomposes more complicated electromagnetic wave fields into simple Gaussian beams [Popov 1982; Bogush and Elkins 1986; Mantica et al. 1986].

The equations for the fields in a Gaussian beam can be derived from the wave equation relatively easily [Yariv 1971]. In the derivation however, some approximations have to be made. Therefore, Gaussian beams do not represent rigorous solutions of Maxwell's equations, but approximations which hold only if the beam diameter at the waist is large in comparison to the wavelength or, stated in other terms, if the beam diverges only slightly.

There are different Gaussian beams with different intensity distributions in the beam cross section. We first discuss only the simplest form, the fundamental Gaussian beam, which is similar to the fundamental mode of a glass fiber. Later, in Sect. 4.8, we add some information on Gaussian beams of higher order.

4.1 Amplitude Distribution

We consider a fundamental Gaussian beam in a medium with refractive index n propagating along the z-axis. The beam waist is located in the plane $z = 0$. For simplicity, the medium is assumed to be lossless, linear, isotropic, and homogeneous.

For describing the electromagnetic field of a Gaussian beam (and, later, of the waves guides by fibers), it is very advantageous to simultaneously use both Cartesian and cylindrical polar coordinates: the field vectors are decomposed into their Cartesian components, i.e. E_x, E_y, and E_z, whereas their spatial variation is described by the cylindrical coordinates ϱ, ϕ, and z of the point of observation.

The wave is assumed to be linearly polarized in the x-direction, i.e. the electric field vector has a dominant component $\underline{E}_x$ in the x-direction. For the fundamental beam, this component depends only on the distance ϱ from the z-axis and on the z-coordinate, but not on the azimuth ϕ [Kogelnik and Li 1966; Marcuse 1982a]:

$$\underline{E}_x(\varrho, z) = \underline{E}_x(0,0) \frac{w_0}{w(z)} \exp\left[\frac{-\varrho^2}{w^2(z)}\right] \exp\left[-jnkz + j\Theta(z) - \frac{jnk\varrho^2}{2R(z)}\right] \quad (4.1)$$

This equation contains several quantities and functions which we have to explain. We will do this while discussing the field distribution described by (4.1).

The factor

$$\underline{E}_x(0,0) = E_0 \quad , \tag{4.2}$$

can be made real by choosing a suitable zero for the time scale. All quantities and functions on the right hand side are then real except for the imaginary unit j. From this, we recognize that the amplitude distribution $E_x(\varrho, z)$, as defined by (3.3), is given by

$$E_x(\varrho, z) = E_0 \frac{w_0}{w(z)} \exp\left[\frac{-\varrho^2}{w^2(z)}\right] \quad , \tag{4.3}$$

and the zero phase angle (3.3) by

$$\Phi_x(\varrho, z) = -nkz + \Theta(z) - \frac{nk\varrho^2}{2R(z)} \quad . \tag{4.4}$$

First, we discuss the radial amplitude distribution (4.3) in an arbitrary cross section with fixed value of z. It is described by the simple bell-shaped Gaussian function $\exp[-\varrho^2/w^2(z)]$ (Fig. 4.1). This Gaussian transverse field distribution is the reason for the name Gaussian beam. On the z-axis, i.e. for $\varrho = 0$, the value of the Gaussian function is unity, and at distance $\varrho = w(z)$ from the axis it is $1/e = 0.37$. Since the intensity S is proportional to the square of the field strength, at the distance $\varrho = w(z)$, the intensity is $1/e^2 = 0.135$ of its maximum value on the axis (Fig. 4.2). Therefore, the function $w(z)$ is a measure for the radius of the beam in the cross section considered. It is the same as the spot size already introduced in Sect. 2.1. The beam diameter is $2w(z)$.

Because of diffraction spreading, the spot size depends on the distance z between the cross section considered and the beam waist [Kogelnik and Li

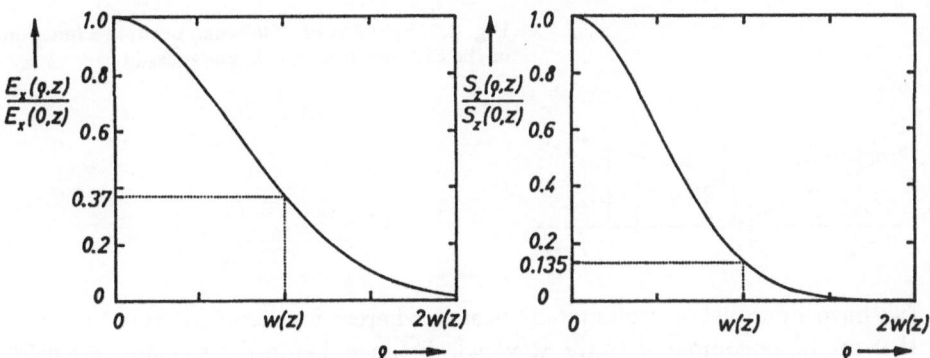

Fig. 4.1. Radial amplitude distribution (4.3) in a Gaussian beam

Fig. 4.2. Radial intensity distribution (4.26) in a Gaussian beam

1966]

$$w(z) = w_0 \sqrt{1 + (z/z_R)^2} \quad . \tag{4.5}$$

In the waist, i.e. for $z = 0$, the spot size has its minimum value w_0.

To give two examples: the spot sizes at the waist for the Gaussian beams emitted by a He-Ne laser at a wavelength of $0.633\,\mu$m and by the end of a typical single-mode fiber at a wavelength of $1.3\,\mu$m are typically $0.5\,$mm and $5\,\mu$m, respectively.

The quantity z_R in (4.5) is an important characteristic length along the beam, the so-called Rayleigh range; it is defined by the equation [Siegman 1971]:

$$z_R = \pi n w_0^2 / \lambda \quad . \tag{4.6}$$

In this equation, the quantity λ is the vacuum wavelength of a uniform plane wave (see Sect. 4.5) of the same frequency. For the beams radiated by a He-Ne laser and from the fiber end into air $(n = 1)$, the values of the Rayleigh range are $1.24\,$m and $60\,\mu$m, respectively.

At a distance of one Rayleigh range from the beam waist, i.e. for $z = z_R$, or $-z_R$, the spot size $w(z_R)$ is larger by a factor of $\sqrt{2}$ than the spot size w_0 in the waist (Fig. 4.3). For values of z with a magnitude very large as compared to the Rayleigh range, the spot size increases in proportion to $|z|$:

$$w(z) = \frac{w_0 |z|}{z_R} = \frac{|z| \lambda}{\pi n w_0} \quad \text{for } |z| \gg z_R \quad . \tag{4.7}$$

Since the beam diameter equals two spot sizes, one sees that in a range of two Rayleigh distances around the waist, the beam diameter is roughly constant

37

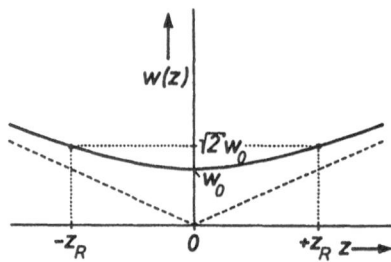

Fig. 4.3. Spot size of a Gaussian beam as a function of the distance from the beam waist (4.5)

(we have a parallel or "collimated" beam), whereas for distances from the waist that are large compared to the Rayleigh distance, i.e. in the so called far-field, the beam boundary (the surface, where the intensity is $1/e^2$ of its value on the beam axis at fixed z) is a circular cone with a divergence angle θ_d given by

$$\tan \theta_d = \frac{w(z)}{|z|} = \frac{\lambda}{\pi n w_0} \quad . \tag{4.8}$$

The beam divergence angle is that angle with respect to the beam axis for which, in the far-field, the intensity is $1/e^2$ of its maximum value. When the spot size w_0 at the waist is large compared to the wavelength λ, the beam divergence is small and the tangent in (4.8) may be approximated by its argument θ_d.

From (4.8) we learn that the diffraction spreading increases with decreasing field diameter at the waist. This fact is well known from antenna techniques: in order to produce a highly collimated antenna beam, one needs an antenna with an aperture which is large compared to a wavelength. In Fig. 4.4, we represent the beam by plotting the beam contour, i.e. the intersections between the beam boundary and the plane $y = 0$. Figure 4.4a corresponds to a Gaussian beam with a large spot size at the beam waist relative to the wavelength, which diverges slowly. In Fig. 4.4b, a small ratio of beam waist and wavelength has been chosen which causes the beam to diverge quickly (strictly speaking, the theory of Gaussian beams must not applied to beams with such a large divergence angle as that sketched in Fig. 4.4b).

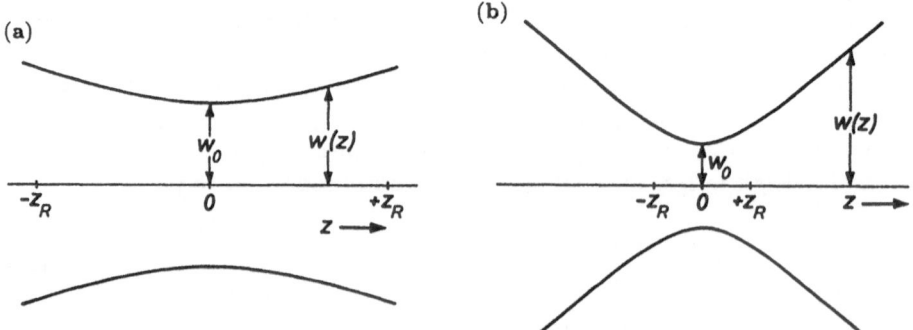

Fig. 4.4a,b. A Gaussian beam with a large beam waist diverges slowly (a) whereas a beam with a small beam waist diverges rapidly (b).

For our numerical example of a Gaussian beam radiated by a He-Ne laser at a wavelength of $0.633\,\mu m$, the beam divergence angle θ_d is only $0.4\,mrad$ or $0.02°$, whereas for the Gaussian beam radiated into air from the end of a single-mode fiber with a spot size of $5\,\mu m$ at a wavelength of $1.3\,\mu m$, it is $82\,mrad$ or $4.7°$.

The far-field radiation pattern, i.e. the normalized electric field strength $E(\theta)/E(0)$ as a function of the polar angle θ measured against the z-axis is given by:

$$\frac{E(\theta)}{E(0)} = \exp\left[-\left(\frac{\theta}{\theta_d}\right)^2\right] \quad . \tag{4.9}$$

Since the far-field distribution is described by a Gaussian function, there are no side lobes.

This concludes the discussion of the field distribution in a beam cross-section. Next, we discuss the amplitude distribution along the beam axis, i.e. for $\varrho = 0$ in (4.3):

$$E_x(0, z) = E_0 \frac{w_0}{w(z)}$$

$$= E_0 \left[1 + \left(\frac{z}{z_R}\right)^2\right]^{-1/2} \quad . \tag{4.10}$$

From this equation, one recognizes that the quantity E_0 introduced in (4.2) is simply the amplitude of the x-component of the electric field vector at the waist $(z = 0)$ on the axis $(\varrho = 0)$. At this point we find the maximum value of the field strength in the beam.

Figure 4.5 represents the z-dependence of the field amplitude on the beam axis. Since we have assumed lossless materials, the power transmitted by the beam does not change along the beam. But because of diffraction spreading, for $z > 0$, the beam diameter increases with increasing value of z. Since the same power is distributed over a larger beam cross section, the intensity and, therefore, the field amplitude on the beam axis will decrease as shown in Fig. 4.5. At a distance of one Rayleigh range from the waist, the amplitude is 0.707 times its value in the waist. For very large values of z, the amplitude is inversely pro-

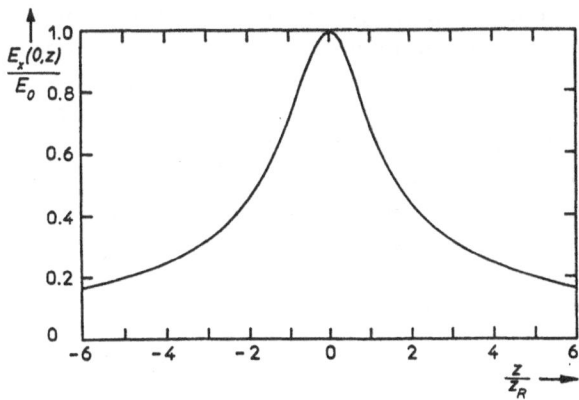

Fig. 4.5. Field amplitude on the axis of a Gaussian beam as a function of the distance z from the beam waist (4.10)

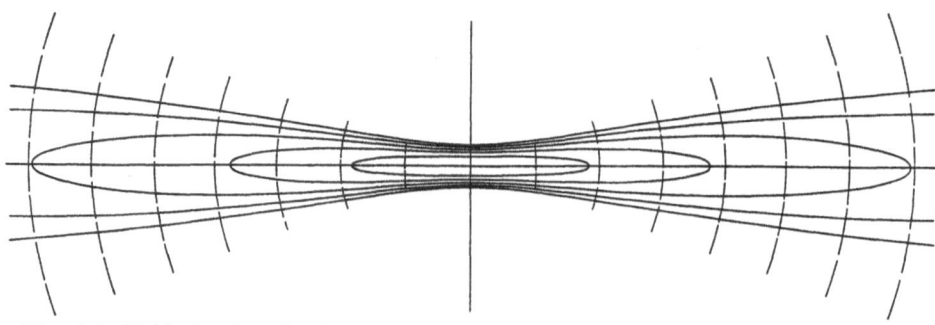

Fig. 4.6. Field of a Gaussian beam in a longitudinal plane represented by lines of constant field strength (*solid lines*) and by wavefronts (*dashed lines*). $w_0/\lambda = 2$

portional to z, and the intensity decreases in inverse proportion to the square of z. This behavior is well known from the far field of any antenna.

In order to further illustrate the Gaussian beam, we use the method of Sect. 3.5.4: In Fig. 4.6 several solid lines of constant amplitude are plotted. The field level $20 \log E_x/E_0$ is related to the maximum electric field strength E_0. Between neighboring constant amplitude contours, the field strength decreases by 5 decibels, i.e. the solid lines represent the loci where the field amplitude is $-5 \, \text{dB}$, $-10 \, \text{dB}, \ldots, -25 \, \text{dB}$ below its maximum value in the center of the waist. For the spot size at the waist, we assumed the special value

$$w_0 = 2\lambda \quad . \tag{4.11}$$

For the Gaussian beam radiated from the end of a single-mode fiber, the ratio of the spot size (e.g. $w_0 = 5 \, \mu\text{m}$) and the wavelength (e.g. $\lambda = 1.3 \, \mu\text{m}$) is larger by a factor of about 2, and the beam divergence is therefore less than that shown in Fig. 4.6.

As we have seen in this section, the transverse field distribution in each beam cross section is described by the same Gaussian function, though its width (the spot size) changes. In waveguide theory, it is usual to call a solution of the wave equation a waveguide "eigenmode" if the transverse field distribution does not change during propagation. If one generalizes this notation slightly by permitting a change of scale in the transverse field distribution along the direction of propagation, one can call the fundamental Gaussian beam (and also the higher order beams to be described in Sect. 4.8) an electromagnetic "mode" of a homogeneous medium. In contrast to waveguides, the parameters describing the free beam can take any value, whereas in waveguides only certain discrete values are allowed. To give an example: the spot size w_0 at the waist of a free Gaussian beam in a homogeneous medium can have any value, whereas in a single-mode fiber, the spot size of the fundamental guided mode has a specific value for which diffraction spreading and focusing just cancel each other, and which is determined by the optical frequency and the refractive index profile of the fiber.

Let us note in passing, that there are optical systems, such as lens wave-guides [Goubau and Schwering 1961] or open optical resonators with spherical mirrors [Boyd G.D. and Gordon 1961; Kogelnik and Li 1966], in which Gaussian beams can propagate. These beams must have a particular width w_0 at the waist, for which the focusing effect of the lens or the mirror just compensates for the diffraction spreading [Siegman 1971]. These Gaussian beams are said to be eigenmodes of the lens waveguide or the optical resonator.

4.2 Phase Distribution

After the spatial distribution of the field strength, we have to discuss the phase distribution in the fundamental Gaussian beam as described by (4.4). First, we study the phase change along the beam axis, i.e. for $\varrho = 0$:

$$\Phi_x(0, z) = -nkz + \Theta(z) \quad . \tag{4.12}$$

In this equation, the quantity k is the vacuum wave number defined by the equation

$$k = \omega\sqrt{\varepsilon_0\mu_0} = \frac{\omega}{c} \quad , \tag{4.13}$$

where the quantities ω, ε_0, μ_0, and c are the optical angular frequency, the permittivity, the permeability, and the velocity of light in vacuum, respectively. The vacuum wave number is equal to the phase constant of a uniform plane wave propagating in free space. The product of the refractive index and the vacuum wave number

$$k_n = nk \quad , \tag{4.14}$$

the wave number of the material, is the phase constant of a uniform plane wave propagating in the material with refractive index n.

Note that in optical communications, the optical frequency is sometimes expressed in terms of the (spectroscopic) wave number $\overline{\nu}$, which is defined (3.46) as the number of wavelengths in vacuo, per unit of length (usually per centimeter)

$$\overline{\nu} = \frac{1}{\lambda} = \frac{f}{c} = \frac{k}{2\pi} \quad . \tag{4.15}$$

The first term, $-nkz$, on the right hand side of (4.12) describes a linear decrease of the zero phase angle in the direction of wave propagation (see the dotted straight line in Fig. 4.7). For a uniform plane wave (Sect. 4.5) propagating in the z-direction, the phase distribution would be described by this term only. The term $\Theta(z)$ in (4.12) can therefore be interpreted as the phase difference between the Gaussian beam and an infinite plane wave. It can be calculated

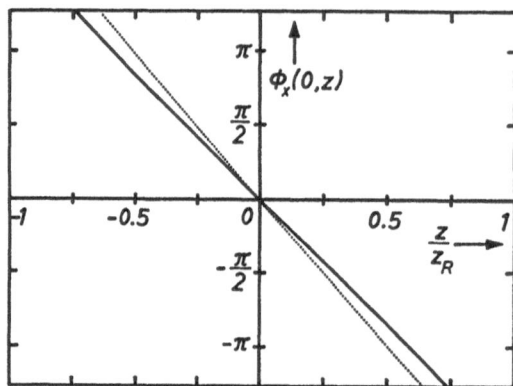

Fig. 4.7. Phase angle of the electric field on the z-axis for a uniform plane wave (*dotted line*) and for the fundamental Gaussian beam mode (4.13) (*solid line*)

from the formula [Kogelnik and Li 1966]:

$$\Theta(z) = \tan^{-1}\frac{z}{z_{\mathrm R}} \ . \tag{4.16}$$

For large negative values of z, the angle $\Theta(z)$ is approximately $-\frac{\pi}{2}$, for $z = 0$ it is zero, and for large positive values of z is approximately $+\frac{\pi}{2}$. By adding the additional term $\Theta(z)$, we get the z-dependence of the zero phase angle represented by the solid curve in Fig. 4.7. In order to show the difference between the phases of a uniform plane wave and of a Gaussian beam, it was necessary to choose an unrealistically small value of the Rayleigh distance $z_{\mathrm R} = \lambda$ in calculating the curves in Fig. 4.7.

Bearing in mind that the Rayleigh distance (where $\Theta(z_{\mathrm R}) = \frac{\pi}{4}$) is many wavelengths, one recognizes that the phase distribution on the axis of a Gaussian beam is very similar to that of a uniform plane wave. Only around the waist is the slope of the phase curve a little smaller. Since the wavelength is the distance between the surfaces of equal phase (Sect. 3.5.3), this means that near the waist, the local value of the wavelength on the beam axis is slightly larger than in a uniform plane wave. On the other hand, the phase velocity is the product of the wavelength and the frequency. Therefore, the local value of the velocity of the wavefronts is slightly increased around the waist. We already found this from simple physical arguments in Sect. 2.1.

In going from large negative values of z through the waist to large positive values of z, the phase change is smaller by π than the phase change in a uniform plane wave. This effect is well known for a converging spherical wave. The field undergoes a phase change of π in passing through the focus. In the optical literature, this effect is called the phase anomaly [Boyd R.W. 1980].

We now consider the phase distribution (4.4) in a cross section of the beam, i.e. for constant z: the phase decrease is proportional to the square of the distance ϱ from the beam axis. From this one can show [Yariv 1971] that the

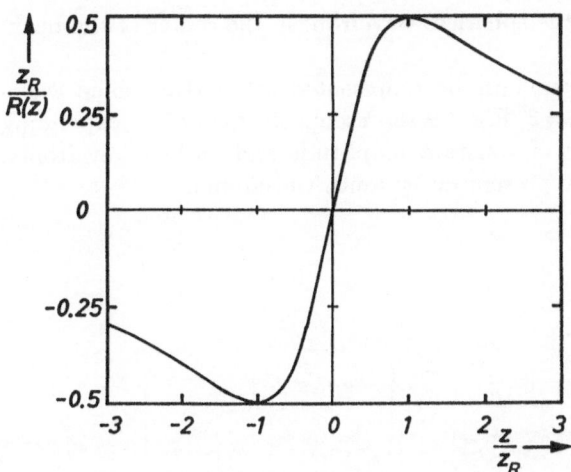

Fig. 4.8. Normalized curvature of the wavefronts in a fundamental Gaussian beam as a function of the distance z from the beam waist (4.17)

wavefronts are spherical surfaces with radius of curvature given by the function $R(z)$ in the denominator of the last term on the rhs of (4.4).

The curvature of the wavefronts, which is the inverse $1/R(z)$ of the radius of curvature, is given by the formula [Kogelnik and Li 1966]:

$$\kappa(z) = \frac{1}{R(z)} = \frac{z}{z^2 + z_{\mathrm{R}}^2} \quad , \tag{4.17}$$

as represented graphically in Fig. 4.8. In the waist, i.e. for $z = 0$, the curvature is zero, which means that the wavefront is a plane. All other wavefronts are spherical surfaces. For positive values of R, the center of curvature lies behind the wavefront, and the beam diverges. For negative values, the center of curvature is situated in front of the wavefront and the beam contracts.

Note, that the radius of curvature $R(z)$ is different from the z-coordinate of the intersection between the wavefront and the beam axis. Thus, the center of curvature of the wavefront is *not* in the center of the beam at $z = 0$. This is in contrast to the case of spherical waves emitted by a point source located at the origin ($\varrho = 0$, $z = 0$) of the coordinate system.

Near the waist, the curvature is proportional to the distance z from the waist. The curvature is maximum at the two points located at one Rayleigh range z_{R} from the waist:

$$\frac{1}{R(z_{\mathrm{R}})} = \frac{1}{2z_{\mathrm{R}}} \quad , \tag{4.18}$$

$$\frac{1}{R(-z_{\mathrm{R}})} = \frac{-1}{2z_{\mathrm{R}}} \quad . \tag{4.19}$$

For very large values of $|z|$, the curvature asymptotically approaches zero and the radius of curvature $R(z)$ can be approximated by the coordinate z of the intersection between the spherical wavefront and the beam axis. Only in this limiting case is the center of the spherical wavefront in the center of the beam waist.

The phase distribution can best be represented by plotting some wavefronts (Sect. 3.5.3). For $w_0/\lambda = 3$, Fig. 4.9 represents the field of an expanding Gaussian beam by solid lines of constant amplitude and dashed wavefronts. (The dashed lines in Fig. 4.6 represent every tenth wavefront).

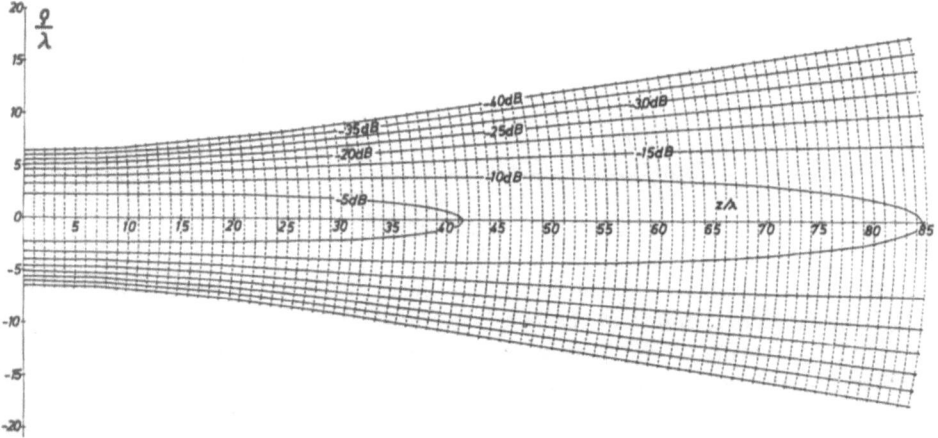

Fig. 4.9. Field of a Gaussian beam in a longitudinal plane represented by lines of constant field strength (*solid lines*) and by wavefronts (*dashed lines*). $w_0/\lambda = 3$

To the right of the waist, i.e. for $z > 0$, R is positive and the beam expands because of diffraction spreading. But as shown in Figs. 4.3–8, to the left of the waist ($z < 0$), there is also a contracting part of the Gaussian beam which is the mirror image of the expanding part of the beam.

Such a contracting beam can be produced simply by putting a convex lens in an expanding Gaussian beam. The transformation of Gaussian beams by a system of lenses or mirrors is an important field in itself (Sect. 4.9).

For large negative values of the coordinate z, we have converging spherical wavefronts. Since the intensity vector is normal to the wavefronts, using arguments from geometrical optics, one would presume that the wave converges to a single point in space, the focus. But during propagation, the beam diameter becomes smaller and as has been explained in Sect. 2.1, the phase velocity near the beam axis increases. Thus the radius of the wavefronts does not decrease to zero at the focus, but has a minimum value at $z = -z_R$ and becomes infinite at $z = 0$. The physical phenomenon of diffraction is thus the reason behind the finite beam diameter w_0 at the focus or waist.

For the fundamental Gaussian beam mode, there is a simple relation between the local curvature $1/R(z)$ of the wavefronts and the increase in spot size per unit length. Remember that the intensity vector is normal to the wavefronts. In the waist, the wavefront is plane, and the intensity vector is parallel to the fiber axis. No (time averaged) power is flowing in the radial direction and therefore, at the waist, the spot size is locally constant. On the other hand, outside the waist, the larger the curvature of the wavefronts the larger is the radial component of the intensity vector and the larger is the radial power flow and the rate of increase in spot size.

These arguments can be quantified to give a simple relation between the derivative of the natural logarithm of the spot size and the curvature of the wavefronts [Arnaud 1969; Neumann 1987]:

$$\frac{d[\ln w(z)]}{dz} = \frac{1}{R(z)} \quad . \tag{4.20}$$

Inserting (4.5) and (4.17) for $w(z)$ and $R(z)$, this equation can easily be verified for a Gaussian beam propagating in a homogeneous medium. For negative z, the radius of curvature is also negative, and the spot size of the beam decreases during propagation, as is illustrated by Fig. 4.6.

The relation (4.20) between the curvature of the wavefronts and the rate of change in spot size also holds for a fundamental Gaussian beam mode propagating in a focusing or defocusing quadratic index medium, e.g. a multimode graded-index fiber or a graded-index lens [Neumann 1987].

Around the beam waist, the z-dependence of the curvature of the wavefronts (4.17) can be approximated by the linear function

$$\kappa(z) = \frac{1}{R(z)} = T_b z \quad , \qquad \text{where} \tag{4.21}$$

$$T_b = \frac{1}{z_R^2} = \frac{\lambda^2}{(\pi n w_0^2)^2} \quad . \tag{4.22}$$

Inserting (4.21) into (4.20) and integrating, gives for the evolution of the spot size around the waist

$$w(z) = w_0 \exp\left(\frac{T_b z^2}{2}\right) \quad . \tag{4.23}$$

The larger the quantity T_b, the faster the increase both in the curvature of the wavefronts and in the spot size. Therefore the quantity T_b may be used as a measure for the tendency of the beam to diverge because of diffraction [Neumann 1987]. T_b is called the diffraction tendency of the beam.

4.3 Electric and Magnetic Field Lines

Until now, we have only considered the x-component of the electric field strength, which is the dominant one. It is now time also to ask about the other two components of the electric field and the three components of the magnetic field.

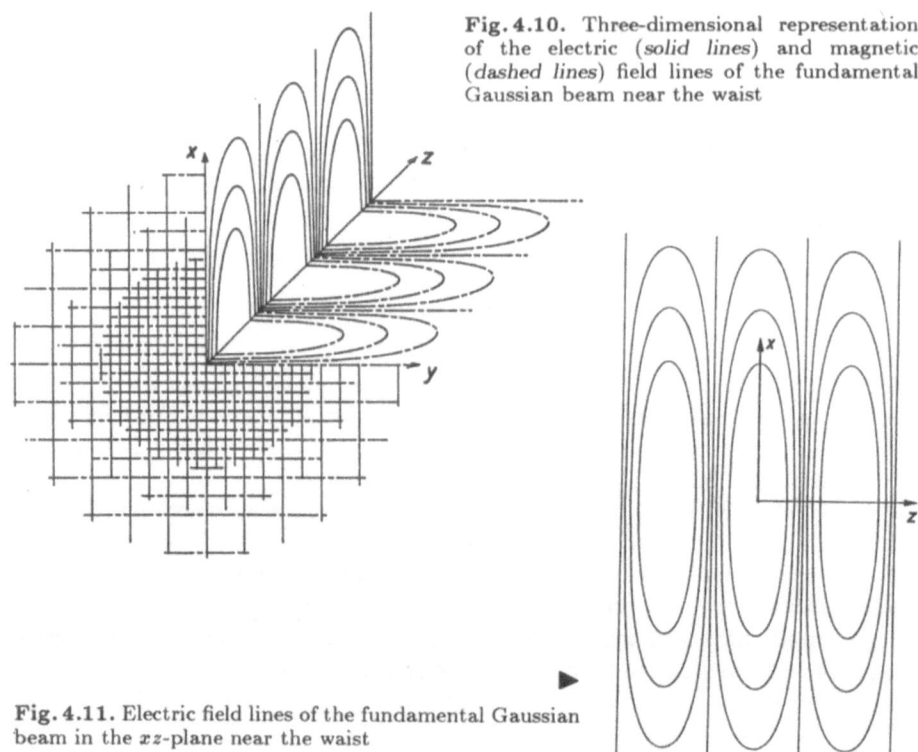

Fig. 4.10. Three-dimensional representation of the electric (*solid lines*) and magnetic (*dashed lines*) field lines of the fundamental Gaussian beam near the waist

Fig. 4.11. Electric field lines of the fundamental Gaussian beam in the xz-plane near the waist

The structure of the electromagnetic field of a Gaussian beam is best illustrated by drawing the electric and magnetic field lines.

In Fig. 4.10, we use the method of Sect. 3.5.1 to represent the electromagnetic field near the beam waist by plotting solid lines for the electric field lines and dashed lines for the magnetic field lines. First we have projected the electric field lines onto three-quarters of the cross section at the waist. In the approximations used here, the other transverse component of the electric field, i.e. $\underline{E}_y$, is zero. Thus the field lines projected on the cross section are parallel to the x-direction. Since the field strength decreases like a Gaussian function with increasing distance ϱ from the beam axis, the density of the field lines must decrease in the radial direction.

In the cross section, it may appear that the electric field lines end freely in space. But since there are no space charges, the field lines have no ends and close in the longitudinal, i.e. the z-direction. In order to illustrate this, we have also plotted the electric field lines stereoscopically in Fig. 4.10 in part of the longitudinal $x - z$-plane and for clarity once more in another figure (Fig. 4.11) in the $x - z$-plane. From these pictures, one sees that in a Gaussian beam there is inevitably also a longitudinal component $\underline{E}_z$ of the electric field [Lax et al. 1975; Simon et al. 1986] which is zero in the plane $x = 0$ and has opposite signs for $x>0$ and for $x<0$. When the beam diameter is large compared to the wavelength, the longitudinal component is usually very small compared to the transverse component.

As can be seen from the pictures of the field lines, the maximum of the longitudinal component is in a cross section a quarter of a wavelength distant from the cross section of maximum transverse component. Considering the time dependence of both components, this means that there is a phase shift of 90° between the longitudinal and transverse field components.

The dominant component of the magnetic field vector $\boldsymbol{H}$ lies in the y-direction, i.e. normal to the transverse component of the electric field. Since no magnetic materials are present, the magnetic field lines are also closed loops lying in longitudinal planes parallel to the $y - z$-plane. The picture of the (dashed) magnetic field lines can be obtained simply by turning the electric field lines through 90° about the beam axis (Fig. 4.10). As is the case for the electric field, the longitudinal component of the magnetic field vector is small compared to the transverse component and in phase quadrature with it.

Since there are longitudinal components of both the electric and the magnetic field, the Gaussian beam is an example of a so-called "hybrid" wave (In hollow metallic waveguides, one has two classes of modes, those with only a longitudinal component of the electric field, the so-called transverse magnetic or TM waves, and those with a longitudinal component of the magnetic field only, so called TE waves).

There is a simple approximate relation between the phasors of the transverse components of the magnetic and electric field vectors:

$$\underline{H}_y(\varrho, z) = \frac{n\underline{E}_x(\varrho, z)}{Z_0} \quad , \tag{4.24}$$

where the quantity

$$Z_0 = \sqrt{\mu_0/\varepsilon_0} = 377\Omega \tag{4.25}$$

is the wave impedance of free space. Since $\underline{H}_y$ equals $\underline{E}_x$ multiplied by the real constant n/Z_0, the transverse components of the electric and magnetic field oscillate in phase and depend in the same manner on the space coordinates ϱ and z.

In order to illustrate the change of the field diameter along the beam axis, Fig. 4.12 shows some electric field lines in the longitudinal $x - z$-plane around the beam waist for a larger range of z-values.

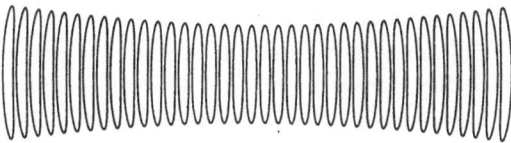

Fig. 4.12. Electric field lines of the fundamental Gaussian beam in the xz-plane showing the beam divergence

4.4 Power Transmitted by a Gaussian Beam

Knowing the electric and magnetic fields of a Gaussian beam, we now can calculate the time average of the longitudinal component of the Poynting vector (3.16) by taking half of the real part of the cross product of the complex electric field vector (4.1) and the complex conjugate of the complex magnetic field vector (4.24):

$$S_z(\varrho, z) = \frac{n w_0^2 E_0^2}{2 w^2(z) Z_0} \exp\left[\frac{-2\varrho^2}{w^2(z)}\right] \quad . \tag{4.26}$$

Since the intensity is proportional to the square of the electric field component E_x, whose distribution in space has been discussed in detail before, we proceed to calculate the power P transmitted by the beam by integrating the power density, i.e. the function $S_z(\varrho, z)$, over the total beam cross section at an arbitrary location z (3.17). Since the indefinite integral of the Gauss function multiplied by its argument is simply the Gauss function itself, one gets a closed form expression for the transmitted power:

$$P = \frac{\pi n E_0^2 w_0^2}{4 Z_0} \quad . \tag{4.27}$$

Since a lossless medium has been assumed, the transmitted power is constant along the beam. Equation (4.27) can be used to calculate the maximum electric field strength E_0 in the waist, which cannot be measured directly at optical frequencies, from the transmitted power P, which can easily be measured with an optical power meter (Sect. 13.2).

One can write the expression for the transmitted power in another form, which is easy to remember:

$$P = \tfrac{1}{2} S_z(0, z) A_b(z) \tag{4.28}$$

where the quantity $S_z(0, z)$ is the intensity on the beam axis ($\varrho = 0$) in the cross section characterized by the coordinate z, and where the quantity

$$A_b(z) = \pi w^2(z) \tag{4.29}$$

is the area of the beam cross section circumscribed by the beam boundary at which the intensity is 13.5 % of the value on the axis. Thus the transmitted power is simply half of the product of the intensity on the beam axis and the area of the beam cross section. This holds independently of the value of z.

Some authors [e.g. Geckeler 1986a] define a different spot size σ_ϱ as that distance from the beam axis for which the *intensity* is $1/e = 0.37$ of its value on the axis, at fixed z. Setting the argument of the Gaussian function in (4.26) equal to unity, gives a relation between our amplitude-based spot size w and the alternate intensity-based spot size σ_ϱ:

$$w_G = 2^{1/2}\sigma_\varrho \quad . \tag{4.30}$$

Using this spot size σ_ϱ, the transmitted power is the product of the intensity on the beam axis and the beam cross section $\pi\sigma_\varrho^2$ without the factor of $\frac{1}{2}$ in (4.28):

$$P = S_z(0,z)\pi\sigma_\varrho^2 \quad . \tag{4.31}$$

Since the standardization organizations recommend that the spot size be defined as the $1/e$-*amplitude* radius, in this book we use w instead of σ_ϱ to characterize the radial field extent.

There is an important application of the relationship between the transmitted power, the beam cross-sectional area, and the maximum intensity: consider the end of a single-mode fiber guiding a fundamental mode with power P. For a plane and normal endface, 4 % of this power is Fresnel reflected (Sect. 5.9.5), and 96 % is radiated from the end as a diverging beam which is very similar to the Gaussian beam discussed in this chapter. In order to avoid damage to the eye, the intensity at the eye must not be larger than a certain value, which depends on the optical wavelength used. For a given distance z between the fiber end and the eye, one can use (4.7) to calculate the spot size $w(z)$ at the location of the eye. Then (4.29) gives the beam area $A_b(z)$ and (4.28) the maximum possible intensity $S_z(0,z)$ at the eye. By comparing this value with the maximum value allowed, one can check whether one is within safety limits [Petersen and Philen 1986].

If we integrate the Poynting vector not over the total cross section of the wave beam, but only over a circular area with radius ϱ, we get for the part of the power transmitted through this area

$$P(\varrho) = P\left\{1 - \exp\left[\frac{-2\varrho^2}{w^2(z)}\right]\right\} \quad . \tag{4.32}$$

Taking for the radius ϱ the spot size $w(z)$, we get

$$P(w) = 0.86P \quad , \tag{4.33}$$

meaning that 86 % of the total power is transmitted within the boundary defined by the $1/e^2$-intensity points.

In a diverging or converging Gaussian beam, besides the dominant longitudinal component $S_z(\varrho, z)$ of the Poynting vector, there is also a small radial component $S_\varrho(\varrho, z)$ which causes the redistribution of the axial intensity along the beam.

4.5 Uniform Plane Wave

A wave with plane wavefronts is called a plane wave. If the amplitude of the field strength is constant on the wavefronts, it is called a uniform or infinite plane wave. In a uniform plane electromagnetic wave, the electric and magnetic field vectors are purely transverse and orthogonal to each other.

In this section, we want to show, that a fundamental Gaussian beam transforms into a uniform plane wave, if the spot size w_0 at the waist increases infinitely. To show this, we choose for the spot size a very large value, for instance $w_0 = 1000\,\lambda\,(n = 1)$. From (4.6), we obtain for the Rayleigh range $z_R = 3\,141\,593\,\lambda$. We consider the electromagnetic field distribution within a cylindrical volume of a radius $\varrho = 0.1\,w_0 = 100\,\lambda$ between the two planes $|z| = 0.1\,z_R = 314\,159\,\lambda$.

From (4.17), it follows that in the volume considered, the radius of curvature of the wavefronts is larger than $10.1\,z_R = 3.2\,10^7\,\lambda$, meaning that the wavefronts are practically planar. Within this volume, the amplitude of the field strength is constant within $1.5\,\%$ (4.3). Thus, the field approximates that of a uniform plane wave.

Since the transverse field extent is very large, the electric field lines are very long in the transverse direction (Fig. 4.11). In a cross section of maximum transverse field component, the density of the field lines is very large, i.e. the transverse field component is very large. In another cross section a quarter wavelength from the first, the field lines are purely longitudinal, and their density is very small. Thus, the longitudinal component of the electric field is vanishingly small as compared to the transverse component. The same holds for the longitudinal component of the magnetic field in relation to the transverse component. Therefore, the electric and magnetic field vectors are approximately normal to the direction of wave propagation. As in the uniform plane wave, they are also orthogonal to each other.

Finally, we compare the wavelength of the wide Gaussian beam with that of a uniform plane wave. From (4.12) one can deduce that the phase change in the beam from $z = -0.10\,z_R$ to $0.1\,z_R$ is $226\,194\,659.7°$, whereas in a strictly uniform plane wave, it is $226\,194\,671.1°$. Since the wavelength is the distance for the phase angle to change by $360°$, we conclude that in the volume considered, the wavelength in the beam differs only very slightly from that in a uniform plane wave, which is related to the wave number k in vacuum (4.13) by:

$$\lambda = 2\pi/k \quad . \tag{4.34}$$

In text books, the uniform plane wave is usually introduced as that solution of the wave equation which is mathematically most simple. But this solution has an infinite field extent in the transverse direction and transmits infinite power. Therefore, the uniform plane wave represents a mathematical fiction and can never be realized in nature. An electromagnetic field similar to a uniform plane wave can exist only in a limited spatial region, e.g. in a small volume far from an antenna.

In contrast to the fictitious uniform plane wave, the Gaussian beam as discussed in this chapter has a finite beam diameter, transmits a finite amount of power, and can therefore be realized physically. Our numerical example shows that the field in a cylindrical volume is very similar to the wave field of the ideal uniform plane wave provided the spot size at the beam waist is very large compared to the wavelength.

Since the electric and magnetic field vectors in a uniform plane wave are transverse to the direction of propagation, it is also called a transverse-electromagnetic wave or TEM wave. We have seen that a wide Gaussian beam is very similar to the TEM wave; therefore, it is sometimes called a "quasi TEM wave". But, strictly speaking, it is a hybrid wave with nonzero longitudinal components of both the electric and the magnetic field!

At this point it is advantageous to review some other properties of a uniform plane wave propagating in a homogeneous medium with refractive index n.

The intensity S of the uniform plane wave is related to the electric field strength by

$$S = \frac{nE_0^2}{2Z_0} \quad , \tag{4.35}$$

as can be seen by approximating $w(z)$ by w_0 in (4.26) and by taking into account that ϱ has been assumed to be very small compared to w_0.

For a uniform plane wave, the phase constant β is the refractive index n multiplied by the wave number k in vaccuum:

$$\beta = nk = k_n \quad . \tag{4.36}$$

as can be seen by neglecting $\Theta(z)$ in (4.12). The quantity k_n is the wave number of the medium (4.14).

Several quantities are used for characterizing the velocities of wave propagation and their wavelength dependency; these are now discussed for the most simple case of a uniform plane wave propagating in an infinite lossless medium with refractive index n. Later, in Sects. 5.6–7, we will introduce the corresponding quantities for the fundamental mode propagating in a single-mode fiber.

By definition, the refractive index n is the ratio of the velocity of light in vacuum c and the phase velocity v_p of a uniform plane wave in the medium considered. Therefore, the phase velocity can be written

$$v_{\mathrm{p}} = \frac{c}{n} \quad . \tag{4.37}$$

The wavelength dependence of the refractive index can be approximated by a Sellmeier expansion of the form [Fleming J.W. 1978]

$$n^2 = 1 + \sum_{i=1}^{3} \frac{A_i \lambda^2}{\lambda^2 - \lambda_i^2} \quad , \tag{4.38}$$

where the coefficients A_i are constants related to material oscillator strengths and the λ_i's are oscillator wavelengths. The values of Sellmeier coefficients A_i and λ_i are tabulated for limited choices of composition of silica and its dopants [Fleming J.W. 1978]. The wavelength dependence of the refractive index of both undoped and doped fused silica has been measured several times [Malitson 1965; Fleming J.W. 1976; Hammond and Norman 1977; Hammond 1978; Fleming J.W. 1978; Shibata N. et al. 1981b; Fleming J.W. and Wood 1983].

One is often interested in the time Δt_{p} needed for the wavefront to propagate a certain distance Δz:

$$\Delta t_{\mathrm{p}} = \Delta z / v_{\mathrm{p}} \quad , \tag{4.39}$$

or in the time

$$\tau_{\mathrm{p}} = \frac{\Delta t_{\mathrm{p}}}{\Delta z} \quad , \tag{4.40}$$

needed to propagate unit distance. Inserting (4.39) into (4.40) gives

$$\tau_{\mathrm{p}} = \frac{1}{v_{\mathrm{p}}} \quad , \tag{4.41}$$

i.e. the phase delay time τ_{p} per unit length simply is the inverse of the phase velocity.

Expressing the phase velocity v_{p} by the vacuum velocity c and the refractive index n (4.37) gives

$$\tau_{\mathrm{p}} = \frac{n}{c} \quad . \tag{4.42}$$

In vacuum, the phase delay time per unit length is

$$\tau_0 = 1/c = 3.33\,\mu\mathrm{s/km} \quad , \tag{4.43}$$

so that one can write for τ_{p}:

$$\tau_{\mathrm{p}} = n\tau_0 \quad . \tag{4.44}$$

Thus the phase delay in the material is simply the delay in vacuum multiplied by the refractive index.

The phase velocity is the velocity of the wavefronts of a wave with sinusoidal time dependence (Sect. 3.5.3). It is related to the phase constant by (3.13)

$$v_{\mathrm{p}} = \frac{\omega}{\beta} \quad . \tag{4.45}$$

When an optical wave is used for communications, its power is modulated with the signal to be transmitted. The wave envelope does not propagate with the phase velocity, but with the group velocity v_{g}. The formula for the group velocity that is simplest to remember is

$$v_g = \frac{d\omega}{d\beta} \quad ,$$ (4.46)

Note that the phase velocity is the ratio of the angular frequency and the phase constant (4.45), whereas the group velocity is the derivative of the angular frequency with respect to the phase constant.

It is now convenient to define a group index n_g by the ratio of the velocity of light in vacuum and the group velocity of a uniform plane wave [Gloge 1971b]. Then, the group velocity can be expressed by the equation

$$v_g = \frac{c}{n_g} \quad ,$$ (4.47)

which is equal in form to the corresponding equation (4.37) for the phase velocity.

The inverse of the group velocity is the time needed for the wave envelop to propagate over unit length, the so called specific group delay time

$$\tau_g = \frac{1}{v_g} \quad .$$ (4.48)

Using (4.47) and (4.43), the specific group delay can also be expressed by the group index

$$\tau_g = n_g/c = n_g\tau_0 \quad .$$ (4.49)

Combining (4.13, 36, 46, 49), the group index can be related to the refractive index n and its derivative $dn/d\lambda$:

$$n_g = \frac{d(nk)}{dk} = n - \lambda\frac{dn}{d\lambda} \quad .$$ (4.50)

For media without dispersion, i.e. with constant refractive index, the group index is equal to the refractive index and the group velocity therefore equals the phase velocity. For the glasses used in optical communications, the refractive index decreases slightly with increasing wavelength [Fleming J.W. 1976; Fleming J.W. 1978].

In the 1.3–1.6 μm region, the group index n_g is about 1 % larger than the refractive index n [Kapron 1977].

For later use, we also give a formula for the product of the group index and the refractive index of the medium, which can be derived by multiplying (4.50) by n:

$$nn_g = n^2 - \frac{\lambda}{2}\frac{dn^2}{d\lambda} \quad .$$ (4.51)

Note that $dn^2/d\lambda (= 2n\, dn/d\lambda)$ is the first derivative of the refractive index squared and neither the second derivative $(d^2n/d\lambda^2)$ nor the square of the first derivative $[(dn/d\lambda)^2]$!

Since distortions may be caused by the wavelength dependence of the group delay time τ_g (Chap. 11), a material dispersion parameter D_m is defined as the derivative of the specific group delay time τ_g of a uniform plane wave in the infinite homogeneous material with respect to wavelength:

$$D_m = \frac{d\tau_g}{d\lambda} \quad . \tag{4.52}$$

Expressing τ_g in (4.52) by the group index (4.49) and using the relation (4.50) between the group index n_g and the refractive index n, one obtains for the material dispersion parameter

$$D_m = -\frac{\lambda}{c}\frac{d^2n}{d\lambda^2} \quad . \tag{4.53}$$

For silica glass, the material dispersion is about -70, 0, and $20\,\mathrm{ps/(km\,nm)}$ at wavelengths of 0.90, 1.30, and $1.55\,\mu m$, respectively [Fleming J.W. 1978]. Doping the silica with GeO_2 or P_2O_5 to increase the refractive index, or with B_2O_3 or F in order to reduce the refractive index, slightly changes the value of the material dispersion parameter D_m.

For pure silica, the material dispersion D_m is zero at a wavelength of $\lambda_{0m} = 1.273\,\mu m$ [Srivastava and Franzen 1985]. Dopants like Ge and P shift this wavelength to slightly higher values.

Material dispersion data for fiber glasses have been reported by Wemple [1979] and Bachmann et al. [1983a].

Table 4.1 presents an overview of the different quantities introduced to characterize the wave velocities and their derivatives.

4.6 Polarization of Gaussian Beams

Until now, we have only considered a fundamental Gaussian beam linearly polarized in the x-direction, i.e. with the transverse electric field component

Table 4.1. Quantities used for characterizing wave velocities. The physical meaning of the inverse of the velocity of light in vacuum is the specific delay time $\tau_0 = 1/c = 3.33\,\mu s/km$ in vacuum

Quantity	Symbol	Definit.	Unif. plane wave	Fund. fiber mode
Phase velocity	v_p	ω/β	c/n	c/n_e
Spec. phase delay	τ_p	$1/v_p$	n/c	n_e/c
Group velocity	v_g	$d\omega/d\beta$	c/n_g	c/n_{ge}
Spec. group delay	τ_g	$1/v_g$	n_g/c	n_{ge}/c
Chromat. dispersion	D	$d\tau_g/d\lambda$	D_m	$D_m + D_w$

$\underline{E}_x$. When we rotate this wave field by 90° around the z-axis, we get another possible form of wave beam, which is linearly polarized in the y-direction, i.e. with a transverse electric field component $\underline{E}_y$.

The second beam with orthogonal polarization is perfectly independent of the first one. We assume the beam waists of both beams to be located in the same plane $z = 0$ and the spot sizes in the waist to be equal. The two beams can be launched with different amplitudes and phase angles, i.e. with different complex amplitudes $\underline{E}_{0x}$ and $\underline{E}_{0y}$. The transverse part of the total electric field vector is obtained by vector addition of the transverse parts of the field vectors of each beam. Considered as a function of time, the endpoint of the resulting field vector will generally describe an ellipse. At each observation point in the beam, the field vector describes an ellipse of the same form: one speaks of an elliptically polarized Gaussian beam.

For special values of the complex amplitudes one obtains linear or circular polarization: When the orthogonal components vibrate in phase or antiphase (i.e. when the phase angles of the phasors are equal or differ by 180°) one obtains linear polarization with the direction of the resulting field vector being determined by the ratio of the moduli of the complex field amplitudes. If there is a phase difference of +90° or −90° between the orthogonal field components and if their real amplitudes are equal, the endpoint of the resulting field vector describes a circle and one has a circularly polarized Gaussian beam.

4.7 Curved Gaussian Beams

Until now, we assumed the Gaussian beam to propagate in a homogeneous medium, i.e. a transparent material with a constant value of the refractive index n. In such a medium, the beam will follow a straight path, i.e. the beam axis will be a straight line.

Now we assume the refractive index to increase linearly in a transverse direction, e.g. in the x-direction:

$$n(x) = n(0) + gx \quad , \quad g > 0 \tag{4.54}$$

where the quantity g is the x-component of the gradient of the scalar function n. It determines the rate of change of the refractive index in the x-direction. We consider a Gaussian beam propagating locally at $z = 0$ in the z-direction with its waist at $z = 0$. Because of the transverse gradient of the refractive index, the local velocity of the wavefronts will be larger for $x > 0$ and smaller for $x > 0$ than at $x = 0$. The wavefronts will thus turn during propagation. Since the power is locally transmitted normal to the wavefronts, the beam will follow a curved path.

In order to illustrate this behavior, Fig. 4.13 shows some electric field lines drawn in the plane of the curved beam axis. The radius of curvature R of the beam axis is simply the inverse of the transverse component g of the index

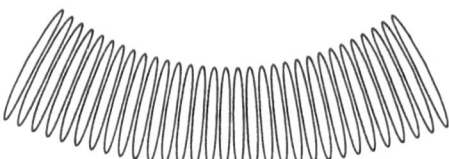

Fig. 4.13. Electric field lines of a bent Gaussian beam in medium with transverse index gradient

gradient multiplied by the refractive index $n(0)$ on the beam axis [Unger 1976; Iizuka 1985]:

$$R = \frac{n(0)}{g} \quad . \tag{4.55}$$

This equation can readily be obtained by considering the local wavelengths, i.e. the distances between two wavefronts, taken at two different values of the transverse coordinate x and by equating the ratios of the sides of two similar triangles.

Equation (4.55) only applies if the beam axis departs only slightly from the z-axis. But we can generalize by stating that a Gaussian beam will bend whenever there is a transverse gradient of the refractive index. The beam bends in the direction of increasing refractive index. The radius of curvature of the beam axis is the inverse of the magnitude of the vector $\mathrm{grad}\, n$ multiplied by the refractive index on the beam axis.

In nature, the bending of light beams can be observed above hot roads or hot sand. It gives rise to mirages and is also called the "Fata Morgana" effect.

Vector wave solutions for the propagation of Gaussian beams in inhomogeneous media with slow spatial variations of the index of refraction (or gain, or loss) have been obtained by Casperson [1973; 1976].

4.8 Gaussian Beams of Higher Order

The fundamental Gaussian beam discussed so far is the most important Gaussian beam. However, there are also other forms that have more than one intensity maximum in the beam cross section; these are the so-called higher order Gaussian beams.

For the beginner, it is a little confusing that the form of the higher order Gaussian beam depends on the coordinate system used to describe the location of the point of observation [Siegman 1971]. In Cartesian coordinates, the transverse field distribution is given by the product of a Gauss function and two Hermite polynomials [Boyd G.D. and Gordon 1961], whereas in cylindrical polar coordinates, it is given by the product of a Gauss function, a power of ϱ, and an associated Laguerre polynomial [Goubau and Schwering 1961]. But each Hermite-Gaussian beam can be regarded at as a superposition of Laguerre-Gaussian beams and vice versa [Siegman 1971; Iga et al. 1984].

In the context of single-mode fibers, we are mainly interested in one of the higher order beams that is very similar to the first higher order mode of a fiber, the LP_{11} mode (Sect. 6.3). Since we are now considering Gaussian beams different from the fundamental beam treated so far, we have to introduce indices.

It is unfortunate that there is no standard notation for numbering the different Gaussian modes. For instance, one can first give the mode number belonging to the azimuthal direction and then the number belonging to the radial direction or vice versa. Since our interest is mainly in fiber waveguides, we use the same numbering for Gaussian beam modes as is generally used for the pertinent LP_{lm} fiber modes (Sect. 6.1): the first integer index, l, is half the number of intensity maxima found in going once round the beam axis in the azimuthal direction, and the second index m is the number of intensity maxima found in going from the beam axis in the radial direction. The same mode numbering has also been used by Facq et al. [1984]. It would make sense to use the same mode designation for a wave beam with a given field structure, independent of the medium through which it propagates, e.g. air, a fiber, or an integrated optic channel waveguide.

For the fundamental Gaussian beam, the intensity does not depend on the azimuth ϕ, therefore the first index l is zero. In the radial direction, we have only one maximum which is located on the axis; therefore the second index m is unity. Therefore, the fundamental Gaussian beam is also called the quasi TEM_{01} beam.

The amplitude distribution of the only higher order beam we are interested in, is given in cylindrical polar coordinates (Laguerre-Gaussian mode) by the equation [Kogelnik and Li 1966]:

$$E(\varrho, \phi, z) = E_0 \frac{w_0 \varrho}{w^2(z)} \cos \phi \exp\left(-\frac{\varrho^2}{w^2(z)}\right) \quad . \tag{4.56}$$

When using Cartesian coordinates (Hermite-Gaussian mode), one gets the same expression but with $\varrho \cos \phi$ replaced by x and ϱ^2 replaced by $x^2 + y^2$.

We compare this expression with the corresponding equation (4.3) for the fundamental Gaussian beam mode. There is an additional factor $\varrho \cos \phi / w(z) = x/w(z)$ which causes the field to be zero in the plane $x = 0$ and to be of opposite sign for $x > 0$ and for $x < 0$. The z-dependence of the spot size $w(z)$ is still given by (4.5).

In order to illustrate the transverse field distribution in this higher order beam, we use the methods of Sect. 3.5.2 and 3.5.7: Fig. 4.14a shows two electric field vectors, together with the boundaries of areas of high intensity. We now find two intensity maxima in the azimuthal and one intensity maximum in the radial direction. The azimuthal and radial beam indices are therefore $l = 1$ and $m = 1$, and the wave beam is designated the quasi TEM_{11} beam.

Instead of the cosine function in (4.56), we can also use the sine function. This gives the field distribution drawn in Fig. 4.14b. Since the numbers of in-

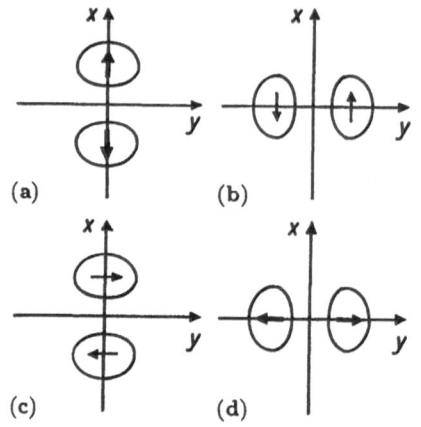

Fig. 4.14. Transverse field distribution of the second order Gaussian beam mode represented by some electric field vectors and lines of constant intensity. The four modes differ in polarization (direction of the electric field vector) and orientation ($\cos \phi$ or $\sin \phi$). This figure also represents the four LP_{11} modes guided by a fiber

tensity maxima in the azimuthal and radial directions are the same as before, this new beam is characterized by the same indices. The two quasi TEM_{11} beams shown in Fig. 4.14a and Fig. 4.14b are said to differ in orientation.

Finally, one can turn the electric field vector by 90° around the beam axis without changing the intensity distribution. This gives two additional possible quasi TEM_{11} beams (Figs. 4.14c and d), which are different in their polarization.

Whereas there were two fundamental quasi TEM_{01} beams of different polarization, we now have four quasi TEM_{11} beams of different orientation ($\cos \phi$ or $\sin \phi$) and polarization (E_x or E_y). The same behavior will be found later for the fundamental LP_{01} modes (Sect. 5.3.5) and the higher order LP_{11} modes of a fiber waveguide (Sect. 6.3.1).

In this section on higher order Gaussian beams, we have only discussed the amplitude or intensity distributions. Concerning the phase distribution it suffices to notice that the wavefront in the beam waist at $z = 0$ is planar. Formulae describing the phase distribution can be found in the literature [Siegman 1971].

4.9 Propagation Through Optical Systems

When a Gaussian beam is transmitted through an optical element such as a single lens, a combination of lenses, a mirror, a length of homogeneous or lens-like medium, it changes its spot size w and the radius R of the wavefronts. In optical communications, it is often necessary to analyze the transformation of Gaussian beams by optical elements, e.g. when optimizing a lens system for coupling a single-mode fiber to a semiconductor laser (Sect. 12.2.2).

It is important to note that the methods of ray optics must be applied with caution. When the distance between a focusing lens and the beam waist is larger than the Rayleigh distance (4.6), one is allowed to apply the well-known lens maker's formula for calculating the location of the waist of the transformed beam behind the lens. However, if the lens is too close to the beam waist, the focal point predicted by the standard thin lens formula will be too far from the lens, i.e. one observes a "focal shift" towards the lens [Williams C.S. 1973; Carter 1982; Self 1983; Sucha and Carter 1984].

An extreme example of the difference in behavior between Gaussian beams and rays emitted from a point source occurs when the waist of the incident beam is at the front focal plane of a positive lens, in which case the emerging beam has its waist at the back focal plane [Self 1983]. This is inexplicable in terms of geometrical optics, which predicts that a point object at the front focus yields a collimated ray bundle in the image space, i.e. the image is at infinity.

In these cases one may not use the simple lens maker's formula derived on the basis of geometric optics, but must turn instead to more accurate wave-optical techniques for analyzing the propagation of Gaussian beams through optical systems [Deschamps and Mast 1964; Tien et al. 1965; Kogelnik 1965a, 1965b; Miller S.E. 1965; Kogelnik and Li 1966; Yariv 1971; Deschamps 1971; Ronchi and Scheggi 1980; Self 1983; Herloski et al. 1983a, 1983b; Tanaka K. 1984].

The most popular method is based on the concept of a 2×2 transfer or $ABCD$-matrix [Kogelnik 1965a, 1965b; Kogelnik and Li 1966; Yariv 1971]. In this powerful technique, a complex beam parameter or complex curvature radius q is defined by the equation

$$\frac{1}{q} = \frac{1}{R} - \frac{jn\lambda}{\pi w^2} \quad . \tag{4.57}$$

The complex beam parameter combines the information about the radius R of the wavefront and the spot size w.

For describing the beam transformation by an optical element, one has to define a transverse input reference plane and a transverse output reference plane. The complex radius q_2 in the output plane of the optical device is related to the complex radius q_1 in the input plane by the so called $ABCD$-law [Kogelnik 1965a]

$$q_2 = \frac{Aq_1 + B}{Cq_1 + D} \quad . \tag{4.58}$$

The letters $A, B, C,$ and D denote the elements of the $ABCD$-matrix.

When the device is analyzed in terms of geometrical optics, the $ABCD$-matrix relates the ray position and ray slope at the output to the ray parameters at the input, and is therefore also called the ray matrix. The matrix elements for thin single lenses, combinations of lenses, spherical mirrors, refracting plane or spherical interfaces and sections of homogeneous, focusing, or defocusing media, can be found in the literature [Kogelnik 1965b; Kogelnik and Li 1966; Yariv 1971].

Using the $ABCD$-law, one can easily calculate the spot size w_2 and wavefront radius R_2 at the output, from the corresponding values w_1 and R_1 at the input of the optical component.

Optical systems are composed of elementary optical devices. The overall $ABCD$-matrix of the optical system can be calculated as the ordered matrix product of the $ABCD$-matrices of the individual elements that constitute the system.

Thus the techniques for analyzing the propagation of Gaussian beams through optical systems are well known. The opposite process of synthesis, i.e. the design of an optical system which will produce some desired transformation of a beam has been studied by Casperson [1981].

A special problem of synthesis that is frequently encountered in optical communications is the transformation of a Gaussian beam with a given waist size w_{01} into another Gaussian beam with waist size w_{02}. A lens inserted into the beam performs the required transformation. Simple design rules for the focal length of the lens and the distances between the lens and the two beam waists have been published by Kogelnik [1964b].

The field of propagation, reflection, and scattering of wave beams is still under active investigation as can be seen from the fact that a recent issue of the Journal of the Optical Society of America has been devoted to "Propagation and Scattering of Beam Fields" [Tamir and Blok 1986].

5. The Fundamental Fiber Mode

In Sects. 2.2, 3, waveguiding by a single-mode fiber was explained in physical terms. In this chapter, the wave beam guided by the fiber will be described in more detail.

We have shown in Sect. 2.2, that the increased refractive index in the core of a single-mode fiber is necessary to compensate for the natural tendency of a wave beam to diffraction spreading. In order to guide the fundamental mode with constant field diameter, the transverse distribution of the refractive index in the core can be chosen rather arbitrarily. For single-mode operation, it is simply necessary to avoid index profiles which can also guide higher order modes (Chap. 6).

Thus one is free to choose the refractive index profile in such a manner as to optimize e.g. the attenuation characteristics or the bandwidth of the fiber. Of course, the special form of the index profile chosen and the operating wavelength determine the properties of the guided fundamental mode, e.g. the transverse field distribution, the wavelength, the group velocity, the chromatic dispersion, etc. As we will see later, it is necessary when designing single-mode systems to know these properties. Therefore one has either to calculate these wave properties from the index profile or to measure them directly (Chap. 13).

For analyzing single-mode fibers, one usually assumes a certain form of the refractive index profile, and solves the wave equation, taking into account the conditions that the field strength is finite on the fiber axis, that it tends to zero for large distances from the fiber axis, and that the tangential components of the electric and magnetic fields are continuous at the core-cladding boundary. This is a complicated mathematical task.

It is a disadvantage of optical waveguide theory that, in general, it is not possible to find rigorous closed-form expressions for the functions and quantities of interest. One has to resort to numerical methods for calculation and to graphical methods for representing the results. Because simple formulas are needed in practice, many approximate relations have been derived in the past. However, the results apply only to the special form of refractive index profile that has been assumed at the beginning of the calculation and for the assumptions made to simplify the theory.

In practice, the index profile often departs from the profile assumed for the calculation. The simplest fiber consists of a circular core with constant refractive index n_1, surrounded by a homogeneous cladding with index n_2. The cladding index n_2 is slightly smaller than the core index n_1. This step-index fiber has been investigated very thoroughly and its properties are well known. However, in real fibers the index profile will depart from the assumed ideal step profile, since there may be an axial index dip, profile ripples, or a rounded index step (Sect. 5.1). Therefore the formulas found in the literature for ideal step-index fibers will not accurately describe the properties of the fundamental mode in a real fiber with profile deviations.

Besides the simple step-index fiber, there are many other types of single-mode fibers, e.g. fibers with multiple claddings which are designed to optimize the fiber bandwidth. The manufacturer of the fiber will not usually specify the data of the index profile. It is of course, possible to measure the refractive index profile of a fiber (Sect. 13.10.2), but it is no easy task to measure it accurately.

The user of single-mode fibers might thus find himself in a difficult situation: usually, he neither knows the exact refractive index profile, nor can he measure it simply. Even if he knows the index profile, he will not readily find formulas for that special index profile. Solving the wave equation numerically in order to deduce the wave properties from the index profile, is not easy, and is rarely done. But, on the other hand, it is very important for the user to know the wave properties in order to design single-mode transmission systems.

In this chapter, we will try to solve some of these problems by first discussing only those general properties of the fundamental fiber mode that apply for fibers with any index profile. We will define the wave properties, but will not relate them to the index profile. Provisionally, we must be satisfied with the possibility of measuring the quantities defined (Chap. 13). The information given is based on generalizing special results.

This chapter concerns the most important properties of the fundamental fiber mode and we start by introducing the quantities used for describing the refractive-index profile (Sect. 5.1). Then, a normalized frequency is defined as a universal fiber parameter that combines the profile data with the operating wavelength (Sect. 5.2). The next section discusses the spatial distribution of the electric and magnetic field of the fundamental fiber mode. The field distribution must be known e.g. for optimizing source to fiber couplers. Section 5.4 gives formulas for the intensity distribution in a fiber cross section and for the power transmitted by the fundamental fiber mode. Section 5.5 describes the phase constant of the fundamental fiber mode which must be known for calculating the phase velocity (Sect. 5.6) and the group velocity (Sect. 5.7). The group velocity of the fundamental fiber mode depends slightly on wavelength. This "chromatic dispersion" (Sect. 5.8) limits the fiber bandwidth. The fundamental fiber mode loses power for several reasons: light scattering, absorption, reflection, nonlinear effects, bending, etc. Fiber attenuation will be reviewed in Sect. 5.9.

The fiber with a simple step-index profile has been thoroughly analyzed [Snitzer 1961; Snyder 1969; Gloge 1971a] and much is known about its properties. The step-index fiber forms a kind of reference fiber. In order to be able to give some quantitative results, we add information for the special case of a step-index profile. In order to show that this information is not generally valid, these paragraphs will appear in small print.

5.1 The Refractive-Index Profile

In this section, we define the quantities and functions that specify the refractive-index profile.

In general, the refractive index depends both on the distance ϱ from the fiber axis, and the azimuth angle ϕ (Fig. 3.1), e.g. for fibers with elliptical core. In this volume, the fiber is assumed to be of circular symmetry, that means that the refractive index does not depend on the azimuthal angle ϕ.

In the past, many different single-mode fibers with various refractive-index profiles have been investigated theoretically or experimentally or have been applied in practice. Figure 5.1 represents some of these index profiles to scale: that of a standard step-index fiber, dispersion-shifted step-index, triangular core, and Gaussian core fibers, depressed cladding fibers, and of double, triple, and quadruple clad fibers, to show just a few profiles out of a large variety.

An arbitrary radial index profile described by the function $n(\varrho)$ is depicted in Fig. 5.2. The core has a dip on axis and an index depression near the core-cladding boundary. It is surrounded by a homogeneous cladding with index n_2, which isolates the region carrying electromagnetic energy from the environment and gives the fiber mechanical strength. The maximum refractive index in the core is denoted n_1.

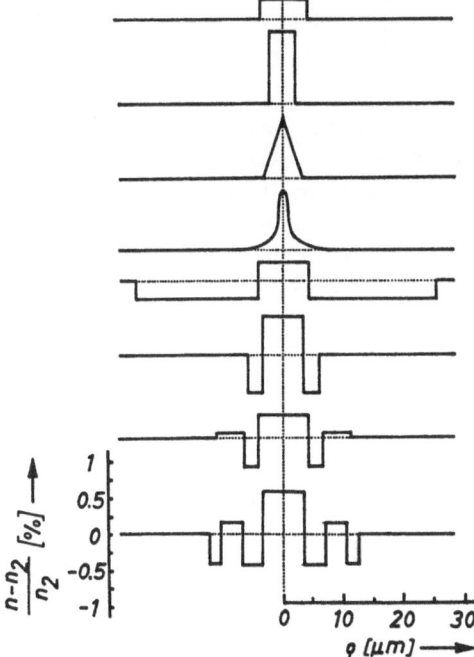

Fig. 5.1a–h. Typical refractive index profiles: (a) matched cladding step-index standard fiber (b) dispersion-shifted step-index fiber (c) dispersion-shifted triangular-core fiber (d) dispersion-shifted Gaussian core fiber (e) depressed-cladding standard fiber (f) dispersion-flattened double-clad fiber (W-fiber) (g) dispersion-flattened triple-clad (TC) fiber (h) dispersion-flattened quadruple-clad (QC) fiber

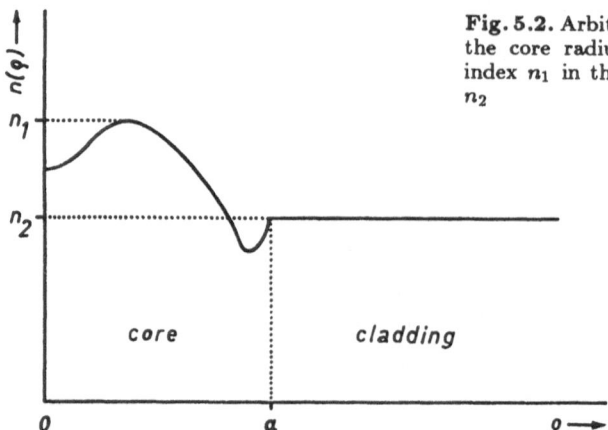

Fig. 5.2. Arbitrary index profile used to define the core radius a, the maximum refractive index n_1 in the core, and the cladding index n_2

An index difference Δn and an relative index difference Δ are defined by

$$\Delta n = n_1 - n_2 \quad \text{and} \tag{5.1}$$

$$\Delta = \frac{n_1^2 - n_2^2}{2n_1^2} \quad , \tag{5.2}$$

respectively.

For the simple step-index fiber, the refractive index in the core is a constant

$$n(\varrho) = n_1 \quad \text{for} \quad \varrho \leq a \quad . \tag{5.3}$$

In optical glass fibers, the refractive index in the core is only slightly higher than that of the cladding for two reasons. Firstly, the refractive index is changed by doping the glass with oxides. Since the scattering losses increase with the doping level (Sect. 5.9.2), high doping levels will usually be avoided. Secondly, the higher the core index relative to the cladding index, the smaller is the field diameter and the larger are splice and connector losses caused by transverse misalignment (Sect. 9.2).

Since the core index is only slightly larger than the cladding index, the relative index difference is very small compared to unity:

$$\Delta \ll 1 \quad , \tag{5.4}$$

and can be approximated by

$$\Delta \approx \frac{n_1 - n_2}{n_1} \quad . \tag{5.5}$$

Typical values of the relative index difference of single-mode fibers are 0.3–1 %.

Fibers with values of the relative refractive index difference Δ which are very small compared to unity, are called weakly guiding [Gloge 1971a], because the radius of curvature of the fiber axis must be large in order to avoid bending losses (Sect. 5.9.7).

For a general index profile, the problem arises of how to define the core radius. In this book, we will define the core radius a as the distance from the axis at which the homogeneous cladding with index n_2 begins (Fig. 5.2).

For later use, we define a normalized index profile function $g(\varrho/a)$ by the equation

$$n(\varrho) = n_1[1 - \Delta g(\varrho/a)] \quad . \tag{5.6}$$

In the cladding, the function $g(\varrho/a)$ equals unity, and at the point of maximum index n_1, it is zero.

In theoretical analyses of single-mode fibers, it is often assumed that the grading of the index can be described by a power-law function

$$g(\varrho/a) = (\varrho/a)^g \quad , \tag{5.7}$$

where the profile exponent g has a value between 1 and infinity. The step-index profile is obtained for $g \to \infty$.

The cladding of standardized single-mode fibers has the same diameter of 125 μm as for multimode fibers for telecommunications [CCITT 1984]. Usually, one is able to assume that the fields of the guided waves are negligibly small at the boundary between the cladding and the coating, so that the form of the index profile outside the cladding is of no concern. The analysis can then be simplified by assuming the cladding to be of infinite extent. In this book, we will make this assumption unless stated otherwise. However, when discussing e.g. the single-mode limit (Sect. 6.3.1,2), it is very important to remember that, in reality, the cladding radius is finite. The influence of a finite cladding diameter on the properties of the modes guided by the fiber core has been studied [Clarricoats and Chan 1973; Kuhn 1974, 1975a].

The index profile of nominal step-index fibers usually deviates from the simple step function. Firstly, there is no index step at the core-cladding boundary, but a steep, continuous transition of the refractive index. This "rounding" of the index profile is caused during fiber drawing by diffusion of the oxides which are used to modify the refractive index.

Secondly, fibers manufactured by the OVD, MCVD, or PCVD techniques intended to have a step-index profile, always have profile ripples, since the core is built up from a few (2–5) layers in the chemical vapour deposition (CVD) fabrication process.

Thirdly, optical fibers made by the MCVD or PCVD techniques usually show a depression of the refractive index on the axis known as the "center dip" [Gambling et al. 1977c, 1978d; Presby 1979]. Formerly, this center dip was attributed to the out-diffusion of dopants during the collapse of the preform in

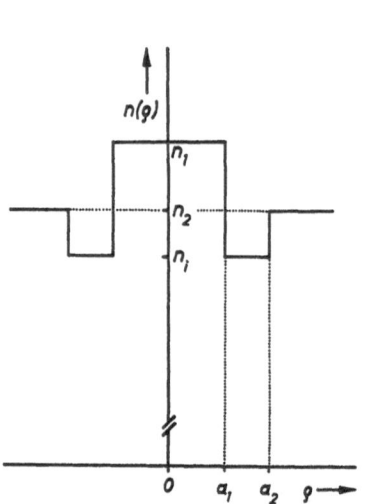

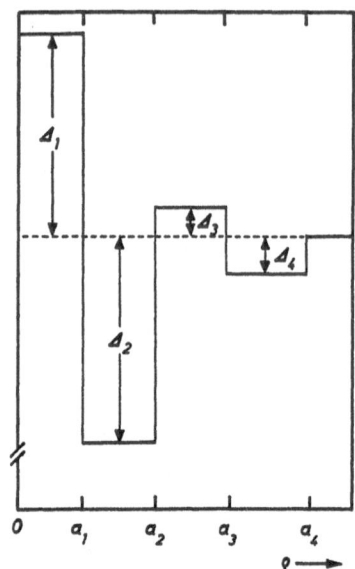

Fig. 5.3. Refractive index profile of a double-clad fiber

Fig. 5.4. Refractive index profile of a qua-druple-clad (QC) fiber

the CVD process. Recently, however, it has been shown [Ritger 1985] that the center dip is largely due to the layer structure inherent to the deposition stage, rather than to "burn off" during collapse.

For later use, Figs. 5.3 and 5.4 define the dielectric and geometric parameters of double- and quadruple-clad single-mode fibers, respectively.

5.2 Normalized Frequency

The properties of the waves guided by a fiber depend on its refractive index profile $n(\varrho)$ and the vacuum wavelength λ used. It turns out that the relations between the properties of a given dielectric optical waveguide and the properties of the waves guided by it, can most easily be described by introducing a dimensionless quantity, the so-called "normalized frequency", or "fiber characteristic term" [Kapany 1967], or "structural parameter" [Snitzer 1961; Snitzer and Osterberg 1961; Kapany and Burke 1961; Biernson and Kinsley 1965; Snyder 1969a; Gloge 1971a]

$$V = \frac{2\pi a}{\lambda} \sqrt{n_1^2 - n_2^2} \quad . \tag{5.8}$$

Since the relative index difference Δ (5.2) is very small, one can approximate the square root by $n_1(2\Delta)^{1/2}$ to obtain

$$V \approx \frac{2\pi a}{\lambda} n_1 \sqrt{2\Delta} \quad . \tag{5.9}$$

The normalized frequency combines in a very useful manner the information about three important experimental design variables: the core radius a, the relative refractive index difference Δ, and the operating wavelength λ. Since the vacuum wavelength is the velocity of light in vacuum divided by the frequency f, one recognizes that the quantity V is proportional to the optical frequency. The reason for using λ instead of f in defining the normalized frequency is simply that the wavelength λ can be measured easily with a prism or grating spectrometer whereas counting optical cycles is still rather complicated [Jennings et al. 1986].

The normalized frequency is the first of several universal parameters (V, b, $d(Vb)/dV, V d^2(Vb)/dV^2$) which play a central role in the theory of single-mode fibers.

The square of the normalized frequency has a definite physical meaning. One method for increasing the refractive index in the core region relative to that in the surrounding cladding is to dope the silica in the core with a metal oxide. It has been shown [Gambling et al. 1978d] that the square V^2 of the normalized frequency is proportional to the quantity of dopant incorporated into the core cross section.

For *multi*mode step-index fibers, the sine of the maximum entrance angle in air that still produces ray trapping is called the numerical aperture (A_N) of the fiber. Applying Snell's law of refraction twice (at the front face and at the core-cladding interface), one finds [Gloge 1979b]

$$A_N = \sqrt{n_1^2 - n_2^2}. \tag{5.10}$$

Therefore, the square root $(n_1^2 - n_2^2)^{1/2}$ in the definition of the normalized frequency (5.8) is called the numerical aperture of the fiber.

For *single*-mode fibers, the angular width of the far-field radiation pattern is proportional to the ratio of the wavelength and the spot size (Sect. 8.1), so that there is no direct correspondence between the numerical aperture (5.10) and the far-field diffraction angle θ_d (8.30)! When the term "numerical aperture" is used in connection with single-mode fibers, it means the square root on the right hand side of (5.8). It must not be used for calculating the angular width of the far-field radiation pattern of the fiber end!

In (5.8), the normalized frequency has been defined using the maximum refractive index n_1 in the core, and the refractive index n_2 in the homogeneous cladding. For fibers with special refractive-index profiles, e.g. depressed cladding fibers (Fig. 5.3) or segmented core fibers (Fig. 5.4), the normalized frequency is sometimes defined somewhat differently.

In principle, the fundamental fiber mode is guided for all values of V. However, only a small range of V values can be used in practice: for small values of V, the fundamental mode ceases to be guided by a curved fiber (Sect. 5.9.7) since it transforms into a uniform plane wave, and for V values larger than a critical cutoff value (which is 2.4 for step-index fibers) the fiber can guide other, higher-order modes besides the fundamental, i.e. it becomes a few-mode fiber (Sect. 6.3.1).

67

5.3 Field Distribution

In general, a fiber waveguide can guide a finite number of waveforms (modes), which have different transverse field distributions. The user must know the structure of the electromagnetic field of the waves propagating in a fiber in order to be able e.g. to optimize the power coupled from a light source into the fiber, to estimate losses at misaligned connectors, or bending losses. Usually, for the user of single-mode fibers, it is sufficient to know the field of the lowest order fundamental mode (this section) and that of the second order mode (Sect. 6.3).

5.3.1 Electric Field

As has been mentioned, the relative refractive-index difference Δ in glass fibers is very small (5.4). This fact causes the structure of the electromagnetic field to be relatively simple as compared to the field structure in dielectric waveguides for microwaves, where there is a large difference between the permittivities of the dielectric rod and the surrounding air [Neumann 1964a].

For describing the electromagnetic field of the waves guided by glass fibers, it is advantageous to use simultaneously both Cartesian and cylindrical polar coordinates: the field vectors are decomposed into their Cartesian components, i.e. E_x, E_y, and E_z, whereas their spatial variation is described by the cylindrical coordinates ϱ, ϕ, and z of the point of observation [Neumann 1964a; Snyder 1969a; Gloge 1971a].

In a glass fiber with a small refractive-index difference, one of the three components of the electric field vector, e.g. E_x, is large compared to the two other components. Neglecting losses, the equation describing the spatial distribution of the dominant component of the electric field in a fundamental mode propagating in the positive z-direction is:

$$\underline{E}_x(\varrho, z) = E(\varrho) \exp(-j\beta z) \quad . \tag{5.11}$$

The quantity $\underline{E}_x$ is the complex phasor (Sect. 3.4) of the x-component of the electric field vector. The instantaneous E_x value of the x-component of the electric field vector is obtained by multiplying the phasor (5.11) by the function $\exp(j\omega t)$ and by taking the real part of the product:

$$E_x(\varrho, z, t) = E(\varrho) \cos(\omega t - \beta z) \quad . \tag{5.12}$$

In general, the function $E(\varrho)$ is complex with an imaginary part which is independent of ϱ. For simplicity, we choose the origin of time in such a way that $E(\varrho)$ is a real function.

Comparing (5.11) with (3.3), the zero phase angle $\Phi_x(z)$ of the x-component of the electric field is seen to be

$$\Phi_x(z) = -\beta z \quad . \tag{5.13}$$

The wavefronts, i.e. the surfaces of equal phase, are obtained by setting Φ_x in (5.13) constant. This gives the condition

$$z = \text{const.} \quad , \tag{5.14}$$

i.e. the wavefronts are planes normal to the fiber axis. In other words, the fundamental fiber mode is a plane wave. It is important to know this if one wants to couple a light source efficiently to the fiber (Sects. 7.2.2 and 12.2.2). The zero phase angle Φ_x, decreases linearly in the direction of wave propagation with a rate given by the phase constant β (Sect. 3.5.3). The distance $2\pi/\beta$ between two neighboring planes of equal phase is the wavelength λ_{01} of the fundamental guided mode (The indices will be explained later).

The phase constant β depends on the refractive-index profile $n(\varrho)$ and the optical operating wavelength. More information about the phase constant can be found in Sect. 5.5.

The amplitude $E(\varrho)$ of the electric field is independent of the azimuth ϕ and the longitudinal coordinate z, i.e. the field is (approximately) circularly symmetric and unattenuated.

In this section, we are interested in wave properties that are independent of the special form of the index profile. Generally, the radial amplitude distribution of the fundamental mode has the form of a bell, i.e. the field amplitude is maximum on axis and decreases monotonically with increasing distance ϱ from the axis.

For any arbitrary refractive-index profile in the core, the amplitude distribution $E(\varrho)$ in the homogeneous cladding is described by a modified Hankel function (modified Bessel function of the second kind) of zero order

$$E(\varrho) = C \mathrm{K}_0 \left(\frac{W\varrho}{a} \right) \quad , \tag{5.15}$$

where C is a constant proportional to the square root of the power transmitted by the fundamental mode.

For large values of its argument, the modified Hankel function can be approximated by [Abramowitz and Stegun 1965]

$$\mathrm{K}_0 \left(\frac{W\varrho}{a} \right) \approx \sqrt{\frac{\pi a}{2W\varrho}} \exp \left(\frac{-W\varrho}{a} \right) \quad . \tag{5.16}$$

When the argument $W\varrho/a$ of the modified Hankel function is larger than 2, the error made in using the approximation (5.16) is smaller than 5 %. Since the exponential function changes much more rapidly than the square root, the electric field in the cladding decays approximately exponentially with increasing distance ϱ from the core-cladding boundary.

At this point, it is important to emphasize that the field of the fundamental mode is not confined to the core, but also extends into the homogeneous cladding. The part of the mode field in the cladding, is called the evanescent field.

From (5.16), one can deduce the physical meaning of the parameter W: the quantity W/a determines the decay rate of the field level in the cladding and, therefore, W can be called the cladding decay parameter [Gloge 1979b]. Other names for W are the eigenvalue in the cladding or the exponential decay constant.

The inverse of W/a, i.e. the quantity a/W, is a measure for the field extent in the cladding: well inside the cladding ($\varrho \gg a/W$), if the distance from the axis is increased from ϱ to $\varrho + a/W$, the field strength decreases by a factor of about $1/e = 0.37$ corresponding to a change in field level of about $8.7\,\mathrm{dB}$ (The small change of the square root in (5.16) can be neglected). Therefore, within a distance of a/W, the power density decays by approximately one order of magnitude. The quantity a/W can be named transverse field extent in the cladding.

There is a simple relation between the quantity W/a, the phase constant β, and the wave number $n_2 k$ (4.14) in the cladding; this is the so-called separation equation in the cladding:

$$(W/a)^2 = \beta^2 - n_2^2 k^2 \quad .\tag{5.17}$$

This equation follows from inserting the field distribution (5.15) into the scalar wave equation (5.20).

The value of the cladding decay parameter W is unequivocally determined by the refractive index profile $n(\varrho)$ of the fiber and the operating wavelength λ. For arbitrary refractive-index profile, it is very complicated to calculate W.

For the special case of a step-index fiber, Fig. 5.5 represents the quantity $1/W$ as a function of the normalized frequency V (5.8). Numerical values of the function $W(V)$ are listed in Table 5.1. For $V = 2.4$, the quantity W is 1.747, and for a typical core radius of $5\,\mu$m, the transverse field extent in the cladding is $a/W = 2.9\,\mu$m.

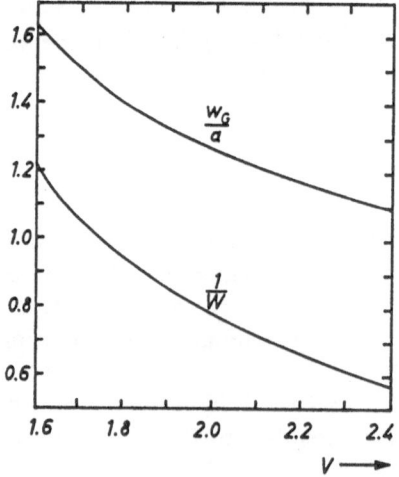

Fig. 5.5. Spot size related to core radius w_G/a (5.25) and normalized transverse field extent $1/W$ (5.15) in the cladding as functions of the normalized frequency V (5.8) for a step-index fiber

Table 5.1. Numerical data for the parameters of step index fibers ($b_1 = d(Vb)/dV$, $b_2 = V d^2(Vb)/dV^2$)

V	U	W	$J_0(U)$	$J_1(U)$	$K_0(W)$	$K_1(W)$	w_G/a	w_e/a	w_d/a	w_a/a	P_1/P	b	b_1	b_2
1.0	0.979	0.202	0.774	0.433	1.742	4.761	5.173	5.854	3.911	5.160	0.167	0.041	0.303	1.346
1.1	1.060	0.293	0.738	0.459	1.395	3.141	3.674	4.129	3.004	3.689	0.255	0.071	0.437	1.458
1.2	1.134	0.392	0.703	0.481	1.132	2.237	2.846	3.166	2.464	2.891	0.336	0.107	0.564	1.440
1.3	1.201	0.497	0.671	0.499	0.929	1.668	2.343	2.574	2.114	2.390	0.411	0.146	0.676	1.343
1.4	1.262	0.606	0.640	0.513	0.769	1.284	2.014	2.184	1.871	2.060	0.479	0.188	0.771	1.209
1.5	1.317	0.718	0.611	0.526	0.642	1.012	1.786	1.913	1.693	1.830	0.539	0.229	0.849	1.063
1.6	1.367	0.831	0.585	0.536	0.539	0.812	1.621	1.722	1.559	1.660	0.592	0.270	0.913	0.919
1.7	1.413	0.946	0.560	0.544	0.455	0.661	1.496	1.569	1.453	1.531	0.637	0.309	0.965	0.785
1.8	1.454	1.060	0.537	0.551	0.386	0.544	1.399	1.454	1.369	1.430	0.677	0.347	1.006	0.664
1.9	1.493	1.175	0.516	0.557	0.329	0.452	1.322	1.364	1.299	1.349	0.711	0.383	1.039	0.556
2.0	1.528	1.290	0.496	0.562	0.282	0.378	1.259	1.291	1.241	1.282	0.741	0.416	1.065	0.462
2.1	1.561	1.405	0.478	0.566	0.242	0.319	1.206	1.231	1.192	1.226	0.767	0.448	1.086	0.380
2.2	1.591	1.519	0.460	0.569	0.209	0.270	1.162	1.181	1.150	1.179	0.789	0.477	1.102	0.309
2.3	1.619	1.634	0.444	0.572	0.180	0.230	1.124	1.139	1.114	1.138	0.809	0.504	1.114	0.248
2.4	1.645	1.747	0.429	0.574	0.156	0.196	1.092	1.103	1.082	1.103	0.827	0.530	1.124	0.195
2.5	1.670	1.861	0.415	0.576	0.135	0.168	1.064	1.072	1.053	1.072	0.842	0.554	1.131	0.150
2.6	1.693	1.974	0.402	0.577	0.118	0.145	1.039	1.045	1.028	1.044	0.856	0.576	1.136	0.110
2.7	1.714	2.086	0.390	0.579	0.102	0.125	1.016	1.021	1.006	1.020	0.868	0.597	1.139	0.077
2.8	1.734	2.198	0.378	0.580	0.089	0.108	0.997	1.001	0.986	0.998	0.879	0.616	1.141	0.048
2.9	1.753	2.310	0.367	0.580	0.078	0.094	0.979	0.982	0.967	0.979	0.889	0.635	1.143	0.023
3.0	1.771	2.421	0.357	0.581	0.068	0.082	0.963	0.965	0.951	0.961	0.897	0.651	1.143	0.001

In the most important range of the normalized frequency

$$1.5 \leq V \leq 2.5 \quad , \tag{5.18}$$

the quantity W can be determined approximately from a simple analytical formula [Rudolph and Neumann 1976]

$$W = 1.1428V - 0.9960 \quad . \tag{5.19}$$

The error in W made by using this approximation is smaller than 0.1 %.

Although (5.19) yields quite accurate values for W, b (5.51, 55), and β (5.54), it leads to significant errors when calculating the second derivative of Vb (5.88) with respect to V for determining waveguide dispersion [Sammut 1979].

In the core, the functional form of the radial amplitude distribution depends on the particular refractive-index profile. From the refractive-index profile $n(\varrho)$, for a given operating wavelength λ, the radial amplitude distribution, $E(\varrho)$, and the phase constant β of the fundamental fiber mode can be obtained by solving the so-called scalar wave equation [Snyder 1983]

$$d^2 E/d\varrho^2 + (1/\varrho)dE/d\varrho + [n^2(\varrho)k^2 - \beta^2]E = 0 \quad . \tag{5.20}$$

subject to the conditions that $E(\varrho)$ is finite for $\varrho = 0$, continuous at ϱ values where the refractive index changes discontinuously, and equal to $C K_0(W)$ at the core-cladding boundary $\varrho = a$.

For arbitrary refractive-index profiles, the solutions $E(\varrho)$ of the scalar wave equation (5.20) may be obtained by direct numerical integration.

Analytical solutions of the scalar wave equation (5.20) are available only for fibers with step-index [Snyder 1969a] and parabolic-index profiles [Streifer and Kurtz 1967; Meunier J.P. et al. 1980].

In a step-index fiber, the radial field distribution in the core is described by the Bessel function of zero order J_0 [Gloge 1971a]:

$$E(\varrho) = C \left[\frac{K_0(W)}{J_0(U)} \right] J_0 \left(\frac{U\varrho}{a} \right) \quad , \tag{5.21}$$

where the "radial phase parameter", "radial propagation constant", or "eigenvalue in the core" U is related to the phase constant β by the separation equation in the core

$$\left(\frac{U}{a} \right)^2 = n_1^2 k^2 - \beta^2 \quad . \tag{5.22}$$

By adding (5.17) and (5.22) and by remembering the definition of the normalized frequency V (5.8), one obtains the important "sum rule" [Garrett and Todd 1982]

$$U^2 + W^2 = V^2 \quad . \tag{5.23}$$

Matching the tangential components of the electric field (E_ϕ, E_z) and of the magnetic field (H_ϕ, H_z) at the core-cladding interface $\varrho = a$, gives the so-called characteristic equation for the fundamental mode of a step-index fiber [Snyder 1969a; Gloge 1971a]:

$$U \frac{J_1(U)}{J_0(U)} = W \frac{K_1(W)}{K_0(W)} \quad , \tag{5.24}$$

which together with (5.23) must be solved graphically or numerically for the eigenvalues U or W. For a given value of the normalized frequency V, the system of transcendental equations (5.23, 24) only contains the two unknown quantities U and W. Therefore, the eigenvalues U and W are functions only of the normalized frequency V.

For V-values larger than 2.405, which is the first zero of the Bessel function $J_0(x)$, the system of equations (5.23, 24) has more than one solution. That solution with the smallest value of the unknown quantity U gives the eigenvalues of the fundamental mode.

A simple approximate solution of the characteristic equation of the step-index fiber is given by (5.19).

Since the step-index fiber is a kind of reference fiber, Table 5.1 lists the most important parameters as functions of the normalized frequency V.

Equation (5.24) is an approximation for the exact eigenvalue equation. However, in the weakly guiding approximation, where $\Delta < 0.01$, it yields very accurate values for U, W, and β. In fact, it has been shown [Rudolph and Neumann 1976] that the errors in the waveguide parameters U and W as a result of using (5.24) instead of the exact eigenvalue equation are less than 1% for $\Delta < 4\%$.

For the special case of $V = 2.2$, the radial field distribution in a step-index fiber is represented in Fig. 5.6 by a solid line.

Up to now, we have only discussed the spatial distribution of the dominant x-component of the electric field vector. We still have to consider the two other components E_y and E_z of the electric field vector. Rigorously, all three components of the electric field are different from zero, so that the field structure is

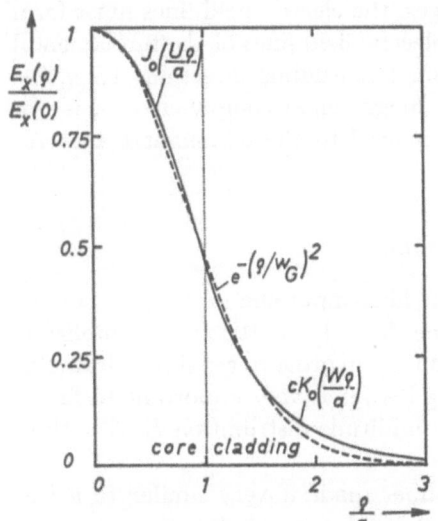

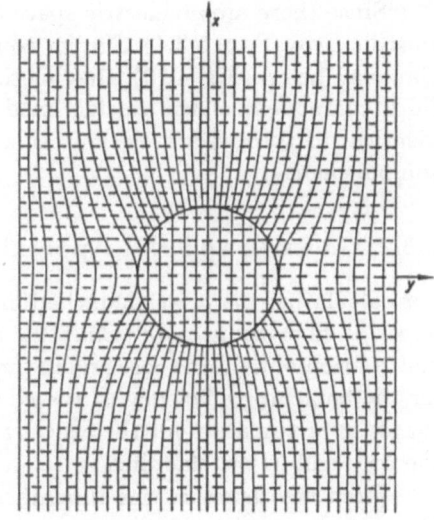

Fig. 5.6. Normalized radial field distribution for a step-index fiber. $V = 2.2$. Exact (5.15, 5.21) (*solid line*); Gaussian approximation (5.25) with $w_G/a = 1.162$ (*dashed line*)

Fig. 5.7. Electric (*solid lines*) and magnetic (*dashed lines*) field lines of the fundamental mode projected onto a transverse plane ($n_1 = 1.6$, $n_2 = 1$, $2a/\lambda = 0.246$)

quite complicated even for the simple step-index fiber [Snitzer 1961]. However, for weakly guiding fibers with a small relative refractive-index difference, the field structure is relatively simple.

The other transverse component of the electric field vector, E_y, cannot be identically zero. This can be seen by looking at Fig. 5.7 which shows the electric field lines of the fundamental mode in a step-index fiber projected onto a transverse plane. In the core, the field lines are approximately parallel to the x-direction, i.e. in the core, E_y is very small. At the core-cladding interface, the tangential components E_ϕ and E_z of the electric field vector and the normal component εE_ϱ of the electric displacement vector are continuous. The change of the refractive index $n = \varepsilon^{1/2}$ causes the electric field lines to refract at the core-cladding interface and to depart from the x-direction. Thus, there is also a nonzero y-component of the electric field. This component is exactly zero in the planes $x = 0$ and $y = 0$ and has alternating signs in the four quadrants of the xy-plane. It is in phase or in antiphase with the dominant component E_x.

In order to demonstrate the refraction of the field lines, a large refractive-index difference between core ($n_1 = 1.6$) and cladding ($n_2 = 1$) has been assumed for plotting Fig. 5.7 [Neumann 1964a]. For weakly guiding fibers with a small relative refractive-index difference Δ, the y-component of the electric field is much smaller than in Fig. 5.7, and the projected field lines deviate only very slightly from the x-direction. In optical fibers, the minor transverse electric field component E_y is very small as compared to the dominant x-component.

Because of the radial decay of the field amplitude, the density of field lines must, in reality, decrease in the radial direction. This is not shown in Fig. 5.7.

Since there are no electric space charges, the electric field lines must form closed curves (Sect. 5.3.4). Similar to the electric field lines of the fundamental Gaussian beam (Fig. 4.10), they close in the longitudinal direction. Thus, the third component of the electric field, the longitudinal component E_z, is not identical to zero. However, it is small compared to the dominant transverse component E_x.

5.3.2 Gaussian and ESI Approximations

For calculating the power launched into the fiber input end (Sect. 7.2.2) or the losses at fiber connectors (Chap. 9), one needs to know the radial amplitude distribution both in the core and in the cladding. In order to get simple formulas for the launching efficiency or the coupling loss, it is very important to find a simple approximation for the exact radial amplitude distribution $E(\varrho)$ both in the core and in the cladding.

From the fact that the fundamental fiber mode is very similar to a free Gaussian beam (Sect. 2.2), one can conclude that a simple function to give a close approximation to the true amplitude distribution is a Gaussian function [Marcuse 1970d, 1977b, 1978a; Gambling and Matsumura 1978a]:

$$E(\varrho) \approx E_0 \exp\left[-\left(\frac{\varrho}{w_G}\right)^2\right] \quad . \tag{5.25}$$

In (5.25), there is just one parameter to match the width of the approximate Gaussian field profile to the width of the true field profile, namely the quantity w_G which is called the spot size or the field radius of the fundamental mode. The mode field diameter is two times the spot size. The mode field diameter can be viewed as the single-mode analogue of core diameter in multimode fibers since both influence the loss of a splice with lateral offset [White K.I. 1985]. For many refractive-index profiles and for typical operating wavelengths, the mode field diameter is slightly larger than the core diameter.

There are at least five definitions for the spot size (Sect. 10.1). As will be explained later in Sect. 10.1.2, one method often used to define the spot size is to consider the problem of launching the fundamental fiber mode by a Gaussian beam incident on the fiber input end (Sect. 7.2.2). The waist of the Gaussian beam is assumed to be located at the fiber input face, and the axes of the beam and of the fiber to coincide. The power launched into the fundamental mode depends on the spot size w_0 of the Gaussian beam at its waist. The spot size w_G of the fundamental fiber mode is then defined as that spot size w_0 of the Gaussian beam which maximizes the launching efficiency.

For the special case of a step-index fiber, the frequency dependence of the spot size is shown in Figs. 5.5 and 10.1. Numerical values are listed in Table 5.1. It is interesting to note that in the range

$$1.6 < V < 2.4 \quad , \tag{5.26}$$

the product $w_G V$ is approximately constant. This means that for a given step-index fiber, in the usual range of operating wavelengths, the spot size increases in proportion to the wavelength.

The amplitude E_0 in the Gaussian approximation (5.25) should be chosen in such a manner that the power transmitted by the substitute Gaussian field (4.27) equals the power transmitted by the true mode (5.35).

The main advantage of using the Gaussian approximation for the radial amplitude distribution is that one does not have to specify two different functions in the core and in the cladding, but only one parameter, namely the spot size w_G.

The Gaussian approximation for the transverse field distribution is very useful for calculating for instance the launching efficiency at the fiber input end or the coupling loss at connectors.

For the fundamental mode guided by a multimode graded-index fiber with untruncated parabolic refractive-index profile [(5.6, 7) with $g = 2$ for all ϱ)], the radial field distribution is described exactly by a Gaussian function [Kogelnik 1965b]. For single-mode fibers with homogeneous cladding, the true field distribution is never exactly Gaussian. Since the field decay in the cladding is more like an exponential function than like a Gaussian function, the Gaussian approximation underestimates the evanescent field in the cladding.

The Gaussian approximation is good for fibers operated at a wavelength near the cutoff wavelength of the second-order mode (Sect. 6.3.2). When the wavelength gets larger, the Gaussian approximation becomes less accurate.

Problems may arise when the decay rate W of the evanescent field in the cladding is of importance, e.g. when analyzing cladding absorption (Sect. 5.9.4), bend losses (Sect. 5.9.7), crosstalk between parallel fibers (Sect. 7.4), or when designing directional couplers (Sect. 12.2.11). In these cases, the Gaussian approximation must be applied with caution and better approximations for the field profile should ideally used, e.g. a Gaussian function near the fiber axis and an exponential function [Sharma A. and Ghatak 1981a; Ghatak et al. 1983], or a modified Hankel function of zero order [Sharma E.K. and Tewari 1984] far from the axis. In contrast to the one-parameter Gaussian approximation, in the Gaussian-exponential and the Gaussian-Hankel approximations, two parameters are necessary to characterize the radial amplitude distribution.

When analyzing the evanescent field in the cladding, it is sometimes better to characterize the transverse field extent of the fundamental fiber mode by the field extent a/W in the cladding instead of by the spot size w_G.

For the example of a step-index fiber, Fig. 5.5 represents the two quantities characterizing the transverse field extent relative to the core radius, i.e. w_G/a and $1/W$, as functions of the normalized frequency V. At the single-mode limit $V = 2.405$ (Sect. 6.3.1), the spot size is 1.91 times the field extent in the cladding (Table 5.1).

For understanding single-mode fibers, it is important to know qualitatively the wavelength dependence of the mode field diameter $2w_G$. In Sect. 2.2, we qualitatively explained waveguiding in a straight single-mode fiber by considering two counteracting effects: diffraction spreading for a beam of finite diameter and focusing by the fiber core. For a given fiber and vacuum wavelength, there is a single value of the field diameter for which the two effects cancel so that the beam diameter does not change during wave propagation.

In Sect. 4.2, a quantity T_b was introduced to characterize the tendency of a Gaussian beam propagating in a homogeneous medium to diverge because of diffraction. The diffraction tendency T_b of a beam with spot size w_0 at the waist propagating in a medium with refractive index n_2 is (4.22)

$$T_b = \frac{\lambda^2}{\pi^2 n_2^2 w_0^4} \quad . \tag{5.27}$$

The smaller the spot size w_0, the larger the diffraction tendency.

The core of a single-mode fiber causes a continuous focusing, which can be described by another quantity $-T_f$ ($T_f < 0$), which is called the focusing tendency [Neumann 1987]. The focusing tendency $-T_f$ is determined by the refractive-index profile.

The fundamental fiber mode is that beam for which the diffraction tendency is compensated for by the focusing tendency of the core:

$$T_b = -T_f \quad , \tag{5.28}$$

so that the spot size $w_0 = w_G$ is constant along the beam.

Since $-T_f$ is approximately independent of wavelength, T_b (5.27) is also nearly constant, and the spot size w_G increases with increasing wavelength. This simple consideration gives qualitative information about the wavelength dependence of the spot size.

The propagation characteristics of a fiber with an arbitrary refractive-index profile are difficult to calculate. On the other hand, the propagation characteristics of step-index fibers are well known. It has therefore been proposed to model a fiber with an arbitrary refractive-index profile by an equivalent step-index (ESI) fiber, hoping that the propagation characteristics of the ESI fiber will be very similar to those of the given fiber. Several different suggestions have been put forward for choosing the core radius a_E and the relative refractive-index difference Δ_E of the ESI fiber.

The concept of the ESI fiber has been fairly useful in specifying standard matched-cladding and depressed-cladding fibers by their equivalent a_E and Δ_E values. However, the ESI methods fail to predict mode-field diameter and waveguide dispersion in dispersion-shifted and dispersion-flattened fibers [Nelson B.P. and Wright 1983, 1984; Samson 1985a]. ESI methods would thus seem to be inadequate for an accurate description of real fibers. For accuracy, it is necessary to use numerical methods for calculating the wave properties from the refractive-index profile.

The intensity is proportional to the square of the field amplitude. This fact opens the possibility of measuring the radial amplitude distribution instead of calculating it from the refractive-index profile: one measures the intensity distribution in the fiber endface (Sect. 13.5) and calculates the amplitude by taking the square root of the intensity.

5.3.3 Magnetic Field

Having described the structure of the electric field of the fundamental fiber mode, we now turn to the corresponding magnetic field. Rigorously, none of the three components of magnetic field vector is identically equal to zero. However, in weakly guiding fibers, one of the components is much larger than the other two. When the dominant component of the electric field is E_x, the dominant component of the magnetic field is H_y, normal to E_x. The other transverse component H_x of the magnetic field is negligibly small compared to the dominant component H_y.

Since there are no magnetic charges, the magnetic field lines must form closed curves (Sect. 5.3.4). Similarly to the magnetic field lines in a Gaussian beam, they also close in the longitudinal direction (Fig. 4.10). Thus, the third component of the magnetic field, the longitudinal component H_z, is not identical to zero. However, the longitudinal component H_z is small in comparison to H_y.

There is a simple approximate relation between the dominant transverse components of the magnetic and the electric field vectors [Snyder 1969a]:

$$\underline{H}_y(\varrho, z) = \frac{n_e \underline{E}_x(\varrho, z)}{Z_0} \quad , \tag{5.29}$$

where $Z_0 = 377\,\Omega$ is the wave impedance of free space (4.25) and $n_e = \beta/k$ the effective refractive index (5.45). The effective index n_e is larger than the cladding index n_2 and smaller than the maximum index n_1 in the core (5.49). Since the relative index difference Δ (5.5) is smaller than 0.01, one can approximate n_e in (5.29) either by n_1 or n_2.

Since $\underline{H}_y$ is proportional to $\underline{E}_x$, the discussion concerning the spatial distribution of the dominant component $\underline{E}_x$ of the electric field also applies to the dominant component $\underline{H}_y$ of the magnetic field. Since the wave impedance is a real quantity, the transverse components of the electric and magnetic field at a fixed location oscillate in phase.

5.3.4 Field Lines

For understanding single-mode fibers, it is useful to know the electric and magnetic field lines of the fundamental fiber mode. The electric field vector has essentialy only an x- and a z-component, and the magnetic field vector only a y- and a z-component. For calculating the field lines, which are tangential to the field vector, besides the formulas for the transverse components of $\mathbf{E}$ (5.11) and $\mathbf{H}$ (5.29), one needs approximate analytical expressions for the longitudinal components E_z and H_z.

The phasors of the longitudinal components, $\underline{E}_z$ and $\underline{H}_z$, can be obtained by inserting the expressions for the transverse components into Maxwell's first and second equation, respectively. Using the Gaussian approximation (5.25) one finds [Simon et al. 1986]

$$\underline{E}_z = \frac{jx\lambda}{\pi n_2 w_G^2} E_0 \exp\left[-\left(\frac{\varrho}{w_G}\right)^2\right] \exp(-j\beta z) \quad, \tag{5.30}$$

and

$$\underline{H}_z = \frac{jy\lambda}{\pi Z_0 w_G^2} E_0 \exp\left[-\left(\frac{\varrho}{w_G}\right)^2\right] \exp(-j\beta z) \quad. \tag{5.31}$$

In the beam cross-section, i.e. for $\varrho < w_G$, the coordinates x and y are smaller then the spot size w_G. Single-mode fibers for the 1.3 μm band have a spot size of about 5 μm. Since the spot size equals about four free-space wavelengths, it can be seen from (5.30, 31) that the longitudinal field components are small compared to the transverse components. Because of the imaginary unit in the expressions for the longitudinal components, there is a time shift of a quarter period and a spatial shift of a quarter wavelength between the maxima of the longitudinal and transverse components.

Since the y-component of the electric field, and the x-component of the magnetic field, are extremely small, the electric and magnetic field lines lie in planes parallel to the xz- and yz-plane, respectively. Since a field line is parallel to the local direction of the vector, the differential equation for an electric field line can be obtained by equating its slope to the ratio of the instantaneous values of the transverse and longitudinal electric field components. Solving this differential equation, gives a formula for the electric field lines:

$$x^2 = w_G^2 \ln\left[\cos\beta(z - z_0)\right] + x_{max}^2 \quad, \tag{5.32}$$

where x_{max} is the x-coordinate of that point of the field line in the cross section $z = z_0$ which is farthest from the fiber axis.

Figure 5.8 shows some electric field lines of the fundamental mode propagating along a straight fiber. Strictly speaking, at surfaces where the refractive index changes abruptly, e.g. at the core-cladding boundary of a step-index fiber, the electric field lines refract, i.e. there is a tilt. But because there are only slight changes of the refractive index in glass fibers, the tilt angles are very small. Therefore, the refraction of the field lines has been neglected both in (5.32) and in Fig. 5.8.

The magnetic field lines lie in planes $x =$ constant. Their form is nearly identical to that of the electric field lines. For an illustration showing both the electric and the magnetic field lines [Snitzer 1961] look at the similar field of a free Gaussian beam near its waist (Fig. 4.10).

At this point, it seems useful to compare the field of the fundamental fiber mode both with a uniform plane wave and with a free Gaussian beam. Like a uniform plane wave, the fundamental fiber mode is a plane wave, i.e. a wave with plane wavefronts. However, in contrast to a uniform plane wave, the field amplitude of the fundamental fiber mode is not constant over a wavefront.

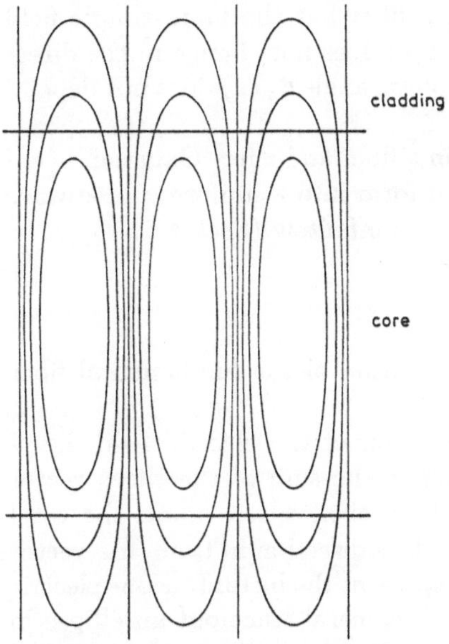

Fig. 5.8. Electric field lines of the fundamental fiber mode in the xz-plane

Therefore, the fundamental fiber mode is an example of a so called inhomogeneous plane wave.

In a uniform plane wave, there are only transverse components of the electric and magnetic fields and these are normal to each other. The uniform plane wave is a so-called transverse electromagnetic (TEM) wave. In contrast to this, the fundamental fiber mode has, besides the main transverse components of the fields, small longitudinal components of both the electric and the magnetic field, i.e. it is a hybrid wave. However, the dominant transverse components E_x and H_y are normal to each other, in the fiber mode, too.

For increasing wavelength, the spot size w_G of the fundamental mode increases infinitely and the field strength in the fiber core becomes approximately constant. The field extends far into the (infinite) cladding (5.25). Thus, the guided wave "sees" mainly the cladding material, and its wavelength approaches that in the bulk cladding material (λ/n_2). The longitudinal components of the fields vanish with increasing spot size (5.30, 31). Therefore, with increasing wavelength, the fundamental mode continuously transforms into a uniform transverse electromagnetic (TEM) wave.

In principle, a straight fiber can guide the fundamental mode for arbitrarily long wavelengths, i.e. its theoretical cutoff frequency (Sect. 6.2) is zero (There are single-mode fibers with special refractive index profiles, for which the fundamental mode has a finite cutoff wavelength (6.22)). However, since for large wavelengths the fundamental mode behaves like a uniform plane wave, it no longer follows the fiber when there is a bend or a tilt. At long wavelengths, a single-mode fiber will therefore suffer large radiation losses.

Comparing the electric field distribution of the fundamental fiber mode (5.11) with that of a free Gaussian beam (4.1), the fields are seen to be very similar. However, the fundamental mode of a fiber has the more simple field structure, since the radial field distribution $E(\varrho)$ does not change in the direction of wave propagation, and since the zero phase angle $\Phi_x(z)$ does not depend on the distance ϱ from the fiber axis.

Both a fundamental mode propagating in a fiber and a free Gaussian beam propagating in a homogeneous medium, transform into a uniform plane wave if the spot sizes w_0 or w_G, respectively, increase infinitely (Sect. 4.5).

5.3.5 Polarization

In this section, we want to discuss the polarization of the fundamental fiber mode.

Assuming a strictly sinusoidal time variation of the electric field, i.e. a monochromatic wave, the electric field vector is the sum of the three vector components $E_x e_x$, $E_y e_y$, and $E_z e_z$. At a fixed observation point, the total electric field vector changes both its length and its direction in time. If a vector is drawn with its origin at the fixed point to represent the instantaneous electric field vector, the end-point of this vector will in general trace out an ellipse in space. This "ellipse of polarization" will be traced periodically with the optical frequency. The field at this point is then said to be elliptically polarized, or to be in an elliptical state of polarization. Under certain conditions, the ellipse will become a circle or a straight line as limiting special cases. The field is then said to be circularly or linearly polarized.

For the particular case of the electric field of the fundamental fiber mode, there is a large x-component and a smaller z-component, which are in phase quadrature (We first neglect the extremely small y-component of the electric field). The polarization is found to be elliptical with the plane of the polarization ellipse being parallel to the longitudinal xz-plane! Thus, strictly speaking, a *single* fundamental fiber mode is elliptically polarized. But usually, when considering polarization in single-mode fibers, the small longitudinal component E_z is neglected. Alternatively, one considers the polarization only on the fiber axis, where the longitudinal component E_z is exactly zero. On the fiber axis, the polarization of a *single* fundamental fiber mode is strictly linear.

We now discuss the influence of the small y-component of the electric field which causes the field lines when projected on a fiber cross section, to deviate from the x-direction (Fig. 5.7). Since the major x- and the minor y-components of the electric field of a *single* fundamental mode oscillate in phase or antiphase at each point of the beam cross section, the resulting transverse electric vector describes a straight line, i.e. the local polarization is linear. But, strictly speaking, because of the nonzero y-component, the direction of the transverse part of the electric field vector varies slightly over the beam cross-section. Thus, the state of polarization of a *single* fundamental fiber mode is linear, but not strictly uniform over the beam cross section. This effect limits the maximum extinction ratio which can be measured with a rotatable linear analyzer when analyzing the state of polarization in a single-mode fiber [Varnham et al. 1984].

In order to avoid these difficulties, we consider the state of polarization only on the fiber axis, where both the longitudinal z-component and the minor transverse y-component are exactly zero. There, the electric field of the *single* fundamental mode has only an x-component, i.e. the state of polarization is linear.

Until now, we assumed the dominant component of the electric field vector of the fundamental mode to be parallel to the x-direction. The same fiber can transmit a second fundamental mode with the dominant component of the electric field vector parallel to the orthogonal transverse direction, i.e. the y-direction. The frequencies, amplitudes, and phases of the two waves linearly polarized in the x- and y-directions are determined by the launching conditions and can be chosen independently.

In general, the source will launch both fundamental modes simultaneously with the same frequency. The total electric field vector on the fiber axis is now the vector sum of the electric vectors of the *two* fundamental modes. Its endpoint will periodically trace an ellipse of polarization which lies in a transverse plane. For arbitrary amplitudes and phases of the two fundamental modes, the total field on the fiber axis will be in an elliptical state of polarization. Hence, the superposition of the two fundamental modes polarized linearly in the x- and y-directions, respectively, gives an elliptically polarized fundamental fiber mode. The magnitude, form, and orientation of the polarization ellipse depend on the amplitudes and the phase difference of the x- and y-polarized fundamental modes. Under certain conditions, the ellipse will become a circle or a straight line as limiting special cases. The fundamental mode is then said to be circularly or arbitrarily linearly polarized.

The two independent fundamental modes will have the same phase constants if the fiber is ideal, i.e. if it has a circular core, is straight, is made of isotropic materials, is free from internal or external mechanical stresses, and is not subjected to electric or magnetic fields. Because of this, the orthogonally polarized waves are said to be degenerate. Since the phase constant is the change of the phase angle per unit length, the phase difference between the two modes will be maintained along an ideal fiber. Usually, both modes have the same attenuation, and the ratio of the amplitudes is also constant along the length of the fiber. Therefore, for an ideal fiber, the form of the polarization ellipse does not change along the length.

In practice, the fiber core is usually slightly noncircular, or it is bent or twisted. It can be subjected to mechanical stresses or electric and magnetic fields. These deviations from the ideal fiber cause the form of the polarization ellipse to change along the fiber; the fiber is said to be birefringent. A birefringent fiber changes the state of polarization of transmitted guided light in a similar manner to a birefringent crystal. If the birefringence changes with temperature or in time or if the laser wavelength drifts, the form of the polarization ellipse at the fiber output will also vary.

For the modulation format most frequently used today, i.e., intensity modulation and direct detection, the change of the polarization caused by the fiber and the fluctuations of the state of polarization do not matter, provided there are no optical components with an insertion loss which depends on the polarization, e.g. grating demultiplexers (Sect. 12.3.3). However, in future coherent systems, which use homodyne or heterodyne receivers, for obtaining maximum receiver sensitivity, it is very important to match the state of polarization of the received wave to that of the local oscillator wave.

There are other applications of single-mode fibers where it is very important to maintain a stable linear polarization at the fiber output, e.g. in interferometric fiber sensors, or in fibers to be connected to integrated optical waveguides (Sect. 12.2.3). One method to maintain a constant state of polarization along the fiber is to use highly birefringent fibers and to excite only one of the orthogonally polarized eigenwaves at the fiber input end. Methods for making these so-called polarization maintaining fibers and their applications have been reviewed recently [Noda et al. 1986].

5.3.6 Fiber and Mode Designation

In this section we describe the nomenclature used for the single-mode fiber and for the fundamental fiber mode.

Since a single-mode fiber can transmit two independent orthogonally polarized waves, the fiber is in fact "bimodal", and the usual names "*single-mode* fiber" or "*monomode* fiber" are misleading. It would be better to use the names "*fundamental* mode fiber" or "*single-spatial-mode* fiber" [Rashleigh 1983]. However, the names "single-mode fiber" or "monomode fiber" are widely used. In this book, we follow the recommendation of the CCITT and use the name single-mode fiber, since we then have unambiguous abbreviations for single-mode (SM) and multimode (MM) fibers and since the prefixes "single" and "multi" are both of Latin origin. In terms of Greek origin, one would have to distinguish between monomode and polymode fibers.

As is the case for the single-mode fiber itself, there are several alternative names for the wave guided by it. Up to now, we have used the name *fundamental mode*.

In microwave techniques, the same wave type guided by a dielectric rod waveguide or by a dielectric image line is called a *dipole mode*, since it can easily be launched by an electric dipole antenna located at the input end of the waveguide.

In the rigorous theory of the step-index fiber, the fundamental fiber mode is called the HE_{11} *wave*. The combination of the letters E and H indicates that there are nonzero longitudinal components of both the electric and the magnetic field.

The most common name for the fundamental fiber mode is the LP_{01} *mode* [Gloge 1971a]. The letters LP are an abbreviation for "linearly polarized", indicating that a single fundamental mode is linearly polarized, as has been discussed in Sect. 5.3.5. The subscripts 01 are necessary to distinguish the fundamental mode from other, so called higher order modes, which can be transmitted by the fiber at sufficiently short wavelengths.

The other, higher order linearly polarized LP_{lm} modes (Sect. 6.1), are characterized by two positive integers, the azimuthal and radial mode numbers l and m, respectively. The azimuthal mode number l equals half the number of intensity maxima found when going in the azimuthal direction around the fiber axis. For the fundamental mode, the intensity does not depend on the azimuth ϕ and therefore its azimuthal mode number l is zero. The radial mode number m equals the number of intensity maxima found when going in the radial direction from $\varrho = 0$ to $\varrho = \infty$ including the possible maximum at $\varrho = 0$. For the fundamental mode, there is only one intensity maximum at the fiber axis and therefore its radial mode number m is one.

At this point, it is useful to summarize some properties of the LP_{01} mode.

i) It is a *plane* wave, since the wavefronts are planes normal to the fiber axis.

ii) It is an *inhomogenous* plane wave, since the amplitude is not constant over a wavefront.

iii) It is a *hybrid* wave [Snitzer 1961], since there are nonzero longitudinal components of both the electric and the magnetic field.

5.4 Intensity Distribution and Transmitted Power

Since the wavefronts of the fundamental fiber mode are planes normal to the fiber axis, the time averaged intensity vector only has a component S_z parallel to the fiber axis. The intensity is proportional to the field amplitude squared (3.16). The amplitude distribution has been discussed in detail in Sect. 5.3.

The power transmitted by the fundamental fiber mode is obtained by integrating the z-component of the intensity vector

$$S_z = \tfrac{1}{2}\mathrm{Re}\{E_x H_y^*\} = \frac{1}{2Z_0} n_2 E_x^2(\varrho) \quad , \tag{5.33}$$

over the total fiber cross section A (3.17)

$$P = \int_A S_z \, dA \quad . \tag{5.34}$$

Since the intensity S_z is independent of the azimuth ϕ, the integration over ϕ yields

$$P = 2\pi \int_0^\infty S_z \varrho \, d\varrho \quad . \tag{5.35}$$

For arbitrary refractive-index profile, S_z depends in a complicated manner on ϱ, and the integral can in general be evaluated only numerically.

However, by using the Gaussian approximation (5.25) for the radial field distribution, the integral in (5.35) can be solved in closed form. Inserting (5.25) into (5.33), the intensity distribution can be approximated by

$$S_z(\varrho) = \frac{n_2 E_0^2}{2Z_0} \exp\left[-2\left(\frac{\varrho}{w_G}\right)^2\right] \quad .$$

(5.36)

At a distance of one spot size w_G from the fiber axis, the intensity is $1/e^2 = 0.135$ of its maximum value on axis.

In a fiber cross section, the contours of constant intensity are circles with the fiber axis as the center. In Fig. 5.9, the dashed circle represents the core-cladding boundary and the solid circles lines of constant intensity, on which the intensity is 5, 10, 15, 20, and 25 dB below its maximum value on the fiber axis. A step-index fiber and a normalized frequency of 2.2 have been assumed.

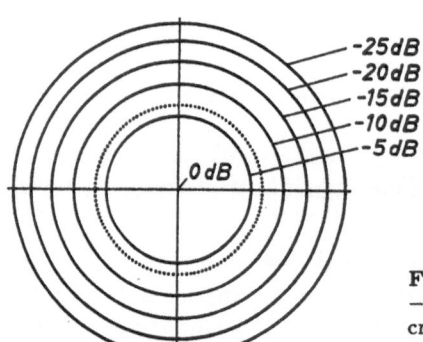

Fig. 5.9. Lines of constant field level $-5\,\mathrm{dB}$, $-10\,\mathrm{dB}\ldots$ of the fundamental fiber mode in a fiber cross section. The field level at the fiber axis is set to $0\,\mathrm{dB}$. The dashed line represents the core-cladding interface of a step-index fiber. $V = 2.2$

Using the Gaussian approximation for the radial intensity distribution (5.36), the integration of the intensity over the total fiber cross section gives for the transmitted power

$$P = \frac{\pi n_2 E_0^2 w_G^2}{4Z_0} \quad .$$

(5.37)

This equation corresponds to (4.27), which gives the power transmitted by a free Gaussian beam. It can be used to estimate the electric field strength E_0 on axis from the measured value of the transmitted optical power P.

As in the case of the fundamental Gaussian beam (4.28), there is a simple relation between the intensity $S_z(0) = n_2 E_0^2/(2Z_0)$ on the fiber axis, the total transmitted power P, and the area $A_b = \pi w_G^2$ (4.29) of the beam cross section defined by the $1/e^2$-intensity surface:

$$P = \tfrac{1}{2}S_z(0)A_b \quad .$$

(5.38)

The part $P(\varrho)$ of the total power P that is transmitted through a circular part of the cross section with radius ϱ, is given by (4.32)

$$P(\varrho) = P\left\{1 - \exp\left[-2\left(\frac{\varrho}{w_G}\right)^2\right]\right\} \quad . \tag{5.39}$$

Inserting the spot size w_G for ϱ, one finds that 86% of the total power is transmitted within the beam boundary defined by the $1/e^2$ intensity surface.

For the special case of a step-index fiber, the transmitted power is [Gloge 1971a]:

$$P = \left(\frac{\pi}{2}\right) a^2 C^2 \left(\frac{n_2}{Z_0}\right)\left(\frac{V}{U}\right)^2 K_1^2(W) \quad , \tag{5.40}$$

where C is the amplitude coefficient introduced in (5.15). The relative power in the core is given by [Snyder 1969a; Gloge 1971a]:

$$\frac{P_1}{P_1 + P_2} = \left(\frac{W}{V}\right)^2\left(1 + \frac{J_0^2(U)}{J_1^2(U)}\right) \quad . \tag{5.41}$$

For a given value of the normalized frequency V, W can easily be obtained from (5.19) and U from (5.23) [Kapron and Lukowski 1977]. The relative core power can also be calculated from the quantities b (5.51) and $d(Vb)/dV$ (5.71) related to the phase and group velocities, respectively [Krumbholz et al. 1980; Hussey 1984]

$$\frac{P_1}{P_1 + P_2} = \frac{1}{2}\left(b + \frac{d(Vb)}{dV}\right) \quad . \tag{5.42}$$

In Fig. 5.10 this power ratio is plotted as a function of the normalized frequency V. At the single-mode limit ($V = 2.4$), 83 % of the power is transmitted in the core and 17 % in the cladding.

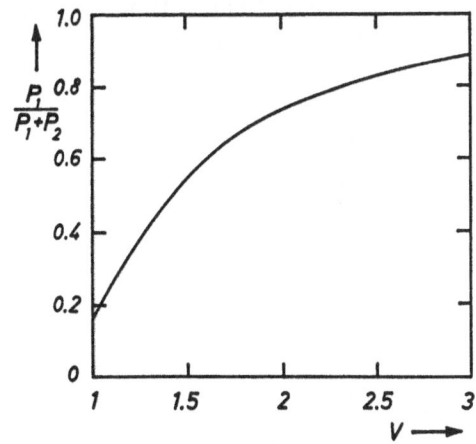

Fig. 5.10. Relative power $P_1/(P_1 + P_2)$ transmitted through the fiber core of a step-index fiber (5.41)

The transmission of significant power in the cladding material implies that the attenuation of the (inner) cladding material must be just as low as that of the core material.

Because of the longitudinal components of the electric and the magnetic field, there is also an instantaneous power flow in transverse directions. But due to the 90° phase shift between the transverse component of the electric (magnetic) field and the longitudinal component of the magnetic (electric) field, the time average of the transverse power flow (3.16) is zero. In the transverse direction there is only an energy oscillation, whereas in the longitudinal direction, there is a nonzero net energy transport.

In a single-mode fiber, the power guided by the fundamental mode is concentrated in a very small circular cross-section with a diameter of about $10\,\mu$m. Thus, even a small optical power of e.g. $1\,$mW causes a high power density of the order of $1\,$kW/cm^2. For such high power densities, the relation between the electric field and the electric polarization becomes nonlinear. Moreover, because of the small attenuation of fiber waveguides, high power densities can be maintained over long propagation lengths. Therefore, nonlinear optical effects, which are hard to detect with ordinary light propagating freely, are easily observed with laser light propagating in a single-mode fiber. Single-mode fibers are a particularly useful medium for the study of nonlinear optics [Shen 1984].

There are many nonlinear effects which occur in fibers, often in complicated and bewildering combinations [Stolen 1979b]. On one hand, nonlinear effects in single-mode fibers have found useful applications in optical amplifiers, frequency converters, tunable sources, and pulse compressors [Shank et al. 1982]. On the other hand, optical nonlinearities place an ultimate practical limitation on the power levels that can be used in fiber transmission systems.

In the past, in designing single-mode transmission systems, optical nonlinearities could be neglected. But modern systems use extremely low-loss fibers, higher transmitter powers, lasers with small linewidths, or dispersion-shifted fibers with small mode field diameter. All these circumstances enhance nonlinear effects. Therefore, for optical communications, it is important to know the maximum power that can be injected into the fiber input end without causing nonlinear effects. Order of magnitude values of the critical powers for some nonlinear processes are:

Raman amplification:	$1\,$mW
Stimulated Brillouin scattering:	$10\,$mW
Self-phase-modulation:	$30\,$mW
Stimulated Raman scattering:	$1000\,$mW .

A *coherent* light input has been assumed. For more details, the reader is refered to reviews of nonlinear effects in fibers [Stolen 1979a,b, 1980; Cotter 1983a,b; Desurvire et al. 1983; Mollenauer 1985; Lin C. 1986] and to papers which estimate the critical powers [Smith R.G. 1972; Gloge 1979b; Uesugi et al. 1981; Chraplyvy 1984; Cotter 1986].

Optical fibers can transmit not only information but also moderate amounts of power over distances of up to a few kilometers. The transmitted optical power can be transformed by a photocell into electrical power which can be used to power electronic circuits or sensors. However, because of their larger core cross-sectional area, multimode fibers are more suitable than single-mode fibers for power transmission applications. A $50\,\mu$m core multimode fiber may transmit high optical powers of up to $250\,$W over a distance of $1\,$km before significant power is transfered to longer wavelengths by Raman scattering [Byron and Pitt 1985].

At extremely high powers, a single-mode fiber will become permanently damaged by the transmitted light. The damage power density limit for pure silica is about $S = 10^{10}\,$W/cm^2, which inserted into (5.38) for a spot size of $5\,\mu$m gives a power limit of $4\,$kW [Cotter 1986].

5.5 Phase Constant

The phase distribution of the fundamental mode propagating along a straight fiber is very simple (5.13): the wavefronts are planes normal to the fiber axis and the phase decreases linearly in the direction of wave propagation. The phase constant β determines the rate of phase change.

In this section, we want to discuss the wavelength dependence of β, since it is necessary to know this function in order to calculate the signal velocity (Sect. 5.7) and the wavelength dependence of the signal velocity (Sect. 5.8), which causes signal distortions (Chap. 11).

The phase constant β is directly related to the wavelength λ_{01} of the fundamental fiber mode. By definition, the wavelength is the distance between two successive cross sections in which the electric field oscillates in phase. Since β gives the increase in phase angle per unit length, the product of β and λ_{01} must equal 2π:

$$\beta\lambda_{01} = 2\pi \quad , \quad \text{or} \tag{5.43}$$

$$\lambda_{01} = \frac{2\pi}{\beta} \quad . \tag{5.44}$$

It is convenient to define an "effective refractive index", "phase index" [Kogelnik and Weber 1974], or "normalized phase-change coefficient" [Clarricoats and Chan 1973] n_{e} by the ratio of the phase constant to the wave number in vacuum (4.13):

$$n_{\mathrm{e}} = \frac{\beta}{k} \quad . \tag{5.45}$$

The wavelength λ_{01} of the LP$_{01}$ mode is smaller than the vacuum wavelength λ by the factor $1/n_e$:

$$\lambda_{01} = \frac{\lambda}{n_e} \quad . \tag{5.46}$$

The fundamental mode propagates in a medium with a refractive index $n(\varrho)$, which depends on the distance ϱ from the fiber axis. It seems plausible that the effective refractive index is a certain average over the refractive index of the medium. Quantitatively, the following relation holds [Stewart 1980b; Geckeler 1986a]

$$n_e^2 = \frac{\displaystyle\int_0^\infty n^2(\varrho) E_x(\varrho) 2\pi\varrho\, d\varrho}{\displaystyle\int_0^\infty E_x(\varrho) 2\pi\varrho\, d\varrho} \quad . \tag{5.47}$$

This formula can be interpreted as follows: decompose the infinite cross section into narrow rings of width $d\varrho$ and radius ϱ and with center at the fiber axis. For each ring, multiply the square $n^2(\varrho)$ of the local refractive index by the area $2\pi\varrho\, d\varrho$ of the ring and by the local amplitude $E_x(\varrho)$ of the electric field as a weighting function. Then add all products. The effective refractive index squared is simply the weighted average of the square of the local refractive index. The area integral in the denominator is necessary for normalization. The formula (5.47) for the effective index can be used to discuss qualitatively the wavelength dependence of the effective index and, therefore, of the phase constant (In the discussion to follow we exclude depressed-cladding fibers).

For long wavelengths, i.e. small values of V, the field diameter is very large compared to the core radius (Sect. 5.3.2). The electric field distribution as a weighting function extends far into the cladding and the rings in the cladding contribute heavily to the integral in the numerator. The effective index will thus be similar to the index n_2 of the cladding, and the phase constant β will approximate the wavenumber $n_2 k$ of the cladding. This can also be stated in physical terms: most of the power is transmitted in the cladding material. The fundamental mode "sees" mainly the cladding material and the phase constant β is approximately that of a uniform plane wave in the infinite cladding material (4.36).

For short wavelengths, on the other hand, the field is concentrated in the region of maximum refractive index and the rings with maximum refractive index will determine the value of the integral. The wave "sees" the region of maximum refractive index and the phase constant β approximates the maximum wave number $n_1 k$ in the core.

From this discussion, we learn that the phase constant and the effective refractive index vary in the intervals

$$n_2 k < \beta < n_1 k \quad , \quad \text{and} \tag{5.48}$$

$$n_2 < n_e < n_1 \quad , \tag{5.49}$$

respectively.

From (5.46), the wavelength λ_{01} can be seen to lie in the interval

$$\frac{\lambda}{n_1} < \lambda_{01} < \frac{\lambda}{n_2} \quad . \tag{5.50}$$

The limiting values, λ/n_1 and λ/n_2, are the wavelengths of a uniform plane wave propagating in a homogenous medium with refractive index n_1 or n_2, respectively.

From the general form of the refractive-index profile (Fig. 5.2) it can be seen that the refractive index in part of the core region can become smaller than the cladding index, i.e. $n_{\min} < n_2$. In these so-called depressed inner cladding fibers, the phase constant β can become smaller than the cladding wave number $n_2 k$. In this case, some power leaks through the depressed cladding, and the wave is not a truly guided wave, but a leaky wave (Sect. 6.4).

Because of the very small values of the relative refractive-index difference Δ used in single-mode fibers, the relative width of the interval (5.48) for β is very small and one has to write down many decimal places to specify the numerical value of β. For this reason, and for reasons of a simple graphical representation of the results of a numerical analysis, it is convenient to consider a normalized phase constant b defined as [Gloge 1971b]:

$$b = \frac{\beta^2 - n_2^2 k^2}{n_1^2 k^2 - n_2^2 k^2} \quad . \tag{5.51}$$

Making use of the mathematical relation

$$A^2 - B^2 = (A + B)(A - B) \quad , \tag{5.52}$$

both in the numerator and the denominator, and of the fact that the phase constant β differs only very slightly from the wavenumber $n_1 k$, one gets a simple approximate expression for the normalized phase constant:

$$b \approx \frac{\beta - n_2 k}{n_1 k - n_2 k} = \frac{n_e - n_2}{n_1 - n_2} \quad . \tag{5.53}$$

The dimensionless quantity, b, can be interpreted graphically as the distance $\beta - n_2 k$ of β from the left boundary $n_2 k$ of the interval relative to the total length of the interval $n_1 k - n_2 k$ for β (Fig. 5.11). For small values of the normalized frequency V, the normalized phase constant b approximates zero and with increasing frequency, it increases monotonically towards the value 1 for very large values of V. The normalized phase constant b is the second of several universal parameters $(V, b, d(Vb)/dV, V d^2(Vb)/dV^2)$ which play a central role in the theory of single-mode fibers.

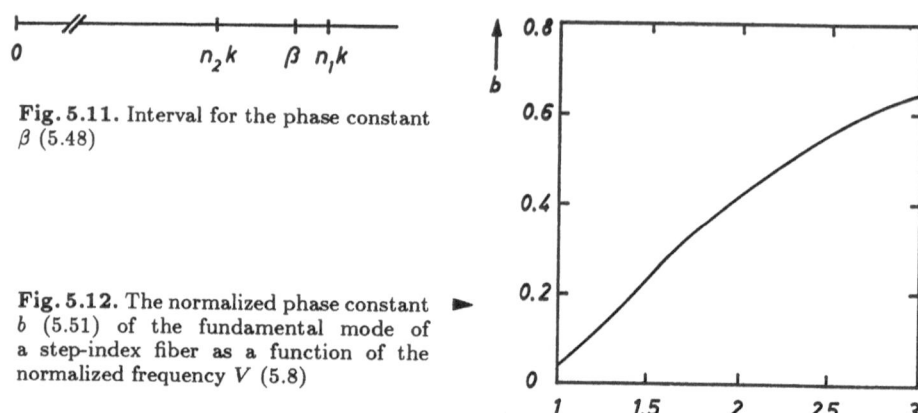

Fig. 5.11. Interval for the phase constant β (5.48)

Fig. 5.12. The normalized phase constant b (5.51) of the fundamental mode of a step-index fiber as a function of the normalized frequency V (5.8)

The normalized phase constant b is determined by the refractive-index profile and the operating wavelength. Many different numerical and approximate methods have been developed for calculating b for arbitrary index profiles.

For the special case of a step-index fiber, Fig. 5.12 shows the function $b(V)$ [Hussey and Pask 1982b]. Numerical values are given in Table 5.1. It is worth noting, that when $V < 2.4$, b is limited to $0 < b < 0.5$ and because of (5.53) this means that the effective index $n_e = \beta/k$ "seen" by the LP_{01} mode is closer to the cladding index throughout most of the single-mode V-value regime [Garrett and Todd 1982].

Knowing the normalized phase constant b, the phase constant β is given in terms of b by the approximate expression

$$\beta = k[n_2 + (n_1 - n_2)b] = k_2(1 + b\Delta) \quad . \tag{5.54}$$

Using in the definition of the normalized phase constant (5.51) the separation equation in the cladding (5.17) for the numerator and the definition of the normalized frequency (5.8) for the denominator, we find the relation

$$b = \frac{W^2}{V^2} \quad , \tag{5.55}$$

which shows that a dispersion curve b(V) can also be used to read the eigenvalue W that determines the radial field distribution in the cladding. For a given index profile, the eigenvalue W and, therefore, the normalized phase constant b, only depend on the normalized frequency V. The preference for using b instead of β lies in the fact that $b(V)$ is a universal function which does not depend explicitly on other fiber parameters.

For step-index fibers, the eigenvalue U, which determines the radial field distribution in the core, can also be obtained from a plot of b versus V, since because of (5.23)

$$U^2 = V^2(1 - b) \quad . \tag{5.56}$$

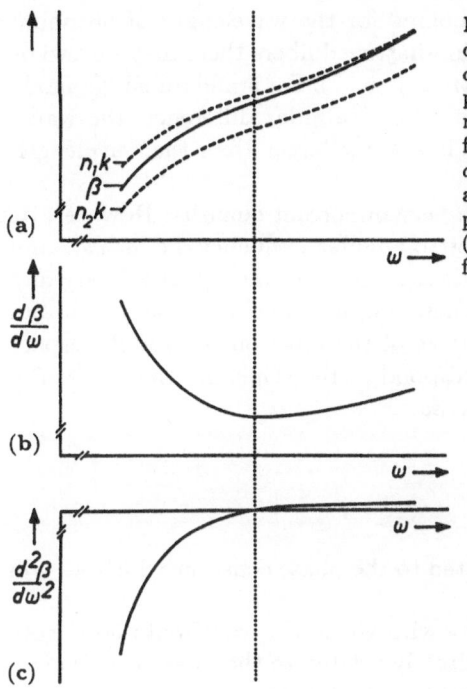

Fig. 5.13. (a) Wavenumbers $n_1 k$ and $n_2 k$ of uniform plane waves in the core and cladding materials, repectively, and of the phase constant β of the fundamental fiber mode as functions of the optical angular frequency ω. (b) Specific group delay $\tau_g = d\beta/d\omega$ (5.66) as a function of the optical angular frequency ω. (c) Chromatic dispersion parameter $d\tau_g/d\omega = d^2\beta/d\omega^2 = -(\lambda/\omega)D$ (5.86) as a function of the optical angular frequency ω

For later use, we also plot qualitatively the phase constant β as a function of the angular frequency ω which is proportional to the wave number $k = \omega/c$ in vacuum (Fig. 5.13a). From (5.48), the quantity β is seen to lie between the wave numbers $n_2 k$ and $n_1 k$. Therefore, it is helpful also to plot these wave numbers as functions of ω. If the refractive indices n_2 and n_1 were constants, these curves would simply be straight lines through the origin with their slopes proportional to the corresponding refractive indices. But in reality, the refractive indices vary slightly with wavelength because of material dispersion (Sect. 4.5).

The dashed curves in Fig. 5.13a schematically represent the frequency dependence of the wave numbers $n_2 k$ and $n_1 k$. In reality, the vertical distance between the curves (for standard single-mode fibers, the relative refractive index difference Δ is typically 0.3%) and the curvature are very small; in Fig. 5.13a the abscissa and the ordinate have been expanded in order to make the separation and the curvature visible. The curvature of the curves for $n_1 k$ and $n_2 k$ changes sign near a certain frequency, which for the silica fibers used today corresponds to a wavelength λ_{0m} in the vicinity of 1.3 μm. This wavelength is called the wavelength of minimum material dispersion (Sect. 4.5).

The dependence of β on $k = \omega/c$ can be calculated from (5.54). In Fig. 5.13a, the function $\beta(\omega)$ is represented by the solid curve, which for small and large frequencies approaches the cladding and core curves, respectively. The curvature of the dispersion curve of the fundamental mode of a fiber also changes its sign at least once. This so-called wavelength of minimum dispersion, λ_0, will be found to be very important when looking for the wavelength of optimum fiber bandwidth (Sect. 11.2) (In dispersion-flattened fibers there may be two or three wavelengths of minimum dispersion, Fig. 5.17). For standard single-mode fibers with a small value of the relative refractive index difference, the wavelength λ_0 of minimum total dispersion is slightly larger than the wavelength λ_{om} of minimum material dispersion.

The phase constant β itself is not a very important quantity. However, its first and second derivatives with respect to angular frequency, i.e. $d\beta/d\omega$ and $d^2\beta/d\omega^2$, are important quantities which are related to the signal velocity and to the fiber bandwidth, respectively. Therefore, for later use, Figs. 5.13b and 5.13c show the first and second derivatives of the function $\beta(\omega)$ with respect to optical angular frequency; these correspond to the slope and the curvature, respectively, of the solid curve in Fig. 5.13a.

5.6 Phase Velocity

There are several quantities closely related to the phase constant which we will discuss next.

The phase velocity v_p is the velocity with which the wavefronts propagate along the fiber. The phase velocity is directly related to the phase constant β: Consider the propagation of a wave crest, for which the argument of the cosine function in (5.12) is a certain integer multiple of 2π:

$$\omega t - \beta z = n2\pi \quad . \tag{5.57}$$

In this relation, z is the time-dependent location of the wave crest. Taking the derivative of (5.57) with respect to time yields

$$\omega - \frac{\beta dz}{dt} = 0 \quad . \tag{5.58}$$

Since dz/dt is the velocity of the wave crest, i.e. the phase velocity v_p, the relation between v_p and β is (3.13)

$$v_p = \frac{\omega}{\beta} \quad . \tag{5.59}$$

Combining (5.48) and (5.59), we have

$$\frac{c}{n_1} < v_p < \frac{c}{n_2} \quad , \tag{5.60}$$

i.e. the phase velocity of the fundamental mode is larger than that of a uniform plane wave propagating in a homogeneous medium with refractive index n_1, but smaller than that of a uniform plane wave propagating in a homogeneous medium with refractive index n_2. Since the refractive indices n_1 and n_2 are near to 1.5, the phase velocity is about 200 000 km/s.

Using the definition of the effective refractive index (5.45), the phase velocity can also be written in the form

$$v_{\mathrm{p}} = \frac{c}{n_{\mathrm{e}}} \quad . \tag{5.61}$$

The refractive index of a medium is defined as the ratio of the phase velocity of a uniform plane wave in vacuum to that in the medium (4.37). In analogy, the effective refractive index n_{e} as defined by (5.45) can be seen to equal the ratio of c to the phase velocity of the fundamental fiber mode.

Expressing the phase constant β by the wavelength λ_{01} of the fundamental fiber mode (5.46), the phase velocity can also be written

$$v_{\mathrm{p}} = f \lambda_{01} \quad . \tag{5.62}$$

A wavefront takes a time L/v_{p} to propagate through a fiber of length L. Usually, the phase delay time L/v_{p} is divided by the fiber length L to give the phase delay time per unit length:

$$\tau_{\mathrm{p}} = \frac{1}{v_{\mathrm{p}}} = \frac{\beta}{\omega} = \frac{n_{\mathrm{e}}}{c} = n_{\mathrm{e}} \tau_0 \quad , \tag{5.63}$$

where $\tau_0 = 1/c = 3.33 \, \mu\mathrm{ms/km}$ (4.43) is the specific delay time in vacuum.

From (5.63), the specific phase delay time is the phase constant divided by the angular frequency. Therefore, in order to graphically determine the phase delay from a dispersion diagram like Fig. 5.13a, one connects the point on the solid $\beta(\omega)$-curve corresponding to the given optical frequency with the origin of the diagram (not shown) by a straight line. The slope of this line equals the specific phase delay τ_{p} and is proportional to the effective index n_{e}.

5.7 Group Velocity

The phase velocity is the velocity of the wavefronts of a wave with sinusoidal time dependence (Sect. 3.5.3). When an optical wave is used for communications, its power is modulated with the signal to be transmitted. The wave envelope propagates not with the phase velocity, but with the so-called group velocity, which, in general, is different from the phase velocity.

Whereas the phase velocity is the quotient of the angular frequency and the phase constant (5.59)

$$v_{\mathrm{p}} = \frac{\omega}{\beta} \quad , \tag{5.64}$$

the group velocity is the derivative of the angular frequency with respect to the phase constant [Ramo et al. 1965; Siegman 1971]

$$v_g = \frac{d\omega}{d\beta} \quad .$$

(5.65)

Of the several equivalent expressions for the group velocity, this form is the most simple to remember.

The values of the phase and the group velocity of the fundamental mode of a single-mode fiber differ only slightly (some few percent, at most), and for a rough estimate one can approximate both velocities by 2×10^8 m/s. However, in some cases, it is important to differentiate between them. For example, when using the backscattering method (Sect. 13.4) to measure the fiber length or to locate a fault, one has to use the *group* velocity to calculate the lengths from the time delays measured. Moreover, it is the wavelength dependence of the *group* velocity that limits the signal transmission capacity of single-mode fibers (Sect. 11.2).

A fiber of length L delays the signal by a time L/v_g. Thus, the group delay time τ_g per unit length is the inverse of the group velocity:

$$\tau_g = \frac{1}{v_g} = \frac{d\beta}{d\omega} = \left(\frac{1}{c}\right)\frac{d\beta}{dk} \quad .$$

(5.66)

The group index of a uniform plane wave propagating in a homogeneous medium has been defined by (4.47)

$$n_g = \frac{c}{v_g} \quad ,$$

(5.67)

where c is the velocity of light in vacuum and v_g the group velocity of the infinite plane wave.

In analogy, for a single-mode fiber, it is usual to define an "effective group index" or "group index of the guide" n_{ge} by [Kogelnik and Weber 1974]

$$n_{ge} = \frac{c}{v_g} \quad ,$$

(5.68)

where the symbol v_g now denotes the group velocity of the fundamental fiber mode. The specific group delay of the fundamental fiber mode can then be written

$$\tau_g = \frac{n_{ge}}{c} = n_{ge}\tau_0 \quad .$$

(5.69)

The effective group index n_{ge} is related to the effective index n_e (5.45) by a relation [Kogelnik and Weber 1974]

$$n_{ge} = n_e - \lambda\frac{dn_e}{d\lambda} \quad ,$$

(5.70)

which is formally equal to the relation (4.50) between the group index and the refractive index of a transparent medium.

Table 4.1 summarizes the symbols and formulas for the wave velocities and time delays both for a uniform plane wave propagating in a homogeneous medium and for the fundamental mode propagating in a single-mode fiber.

The specific group delay time τ_g (5.66) equals the derivative of the function $\beta(\omega)$. Therefore, from a dispersion curve $\beta(\omega)$ such as the one plotted in Fig. 5.13a, one can graphically determine the group velocity by drawing the tangent to the curve at the operating point. Its slope equals the specific group delay τ_g and is proportional to the effective group index n_{ge}.

The specific group delay of the fundamental mode in a single-mode fiber is not much different from the specific group delay of a uniform plane wave in glass and is therefore of the order of 5 μs/km. At the wavelength λ_0 of minimum dispersion, where the dispersion curve $\beta(\omega)$ has its point of inflection, the slope of the tangent is minimum and thus there is a minimum of the group delay time and a maximum of the group velocity.

Figure 5.13b is a schematic plot of the frequency dependence of the specific group delay obtained by differentiating the solid curve in Fig. 5.13a.

Inserting (5.54) for β in (5.66) and approximating the relative difference of the refractive indices $(n_1 - n_2)/n_2$ by the relative difference of the group indices $(n_{g1} - n_{g2})/n_{g2}$, gives the group delay per unit distance [Gloge 1971b] as

$$\tau_g = \left(\frac{1}{c}\right)\left[n_{g2} + (n_{g1} - n_{g2})\frac{d(Vb)}{dV}\right] \quad . \tag{5.71}$$

Usually, the dispersive properties of the core and the cladding glass are approximately the same, so that the wavelength dependence of Δ can be ignored. Then

$$\tau_g = \left(\frac{1}{c}\right)\left[n_{g2} + n_2\Delta\frac{d(Vb)}{dV}\right] \quad . \tag{5.72}$$

The first term in (5.72) describes the wavelength dependence of the group delay which would arise from propagation of a uniform plane wave in an infinitely extended medium whose refractive index equals that of the fiber cladding. This term does not depend on mode properties. The second part, which is due purely to the waveguide properties of the fiber, is governed by the derivative $d(Vb)/dV$ and describes the change in group delay caused by the change in power distribution between core and cladding.

The "mode delay factor" [Unger 1977a] $d(Vb)/dV$ is the third of several universal parameters $(V, b, d(Vb)/dV, Vd^2(Vb)/dV^2)$ which play a central role in the theory of single-mode fibers. For a fiber with arbitrary refractive index profile, the derivative $d(Vb)/dV$ can be obtained from the normalized phase constant b and the spot size w_d related to dispersion (10.10), and the core radius a [Hussey 1984; Hussey and Martinez 1985]

$$\frac{d(Vb)}{dV} = \frac{4a^2}{V^2w_d^2} + b \quad . \tag{5.73}$$

For a step index fiber, the V-dependence of the function $d(Vb)dV$ is [Hung-chia and Zi-Hua 1981]

$$\frac{d(Vb)}{dV} = 1 - \frac{U^2}{V^2}\left(1 - 2\frac{\mathrm{K}_0^2(W)}{\mathrm{K}_1^2(W)}\right) \quad . \tag{5.74}$$

This function is plotted in Fig. 5.14. Numerical values are listed in Table 5.1.

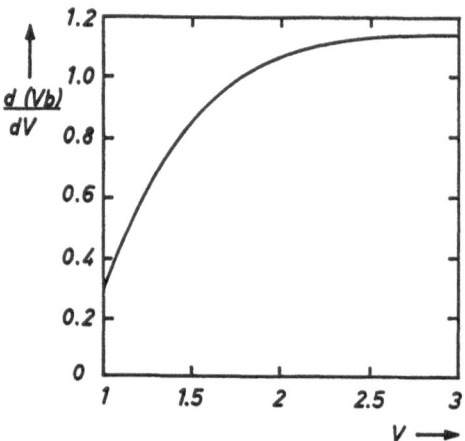

Fig. 5.14. The function $d(Vb)/dV$ (5.71) determining the group delay of the fundamental mode in a step-index fiber as a function of the normalized frequency V (5.8)

For small values of V, $d(Vb)/dV$ is very small and for large values of V, it approaches unity. Therefore, for small and large V the group delay (5.71) equals the values corresponding to those of a uniform plane wave propagating in the cladding and core material, respectively. For small V, most of the power is transmitted in the cladding, and the guided mode "sees" the cladding material, whereas for large V, most of the power flows in the core and the fundamental mode "sees" the core material.

Having discussed a method for obtaining the specific group delay time from the frequency dependence of the phase constant of the fundamental mode, we now describe another method which determines τ_g directly from the dispersive properties of the glass of which the fiber is made.

Since the effective refractive index n_e of the fundamental mode is a certain average of the refractive index $n(\varrho)$ of the glass (5.47), one might presume that the effective group index n_{ge} of the mode is, likewise, an average of the group index $n_g(\varrho)$ of the glass. In fact, this is not the case, but there is a formula for the product $n_{ge}n_e$ of the effective group index and the effective index [Brown 1966; Case 1972; Eaves 1973; Kawakami 1975; Nemoto and Makimoto 1976; Krumbholz et al. 1980; Stewart W.J. 1980b; Geckeler 1986a]:

$$n_{ge}n_e = \frac{\int\limits_0^\infty n(\varrho)n_g(\varrho)S_z(\varrho)2\pi\varrho\,d\varrho}{\int\limits_0^\infty S_z(\varrho)2\pi\varrho\,d\varrho} \quad . \tag{5.75}$$

This formula is sometimes called the partial-power law or Brown's identity. It is also applicable to lossy waveguides [Kuester 1984]. The interpretation of this formula is analogous to the corresponding for the effective refractive index (5.47): the product of the effective group index and the effective index of the fundamental mode is the area average of the product of the local group index and the local refractive index of the glass. This time, the weighting function is not the electric field, but the intensity distribution $S_z(\varrho)$, which is proportional to the square of the electric field. Since the effective index n_e is known from (5.47), we can calculate n_{ge} from (5.75) and, from (5.69), the specific group delay τ_g.

The product of the effective index n_e and the effective group index n_{ge}, which appears on the left hand side of (5.75), can also be expressed by the *first* derivative of the square β^2 of the phase constant with respect to the square ω^2 of the angular frequency, as can be derived by applying the rules of differential calculus:

$$n_e n_{ge} = c^2 \frac{d\beta^2}{d\omega^2} \quad . \tag{5.76}$$

Be careful not to confuse $d\beta^2/d\omega^2$ with the *second* derivative $d^2\beta/d\omega^2$ of β with respect to ω!

The group velocity equals the velocity of energy propagation, which, for a guided wave, is defined by the ratio of the transmitted power to the energy stored in the electromagnetic field per unit length. Therefore, alternatively, the group velocity can be obtained from the field distribution by calculating the power P transmitted and the electromagnetic field energy W contained in a fiber section of length Δz. Then [Clarricoats and Chan 1973; Kogelnik and Weber 1974; Kogelnik 1975]:

$$v_g = \frac{P \Delta z}{W} \quad . \tag{5.77}$$

The stored energy W is the sum of the energy W_e stored in the electric field and the energy W_m stored in the magnetic field:

$$W = W_e + W_m \quad . \tag{5.78}$$

In dielectric waveguides, the electric energy stored in the waveguide is equal to the magnetic energy [Kogelnik and Weber 1974]:

$$W_e = W_m \quad . \tag{5.79}$$

In a uniform plane wave propagating in a homogeneous medium, the energy densities, i.e. the energies per unit volume, are equal:

$$w_e = w_m \quad . \tag{5.80}$$

But in a dielectric waveguide, we have equality only between the integrated energy contributions over the whole cross section of the beam; the local energy densities at an arbitrary point in the cross section are not necessarily equal.

In the absence of material dispersion, the physical reason for the difference between group and phase velocity in dielectric waveguides is the existence of longitudinal components of the electric and magnetic fields [Kogelnik and Weber 1974; Kogelnik 1975]. If there is no material dispersion, the group velocity is always less than the phase velocity

$$v_g \leq v_p \quad . \tag{5.81}$$

In the presence of anomalous material dispersion ($dn/d\lambda > 0$, the inequality (5.81) is no longer guaranteed [Kogelnik and Weber 1974].

The group delay time

$$t_g = \tau_g L = \frac{n_{ge} L}{c} \quad , \tag{5.82}$$

in a fiber of length L depends on temperature, since both the effective group index n_{ge} and the fiber length L change with temperature. The change in group delay Δt_g for a temperature change ΔT is

$$\Delta t_g = \frac{1}{c} \left(\frac{dn_{ge}}{dT} L + n_{ge} \frac{dL}{dT} \right) \Delta T \quad . \tag{5.83}$$

For *bare* doped fused silica fibers, the thermal expansion coefficient α is very small

$$\alpha = \left(\frac{1}{L} \right) \frac{dL}{dT} = 8 \times 10^{-7}/\text{deg} \quad . \tag{5.84}$$

Since the temperature coefficient of the effective group index dn_{ge}/dT is of the order of $10^{-5}/\text{deg}$ [Cohen L.G. and Fleming 1979], the increase of group delay in bare fibers with temperature is primarily due to an increase of the refractive index.

Plastic materials have a coefficient of linear expansion which is much larger than that of glass. Therefore, the temperature coefficient of group delay dt_g/dT is larger for jacketed fibers than for bare fibers and becomes greater with increasing thickness of the jacketing layer [Tateda et al. 1980; Namihira et al. 1981; Nakahara et al. 1983]. However, the temperature coefficient of fibers coated with fiber reinforced Kevlar has been reported to be smaller than that of the bare fiber [Tanaka S. and Ono 1984].

The group delay increases with increasing microbending loss (Sect. 5.9.9). A theoretical estimation gave a change of relative delay of 3.4×10^{-7} per decibel of excess microbending loss [Das S. et al. 1985].

5.8 Chromatic Dispersion

For a hypothetical fiber with a specific group delay that is independent of optical frequency, the time dependence of the wave envelope at the far end of the fiber would be identical to that at the fiber front end (neglecting a decrease in the amplitude and a time delay). The wave envelope would not be distorted by the fiber. Transmitted pulses would not change their form or duration. This ideal fiber would have infinite baseband bandwidth.

But in reality, the specific group delay varies with wavelength (Fig. 5.13b). In combination with the finite width of the source spectrum, this effect causes a distortion of the wave envelope: pulses will be temporally broadened and higher modulation frequencies will be attenuated more strongly than lower frequencies. When designing single-mode fibers for telecommunication applications, it is essential that the temporal broadening of the launched pulse over a repeater section length remains much smaller than the time interval between two pulses in order to avoid intersymbol infererence.

In Chap. 11 we will discuss quantitatively pulse broadening and the bandwidth of single-mode fibers. Here, we discuss the physical reason for these effects, namely the wavelength dependence of the specific group delay, which is called chromatic dispersion. Sometimes, it is also called "intramodal" dispersion, whereas the variation of group velocity with mode number in multimode fibers is called "intermodal" dispersion.

The derivative of the specific group delay τ_g with respect to the vacuum wavelength is called the total first-order dispersion parameter, or for short "chromatic dispersion"

$$D = \frac{d\tau_g}{d\lambda} \quad . \tag{5.85}$$

Since the change of the specific group delay is usually measured in picoseconds per kilometer and the change of the wavelength in nanometers, the common unit for the first-order dispersion parameter is ps/(km nm).

Replacing the variable λ by ω, the dispersion parameter D can also be written in the form

$$D = -\left(\frac{\omega}{\lambda}\right)\frac{d\tau_g}{d\omega} = -\left(\frac{\omega}{\lambda}\right)\frac{d^2\beta}{d\omega^2} \quad . \tag{5.86}$$

From (5.86), we see that the dispersion parameter D is essentially (neglecting the change of the factor ω/λ) is the negative slope of the $\tau_g(\omega)$-curve in Fig. 5.13b or the negative curvature of the $\beta(\omega)$-curve in Fig. 5.13a. In Fig. 5.13c, the second derivative $d^2\beta/d\omega^2$ of the phase constant β with respect to ω is plotted as a function of ω. Remember that the curvature of the curves plotted in Figs. 5.13a–c, have been exaggerated in order to illustrate the main points. In reality, the function $\beta(\omega)$ is nearly linear, $\tau_g = d\beta/d\omega$ nearly constant, and $d^2\beta/d\omega^2$ extremely small. Therefore, in order to calculate D,

$\beta(\omega)$ needs to be known extremely accurately. If absolute accuracy for D is the major concern, all approximations can and should be avoided in the analysis [Garrett and Todd 1982].

The dispersion parameter D is zero at the wavelength λ_0, where the curvature of the $\beta(\omega)$-curve changes sign, and where the group delay is minimum. This is the wavelength of minimum chromatic dispersion mentioned before. In standard fibers, for wavelengths smaller than λ_0, i.e. higher frequencies, $d^2\beta/d\omega^2$ is positive and D negative. For wavelengths larger than λ_0, the dispersion parameter D is positive.

There are two reasons for the curvature of the $\beta(\omega)$-curve, i.e. for chromatic dispersion:

1. the refractive indices n_1 and n_2 depend on the wavelength (material dispersion),
2. the field distribution of the fundamental mode changes with wavelength, causing the $\beta(\omega)$-curve in Fig. 5.13a to approximate the $n_2 k$-curve for small frequencies and the $n_1 k$-curve for large frequencies (waveguide dispersion).

First we consider the two causes of dispersion separately. Pure material dispersion in the absence of waveguide dispersion has already been discussed in Sect. 4.5. and has been found to be substantially proportional the second derivative of the refractive index with respect to wavelength (4.53):

$$D_{\mathrm{m}} = -\left(\frac{\lambda}{c}\right)\frac{d^2 n}{d\lambda^2} \quad . \tag{5.87}$$

Waveguide dispersion in the absence of material dispersion can be studied by considering the fundamental mode propagating in a fiber with a refractive index profile which is independent of wavelength. The curves of $n_2 k$ and $n_1 k$ versus ω are then straight lines and the curvature of the $\beta(\omega)$-curve is due solely to the frequency dependence of the normalized phase constant b (5.51). Inserting (5.72) into (5.85), one obtains for the waveguide dispersion parameter D_{w}

$$D_{\mathrm{w}} = -\left[\frac{n_1 - n_2}{\lambda c}\right] V \frac{d^2(Vb)}{dV^2} \quad . \tag{5.88}$$

Since the normalized phase constant b for a given fiber only depends on the normalized frequency V, the function $V d^2(Vb)/dV^2$ also only depends on V. The normalized waveguide dispersion coefficient $V d^2(Vb)/dV^2$ is the fourth of several universal parameters $(V, b, d(Vb)/dV, V d^2(Vb)/dV^2)$ which play a central role in the theory of single-mode fibers.

Since the waveguide dispersion is related to the relative powers transmitted in the core and the cladding, which depend on the spot size, one presumes a relation between waveguide dispersion and spot size. Indeed, it has been shown [Sansonetti 1982a,b; Petermann 1983; Pocholle et al. 1983a,b; Pask 1984] that the pure waveguide dispersion parameter D_{w} can be obtained from the wavelength dependence of the spot size w_{d} (10.10), which will be introduced later in Sect. 10.1.4:

$$D_w = \frac{\lambda}{2\pi^2 c n_1} \frac{d}{d\lambda}\left(\frac{\lambda}{w_d^2}\right) \quad . \tag{5.89}$$

Comparing the right hand sides of (5.88) and (5.89) and replacing the derivative with respect to λ by the derivative with respect to V, gives [Hussey 1984; Hussey and Martinez 1985]

$$V\frac{d^2(Vb)}{dV^2} = \frac{d}{dV}\left(\frac{4a^2}{Vw_d^2}\right) \quad . \tag{5.90}$$

For the special case of a step-index fiber a long, but closed form expression for the function $Vd^2(Vb)/dV^2$ can be found in a paper by Hung-Chia and Zi-Hua [1981]. Figure 5.15 represents the function $Vd^2(Vb)/dV^2$ and shows, that the waveguide dispersion parameter D_w is negative in the single-mode regime, has maximum magnitude at $V = 1.14$, and becomes zero outside the region for single-mode operation for $V = 3.01$ [Gambling et al. 1979b; Snyder 1981]. Numerical values for the function $Vd^2(Vb)/dV^2$ can be found in Table 5.1.

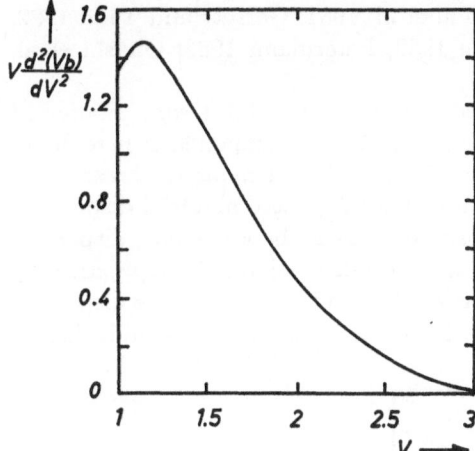

Fig. 5.15. The function $Vd^2(Vb)/dV^2$ (5.88) determining waveguide dispersion in a step-index fiber as a function of the normalized frequency V (5.8)

In a real fiber, there will be both material dispersion and waveguide dispersion at the same time. The total dispersion parameter is obtained by adding the material dispersion D_m, the waveguide dispersion D_w, and a third term D_p:

$$D = D_m + D_w + D_p \quad . \tag{5.91}$$

Strictly speaking, in fibers with a power-law index profile (5.6, 7), for D_m one has to take an average of the material dispersion of the core and cladding materials, the so-called composite material dispersion [Gambling et al. 1979b]. Similarly, for the second term in (5.91) one has to insert a so-called composite waveguide dispersion, which differs from (5.88) by a factor near unity which contains $dn_2/d\lambda$. The third term D_p is the so-called composite profile dispersion which is proportional to the derivative of Δ with respect to wavelength. This last term D_p is very small

$$D_p < 0.5\,\mathrm{ps/(km\,nm)} \quad , \tag{5.92}$$

especially at long wavelengths, and can be neglected in a rough estimate of total dispersion.

For inhomogeneous materials, e.g. the core of a graded-index single-mode fiber, one has to use some interpolation scheme to deduce the approximate variation of index both with wavelength and with radial position within the fiber. One such method takes the form of a four-term polynomial fit [Paek et al. 1982]. An interpolation method which is both simpler and has theoretical justification, is based on the Clausius-Mossotti equation [Melman and Davies 1985], which relates the refractive index to the polarizabilities associated with the components.

For a more exact discussion of total chromatic dispersion, the reader is referred to the original literature [Jürgensen 1975; Gambling et al. 1979b,c; Marcuse 1979b; Sugimura et al. 1980; Pal et al. 1981; Garrett and Todd 1982; Imoto and Tsuchiya 1982; Jeunhomme 1982; Petermann 1983; Ohashi et al. 1985].

As Jeunhomme [1979] has pointed out, there has been some confusion in the literature regarding the signs of the dispersion components; here Jeunhomme's convention has been adhered to, since it is based on physical reasoning.

For wavelengths longer than the wavelength λ_{0m} zero material dispersion, the material dispersion parameter is positive whereas the waveguide dispersion parameter is negative (Fig. 5.16). Since the total dispersion D is approximately equal to the sum of the material dispersion D_m and the waveguide dispersion D_w, for a certain wavelength λ_0 slightly larger than λ_{0m} the waveguide disper-

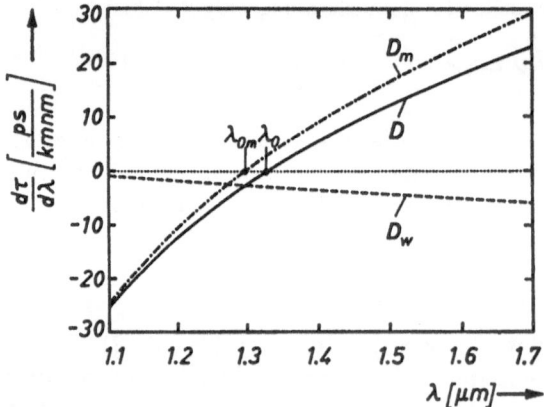

Fig. 5.16. Material (D_m), waveguide (D_w), and total dispersion ($D = D_m + D_w$) for a conventional non-dispersion-modified single-mode fiber as functions of the wavelength λ (5.91). The symbols λ_{0m} and λ_0 denote the wavelengths of zero material dispersion and zero total dispersion, respectively

sion compensates the material dispersion and the total first-order dispersion parameter D is zero.

Thus, in standard fibers, waveguide dispersion shifts the wavelength of minimum chromatic dispersion slightly away from the wavelength λ_{om} of zero material dispersion to longer wavelengths [Kapron 1977].

Several authors [Dyott and Stern 1970; Kapron and Keck 1971; Payne D.N. and Gambling 1975] proposed to operate at this wavelength λ_0 of minimum dispersion in order to minimize the pulse spreading and to increase the bandwidth.

For single-mode fibers optimized for operation at 1300 nm, CCITT [1984] recommends that the maximum magnitude of chromatic dispersion D shall not exceed 3.5 ps/(km nm) in the wavelength range 1285–1330 nm and shall be smaller than 20 ps/(km nm) at 1550 nm [Kapron 1987].

At the wavelength λ_0, the derivative of group delay with respect to wavelength is zero. But looking at the $\tau_g(\omega)$-curve (Fig. 5.13b) we see, that the group delay still depends on the wavelength around λ_0. Therefore, even if we choose the operating wavelength to coincide with λ_0, the signal will be slightly distorted because of the wavelength dependence of the group velocity.

The variation of the chromatic dispersion with wavelength is commonly characterized by the "second-order dispersion parameter", "dispersion slope", or "differential dispersion" [Kapron 1984b]

$$S = \frac{dD}{d\lambda} = \frac{d^2\tau_g}{d\lambda^2} \quad . \tag{5.93}$$

Whereas the chromatic dispersion D is related only to the second derivative of the phase constant with respect to angular frequency (5.86), the dispersion slope is generally related to both the second and third derivatives (11.27, 28):

$$S = \frac{(2\pi c)^2}{\lambda^4} \frac{d^3\beta}{d\omega^3} + \frac{4\pi c}{\lambda^3} \frac{d^2\beta}{d\omega^2} \quad . \tag{5.94}$$

Most important is the value of this function $S(\lambda)$ at the wavelength λ_0 of minimum chromatic dispersion, the so called zero-dispersion slope

$$S_0 = S(\lambda_0) \quad , \tag{5.95}$$

which is determined only by the third derivative of β.

Typical values of the dispersion slope S at λ_0 are $S_0 = 0.085\,\text{ps/(km nm}^2)$ for standard single-mode fibers and $S_0 = 0.05\,\text{ps/(km nm}^2)$ for dispersion shifted fibers with $\lambda_0 = 1550$ nm, respectively [Marcuse and Stone 1985]. For standard fibers CCITT has recently proposed that minimum and maximum values for λ_0 and a maximum value for S_0 are specified (two-parameter representation) instead of recommending $|D| < 3.5\,\text{ps/(km nm)}$ in the wavelength interval 1285–1330 nm (window specification) [Kapron 1987].

When the two parameters λ_0 and S_0 are specified by the fiber manufacturer, the user can estimate the fiber dispersion at an arbitrary wavelength by using (13.109).

103

Often, the wavelength of zero first order dispersion $(D = 0)$ is called the wavelength of *zero* dispersion. Because of second-order dispersion effects, it is perhaps better to call it the wavelength of *minimum* dispersion.

By tailoring the index profile, one can increase the waveguide dispersion relative to the material dispersion and thus either shift the minimum dispersion wavelength from $1.3\,\mu m$ to e.g. $1.55\,\mu m$ [Chang C.T. 1979a,b; Cohen L.G. et al. 1979], where silica fibers have a loss minimum, or even obtain two or three wavelengths of zero first-order dispersion parameter and small values of D in between [Cohen L.G. et al. 1982b, 1983a]. Figure 5.17 shows the wavelength dependence of the chromatic dispersion of typical standard (non-dispersion-modified), dispersion-shifted, and dispersion-flattened fibers. All three types of single-mode fibers are commercially available.

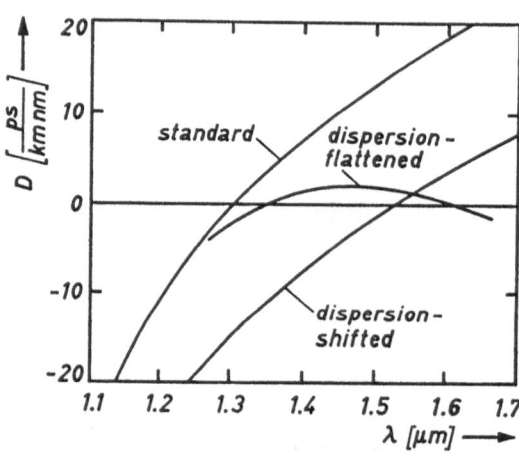

Fig. 5.17. Total chromatic dispersion $D = d\tau_g/d\lambda$ (5.85) for a standard non-dispersion-modified, a dispersion-shifted, and a dispersion-flattened fiber as functions of the wavelength λ

In general, the chromatic dispersion of fibers fabricated to have dispersion minimum near $1.55\,\mu m$ depends sensitively on the core diameter and the refractive-index profile function. Reviews of dispersion-modified fibers have been published [Cohen et al. 1983b; Bhagavatula 1985, 1986; Bachmann 1986; Ainslie and Day 1986].

The wavelength λ_0 of minimum dispersion is slightly dependent on temperature. For silicon-coated, germanium-doped, dispersion-unshifted fibers, the relationship between λ_0 and the temperature was nearly linear with a coefficient of $0.025\,nm/°C$ [Hatton and Nishimura 1986a,c]. The dispersion slope S_0 at zero dispersion wavelength λ_0 was found to be independent of temperature.

5.9 Attenuation

There are several reasons why the power transmitted by the LP_{01} mode will decrease along the fiber. In this section, we introduce the quantities used for specifying the fiber attenuation and discuss the different loss mechanisms both in straight and curved fibers.

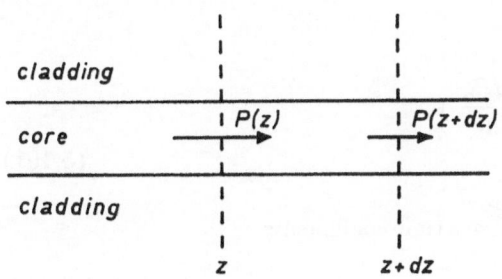

Fig. 5.18. Sketch used to derive the relation (5.99) between the power loss per unit length $-dP/dz$ and the attenuation constant α'

5.9.1 Attenuation Coefficient

Consider a short section of a uniform single-mode fiber with length dz at the location z (Fig. 5.18). The fundamental mode will transmit a power $P(z)$ through the infinite cross section at z and a power $P(z + dz)$ through the cross section at $z + dz$. Because of power losses, the transmitted power at $z + dz$ will be smaller than at z. The power difference

$$P(z) - P(z + dz) = dP_t = dP_a + dP_s + dP_b + \dots \tag{5.96}$$

is the power dP_t lost in the length element dz. The total power loss dP_t is the sum of the power losses caused by absorption (dP_a), scattering (dP_s), bending (dP_b), and other reasons.

For optical powers smaller than about $100\,\text{mW}$, nonlinear scattering processes (Sect. 5.9.6) can be neglected, and the power loss increases in proportion to the power transmitted and the length dz of the section of fiber:

$$dP_t = 2\alpha' P(z)\, dz \quad . \tag{5.97}$$

The reasons for separating a factor 2 from the constant of proportionality and for adding a prime to the letter α will be explained later.

Now, the change dP of transmitted power from z to $z + dz$ is

$$dP = P(z + dz) - P(z) = -dP_t = -2\alpha' P(z)\, dz \quad . \tag{5.98}$$

Thence, for the length dependence of the transmitted optical power P, we obtain a very simple differential equation

$$\frac{dP}{dz} = -2\alpha' P(z) \quad , \tag{5.99}$$

for which the exponential function

$$P(z) = P(0) \exp(-2\alpha' z) \tag{5.100}$$

is a solution. Thus, for a uniform fiber, the transmitted power decays exponentially in the direction of wave propagation.

The rate of power decay is determined by the parameter α' which is therefore called the attenuation coefficient. Combining (5.96) and (5.97) one obtains

$$\alpha' = \frac{1}{2P}\frac{dP_t}{dz}$$
$$= \frac{1}{2P}\frac{dP_a}{dz} + \frac{1}{2P}\frac{dP_s}{dz} + \frac{1}{2P}\frac{dP_b}{dz} + \cdots$$
$$= \alpha'_a + \alpha'_s + \alpha'_b + \cdots \tag{5.101}$$

where we introduced the partial attenuation coefficients

$$\alpha'_a = \frac{1}{2P}\frac{dP_a}{dz} \tag{5.102}$$

$$\alpha'_s = \frac{1}{2P}\frac{dP_s}{dz} \tag{5.103}$$

$$\alpha'_b = \frac{1}{2P}\frac{dP_b}{dz} \tag{5.104}$$

caused by absorption, scattering, and bend radiation, respectively. Thus the total attenuation coefficient α' simply is the sum of the partial attenuation coefficients [Walker 1986]. In Sects. 13.3.1 and 13.4, we will discuss methods for measuring the total attenuation coefficient. Methods for measuring selectively the partial attenuation coefficients for absorption, scattering, macrobending, and micro-bending will be described in Sects. 13.3.2–5.

Since the Poynting vector is proportional to the square of the electric field amplitude (5.33) and since the power transmitted is given by the integral of the Poynting vector over the infinite fiber cross section (5.35), the electric field amplitude is proportional to the square root of the optical power. Whereas the power decays as $\exp(-2\alpha'z)$ (5.100), the electric field amplitude is proportional to $\exp(-\alpha'z)$. Taking attenuation into account and using the Gaussian approximation (5.25), the phasor of the electric field of the fundamental mode (5.11) becomes

$$\underline{E}_x(\varrho, z) = E_0 \exp(-\alpha'z)\exp\left[-\left(\frac{\varrho}{w_G}\right)^2\right]\exp(-j\beta z) \quad . \tag{5.105}$$

In classical transmission-line techniques, the attenuation coefficient is defined from the exponential decay of the line voltage or electric field strength and not from the decay of the power transmitted. By introducing the factor 2 in the equation (5.97) for the power loss, our definition for the attenuation coefficient conforms to the common definition at lower frequencies. Note, that some authors assume that the optical power decays as $\exp(-\alpha''z)$, so that their α'' is twice our α'.

For completeness, we also define the complex propagation coefficient of the LP$_{01}$ mode

$$\gamma = \alpha' + j\beta \quad , \tag{5.106}$$

which combines the information about the attenuation and the rate of phase change of the guided wave.

The attenuation coeffient α' introduced here is used mainly in theoretical papers. Its unit is neper/km. In practice, the attenuation is expressed in decibels per kilometer. Therefore, we have to relate the theoretical and practical measures for the fiber attenuation.

If we locate the fiber input end at $z = 0$ and the fiber output end at $z = L$, the output power $P(L)$ is related to the input power $P(0)$ by (5.100)

$$P(L) = P(0)\,e^{-2\alpha'L} \quad . \tag{5.107}$$

The optical power level difference a between the input and output powers, i.e. the fiber attenuation a in decibels, is thus, by the definition of the decibel,

$$a = 10\log\left[\frac{P(0)}{P(L)}\right] \quad . \tag{5.108}$$

Inserting (5.107) gives

$$a = 8.69\alpha'L \quad . \tag{5.109}$$

The attenuation a in decibels divided by the fiber length L in kilometers is the attenuation coefficient α in decibels per kilometer

$$\alpha = a/L = 8.69\alpha'. \tag{5.110}$$

In terms of attenuation coefficients α' [nepers/km] and α [dB/km], the power decay along the fiber can either be written

$$P(z) = P(0)\,e^{-2\alpha'z} \quad , \quad \text{or} \tag{5.111}$$

$$P(z) = P(0)10^{-(\alpha z/10)} \quad . \tag{5.112}$$

Both the IEEE [1984] and CCITT [1983] recommend the symbol α for the attenuation coefficient measured in dB/km. Following Danielson [1980], I have introduced the prime in α', in order to discriminate between the two quantities.

At this point, it is necessary to explain the difference between optical and electrical decibels. At present, almost all photoreceivers use direct detection, i.e. the photocurrent i received is proportional to the incident optical power:

$$i = RP \quad . \tag{5.113}$$

For a PIN-photodiode, the constant of proportionality R is called responsivity and is given by

$$R = \frac{\eta q}{hf} \quad , \tag{5.114}$$

with the quantities η, q, h, and $f = \omega/(2\pi)$ being the quantum efficiency, the charge of an electron, Planck's constant, and the optical frequency, respectively. The responsivity is of the order of $0.5\,\text{mA/mW}$.

Now, consider two optical powers P_1 and P_2, e.g. the powers at the input and output ends of a fiber link. By definition, the optical level difference in decibels is

$$a = 10 \log\left(\frac{P_1}{P_2}\right) \quad . \tag{5.115}$$

For determining the link loss a, the powers P_1 and P_2 are measured with a photodiode connected to an ammeter. The photocurrents i_1 and i_2 corresponding to the optical powers P_1 and P_2 are obtained from (5.113). The electrical level difference a_{el} in electrical decibels between the two photocurrents is defined by

$$a_{\text{el}} = 20 \log\left(\frac{i_1}{i_2}\right) \quad . \tag{5.116}$$

Replacing the currents in (5.116) by the optical powers from (5.113) gives

$$a_{\text{el}} = 20 \log\left(\frac{P_1}{P_2}\right) \quad , \tag{5.117}$$

or, because of (5.115)

$$a_{\text{el}} = 2a \quad . \tag{5.118}$$

Hence, for direct photodetection, the electrical level difference is twice the optical level difference. For instance, an optical level difference of 10 optical decibels corresponds to an electrical level difference of 20 electrical decibels. This difference must be taken into account when one measures optical attenuation using a photodiode connected to a voltmeter calibrated in electrical decibels!

It is usual to relate optical powers to a reference power of

$$P_0 = 1\,\text{mW} \quad , \tag{5.119}$$

by defining a relative optical power level

$$a_{\text{r}} = 10 \log\left(\frac{P}{P_0}\right) \quad , \tag{5.120}$$

which has the unit of decibels relative to one milliwatt (dBm). In a fiber link, it is common to express the power launched into the fiber input end, the power at the fiber end, and the receiver sensitivity in dBm's.

5.9.2 Attenuation by Rayleigh Scattering

The attenuation behavior of single-mode fibers is not very different from that of multimode fibers.

Intrinsic scattering is the property of the fiber that determines the minimum possible attenuation. All transparent materials scatter light because of microscopic inhomogeneities. The scattering decreases rapidly with increasing wavelength. The most familiar example of this type of scattering is the blue sky: in the atmosphere, blue light from the sun rays is scattered more strongly than red light. Since the light is scattered approximately equally in all directions, the clear sky shines blue. In the rays transmitted through the atmosphere, the blue light is lacking; therefore, at sunrise and sunset the sun appears red.

The same Rayleigh scattering is observed in glass fibers. During fiber drawing, the glass melts. In molten glass, the density of matter fluctuates randomly in time and space. As the glass melt is cooled through the glass transition temperature, these fluctuations are frozen in and become static in time [Miller and Chynoweth 1979]. The density fluctuations lead to irregular spatial variations in the refractive index, the dimensions of which are small in relation to the optical wavelength. Part of the light traversing such a medium will be scattered.

Rayleigh scattering in fibers has been reviewed many times [e.g. Garrett and Todd 1982]. Light scattering in single-mode fibers has been analyzed theoretically by O'Connor and Tauc [1978].

For understanding the directional characteristics of the scattered light, each volume element of the medium can be considered to contain small electric dipole antennas with random orientations. A *linearly polarized* incident wave excites the dipoles which are parallel to the electric field vector. Because of the radiation characteristics of an electric dipole, power is scattered isotropically in the plane normal to the direction of the electric field vector but no power is scattered in the direction of the electric field.

For *unpolarized* light, the directional characteristic of the scattered intensity is obtained by adding the intensities emitted by two orthogonal dipole antennas carrying equal currents: the scattered intensity is independent of the azimuth angle ϕ (Fig. 3.1) and is proportional to the factor $(1 + \cos^2 \theta)$ where θ is the polar angle measured between the fiber axis and the direction pointing to the point of observation (Fig. 5.19). Thus, the forward ($\theta = 0$) and backward

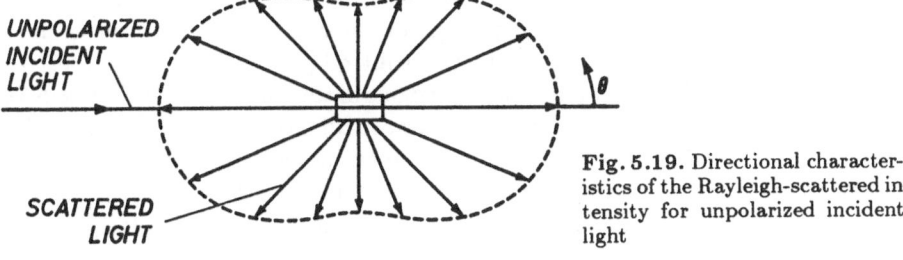

Fig. 5.19. Directional characteristics of the Rayleigh-scattered intensity for unpolarized incident light

$(\theta = \pi)$ scattered intensities are the same and are equal to twice the intensity scattered at right angles $(\theta = \frac{\pi}{2})$. For a very rough estimate, one can assume the power to be scattered isotropically in all directions.

The light scattered into the backward direction and coupled into a fundamental fiber mode guided back to the optical source is used in the very important backscattering technique for measuring fiber attenuation, splice loss, fiber length, and fault location (Sect. 13.4). The light scattered in the perpendicular directions can be observed in order to determine the beat length as a measure for the difference of the phase constants of the orthogonally polarized fundamental fiber modes.

The attenuation caused by Rayleigh-scattering in pure bulk silica glass is about 0.63 dB/km at a wavelength of $1\,\mu$m [Garrett and Todd 1982]. With increasing wavelength, it rapidly becomes smaller: it is inversely proportional to the fourth power of wavelength

$$\alpha_s = \frac{A}{\lambda^4} \quad . \tag{5.121}$$

The Rayleigh scattering coefficient A of pure silica is about 0.63 dBμm^4/km.

For producing a dielectric waveguide, the refractive index of the core is increased, e.g. by adding germanium oxide to the pure silica glass, and/or the cladding index is reduced e.g. by adding fluorine. Since the concentration of these dopants fluctuate randomly in the glass, the refractive index fluctuates correspondingly; the attenuation caused by Rayleigh scattering rapidly increases with increasing dopant concentration in the core. Therefore for doped silica, a larger Rayleigh scattering coefficient A must be inserted in (5.121) [Matsumura et al. 1980a].

The relative refractive-index difference Δ (5.2) and the concentration of the dopants are smaller in single-mode fibers than in multimode fibers. Because of the lower doping level, the scattering loss and, therefore, the total losses in single-mode fibers are somewhat lower than in multimode fibers [Kapron and Lazay 1983; Kapron 1984a].

From (5.103), we can calculate the power loss dP_s caused by Rayleigh scattering. For a wavelength of $1.3\,\mu$m, assuming a Rayleigh scattering coefficient of

$$A = 0.8\,\text{dB}\mu\text{m}^4/\text{km} \quad , \tag{5.122}$$

(5.121) gives a partial attenuation coefficient caused by scattering of $\alpha_s = 0.28$ dB/km or $\alpha_s' = 0.032$ Np/km. For a fiber element of length $10\,$m, (5.103) gives $dP_s = 6.4 \times 10^{-4} P$, showing that the power dP_s scattered from this fiber element is very small compared to the transmitted power P.

Most of the scattered power is scattered out of the fiber and is lost (Fig. 5.19). Only a very small part of the scattered power will excite fundamental modes propagating in the forward and backward directions. A capture fraction S can be defined as the ratio of the power of the fundamental mode exited by backscattering to the total scattered power dP_s. The capture fraction S is found to be mainly determined by the ratio of the spot size to the wavelength [Brinkmeyer 1980a,b]:

$$S = \frac{3}{8\pi^2 n_2^2 (w_G/\lambda)^2} \quad .$$ (5.123)

This relation has been confirmed both theoretically [Nakazawa1983b] and experimentally [Gold and Hartog 1981]. A more general expression for the capture fraction for arbitrary refractive-index profile and scattering-loss distribution in the fiber cross section has been published by Hartog and Gold [1984].

Choosing typical values for the refractive index ($n_2 = 1.5$), the wavelength ($\lambda = 1.3\,\mu m$), and the spot size ($w_G = 5\,\mu m$), the capture fraction S is found to be 0.001, i.e. only 0.1% of the very small total power scattered is coupled into the fundamental mode propagating back to the source.

On the one hand, this is a fortunate circumstance, since cross talk in bidirectional (duplex) systems [Wells 1978] and multiple backscattering [Dyott and Stern 1971; Eickhoff 1981] usually will not disturb signal transmission. On the other hand, in the backscattering method, the backscattered power received at the source end of the fiber is extremely small and signal averaging must be employed to obtain backscattering curves which can be evaluated (Sect. 13.4).

Since the losses caused by scattering decrease dramatically with increasing wavelength, a fiber transmission link at $1.3\,\mu m$ or $1.55\,\mu m$ can be much longer than at $0.85\,\mu m$ wavelength. This is one of the reasons for the tendency to use longer wavelength systems. The other is the smaller chromatic dispersion and larger bandwidth at longer wavelengths (Sect. 5.8).

5.9.3 Attenuation by Intrinsic Absorption

There are several mechanisms by which light power is transformed into heat power, i.e. absorbed. We first consider intrinsic absorption, which is found in pure silica glass.

The knowledge about intrinsic absorption has been reviewed several times [e.g. Garrett and Todd 1982].

At optical wavelengths smaller than about $0.5\,\mu m$, the energy hf of a photon is sufficiently large to excite bound electrons in the glass. By this process, photons are absorbed, causing an exponential increase of the attenuation with decreasing wavelength [Stone F.T. 1982; Walker 1986]. Since this intrinsic ultraviolet absorption increases very rapidly, one speaks of an uv-absorption edge. For wavelengths greater than about 700 nm, intrinsic uv-absorption will have no effect [Keck et al. 1973].

At optical wavelengths longer than about $1.6\,\mu m$, the frequency of light approaches the resonant frequencies of the internal vibrations of the SiO_2 molecules and of the dopant oxides, causing an exponential increase of loss, the so-called intrinsic ir-absorption edge [Izawa et al. 1977; Walker 1986]. For pure silica, the ir-absorption edge starts at the largest wavelength. For GeO_2

or F_2 doped silica, the ir loss increase is similar to that in pure silica. However, doping with P_2O_5 or B_2O_3 shifts the ir-edge to shorter wavelengths. The addition of F_2, GeO_2, P_2O_5, or B_2O_3 causes a minimal, a slight, an intermediate, and a substantial loss increase below $\lambda = 2\,\mu m$, respectively [Garrett and Todd 1982].

Between the uv-absorption edge and the region of ir-absorption, i.e. for wavelengths between about $0.7\,\mu m$ and $1.6\,\mu m$, the partial attenuation constant for intrinsic absorption is small compared to that for Rayleigh scattering. Optical fiber transmission systems based on silica thus operate in this wavelength region.

There are other materials, e.g. Zr-based fluoride glasses, which have smaller resonant frequencies of the molecular vibrations. For these, the attenuation minimum will be found at wavelengths larger than $1.55\,\mu m$. Very small attenuation has been predicted for fibers made from these materials, $10^{-3}\,dB/km$ at $3.4\,\mu m$ [Shibata S. et al. 1981], and, more realistically, $0.03\,dB/km$ at $2.55\,\mu m$ [Moore et al. 1986]. But up to now, the measured minimum attenuation of $0.7\,dB/km$ [Takahashi 1987] is still five times larger than the minimum attenuation in conventional silica glass fibers.

5.9.4 Attenuation by Impurity Absorption

Impurities such as metallic, hydroxyl, or hydrogen ions cause additional absorption losses, which show up as peaks in the attenuation vs wavelength curve.

Nowadays, very high purity starting materials are available, by distillation, for chemical-vapor-phase (CVD) fabrication of single-mode fibers. As a consequence, the concentration of transition metals can be held at a level that is so small that the absorption losses due to metal ion impurities are negligibly small.

Unfortunately, absorption by hydroxyl ions is a more difficult problem. If the fiber is contaminated with hydroxyl (OH^-) ions, sharp attenuation peaks are produced at wavelengths of 1.39, 1.25, and $0.95\,\mu m$. The absorption peak at 1390 nm is the first overtone of the OH stretch vibration. For a weight concentration of 1 ppm (parts per million) of OH, the highest peak at $1.39\,\mu m$ is $35-55\,dB/km$ [Garrett and Todd 1982; Krause D. et al. 1985]. The attenuation at the $1.25\,\mu m$ peak is about $1/20$ of that at $1.39\,\mu m$. The heights of the absorption peaks indicate the concentration of OH-groups in the glass.

The absorption peak at $1.39\,\mu m$ separates two low-loss windows at $1.3\,\mu m$ and $1.55\,\mu m$. The peak attenuation at $1.39\,\mu m$ must be less than $2\,dB/km$ if the additional loss due to OH at $1.3\,\mu m$ is to be kept smaller than $0.05\,dB/km$ and if one is interested in a wide second window around $1.3\,\mu m$ in order to make the location of the source center wavelength uncritical on loss grounds [Garrett and Todd 1982]. The height of the peak at $1.39\,\mu m$ can thus be used to estimate the quality of the glass fiber.

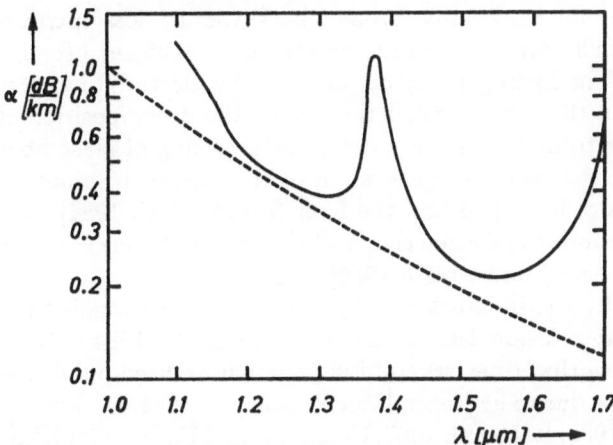

Fig. 5.20. Attenuation constant α versus wavelength λ curve for a good single-mode fiber (*solid line*). The dashed curve represents the attenuation constant α_s caused by Rayleigh scattering only (5.121), assuming that $\alpha_s = 1\,dB/km$ at $\lambda = 1\,\mu m$

By flushing a porous preform with oxygen and $SOCl_2$ vapour [Sudo et al. 1978] or by using chlorine during collapse of a MCVD-preform [Nagel 1984], the OH-concentration can be reduced to below 0.01 ppm, so that no absorption peaks can be seen on the attenuation vs wavelength curve, but the additional processing is costly.

Figure 5.20 shows the attenuation coefficient vs wavelength for a typical good silica glass single-mode fiber. The dashed curve represents the attenuation caused by Rayleigh scattering only (5.121), assuming $\alpha_s = 1.0\,dB/km$ at $\lambda = 1.0\,\mu m$. In the loss minima, the attenuation is determined mainly by Rayleigh scattering. The curve also shows the loss increase for wavelengths larger than $1.6\,\mu m$ caused by intrinsic ir-absorption and the OH-peak at $1.39\,\mu m$. The loss increase for wavelengths smaller than $1.2\,\mu m$ is due to the attenuation of the LP_{11} mode near cutoff (Sect. 13.8.1). The minimum loss values for silica single-mode fibers reported up to now are $0.157\,dB/km$ at a wavelength of $1.57\,\mu m$ [Csencsits et al. 1984] and $0.154\,dB/km$ at $1.55\,\mu m$ [Kanamori et al. 1986].

The dopants added to silica to increase or decrease the refractive index can also be considered as impurities. They too introduce absorption losses. For GeO_2 doped fibers, the loss amounts to about 20%–30% of the total loss at 1.3 and $1.55\,\mu m$, respectively [Garett and Todd 1982].

Hydrogen ions diffusing into the fiber core cause a loss increase at long wavelengths. This unexpected effect was observed for the first time in 1983 [Mochizuki et al. 1983; Uesugi et al. 1983]. Silica glass is completely permeable to hydrogen. Hydrogen ions primarily produce a characteristic absorption peak at $1.24\,\mu m$, which vanishes when the hydrogen molecules diffuse out of the waveguide. But over a longer time and at elevated temperatures, chemical reactions take place in the fiber material, whereby hydroxyl bonds are formed. These cause an irreversible loss increase with a characteristic absorption peak at $1.88\,\mu m$.

113

Hydrogen also causes an irreversible broad "background" loss increase which increases exponentially for wavelengths shorter than $0.95\,\mu$m [Tomita and Lemaire 1984, 1985]. The hydrogen can be produced by electrolytic corrosion of metals contained in the cable construction, or by the decomposition of plastic materials. In order to avoid this very detrimental ageing of glass fiber waveguides, one has either to avoid the generation of H_2 ions, or to build up a barrier against diffusion of hydrogen into the fiber [Beales et al. 1984]. Another countermeasure is to insert hydrogen chemical absorbers (getters) in fiber cables [Anelli et al. 1985; Anelli and Grasso 1986].

The ageing of fibers by the indiffusion of hydrogen has been studied extensively and about twenty relevant letters can be found in the 1984 volume of Electronics Letters. Since that time, several full papers have been published considering the loss increase due to hydrogen [Mochizuki et al. 1984a,b; Stone J. et al. 1984; Noguchi et al. 1985; Itoh et al. 1985; Uchida N. and Uesugi 1986]. Today, the problem of hydrogen contamination seems to be under control [Costa 1985].

Another reason for absorption losses is nuclear radiation. Information about radiation effects can be found in a conference proceedings [Boucher 1983].

Since the field of the fundamental mode also extends into the cladding, the cladding is usually composed of a low loss deposited (synthetic) inner cladding and a high loss outer cladding, which originates from the substrate tube in the MCVD technique or a jacketing tube in the VAD technique. The jacket loss is typically of the order of a dB/μm or nine orders of magnitude higher than the attenuation coefficient in the fiber [Gloge 1979b]. Since it is costly to deposit the inner cladding, the minimum thickness required for avoiding impurity absorption has been studied [Gloge 1975; Marcuse 1977a].

The OH-absorption loss at the $1.24\,\mu$m peak has been measured as a function of cladding to core diameter ratio d_s/d, where d_s is the diameter of the inner, deposited cladding and $d = 2a$ the core diameter [Miya et al. 1981]. The diameter ratio must be greater than seven, to obtain a loss increase of less than $0.1\,$dB/km at $1.3\,\mu$m. Another reason for having a thick deposited cladding is to avoid a loss increase at $\lambda = 1.39\,\mu$m due to OH ions which diffuse from the substrate tube toward the core region. For a core diameter of $9\,\mu$m, a synthetic cladding diameter of $63\,\mu$m, and outer cladding standard diameter of $125\,\mu$m, less than 1% of the fiber material is in the core, 25% in the synthetic cladding and 74% in the substrate tube.

As has been stated before (5.101), the total attenuation coefficient is the sum of the partial attenuation constants. In practice, the wavelength dependence of the total attenuation coefficient can often be written in the form [Krause D. et al. 1985]

$$\alpha = B + \frac{A}{\lambda^4} + C\alpha_{OH}(\lambda) \quad , \tag{5.124}$$

where B is the wavelength independent loss, A the Rayleigh scattering coefficient at $\lambda = 1\,\mu$m in $\,$dB$\,\mu$m^4/km, C the OH-concentration in ppm, and

$\alpha_{OH}(\lambda)$ the wavelength-dependent attenuation coefficient caused by OH groups. At the wavelength of maximum absorption, $\lambda = 1.39\,\mu m$:

$$\alpha_{OH}(1.39\,\mu m) = 35\,dB/(km\,ppm) \quad . \tag{5.125}$$

The characteristic wavelength dependence of the scattering loss (5.121) can be used for determining the various loss components separately from the total attenuation measured as a function of wavelength (Sect. 13.3.1).

Absorption losses can also be due to traces of phosphorus in the fiber core [Walker 1986].

5.9.5 Reflection Losses

When a light beam passes through the interface between two transparent media with different refractive indices n_1 and n_2, some light power is lost by reflection. We only consider the most important case of a uniform plane wave transmitted through a plane interface normal to the direction of wave propagation. From Fresnel's equations [Born and Wolf 1975] one finds for the amplitude reflection factor r as defined by the ratio of the electric field of the reflected wave (E_r) and incident wave (E_1) at the interface

$$r = \frac{E_r}{E_1} = \frac{n_1 - n_2}{n_1 + n_2} \quad , \tag{5.126}$$

and for the amplitude transmission factor t as defined by the ratio of the electric field of the transmitted wave (E_2) and incident wave (E_1) at the interface

$$t = \frac{E_2}{E_1} = \frac{2n_1}{n_1 + n_2} \quad . \tag{5.127}$$

The power reflection factor R as defined by the ratio of the intensities of the reflected and the incident wave simply is the square of the amplitude reflection factor

$$R = r^2 = \frac{(n_1 - n_2)^2}{(n_1 + n_2)^2} \quad . \tag{5.128}$$

This relation holds for any polarization and for both directions of wave propagation.

The power transmission factor T as defined by the ratio of the intensities of the transmitted and incident wave is not equal to the square of the amplitude transmission factor, since the incident and the transmitted waves propagate in two media with different wave impedances Z_0/n_1 and Z_0/n_2, respectively.

Since the Poynting vector is the electric field squared divided by the wave impedance Z/n (4.35), the power transmission factor is

$$T = \left(\frac{n_2}{n_1}\right)t^2 = \frac{4n_1 n_2}{(n_1 + n_2)^2} \quad . \tag{5.129}$$

Because of the conservation of energy, the sum of the power reflection coefficient R (5.128) and of the power transmission coefficient T (5.129) is, of course, unity:

$$T + R = 1 \quad . \tag{5.130}$$

As an important example, we calculate the power reflection factor R of a single interface between glass with refractive index n and air

$$R = \frac{(n-1)^2}{(n+1)^2} \quad . \tag{5.131}$$

For the typical value of $n = 1.46$ for silica glass, we find $R = 0.035$, i.e. 3.5% of the incident power is reflected from the interface and 96.5% is transmitted through the interface. Thus, a single glass-air interface attenuates the transmitted wave by 0.155 dB.

Though the fundamental fiber mode is an inhomogeneous plane wave instead of a uniform plane wave, and though the fiber is not a homogeneous medium, the power reflection factor of an ideal flat fiber end normal to the fiber axis can be estimated by using (5.131). Since there are only slight index variations in the fiber, one can insert the cladding index n_2 for n in (5.131). Often, a broken fiber end face is neither plane nor normal to the fiber axis. The reflection factor can then be much smaller than the value of 3.5% estimated from (5.131) [Marcuse 1975c].

In single-mode fiber connectors without an index matching liquid, there is a small air gap between the fiber ends to be connected. Usually one is not allowed simply to add the losses caused at the two interfaces to obtain a Fresnel loss of 2×0.155 dB $= 0.31$ dB for a connector. Instead one must take into account the phase difference of the waves reflected from the two interfaces. The phase difference depends on the distance between the interfaces. Therefore, the connector loss depends on the length d of the air gap. Assuming that the interfaces are perfectly flat and normal to the direction of wave propagation, that the gap width is so small that diffraction spreading of the beam can be neglected, and that the coherence length (3.36) of the wave is larger than the gap width, the reflection factor is given by

$$R = \frac{(n^2 - 1)^2 \tan^2(kd)}{4n^2 + (n^2 + 1)^2 \tan^2(kd)} \quad . \tag{5.132}$$

Figure 5.21 represents the power reflection factor R of a connector with air gap as a function of the width d of the gap. If the gap width is an integer number of half wavelengths in air, there is no reflection. On the other hand, the maximum value of the reflection factor is $R = 13\%$ corresponding to a reflection loss of 0.6 dB, when the gap width is a quarter of a wavelength plus an integer multiple of half the wavelength. The average of R over all values of the gap width is about 7% and the average loss about 0.31 dB.

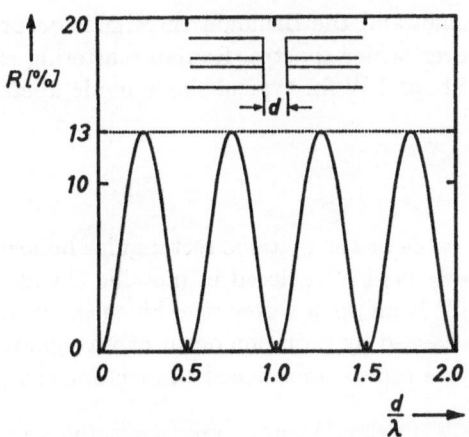

Fig. 5.21. Power reflection factor (5.132) of an air gap in a fiber connector as a function of normalized gap width d/λ

When the power transmitted through a gap is measured as a function of length of the gap, the power oscillations are found to be smaller than is expected theoretically and to decrease with increasing width of the gap. Moreover, the average transmitted power decreases with d (Sect. 9.3). This is due to the divergence of the beam in the air space and the finite coherence length of the light [Wagner and Sandahl 1982].

Fluctuations of the transmitted power have been observed that are also larger than expected theoretically from (5.130) and (5.132) [Shah et al. 1987]. These larger fluctuations can be explained by the presence of a thin region at the endface, which has a higher index than that of the fiber. The higher index region is created when the endface is polished with red iron oxide. Stresses induced during polishing increase the density of the glass surface causing the refractive index to increase from 1.46 to 1.6.

The gap width can be determined interferometrically by measuring the transmitted power as a function of wavelength in a wide wavelength range. Then, the gap width can be calculated uniquely from the wavelengths of two adjacent transmission maxima [Tomita 1982].

5.9.6 Attenuation Caused by Nonlinear Effects

In a single-mode fiber, the power guided by the fundamental mode is concentrated in a very small cross section. As a result, nonlinear effects occur at relatively small values of the transmitted power (Sect. 5.4). Two of these nonlinear effects cause a loss increase at high powers; these are stimulated Raman scattering and stimulated Brillouin-scattering. Both effects extract power from the wave being transmitted and generate waves of longer wavelengths.

It is therefore very important to know the maximum power that can be transmitted, before the fiber attenuation begins to rise.

For a typical single-mode fiber and a *coherent* light input, stimulated Brillouin scattering occurs at power levels as low as 10 mW [Gloge 1979b]. This threshold increases linearly with increasing source spectral width to about 1 W

117

at a width of 0.1 nm for which the Raman and the Brillouin thresholds occur at about the same power density. For even wider spectra, Raman scattering is the limiting effect, with a threshold of about 1 W for typical single-mode fibers (Sect. 5.4).

5.9.7 Pure Bend Loss

In contrast to the conventional coaxial cable or the metallic rectangular hollow waveguide, for which the electromagnetic field is enclosed in metallic shields, the optical fiber as a dielectric waveguide is an open waveguide. Since the field is not enclosed in a shield, power losses caused by radiation occur at waveguide discontinuities such as changes of the core radius, misaligned connectors, tilts, or bends [Neumann 1975].

First, we discuss the radiation losses of bends. In Sect. 2.3, waveguiding by curved single-mode fibers was explained in physical terms: at the beginning of the bend, the wave beam is shifted to the outside of the bend [Miyagi and Yip 1976; Gambling et al. 1977a], causing a decrease in the phase velocity on the inner side and an increase on the outer side of the beam. This beam shift forces the beam to follow a curved path. Nature chooses the beam offset in such a way that the beam tracks the curved fiber. In Fig. 5.22, the shifted beam in a curved fiber is illustrated by plotting some electric field lines.

In propagating along a bend of constant radius of curvature, a so called "macrobend", the fundamental mode continuously loses power by radiation. This power loss is called pure bend loss. First, we explain the reason for the pure bend loss using an analogy in sport.

Consider five runners forming a straight line and running along a straight path with a velocity slightly below the maximum velocity possible (Fig. 5.23a).

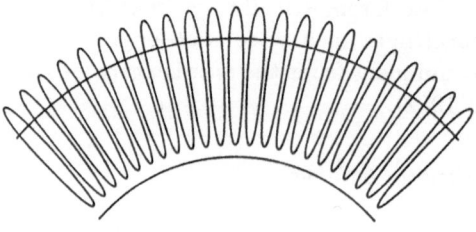

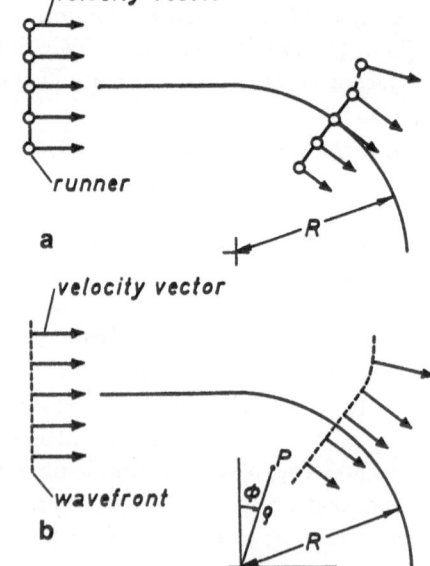

Fig. 5.22. Electric field lines of the fundamental mode propagating in a curved single-mode fiber

Fig. 5.23a,b. Model of five runners (a) to explain radiation losses at fiber bends (b), see text

Beginning at a certain point, the central runner is asked to follow a curved line of radius R without changing his velocity. The other runners are asked to preserve both the straight line normal to the direction of movement of the central runner and the distances to their neighbors. To achieve this, the runners on the inside of the bend must reduce their velocities, whereas those on the outside must speed up. The arrows in Fig. 5.23a represent the velocities of the runners. If the radius of curvature is small, the outermost runner will experience difficulties in gaining the necessary velocity and will eventually lag behind and finally dissociate himself from the others. The reason for the "loss" of the outermost runner is that the velocity cannot become greater than a maximum value.

Now, coming back to the problem of the fundamental fiber mode traversing a curved fiber: Consider the fundamental mode first propagating along a straight and then along a curved fiber section. In Fig. 5.23b, the electromagnetic field is represented by dashed wavefronts and arrows indicating the local phase velocity. The photons correspond to the runners and the wavefronts correspond to the straight line formed by the runners. In order to have no power loss, the photons in the bent wave beam must move on circular paths concentric with the axis of curvature. For that purpose, the local velocities of the wavefronts and of the photons have to increase proportional to the distance ϱ from the axis of curvature (Fig. 5.24). For describing wave propagation in bent fibers, we introduce cylindrical polar coordinates with the z-axis coinciding with the axis of curvature (Fig. 5.23b).

In principle, one has to differentiate between the phase velocity and the velocity of energy propagation, which is defined by the ratio of the magnitude of the Poynting vector to the density of field energy [Jackson J.D. 1975]. But in this intuitive picture, let us assume for a moment that they are equal and cannot become larger than c/n_2, i.e. the velocity of light in the cladding. Arguments for using the local phase velocity can be found in papers by Böhme [1963] and Heyman [1984].

On the fiber axis, i.e. for $\varrho = R$, the wave velocity will be approximately equal to the phase velocity (5.59) $v_p = \omega/\beta$ of the fundamental mode propagat-

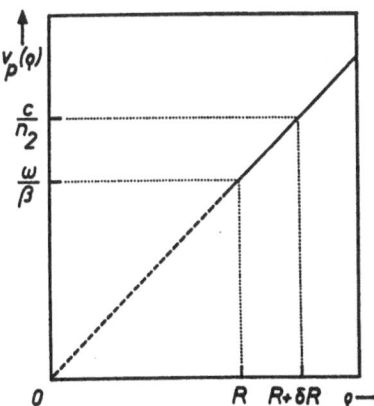

Fig. 5.24. Local phase velocity $v_p(\varrho)$ in the fundamental mode circulating in a bent fiber as a function of the distance ϱ from the axis of curvature

ing in the straight fiber. At a certain distance $R+\delta R$ from the axis of curvature, the local velocity reaches the maximum possible value of c/n_2. Therefore, photons propagating in the region $\varrho > R + \delta R$ cannot keep up with the others and will be lost by tangential radiation [Larsen 1965; Marcatili 1969b; Marcatili and Miller 1969; Marcuse 1974, 1976a].

Comparing two similar triangles in Fig. 5.24, the critical distance δR from the fiber axis is found to be [Marcatili and Miller 1969]

$$
\begin{aligned}
\delta R &= R\left[\frac{c\beta}{n_2\omega} - 1\right] \\
&= R\left[\frac{\beta}{n_2 k} - 1\right] \\
&= R\left[\frac{n_e}{n_2} - 1\right] \\
&\approx R\frac{(W/a)^2}{2n_2^2 k^2} \ .
\end{aligned}
\tag{5.133}
$$

For obtaining the last approximate expression for δR, we have used (5.17) to introduce the the transverse field extent a/W in the cladding (5.15). In the sum $\beta + n_2 k$, the phase constant β has been approximated by $n_2 k$.

For distances ϱ from the axis of curvature, which are larger than $R + \delta R$, the wavefronts are no longer planes normal to the local direction of the fiber axis. The intersection between the wavefront and the plane of the curved fiber axis becomes approximately a section of an Archimedean spiral [Neumann and Rudolph 1975b] as shown schematically in Fig. 5.25. Since the Poynting vector is normal to the wavefronts [Neumann and Rudolph 1974], there is a component of the power flow away from the waveguide which results in an exponential decay

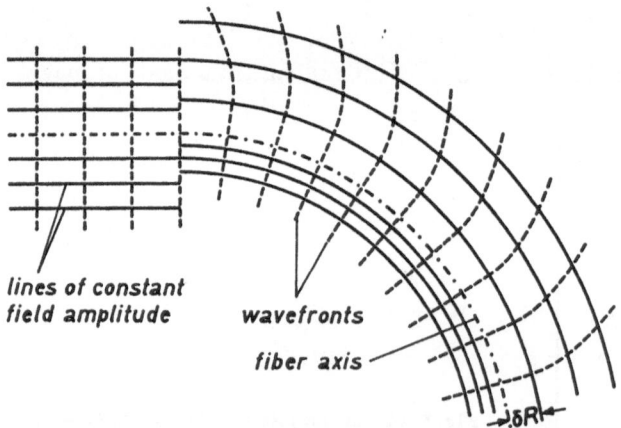

lines of constant
field amplitude

wavefronts

fiber axis

Fig. 5.25. Electromagnetic field distribution of the fundamental modes at a straight and at a bent single-mode fiber section represented schematically by lines of constant field amplitude (*solid lines*) and by wavefronts (*dashed lines*)

$$P(s) = P(0) \exp(-2\alpha'_b s) \tag{5.134}$$

of the azimuthal power flow as a function of the length s of the bent fiber axis. In (5.134) it has been tacitly assumed, that there are no other sources of attenuation besides radiation from the bend.

The principal reason for the pure bend loss is the finite maximum speed of light in the cladding. The quantity α'_b in (5.134) is the partial attenuation coefficient due to pure bend loss in nepers/km. In general, it adds to the partial attenuations constants due to absorption, scattering, etc. (5.101). In the bent section of fiber, the axial coordinate z is replaced by the length of arc

$$s = R\phi \quad , \tag{5.135}$$

along the fiber axis. Corresponding to (5.110), the attenuation coefficient α_b due to bend loss in decibels per kilometer is related to α'_b by

$$\alpha_b = 8.69\alpha'_b \quad . \tag{5.136}$$

Since power continuously leaks from the fundamental mode, a curved single-mode fiber is an example for a "leaky" waveguide (Sect. 6.4).

It can be shown [Neumann and Rudolph 1975b] that at the critical distance $R + \delta R$ from the axis of curvature, the field decay in the ϱ-direction changes from an approximately exponential one, which is characteristic for a wave guided by a dielectric waveguide, into a slower one, which is characteristic for a wave radiated by an antenna.

In order to illustrate wave propagation around a curved fiber section, we measured the field distribution around a bent dielectric waveguide at microwaves frequencies [Neumann and Rudolph 1975b]. The model fiber consisted of a polyethylene rod with a rectangular cross section of 9.5 mm × 12 mm (The fundamental mode guided by a fiber with a rectangular core is very similar to that guided by a fiber with a circular core). The radius of curvature was 248 mm and the frequency used 12.71 GHz. Since the cladding was air, the field distribution in the cladding could easily be measured using a small electric field probe, consisting of the axially protruding inner conductor of a miniature coaxial line. A microwave network analyzer was used as a two-channel superheterodyne receiver to measure the electric field strength with respect to both amplitude and phase. Points of equal amplitude were marked on a sheet of paper and were connected by solid lines (Fig. 5.26). Points of equal phase were connected by dashed curves to represent the wavefronts. Between neighboring lines of equal field amplitude, the level changes by 5 dB. Between neighboring wavefronts, the phase changes by 2π. Thus the normal distance between two wavefronts is the local wavelength, which is proportional to the local phase velocity.

The fundamental mode is incident along the upper straight section from the left. For the straight guide, the field distribution is symmetrical about

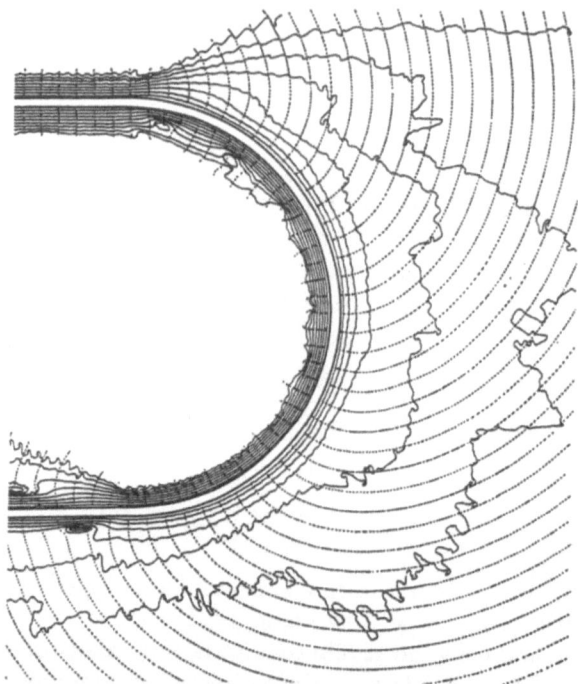

Fig. 5.26. Lines of constant magnitude (*solid lines*) and constant phase angle (*dashed lines*) of the electric field of the fundamental mode propagating around an *H*-plane 180° bend in a dielectric rod transmission line. Model experiment with microwaves of frequency $f = 12.71$ GHz. The dielectric waveguide had a rectangular cross section of 9.5 mm × 12 mm and a permittivity of $\varepsilon = 2.30$. The cladding was air and the radius of curvature of the bent waveguide $R = 24.8$ cm

the waveguide axis. In the air cladding, the field decays approximately exponentially in the radial direction. For a pure traveling wave with negligible attenuation, the lines of constant amplitude are straight lines parallel to the guide with the distances between neighboring lines being almost equal. Since the attenuation by dielectric losses is very small, the average power is flowing in the axial direction and, therefore, the wavefronts are planes normal to the guide axis. The distance between neighboring wavefronts equals the wavelength of the fundamental wave in the straight waveguide.

Behind the transition from the straight to the curved section, there is some field disturbance because the field radiated from the transition (Sect. 5.9.8) superimposes on the field of the wave propagating along the curved dielectric rod waveguide. But for bend angles greater than about 60°, near the waveguide, the field of the wave guided by the bent waveguide dominates, as can be seen from the fact that the field distribution is the same for each cross section apart from a reduction in amplitude and a phase delay. Therefore, one can speak of a "leaky mode" propagating along the bend although its radiation field principally extends to infinity and although its dependence on the longitudinal coordinate is not merely a phase change [Baets and Lagasse 1983].

In the bent dielectric waveguide, the beam is displaced to the outside, as can be seen from the lines of constant amplitude next to the waveguide axis. Moreover, the amplitude distribution is no longer symmetric. On the inside of the bend, the distances between lines of constant amplitude are smaller than for the straight guide. Thus, the field decreases more rapidly with increasing distance from the fiber axis. On the outside of the bend, the distances are larger than for the straight guide indicating a slower field decay with distance from the fiber axis. On the outer side, the field extends far into the air showing that part of the guided power is continuously transformed into radiation. In the radiation field, for $\varrho > R + \delta R$, the normal distance between the wavefronts, i.e. is the wavelength, is a constant, namely the wavelength of a uniform plane wave in air. Since the wavelength is proportional to the phase velocity, it can be seen from the measured wavefronts, that in the radiation field, the phase velocity equals the light velocity c.

Near the curved guide, the wavefronts are planes almost normal to the local direction of the waveguide axis, i.e. the local phase velocity increases linearly with the distance ϱ from the axis of curvature as has been assumed in Fig. 5.24. But, on the outside of the bend, beyond the critical distance δR from the waveguide axis, the wavefronts become curved as sketched in Fig. 5.25. Since the Poynting vector is approximately normal to the wavefronts, one sees that power is radiated tangentially from the outer side of the bend.

At the end of the 180° bend, most of the power of the mode in the curved guide is coupled into the fundamental mode of the second straight waveguide section and only a small part is radiated. At short distances behind the second transition, the field of the guided wave is disturbed by the superimposed radiation from the curved section and from the transition itself.

The total attenuation of the bend was measured to be 0.5 dB, from which 0.4 dB were caused by dielectric losses. Thus, the sum of the pure bend loss and the transition losses occurring at the two ends of the bend (Sect. 5.9.8) was only 0.1 dB. Though only 2.3 % of the incident power was lost by radiation, the radiation can clearly be seen by plotting the field distribution near the bend.

By considering the physical reason for the bend losses, a simple rule for minimizing the radiation losses can be deduced: one should minimize the number of photons propagating in the region beyond the critical distance δR, i.e. reduce the field strength at the critical boundary. Then the radial component of the Poynting vector, which determines the power dP_b radiated from a length element with length ds, will be small.

This can be realized by increasing the critical distance δR relative to the transverse field extent. The designer of a fiber can minimize the bend losses by choosing a relatively large value for the refractive index difference and therewith a small value for the transverse field extent. The fiber user cannot change the transverse field extent and can reduce the bend losses only by increasing the radius of curvature in order to increase the critical distance δR (5.133).

It is the ratio of the critical distance δR and the radial field extent which mainly determines the pure bend loss. There are two measures for the transverse field extent: the spot size w_G which describes the width of the approximate Gaussian field distribution (5.25) in the core, and the transverse field extent a/W in the homogeneous cladding (5.15). These two measures for the transverse field extent are of the same order of magnitude (Fig. 5.5). Usually, one is interested in small radiation losses. In this case, the critical boundary is in the cladding and, therefore, the condition for small radiation losses should be formulated using a/W as a measure for the transverse field extent. Macrobending losses are related to a/W [Petermann 1987].

In order to obtain small pure bend losses, the critical distance δR (5.133) must be very large compared to the transverse field extent a/W in the cladding:

$$\delta R = \frac{R(W/a)^2}{2n_2^2 k^2} \gg \frac{a}{W} \quad . \tag{5.137}$$

The pure bend attenuation coefficient α_b depends essentially exponentially on the ratio of δR and a/W. Therefore, all formulae for the pure bend loss of different dielectric waveguides contain the exponential function [Neumann and Rudolph 1975b]:

$$\exp\left[-\frac{4W\delta R}{3a}\right] = \exp\left[-\frac{W^3 \lambda^2 R}{6\pi^2 a^3 n_2^2}\right] \quad , \tag{5.138}$$

which, by introducing the normalized frequency V (5.9), can also be written in the form

$$\exp\left[-\frac{4W^3 R \Delta}{3aV^2}\right] \quad . \tag{5.139}$$

where Δ (5.5) is the relative refractive index difference. This exponential factor is decisive for the pure bend loss. The bending loss is insensitive to all factors apart from this exponential factor in the bend loss formulae [Sakai and Kimura 1978a]. In a crude approximation, the exponent can be regarded as constant in order to give a constant bending loss.

The bending sensitivity of a single-mode fiber increases with increasing mode field diameter $2w_G$. However, different fibers with the same value of the mode field diameter do not necessarily have the same pure bend loss, general, since it is the ratio of the field extent and the critical distance δR which determines the radiation loss, and since δR can be different for fibers with the same field extent. For a given radius of curvature, the pure bend loss is mainly determined by the third power W^3 of the eigenvalue in the cladding. Therefore, for reducing the bend loss, one has to maximize W. Since

$$W^2 = bV^2 \quad , \tag{5.140}$$

(5.55), and since the normalized phase constant b increases with normalized frequency V (Fig. 5.12), V should be as large as possible. Note however, that

V must not be larger than the normalized cutoff frequency of the second order mode, or else the fiber will become bimodal.

The pure bend attenuation depends very strongly on the radius of curvature. Practically, one can remember that for R somewhat larger than a critical value R_c, bend losses are negligibly small, whereas for R somewhat smaller than R_c, bend losses become prohibitively high. In a cable, the fiber is continuously bent with a radius of typically 10–20 cm, whereas in the splicing pots, a short section of fiber is bent with a radius of some centimeters. The minimum bend radius in joint housings widely accepted for long term deployment of fibers in practical system installations to avoid static-fatigue failure is 3.75 cm [CCITT 1986].

If one wants to know the bend losses more exactly, one has to resort to the theory of wave propagation in curved dielectric waveguides. This is a very complicated mathematical problem. Rigorous solutions for the electromagnetic field distribution, the phase constant, and the attenuation constant of the leaky fundamental mode propagating along a uniformly bent single-mode fiber are not known.

Because radiation losses from curved fibers represent a very important practical problem, many approximate methods have been developed for analyzing wave propagation in curved single-mode fibers [Böhme 1963; Marcatili 1969b; Marcatili and Miller 1969; Shevshenko 1971; Lewin 1973, 1974, 1975, 1976; Arnaud 1974b, 1975; Snyder et al. 1975; Kuester and Chang 1975b; Marcuse 1976a,b, 1982b,c; Miyagi and Yip 1976, 1977; Chang D.C. and Kuester 1976; Lewin et al. 1977; Howard A.Q. 1977; Sammut 1977; Kuester 1977; Gambling and Matsumura 1978a; Miyagi and Nishida 1978; Sakai and Kimura 1978a,c, 1982; White I.A. 1979; Sakai 1979; Cohen et al. 1982c; Snyder and Love 1983; Danielsen and Yevick 1983; Baets and Lagasse 1983; Saijonmaa et al. 1983; Kuznetsov and Haus 1983; Saijonmaa and Yevick 1983; Vassallo 1985; Frantsesson et. al. 1986].

The different techniques for calculating the bending loss have been reviewed in the literature [Neumann and Rudolph 1975b; Baets and Lagasse 1983].

In the approximate methods for analyzing curved fibers, several simplifying assumptions are made, e.g.:

a)   the cladding material extends to infinity,
b)   the relative refractive index difference Δ is very small compared to unity,
c)   one component of the electric field vector is much larger than the other, so that a simple scalar analysis can be made instead of a complicated vectorial theory,
d)   the radius of curvature is very large compared to the core radius, so that the transverse field distribution and the phase constant in the curved fiber are similar to those in the straight fiber,
e)   the curved fiber can be coordinate-transformed into an equivalent straight fiber which propagates a mode with the same transverse field distribution. In the equivalent straight fiber, the influence of curvature is taken into

125

account by introducing an effective refractive index profile n_{eff} [Heiblum and Harris 1975, Arnaud 1975; Kawakami et al. 1975; Petermann 1976a,b; Marcuse 1976b; Sakai 1980a]

$$n_{\text{eff}} = (\varrho_{\text{a}}/R)n_{\text{c}} \quad , \tag{5.141}$$

where n_{c} is the index profile of the curved fiber and ϱ_{a} is the distance between a point in the fiber cross section and the axis of curvature (not the fiber axis!). Please do not confuse the index profile n_{eff} of the equivalent straight fiber with the effective index $n_{\text{e}} = \beta/k$ (5.45).

A single analysis will often only use some of these simplifying assumptions.
For fibers with arbitrary refractive index profile, the pure bend loss of the fundamental fiber mode is given by [Gambling and Matsumura 1978a; Sakai and Kimura 1978a, 1981, 1982]

$$\alpha_{\text{b}}' = \frac{1}{8}\sqrt{\frac{\pi}{aRW^3}}S(V,W)\exp\left[-\frac{4RW^3\Delta}{3V^2 a}\right] \quad , \tag{5.142}$$

where the function $S(V,W)$ is given by

$$S(V,W) = \frac{a^2}{K_0^2(W)}\left[\int_0^\infty \frac{E^2(\varrho)}{E^2(a)}\varrho\, d\varrho\right]^{-1} \quad , \tag{5.143}$$

and where $E(\varrho)$ is the radial field distribution in the *straight* fiber and α_{b}' the attenuation coefficient in Np/km defined using the *field* decay (5.134). The cladding decay parameter can be expressed by the normalized phase constant using (5.55)

$$W = b^{1/2}V \quad . \tag{5.144}$$

Approximating the radial field distribution by a Gaussian function (5.25), the integral in (5.143) can be evaluated in closed form to give

$$\alpha_{\text{b}}' = \frac{1}{2}\sqrt{\frac{\pi}{aRW^3}}\left(\frac{a}{w_{\text{G}}}\right)^2\frac{\exp[-2(a/w_{\text{G}})^2]}{K_0^2(W)}\exp\left[-\left(\frac{W}{a}\right)^3\frac{\lambda^2 R}{6\pi^2 n_2^2}\right]. \tag{5.145}$$

This approximate formula shows that two measures for the transverse field extent of the fundamental mode influence the bending loss: the transverse field extent a/W in the cladding and the spot size w_{G}. But, since the factor

$$\left(\frac{w_{\text{G}}}{a}\right)^2 K_0^2(W)\exp\left[2\left(\frac{a}{w_{\text{G}}}\right)^2\right] \tag{5.146}$$

in the denominator changes relatively little with normalized frequency V (the factor is of the order of unity), it is not the spot size w_G, but the transverse field extent a/W in the cladding, which is crucial for the bend loss. As can be seen from Fig. 5.5, a/W increases more rapidly with decreasing V than w_G.

For the special case of a step-index fiber, the function $S(V, W)$ becomes [Gambling and Matsumura 1978a]

$$S(V,W) = \frac{2U^2}{V^2 K_1^2(W)} \quad , \tag{5.147}$$

and the pure bend loss [Marcuse 1976a]

$$\alpha_b' = \frac{1}{2}\left[\frac{\pi}{4aRW^3}\right]^{1/2}\left[\frac{U}{VK_1(W)}\right]^2 \exp\left[-\frac{1}{6\pi^2}\frac{W^3\lambda^2 R}{a^3 n_2^2}\right] \quad . \tag{5.148}$$

For given values of the core radius a and of the normalized frequency V, the field extent a/W in the cladding can be obtained by the simple approximation (5.19). The radial phase parameter U determining the radial field distribution in the homogeneous core can then be obtained from (5.23).
Several engineering versions of (5.148) have been published [Jeunhomme 1982, 1983; Geckeler 1986a; Walker 1986].
For a standard step-index fiber with a core diameter of $2a = 8\,\mu m$, a relative refractive-index difference of $\Delta = 0.3$, and a theoretical cutoff wavelength of the LP_{11} mode at $\lambda_c = 1.18\,\mu m$, the attenuation of the LP_{01} mode (and that of the unwanted LP_{11} mode, see Sect. 6.3) has been calculated from (5.148) as a function of wavelength for three different values of the radius of curvature (Fig. 5.27). Note, that a slight increase in wavelength or curvature can increase the bend loss by many orders of magnitude.

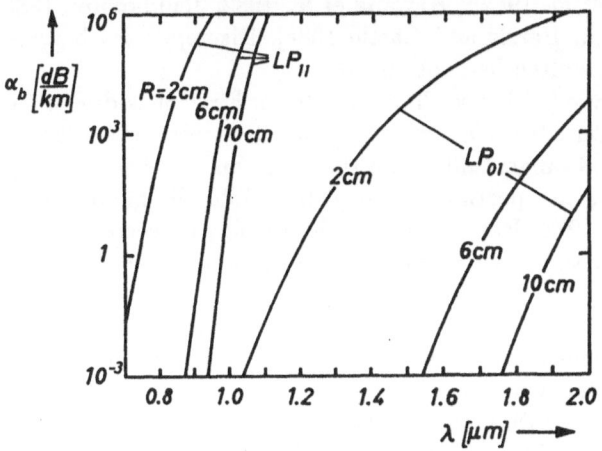

Fig. 5.27. Pure bend loss of the LP_{01} and LP_{11} modes of a step-index fiber with $2a = 8\,\mu m$ and $\Delta = 0.003$ for different values of the radius of curvature R

There is another formula for the pure bend loss in fibers with arbitrary refractive-index profiles [White I.A. 1979; Snyder and Love 1983; Kuznetsov and Haus 1983]. Since it contains an integral over the refractive-index profile multiplied by the amplitude distribution in the fiber core, it looks quite different. Nevertheless, the numerical values obtained from this alternative formula equal those obtained from (5.142) [Schwierz 1986].

127

Since the bend loss depends very sensitively upon the radius of curvature, approximate formulas for α_b are generally sufficient. For this reason, one usually neglects small corrections to the relatively simple bend loss formula (5.142). These corrections take into account the field deformation at the curved fiber [Marcuse 1976b; Sakai and Kimura 1982] or the small component of the Poynting vector normal to the plane of curvature [Sakai and Kimura 1978a, 1982].

One practical method to characterize the bending sensitivity of a fiber, is to determine that radius $R_{3\,dB}$, for which the attenuation a_{360} of a single fiber loop equals 3 dB [Lazay and Pearson 1982]. For a W-type fiber, this so called "3 dB loop test" gave radii as small as 2 mm [Ainslie et al. 1985]. Another critical radius of curvature R_c has also been defined as that radius which leads to a pure bend attenuation of 10^{-3} dB/km [Francois 1982].

For fibers with arbitrary index profile, in order to avoid the integration over the radial intensity distribution in (5.143), some authors have tried to estimate the bend attenuation by first determining the core radius a_E and the relative index difference Δ_E of an equivalent step-index fiber (Sect. 5.3.2) and then calculating the bend loss using the simpler formula (5.148) for a step-index fiber. But depending on the special ESI technique used, there can be extremely large errors in the prediction of bend loss [Saijonmaa and Yevick 1983; Samson 1985a].

A number of papers report the measurements of the pure bend loss and compare the measured bend loss with the calculated value [Kapron et al. 1970a,b; Neumann and Rudolph 1975b; Gambling et al. 1978g; Gambling and Matsumura 1978b; Alard et al. 1982b; Kawana et al. 1982; Jeunhomme 1983; Sharma A.B. et al. 1984a,b; Harris and Castle 1986]. Discrepancies between theory and experiment have often been observed.

These discrepancies may be due to different effects. First, it is difficult to experimentally isolate the pure bend loss from the transition losses (Sect. 5.9.8). Moreover, the wave radiated tangentially from the fiber can be reflected at the cladding-coating boundary, and partially coupled back into the guided mode [Murakami and Tsuchiya 1978; Harris and Castle 1986]. For certain values of the radius of curvature, the wave which has been radiated, reflected, and reconverted adds in phase to the wave guided along the curved core axis, causing the pure bend loss to be smaller than predicted by all theories, which assume an infinite cladding. This effect causes oscillations superimposed on the pure bend loss versus radius of curvature curve.

Other reasons for observed differences between the measured and the calculated values of the pure bend loss may be, that the elasto-optic effect to be described later in this section, or that the transition losses (Sect. 5.9.8) have not been taken into account in the calculation. For fibers tightly wound onto a rough drum, there may be also a loss component caused by microbending (Sect. 5.9.9).

As a consequence of the difficulties in measuring the pure bend loss, there is little convincing evidence of agreement between theory and experiment [Sharma A.B. et al. 1984a,b;]. Moreover, macrobending theories are very complicated, somewhat empirical, and not always in agreement amongst themselves [Kapron and Lazay 1983].

The pure bend loss has been measured in a polarization maintaining linearly birefringent side-pit fiber and has been found to be the same for both normal modes. It was also confirmed that the dependence of the bending loss on the radius of curvature is almost the same as that of conventional fibers without stress birefringence [Hosaka et al. 1981].

Bending losses are a major disadvantage of optical fibers. It therefore seems reasonable to ask, whether a curved dielectric waveguide inevitably loses power by radiation or whether there is a method for constructing low-loss dielectric waveguides with small radii of curvature.

It is well known that a ray of light propagating through a medium with a refractive index that changes in a direction transverse to the direction of propagation, will follow a curved path (Sect. 4.7). Sometimes, this "Fata Morgana" effect can be observed over the hot surface of highways: the hot air near the surface has a smaller density and a smaller refractive index than higher air layers. The bending of the light rays gives the road surface a mirror-like appearance. Quantitatively, the curvature of the ray path, i.e. the inverse of the radius of curvature, is equal to the transverse gradient, i.e. the derivative with respect to the transverse coordinate, of the refractive index (4.55).

It has been proposed this effect be utilized for constructing curved low-loss dielectric waveguides with very small radii of curvature [Neumann and Richter 1983; Marcatili 1985]. For a given refractive index profile n_s in the straight guide, a modified index profile n_b in the bent section of radius R has to be used. The optimum refractive index profile of the bent waveguide is:

$$n_b = (R/\varrho_a)n_s \quad , \tag{5.149}$$

where ϱ_a is the distance between a point in the waveguide cross section and the axis of curvature.

The first factor R/ϱ_a on the right hand side describes an increase of the refractive index on the inner side of the bend and a decrease on the outer side, i.e. a transverse index gradient. Thereby, the local phase velocity is reduced on the inner side and increased on the outer side. This forces the beam to follow the curved fiber. The second factor n_s describes the increased refractive index around the waveguide axis which, as for the straight waveguide, cancels the tendency of the beam to diverge because of diffraction.

Inserting the optimum refractive-index profile n_b (5.149) of a bent dielectric waveguide into the formula (5.141) for the index profile n_{eff} of the equivalent straight fiber, gives

$$n_{eff} = n_s \quad , \tag{5.150}$$

showing that the straight fiber equivalent to the curved fiber with the modified index profile is identical to the unbent fiber. Therefore, theoretically, the transverse field distribution, the phase constant, and the attenuation constant of the fundamental mode propagating in a bent waveguide with the optimum profile are the same as in the straight waveguide. There are no pure bend losses and no losses at the transition between straight and curved fiber sections (Sect. 5.9.8).

This method for reducing the bend losses assumes that the radius of curvature R is known in advance. Thus it cannot be applied to the flexible single-mode fibers, but may perhaps be used for the rigid dielectric channel waveguides in integrated optics. Since smaller radii of curvature are made possible, this technique for reducing bend losses can increase the packing density of optical components.

In a similar scheme [Korotky et al. 1985, 1986], the beam is not bent continuously by a waveguide with the optimum index profile given by (5.149), but is deflected discontinuously by a series of micro-prisms of raised refractive index placed in the beam path in the bend region. For an S-bend transition between two parallel straight waveguides offset laterally by $100 \, \mu$m, the bend loss was reduced from $10 \, \mathrm{dB}$ to $0.1 \, \mathrm{dB}$.

Although the effect of beam bending in an inhomogeneous medium is most important for the dielectric channel waveguides of integrated optics, it also plays a certain role in single-mode fibers. Bending a fiber increases the density of the glass on the inner side of the bend and decreases it on the outer side. Because of the photoelastic effect, the refractive index is proportional to the density [Primak and Post 1959]. Therefore, nature itself introduces a transverse index gradient of the correct sign for reducing the radiation losses. However, the magnitude of the gradient is smaller than the optimum value required by (5.149). Therefore, the photoelastic effect reduces bend losses a little, but not completely.

In calculating the bend losses, the index gradient due to elastic deformation must be accounted for by replacing the true radius of curvature in the formulas (5.142, 145, 148) for the pure bend loss, which do not take into account the transverse index gradient, by an effective radius of curvature, which depending on the refractive-index profile and the fiber material is 25–31 % larger than the true radius of curvature [Hannay 1976; Nagano et al. 1978; Marciniak 1982].

Coating a fiber to preserve the mechanical strength causes a loss increase at low temperatures. In order to reduce microbending losses (Sect. 5.9.9), the coating often consists of a soft (low Young's modulus) inner layer (e.g. silicone) surrounded by a harder outer coating (e.g. Nylon). Because the thermal expansion coefficient of plastic is much greater than that of glass, when the temperature is reduced the hard coating contracts more than the glass and subjects the fiber to both an axial and a radial stress. The axial compressive stress causes the fiber to buckle, i.e. to bend helically in the soft primary coating about the central axis. This uniform bending introduces a bend loss. With decreasing temperature, the bend radius decreases and the losses increase. The loss increase is maximum for long wavelengths. It has been found, that the experimental loss increase agrees well with the values calculated on the basis of a spiral bend model [Katsuyama et al. 1980].

The wave radiated from a bent single-mode fiber can be used to tap the fiber (Sect. 12.2.13), e.g. for local power detection when splicing fibers (Sect. 12.2.4), or for eavesdropping [Horgan 1985]. Because of reciprocity [Ramo et al. 1965], the direction of wave propagation can be reversed. Thus by tangentially irradiating a bent fiber section, one can couple power into the fundamental mode (local power injection). In order to obtain maximum launching efficiency, it is necessary to match both the amplitude distribution and the wavefronts of the incident beam to those of the wave which would be radiated from the bent fiber (Fig. 5.26). The pattern of the wave radiated from a curved optical fiber guiding the LP_{01} mode has been investigated both theoretically and experimentally [Thylén 1980].

Since single-mode fibers are sometimes operated at wavelengths shorter than the theoretical cutoff wavelength of the LP_{11} mode, it is interesting to know the pure bend loss of this mode, too. This information will be given in Sect. 6.3.1.

5.9.8 Transition Loss

In a straight single-mode fiber, the beam axis and the waveguide axis coincide. As has been explained in Sect. 2.3, at a bent fiber the beam shifts to the outside of the bend, i.e. the beam axis and the waveguide axis do not coincide. Figure 5.28 schematically shows the amplitude distributions, wavefronts, and time averaged Poynting vectors of the fundamental modes propagating along a straight and a curved optical waveguide with the axes aligned at the junction. Note that the fields of the radiated waves are not shown in Fig. 5.28.

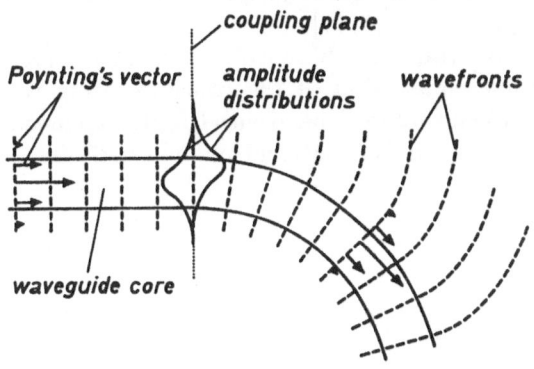

Fig. 5.28. Sketch used for explaining the transition loss: Transverse amplitude distribution, wavefronts, and time averaged Poynting vectors of the fundamental modes on a straight and a bent fiber section

In the plane at the junction normal to the fiber axis, because of the beam shift in the bent fiber, there is an amplitude mismatch between the field distributions of the fundamental mode incident along the straight fiber and the fundamental mode launched in the curved waveguide. Because the wavefronts of the fundamental mode propagating along the curved guide lag outside the critical boundary, the phase distributions also are different. As will be explained in Sect. 7.2.1, both a mismatch in the amplitude distributions and a mismatch of the wavefronts will result in radiation losses. The radiation losses occurring at the junction between a straight and a curved dielectric waveguide are called transition losses [Miyagi and Yip 1977; Gambling et al. 1978g, 1979a].

The displacement d between the beam axis and the bent waveguide axis can be approximated by a convenient formula [Gambling et al. 1978b]:

$$d = \frac{2\pi^2 n_e^2 w_G^4}{\lambda^2 R} \quad . \tag{5.151}$$

In this formula, the spotsize w_G related to launching (Sect. 10.1.2) has to be inserted.

The power coupling efficiency η between the mode on the straight section and the mode on the bent section, or the transition loss

$$a_t = -10 \log \eta \quad , \tag{5.152}$$

can be calculated by evaluating the field overlap integral (7.11) in the cross section at the discontinuity. Using the Gaussian approximation for the field distribution (5.25), and neglecting the field asymmetry, the increase in spot size, and the curvature of the wavefronts beyond the critical boundary, the formula for the transition loss reduces to that for a connector with transverse offset s (9.5). Therefore, the transition loss in decibels between a straight fiber and a bent fiber equals

$$a_t = 4.343 \left(\frac{d}{w_G} \right)^2 \quad , \tag{5.153}$$

with the offset d given by (5.151). A more accurate analysis of the transition losses has been published by Miyagi and Yip [1977].

Due to reciprocity, the loss at the transition between the end of a curved guide and a following straight guide equals that at the transition from the straight to the bent section. Hence, when a uniform bend with length L connects two straight fibers, the total loss is expected to be [Miyagi and Yip 1976, Baets and Lagasse 1983]

$$a = a_t + \alpha_b L + a_t = 2a_t + \alpha_b L \quad . \tag{5.154}$$

This relation only holds when the bent section is long [Gambling et al. 1978c]. For short lengths, the wave radiated from the first junction is incident on the second. There, part of its power is reconverted into guided power, i.e. coupled back into the fundamental mode [Gambling et al. 1979a]. Therefore, for short lengths, the total loss differs from the value predicted by (5.154) [Baets and Lagasse 1983].

Since the component of the transmitted fundamental mode that is caused by reconversion of radiation, differs in phase from the component that is guided around the bend, and since this phase difference changes with the length L of the bend, the total loss does not increase monotonically with L, but shows an oscillatory behaviour for small bend lengths [Gambling et al. 1976c; Gambling and Matsumura 1978b]. A similar effect is known from a succession of two waveguide tilts. By optimizing the waveguide length between the tilts, the radiation loss can be minimized. The two tilts are then said to be coherently coupled [Johnson L.M. and Leonberger 1983; Johnson L.M. and Yap 1984].

After a certain length corresponding to a bend angle of e.g. 100°, the radiated field at the second transition will be negligibly small, and for larger lengths the total bend loss increases linearly with bend length L as described by (5.154). It has been observed experimentally, that the modal field distribution in the pure bending region is constant along the fiber length [Gambling et al. 1977a].

At a transition between two fibers with different radii of curvature R_1 and R_2, there is also a transition loss which can be obtained from (5.153) by replacing the quantity d by the offset between the axes of the two beams. The transition loss has been measured and been found in good agreement with theoretical predictions [Gambling et al. 1978g].

Since it is the field shift at the bent waveguide that causes the transition loss, there is a simple method for reducing this loss: align the axes of the wave beams to be connected instead of the waveguide axes [Neumann 1982b]. For this purpose, the axis of the curved waveguide has to be offset toward the center of curvature by an amount which equals the field offset d (5.151) (Fig. 5.29).

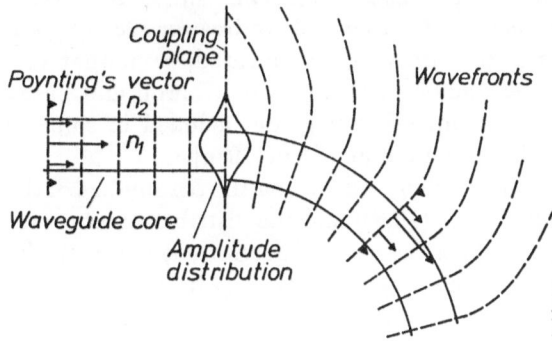

Fig. 5.29. A core offset reduces the field mismatch, and, therefore, the transition loss

By this method, the transition loss cannot be eliminated completely, because the transverse field distribution of the wave guided by the curved fiber is slightly asymmetrical and because its wavefronts are curved far from the axis of curvature. Thus there is both a residual amplitude mismatch and a residual wavefront mismatch, resulting in a residual transition loss. Of course, this method of offsetting the waveguide axes at a sudden change in curvature is not useful for flexible glass fibers, but may be applied with the rigid optical channel waveguides of integrated optics [Döldissen et al. 1985].

It is interesting to note, that the method for reducing the pure bend loss by introducing an additional transverse refractive index gradient (5.149), automatically avoids transition losses, too.

5.9.9 Microbending Loss

Bends with constant radius of curvature are also called macrobends. Microbends are irregular bends in the fiber caused by local transverse axis offsets of a few micrometers with spatial wavelengths in the millimeter and centimeter region. Microbends occur for instance, when the fiber under tension is wound on a rough drum or when a primary-coated fiber is in mechanical contact with a microscopically rough secondary coating. Minute irregularities in the machined surface of a metal drum suffice to cause substantial loss in fibers wound on this drum with only a few grams of tension. Like any discontinuity in a fiber waveguide, microbends cause radiation losses.

Microbending loss was first observed by winding multimode fibers under constant tension onto a drum surface that was not perfectly smooth [Gardner and Gloge 1975; Gardner 1975]. Random microbends can result in an increase in the optical loss on the order of 100 dB/km! Microbends cause a loss increase both in multimode and in single-mode fibers.

There have been a number of theoretical studies of the attenuation of the LP_{01} mode caused by microbending, both rigorous theories [Kuhn 1975b, 1976; Olshansky 1976; Marcuse 1976c], as well as simplified approximate approaches [Petermann 1976a,b, 1977a; Petermann and Storm 1976; Kawakami 1976; Tanaka T. et al. 1977; Sakai and Kimura 1978a, 1979; Gambling et al. 1979a; Sakai 1980a,b; Furuya and Suematsu 1980; Garrett and Todd 1982; Arnold J.M. and Allen 1983; Marcuse 1984; Kühne et al. 1985; Petermann and Kühne 1986; Artiglia et al. 1986a,b; Bjarklev 1986a,b; Povlsen and Andreasen 1986a,b; Andreasen 1986b]. Microbending theories are very complicated, somewhat empirical, and not always in agreement amongst themselves [Kapron and Lazay 1983]. Short reviews of the theory of microbending are included in some of these papers [e.g. Marcuse 1984; Petermann and Kühne 1986].

The attenuation caused by microbending depends on the parameters of the fiber [essentially the effective spot size w_e (10.8)] and the distribution of curvature $\kappa(z) = 1/R(z)$ of the fiber axis along the waveguide. Since it is practically impossible [Marcuse 1984] to measure the irregular curvature function $\kappa(z)$, for analyzing microbend losses, one assumes a special form of the autocorrelation function of the curvature function $\kappa(z)$, e.g. a Gaussian form [Gambling et al. 1978c]. This assumed autocorrelation function can be characterized by two parameters: its height, which is the RMS-value of the curvature, and its width, which is the correlation length of the curvature function.

The Wiener-Khintchine theorem [Carlson 1975] states that the autocorrelation function and the spectral density $\Phi_p(\Omega)$ of the curvature $\kappa(z)$ form a Fourier transforms pair, with Ω being the spatial frequency of a given curvature component along z. Thus the distribution of microbends can also be characterized by the power spectrum $\Phi_p(\Omega)$ of the curvature [Gloge 1979b].

For a Gaussian autocorrelation function, the power spectrum is also Gaussian. Another form of the power spectrum, which is often assumed, is a power law:

$$\Phi_{\mathrm{p}}(\Omega) = A/\Omega^{2p} \quad . \tag{5.155}$$

Since the actual curvature power spectrum strongly decreases with increasing spatial frequency Ω, the characteristic exponent p will be large. A value of e.g. $p = 4.3$ has been experimentally determined [Hornung et al. 1982]. There are other possible choices for the curvature power spectrum, e.g. an exponential form, or more complex functions. The question of the choice of the autocorrelation function which gives the best agreement between theoretical and experimental microbending loss has been investigated by several authors [Furuya and Suematsu 1978, 1980; Blow et al. 1982; Danielsen P. 1983; Arnold J.M. and Allen 1983].

The most widely used approximate theory for calculating microbending loss in single-mode fibers is that of Petermann [1976b, 1977a]. According to this theory, the microbending attenuation is given by

$$\alpha'_{\mathrm{m}} = \tfrac{1}{8}(knw_{02})^2 \Phi_{\mathrm{p}}\left[\frac{2}{knw_{01}^2}\right] \quad , \tag{5.156}$$

where w_{01} and w_{02} are two spot sizes. For calculating these spot sizes, one first has to determine an auxiliary function $f(\varrho)$ by solving an inhomogeneous second order differential equation, which comprises the refractive index profile $n(\varrho)$, the radial field distribution $E(\varrho)$, and the phase constant β of the fundamental fiber mode. Thereafter, one must insert Petermann's auxiliary function $f(\varrho)$ and the radial field distribution $E(\varrho)$ into two integrals which are used to define the spot sizes w_{01} and w_{02}. In general, microbending losses must be calculated numerically.

For the special case of step-index fibers, the auxiliary function $f(\varrho)$ and the spot sizes w_{01} and w_{02} have analytic solutions [Wright J.V. 1983].

When the radial field distribution can be approximated by a Gaussian function, the spot sizes w_{01} and w_{02} are equal to the effective spot size w_{e} (10.8) [Petermann 1976b]:

$$w_{01} = w_{02} = w_{\mathrm{e}} \quad . \tag{5.157}$$

The Gaussian approximation is not a very accurate representation for the field distribution at small V-values, nor for dispersion-modified fibers. By describing the radial field distribution in a fiber with arbitrary index profile by a Gaussian function near the axis and a modified Hankel function of zero order far from the axis, analytical expressions have been derived for w_{01} and w_{02} in excellent agreement with exact numerical results in the case of step-index and parabolic-core fibers [Sarkar et al. 1985]. In another method for calculating the spot sizes w_{01} and w_{02}, an arbitrary refractive-index profile is approximated by a staircase function [Andreasen 1986b]. Twenty layers are sufficient to ensure satisfying accuracy of the loss calculation.

Assuming the curvature power spectrum $\Phi_p(\Omega)$ to be described by a power law (5.155) and by using the Gaussian approximation (5.25), Petermann's microbending formula becomes [Garrett and Todd 1982]

$$\alpha_m = \frac{C w_e^{(4p+2)}}{\lambda^{(2p+2)}} \quad , \tag{5.158}$$

where C is a constant which, with p, governs the curvature statistics.

Since the effective spot size w_e varies with wavelength, the wavelength dependence of microbending loss is rather complex [Sakai 1980b; Marcuse 1984]. As compared to multimode fibers, in which the microbending loss is, to the first order, wavelength independent, microbending loss in single-mode fibers usually increases with wavelength. However, the loss increase at long wavelengths due to microbending is much slower than that due to macrobending [Hordvik and Eriksrud 1985, 1986]. Taking the microbending induced loss α_{mMM} for a multimode fiber with a 50 μm core diameter and a numerical aperture of 0.2 as a reference, the microbending attenuation of the single-mode fiber is [Jeunhomme 1982]

$$\alpha_m = 2 \times 10^{-4} w_G^6 \lambda^{-4} \alpha_{mMM} \quad , \tag{5.159}$$

where w_G and λ should be expressed in μm and both the multimode and single-mode fibers should have the same outside diameter. This relation has been verified experimentally [Auge et al. 1984].

When fiber parameters are properly chosen, a single-mode fiber is less sensitive to microbending than a multimode fiber [Furuya and Suematsu 1980; Kaiser 1983].

Since the spot size increases with increasing wavelength, microbending losses will cause a rise of the attenuation at long wavelengths. The wavelength range which can be utilized in a single-mode fiber is thus limited at short wavelengths by the effective cutoff wavelength λ_{ce} of the LP_{11} mode (Sect. 6.3.3) and at long wavelengths by microbending, macrobending, or intrinsic ir-attenuation.

As long as the shape of the fundamental mode field is nearly Gaussian, the simplified theory of Petermann [1976b, 1977a] is very accurate [Petermann and Kühne 1986]. However, when the radial intensity distribution of the fundamental mode deviates considerably from a Gaussian function, e.g. for dispersion-flattened fibers (Fig. 5.17), one must not apply the 1976 theory of Petermann with the Gaussian approximation [Sakai 1980b], but a new approximate one [Kühne et al. 1985; Petermann and Kühne 1986]. In this analysis, a power law (5.155) with the characteristic exponent p is assumed for the power spectrum of curvature. For an arbitrary single-mode fiber, the microbending loss can always be written as [Petermann and Kühne 1986]

$$\alpha_m = \frac{A}{4}(kn_1 w_e)^2 \left[\frac{kn_1 w^2(p)}{2} \right]^{2p} \quad , \tag{5.160}$$

with the free space wavenumber k, the maximum refractive index n_1 in the core, the effective spot size w_e (10.8), and a parameter $w(p)$ which is characteristic of the specific fiber. The following inequality holds for any arbitrary dispersion-flattened single-mode fiber:

$$w_\infty^2 \geq w^2(p) \geq w_e w_d \geq w_d^2 \quad , \tag{5.161}$$

with w_d the spot size related to dispersion (10.10), and

$$w_\infty^2 = \frac{2}{kn_1(\beta - n_2 k)} \quad . \tag{5.162}$$

Using (5.17) and approximating $n_1 k$ by β, the parameter w_∞ can be related to the transverse field extent a/W in the cladding [Petermann 1987]

$$w_\infty \approx \frac{2a}{W} \quad . \tag{5.163}$$

For large values of the characteristic exponent p, $w(p)$ approaches w_∞.

Whereas macrobending losses are related to w_∞, microbending losses are related to the parameter $w(p)$ [Petermann 1987]. Inserting the upper $(2a/W)$ or the lower (w_d) limit for $w(p)$ in (5.160), one gets simple estimates for upper and lower limits for the microbending loss in arbitrary single-mode fibers. Since the spot size w_d determines the joint losses caused by a small transverse axis offset (Sect. 10.1.4), its value is fixed by the joint losses that can be tolerated. Then, from (5.161) it can be seen, that for small microbending losses, the parameter w_∞ has to be minimized.

The design goal for simultaneously achieving low bending loss (w_∞ small) and low offset splice loss (w_d large) thus corresponds to achieving a ratio w_∞/w_d as close to unity as possible [Petermann 1987]. An ordinary step-index fiber, operated at $V = 2.4$ near the LP$_{11}$ mode cutoff, already exhibits a satisfactory $w_\infty/w_d = 1.06$ (Table 5.1). Dispersion-shifted and dispersion-flattened fiber designs yield a ratio w_∞/w_d far from unity [Povlsen 1985b; Petermann 1987]. These fibers thus exhibit large microbending losses at long wavelengths [Kühne et al. 1985]. The microbending loss of dispersion compensated fibers with triple or quadruple claddings is much higher than that of simple step-index fibers [Petermann and Storm 1976; Yang and Unger 1985].

There are two papers comparing different microbending theories [Francois and Vassallo 1986; Bjarklev 1986a]. The loss values predicted by the theories of Petermann [1977a] and Gambling et al. [1979a] can differ by a factor of 10 or more in realistic cases [Francois and Vassallo 1986]. Microbending losses calculated by using a multilayer approximation for the core have been found to differ only slightly from the results found from Petermann [1976b] theory, but to differ substantially from Marcuses's [1984] results [Bjarklev 1986a].

Several experimental investigations of microbending loss in single-mode fibers have been reported [Tanaka T. et al. 1977; Gambling et al. 1978g; Furuya and Suematsu 1979; Katsuyama et al. 1980; Hornung and Reeve 1981; Hornung et al. 1982; Yabuta et al. 1983; Mitsunaga et al. 1984; Auge et al. 1984; Hordvik and Eriksrud 1985]. In one experimental investigation [Hornung et al. 1982], it was found that a reasonable fit to the experimental loss data could be obtained by a suitable choice of the parameter p in (5.155). For the best fit over a range of fibers for fixed V, the optimum value of the index of the microbending power spectrum was $p = 4.3$. Inserting this value in (5.158) shows that microbending loss is proportional to $w_e^{19.2}$, i.e. very sensitive to an increase in spot size.

In another study, the wavelength dependence of microbending loss has been calculated and compared with experimental results [Arnold J.M. and Allen 1983]. An inverse power spectrum of curvature $\Phi_p(\Omega)$ (5.155) with p somewhere between 3 and 4 appears to give good agreement with the experimental data.

In the two papers mentioned above, good agreement between experimental and theoretical values of microbending loss have been reported. However, in another experimental study of microbending losses [Garg and Eoll 1986b], attempts to relate the experimental results to theory were disappointing. Agreement with theory was at best qualitative.

For obtaining small microbending losses, the spot size must not be too large. The spot size w_e must not become larger than 4λ, lest the microbending loss becomes higher than is normally tolerated in multimode fibers [Petermann 1977a]. The spot size increases with decreasing relative refractive-index difference Δ. For a given cable structure and both for loose-tube coating and tight secondary fiber coating, microbending losses increase rapidly for refractive index differences below 0.3 % [Hooper et al. 1985]. When Δ is reduced further, the microbending loss suffers a catastrophic, i.e. extremely steep increase, at about $\Delta = 0.2\,\%$ [Katsuyama et al. 1979; Hornung and Reeve 1981; Garrett and Todd 1982].

In a fiber cable, loss increase can be caused either by micro- or macrobends. We consider a fiber with loose tube secondary coating. The thermal expansion coefficient of the plastic tube is much larger than that of the glass fiber. In a cable with a loose tube construction, the fibers are usually stranded around a central strength member. Normally, the tube and the fiber have the same length at room temperature. When the cable is heated, the tube will become longer than the fiber. The fiber will move towards the cable axis and eventually be pressed against the tube wall (inner curve). The tube and the primary fiber coating can be expected to have microscopically rough surfaces. The fiber axis will deviate from the helix with constant curvature around the strength member, and microbending losses are induced.

When the fiber is cooled, the tube will shrink. The fiber will move away from the cable axis, and eventually it will be pushed against the tube wall (outer curve). When the tube is jelly filled, the frictional resistance between primary coating and tube is small. Further cooling will result in buckling of the fiber [Stueflotten 1982] to allow it to fit helically into the shortened tube. Buckling causes additional uniform bending of the fiber, i.e. macrobending losses. Thus one expects a fiber enclosed in a jelly filled loose tube to exhibit predominantly microbending losses at high temperatures and predominantly macrobending losses at low temperatures. This has been experimentally verified [Hordvik and Eriksrud 1985, 1986]. The cable designer has to choose the excess fiber length in such a manner, that in the range of operating temperatures, the loss due to micro- or macrobending is minimized.

In plastic-jacketed fibers, the fiber also buckles at low temperatures. The loss increase due to buckling has been measured and found to approximately coincide with values calculated from buckling theory and macrobending theory [Yabuta et al. 1983; Mitsunaga et al. 1984].

Summarizing, at present there is no satisfactory method for predicting microbending loss in single-mode fibers. Even though it is relatively easy to derive formulas for these losses, prediction of actual loss values is hampered by the lack of information about the statistics of the fiber deformations [Marcuse 1984]. Moreover, as has been reported before, different theories can yield different results. Thus, it is more difficult to predict microbending losses than macrobending losses.

5.9.10 Other Causes of Radiation Losses

Neglecting Rayleigh scattering, an ideal straight uniform single-mode fiber does not radiate light into its surroundings. But any imperfection of the dielectric waveguide, such as a local change of its index of refraction, or a deviation from perfect straightness, or an imperfection of the interface between core and cladding, couples the fundamental fiber mode to the radiation field, thus increasing the loss from the desired guided mode. Radiation due to different waveguide imperfections has been demonstrated using microwave model experiments [Neumann 1975].

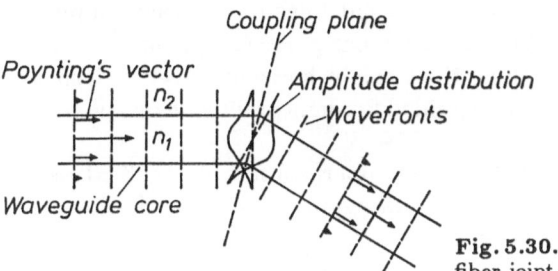

Fig. 5.30. Wavefront mismatch at a tilted fiber joint

Macrobends and microbends have already been considered as causes of radiation. Another example of a radiating discontinuity is an abrupt bend or corner, which can be considered as a bend with zero radius of curvature. In Fig. 5.30, the fields of the fundamental modes in the two fibers are represented by wavefronts, Poynting vectors, and transverse amplitude distributions. At the junction, there is an angular misalignment between the wavefronts of the incident and outgoing LP_{01} modes. This wavefront mismatch causes radiation losses. The power losses occurring at fiber joints with tilted fiber axes will be described in Sect. 9.4.

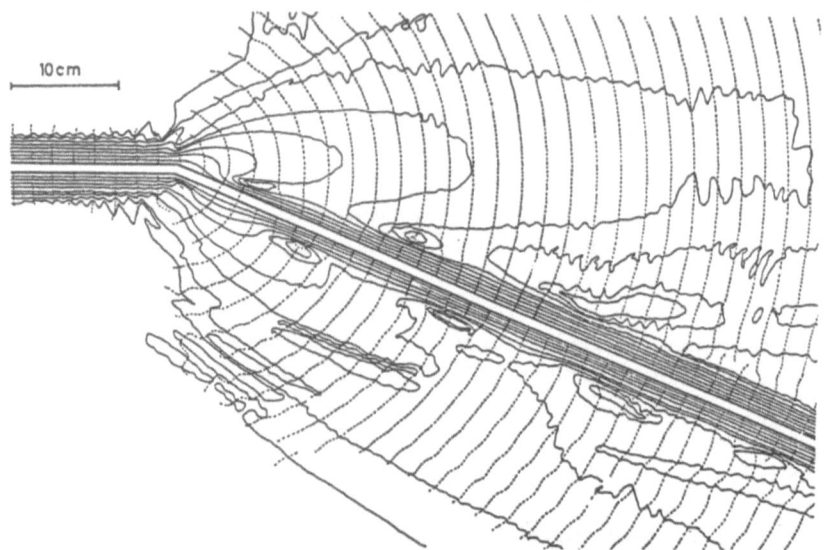

Fig. 5.31. Lines of constant magnitude (*solid lines*) and wavefronts (*dashed lines*) of the electric field of the fundamental mode passing an *H*-plane 22°-corner in a circular dielectric rod transmission line. Model experiment with microwaves at a frequency of 13.92 GHz. The dielectric rod waveguide had a core diameter of $2a = 10$ mm and a permittivity of $\varepsilon = 2.50$

Here, for illustrating the radiation from a tilt, Fig. 5.31 shows the distribution of the electric field measured with microwaves in the H-plane of a corner in a dielectric rod waveguide [Neumann and Rudolph 1975a; Neumann 1975]. As in Fig. 5.26, the field is represented by solid lines of constant amplitude (5 dB level difference) and dashed lines of constant phase (2π phase difference). A fundamental mode is incident along the straight horizontal section from the left-hand side. At the tilt, it launches a guided fundamental mode on the second straight section and a radiated wave.

The observed total electric field is the superposition of the field of the fundamental mode guided by the second waveguide and the radiation field. The radiation from the tilt is very similar to the radiation from the free end of a dielectric waveguide (Sect. 8.1), i.e. to a Gaussian beam with the beam axis coinciding with the axis of the first waveguide. Since the strength of the radiation field decreases with increasing distance from the tilt, whereas the field of the mode guided by the second waveguide is attenuated only very slightly by dielectric losses, far from the tilt, the field of the guided wave dominates near the axis of the second waveguide. For the experimental data chosen, circular rods with a permittivity of 2.5 and a diameter of 10 mm, a frequency of 13.92 GHz, and a tilt angle of 22°, the insertion loss was 1.6 dB.

The field in the second fiber can be expressed in terms of the guided fundamental mode and a continuum of radiation modes. Although this approach is basically rigorous, it gives hardly any insight into the physical situation. However, a simple model has been presented for the radiation field [Kawakami and Baba 1985]: it can be approximated by a diffracting first-order Hermite-Gaussian beam in a uniform medium (Sect. 4.8).

The radiation losses of abrupt bends in optical dielectric waveguides can be reduced either by retarding the wavefronts on the inner side [Neumann 1981] or by accelerating them on the outer side of the tilt [Shiina et al. 1986]. The local phase velocity is increased or decreased by respectively decreasing or increasing the refractive index.

Finally, for completeness, we list some other examples of radiating discontinuities in single-mode fibers:

a)   single steps in core diameter [Marcuse 1970a,c; Guttmann and Krumpholz 1973; Rawson 1974; Neumann and Opielka 1977];

b)   joints between fibers with equal core radius but different refractive indices of the core and the cladding [Neumann and Opielka 1977];

c)   sinusoidal variations of the core radius [Rawson 1974; Marcuse 1975d, 1978b; White I.A. and Snyder 1977; Snyder 1980];

d)   sinusoidal meander of core axis [Rawson 1974];

e)   periodic deformations of the core axis [Wlodarczyk and Seshadri 1987];

f)   helical meander of core axis [Rawson 1974];

g)   a deformation in which the core cross section alternates periodically along the fiber between an ellipse extended along the x-axis and an ellipse extended along the y-axis [Rawson 1974].

h)   continuous changes in core diameter, so called tapers (Sect. 12.2.6) [Marcuse 1970a,c];

i)   random changes in core diameter [Marcuse 1970a,c, 1984; Inada et al. 1982; Bjarklev 1986a,b],

j)   random fluctuations of the refractive index [Inada et al. 1982].

6. Higher-Order Modes

An optical fiber waveguide can guide a finite number of distinct waves or modes, each with a characteristic transverse field distribution. For a given refractive-index profile, the number of guided modes depends on the core radius, the relative refractive-index difference, and the operating wavelength. These data are combined into the normalized frequency V (5.8). The higher the value of V, the larger the number of guided modes. Although a single-mode fiber is designed for transmitting only the fundamental mode, it is necessary to have some knowledge about the other guided modes, the so-called higher-order modes.

If one decreases the operating wavelength of a single-mode fiber, higher-order modes begin to be guided along with the fundamental LP_{01} mode. Usually, the first higher-order mode to appear, and which will eventually disturb system performance (Sect. 11.6.1), is the LP_{11} mode. But note that there are fibers with special refractive index profiles, so-called multiple-clad fibers, in which it is possible, that the LP_{02} mode appears before the LP_{11} mode [Francois et al. 1984].

Therefore, in this chapter, we first describe an arbitrary higher-order mode (Sects. 6.1; 6.2), and then the most important LP_{11} mode (Sect. 6.3). As well as the guided modes, there are also leaky modes (Sect. 6.4), cladding modes (Sect. 6.5), and radiation modes (Sect. 6.6). Finally, we mention the dual-mode fiber (Sect. 6.7).

6.1 Field Distribution

Rigorously, the modes guided by a dielectric waveguide have three nonzero components of the electric field and three nonzero components of the magnetic field. These modes are called exact, ideal, or vector modes. Unfortunately, the designation of the vector modes is not uniform [Morishita 1983b]. The most common mode names are $TE_{0\mu}$, $TM_{0\mu}$, $HE_{\nu\mu}$, and $EH_{\nu\mu}$ modes [Snitzer 1969]. The calculation of the field components and phase constants of the vector modes is "overwhelmingly complex" [Garside et al. 1980]. The modal field patterns of the exact $TE_{0\mu}$, $TM_{0\mu}$, $HE_{\nu\mu}$, and $EH_{\nu\mu}$ modes have been represented graphically both by their radial field distributions [Biernson and Kinsley 1965], and by electric and magnetic field vectors in a waveguide cross section [Snitzer 1969; Kajfez 1983].

Since the formulas for the exact vector modes are very complicated, we only describe the so-called linearly polarized LP$_{lm}$ modes of weakly guiding fibers [Gloge 1971a; Snyder and Young 1978], which represent approximations to combinations of nearly degenerate exact TE$_{0\mu}$, TM$_{0\mu}$, HE$_{\nu\mu}$, and EH$_{\nu\mu}$ modes [Qian and Huang 1986]. This "linearly polarized", "weak-guidance", or "scalar" approximation is very attractive, since for weakly guiding fibers ($\Delta \ll 1$), it provides simple and yet sufficiently accurate formulas for the practical design calculation for single-mode fibers [Tjaden 1978; Yeh et al. 1979b; Garside et al. 1980; Morishita et al. 1980; Hosain et al. 1983].

We use the same coordinate systems as for the fundamental mode: x, y, z and ϱ, ϕ, z are Cartesian and cylindrical polar coordinates (Fig. 3.1), where the z-coordinate is along the fiber axis. First, we assume the electric field to be polarized linearly in the x-direction.

The phasor $\underline{E}_x(\varrho, \phi, z)$ of the x-component of the electric field of the mode with arbitrary mode numbers l and m is given by the equation

$$\underline{E}_x(\varrho, \phi, z) = E(\varrho) \cos(l\phi) \exp(-\mathrm{j}\beta z) \quad . \tag{6.1}$$

The transverse amplitude distribution is described by the function $|E(\varrho) \cos(l\phi)|$ and the zero phase angle by

$$\Phi_x(\varrho, \phi, z) = -\beta z (+\pi) \quad . \tag{6.2}$$

The angle π has to be added in (6.2) for those regions in the cross section for which the product $E(\varrho) \cos(l\phi)$ is negative.

The wavefronts are sections of planes normal to the fiber axis. The longitudinal distance $2\pi/\beta$ between two planes of equal phase is the wavelength λ_{lm} of the higher-order mode considered. Note, that the wavelengths λ_{lm} and phase constants β of the higher-order modes differ from the corresponding values for the fundamental mode. Strictly speaking, one has to add the subscripts l and m to each parameter belonging to a particular fiber mode, e.g. to $\lambda, \beta, W, \alpha, \ldots,$. Usually, we omit these subscripts, and only add them, if two different modes are considered at the same time. But one must bear in mind that for a given fiber and a given operating wavelength, different modes will have different wavelengths, phase constants, cladding decay parameters, attenuation coefficients, etc.

Since the symbol for the vacuum wavelength is λ, the symbol λ_{lm} is used for the wavelength of the guided LP$_{lm}$ mode.

A simple cosine function describes the field amplitude as a function of the azimuthal angle ϕ. Since the integer index l determines the number of full periods of the cosine function passed through when ϕ increases by 2π, it is called the azimuthal mode number (order, index) of the guided mode. The cosine-function in (6.1) can also be replaced by the function $\sin(l\phi)$ giving (for $l \neq 0$) a mode with the same azimuthal order but with a different "orientation".

The real function $E(\varrho)$ describing the dependence of the field amplitude on the distance ϱ from the fiber axis is a solution of the so-called scalar wave equation [Snyder and Love 1983]:

$$\frac{d^2E}{d\varrho^2} + \left(\frac{1}{\varrho}\right)\frac{dE}{d\varrho} + \left[n^2(\varrho)k^2 - \beta^2 - \left(\frac{l}{\varrho}\right)^2\right]E = 0 \quad . \tag{6.3}$$

For $l = 0$, (6.3) transforms into the scalar wave equation (5.20) for the fundamental mode.

For the higher-order modes with $l = 0$, the field $E(0)$ on the fiber axis is finite, whereas for $l>0$ it is zero.

For any refractive-index profile in the core, the radial field distribution in the homogeneous cladding is described by the modified Hankel function K_l of order l:

$$E(\varrho) = CK_l\left(\frac{W\varrho}{a}\right) \quad , \tag{6.4}$$

where the quantity W is the cladding decay parameter of the higher-order mode considered. It is different from that of the fundamental mode (5.19). The cladding decay parameter W, the core radius a, the phase constant β, and the wavenumber n_2k in the cladding are related by the separation equation in the cladding (5.17):

$$\left(\frac{W}{a}\right)^2 = \beta^2 - (n_2k)^2 \quad . \tag{6.5}$$

For large values of the argument of the Hankel function, i.e. far into the cladding, the radial amplitude distribution can be approximated by

$$E(\varrho) = C\sqrt{\frac{\pi a}{2W\varrho}}\exp\left(-\frac{W\varrho}{a}\right) \quad . \tag{6.6}$$

In the core, the radial field distribution $E(\varrho)$ depends on the particular index profile and the mode numbers l and m. In general, it can only be determined from the index profile by solving the scalar wave equation (6.3) using numerical methods. Here, it is sufficient to state that the amplitude $E(\varrho)$ oscillates as a function of the distance ϱ from the fiber axis. The second subscript m in the mode name LP_{lm} gives the number of intensity maxima in the core between $\varrho = 0$ and $\varrho = a$ and is thus called the radial mode number (order, index).

As for the fundamental mode, it is very useful to find a single simple function that approximates the radial field distribution both in the core in the cladding. Whereas a Gaussian function can be used to approximate the radial field distribution of the fundamental mode (5.25), a product of the lth power of ϱ, of a generalized Laguerre polynomial $L_{m-1}^{(l)}$ of order l and degree $m - 1$ [Abramowitz 1965], with a Gaussian function can be used to approximate the radial field distribution of higher-order modes [Love and Hussey 1984; Facq et

144

al. 1984]:

$$E(\varrho) = E_0 \left(\frac{\sqrt{2}\varrho}{w_\mathrm{G}} \right)^l L_{m-1}^{(l)} \left(\frac{2\varrho^2}{w_\mathrm{G}^2} \right) \exp\left(-\frac{\varrho^2}{w_\mathrm{G}^2} \right) \quad , \tag{6.7}$$

where w_G is the spot-size of the fundamental fiber mode.

For the special case of a step-index fiber, the radial amplitude distribution in the core is described by a Bessel function J_l, the order of which equals the azimuthal mode number l:

$$E(\varrho) = C \frac{\mathrm{K}_l(W)}{\mathrm{J}_l(U)} \mathrm{J}_l \left(\frac{U\varrho}{a} \right) \quad . \tag{6.8}$$

By adding the factor $\mathrm{K}_l(W)/\mathrm{J}_l(U)$ in (6.8), continuity of the tangential component of the electric field at the core-cladding boundary is guaranteed.

The eigenvalues U and W in the core and cladding are obtained by solving the characteristic equation [Gloge 1971a]

$$U \frac{\mathrm{J}_{l-1}(U)}{\mathrm{J}_l(U)} = -W \frac{\mathrm{K}_{l-1}(W)}{\mathrm{K}_l(W)} \quad , \tag{6.9}$$

taking into account that

$$U^2 + W^2 = V^2 \quad , \tag{6.10}$$

where V is the normalized frequency (5.8). For a given azimuthal mode number l and for larger V values, the system of equations (6.9, 10) has more than one solution. The solutions are ordered according to increasing value of the eigenvalue U. The number 1 is assigned to the first solution and the number m to the mth solution. The integer m is the radial mode number introduced before.

Using the recurrence relations for the Bessel and modified Hankel functions [Abramowitz and Stegun 1965], the characteristic equation of the LP_{lm} modes can also be written in the form

$$U \frac{\mathrm{J}_{l+1}(U)}{\mathrm{J}_l(U)} = W \frac{\mathrm{K}_{l+1}(W)}{\mathrm{K}_l(W)} \quad . \tag{6.11}$$

The dominant component $\underline{H}_y$ of the magnetic field is orthogonal to the dominant electric field component $\underline{E}_x$ and the direction of propagation. Its amplitude is simply the amplitude of the transverse component of the electric field divided by a mean value Z_0/n_e of the wave impedance (5.29), where $n_e = \beta/k$ is the effective index (5.45) of the guided mode considered. Thus the transverse distribution of the magnetic field equals that of the electric field.

The higher-order LP_{lm} modes have nonzero longitudinal components $\underline{E}_z$ and $\underline{H}_z$ of both the electric and the magnetic field, like the fundamental fiber mode, and, therefore, are also hybrid waves. But for weakly guiding fibers with a small relative refractive-index difference Δ, the longitudinal components are small compared to the dominant transverse components. Strictly speaking, the fields also have transverse components $\underline{E}_y$ and $\underline{H}_x$ normal to the main components. In the theory of weakly guiding fibers [Snyder 1969a; Gloge 1971a], these very small transverse components are neglected.

Since the guided modes described here are essentially linearly polarized, they have been named linearly polarized modes, or, LP modes.

Instead of assuming the electric field to be linearly polarized in the x-direction, one can also launch a wave linearly polarized in the y-direction. Therefore, for a given combination of the mode numbers $l(\neq 0)$ and m, there are four LP_{lm} modes with different polarizations (E_x or E_y) and different orientations [$\cos(l\phi)$ or $\sin(l\phi)$]. For the modes with the azimuthal order $l = 0$, the field does not depend on the azimuth ϕ and, therefore, one has only two LP_{0m} modes with different polarizations.

For an ideal fiber with a circular core made of isotropic materials the four LP_{lm} modes with different orientations and polarizations (or the two LP_{0m} modes with different polarizations) have equal values of the phase constant β, the specific group delay $d\beta/d\omega$, the dispersion $d^2\beta/d\omega^2$, the normalized cutoff frequency V_c, etc. They are called degenerate modes. An elliptical core or a birefringent core material lifts the degeneracy, i.e. the phase constants and the other parameters of the four (or two) modes with equal mode indices become slightly different.

The time averaged intensity (Poynting) vector (3.16) points in the direction of the fiber axis and its magnitude is given by

$$S_z(\varrho, \phi) = [n_e/(2Z_0)]E^2(\varrho)\cos^2(l\phi) \quad . \tag{6.12}$$

On a circle around the fiber axis, there are $2l$ intensity maxima, and, starting at the fiber axis, there are m intensity maxima in the radial direction. This knowledge can be used to determine the mode numbers l and m: Provided that only one mode is excited, one can observe the near-field intensity distribution in the fiber endface (Sect. 13.5), and count the intensity maxima in the azimuthal and radial directions. The azimuthal mode number l is the first value divided by two, and the radial mode number m is the second value. Of course, this method only works when the field of the mode one is interested in prevails. Methods for selectively exciting single guided modes are described in Sect. 7.6.

When a fiber can support several modes, a source will generally launch power into all modes. The relative powers guided by the different modes, described by the so-called mode spectrum, depend on the launching conditions and the attenuation of the modes, which will generally be different.

When a coherent source excites a fiber that can guide more than one mode, one will observe a very irregular transverse intensity distribution caused by the interference of the fields of the different modes, a so called speckle pattern. Since the relative phases between the modes change in time due to frequency drift of the laser or slight movement of the fiber, the speckle pattern will vary with time. If only part of the total light field is coupled to a second fiber or to a photodetector, this part will also fluctuate randomly in time. This effect is called "modal noise" [Epworth 1978]. It is a great problem for multimode fibers, but can also be observed with single-mode fibers, since they can propagate two fundamental mode with orthogonal polarization [Heckmann 1981a] or two LP_{01} and four LP_{11} modes, if they are inadvertently operated at a wavelength slightly shorter than the cutoff wavelength of the second-order mode [Heckmann 1981b]. These modal noise effects in single-mode fibers will be described in Sects. 11.6.10 and 11.6.1, respectively.

The pure bend loss of higher-order modes in fibers with arbitrary refractive-index profile has been analyzed by several authors [Marcuse 1976a,b; Sakai and Kimura 1978a,1982; White I.A. 1979; Snyder and Love 1983].

6.2 Cutoff Wavelength

For higher-order fiber modes, the electric field in the core changes more rapidly in the radial and azimuthal directions than for the fundamental mode. Therefore, the tendency for diffraction spreading is stronger and a larger core or a larger refractive index difference are needed to compensate for the diffraction spreading. Thus, the normalized frequency V has to be larger than a minimum value in order that a particular higher-order mode is guided by the fiber. This minimum value is called the normalized cutoff frequency V_c of the higher-order mode.

At the cutoff frequency, i.e. for

$$V = V_c \quad , \tag{6.13}$$

the field extends infinitely into the cladding, i.e. the condition for cutoff is that the cladding decay parameter W is zero [Gloge 1971a]:

$$W = 0 \quad . \tag{6.14}$$

For arbitrary refractive index profile, the normalized cutoff frequency can only be calculated by using numerical methods, in general. The higher the mode numbers l and m of a mode, the higher its cutoff frequency V_c.

In a step-index fiber, the normalized cutoff frequency of the LP_{lm} mode follows by setting $W = 0$ in the characteristic equation (6.9)

$$J_{l-1}(U) = 0 \quad . \tag{6.15}$$

Thus, at cutoff, the radial phase parameter U in the core is equal to a zero $U_{l-1,m}$ of the Bessel function $J_{l-1}(U)$ of number $l-1$. In the symbol $U_{l-1,m}$ for the zero, the first subscript $l-1$ denotes the order of the Bessel-function, and the second subscript m is the number of the zero. For $l \neq 0$, a zero at U is not being counted. However, for $l = 0$, the zeros of the Bessel function

$$J_{-1}(U) = -J_1(U) \tag{6.16}$$

are counted so as to include $U = 0$ as the first zero [Gloge 1971a].

On account of (6.10) and (6.14), at cutoff, the parameter U is equal to the normalized frequency V, i.e.:

$$V_c = U_{l-1,m} \quad . \tag{6.17}$$

In a step-index fiber, the second-order mode is the LP_{11} mode with a normalized cutoff frequency of

$$V_c = U_{0,1} = 2.405 \quad .$$ (6.18)

The next highest order modes are the LP_{02} and LP_{21} modes with

$$V_c = U_{1,1} = 3.832 \quad ,$$ (6.19)

and the LP_{31} mode with

$$V_c = U_{2,1} = 5.136 \quad .$$ (6.20)

The V regions, in which the different modes can propagate are illustrated in Fig. 6.1.

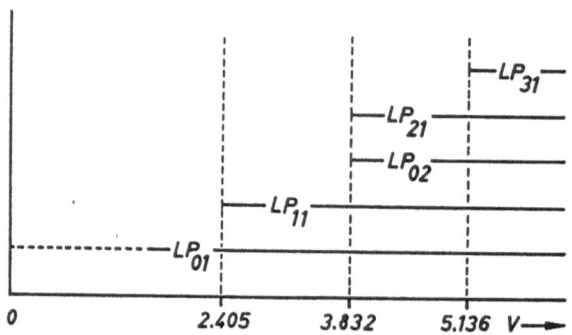

Fig. 6.1. Regions of the normalized frequency V (5.8), for which some low-order modes can be propagated by a step-index fiber. The cutoff condition is described by (6.15)

From the definition of the normalized frequency (5.8), the relation between the cutoff value of the normalized frequency and the corresponding cutoff wavelength λ_c is

$$\lambda_c = \frac{2\pi a \sqrt{n_1^2 - n_2^2}}{V_c} \quad .$$ (6.21)

For wavelengths smaller than the cutoff value, the mode is guided by the fiber. When one tries to launch a fiber mode with a wavelength longer than the cutoff wavelength, the incident beam will not be guided by the fiber and the power will be lost by radiation.

For most refractive-index profiles, the cutoff frequency of the *fundamental* mode is theoretically zero. However, this does not mean that a single-mode fiber can be used to guide microwaves or radio waves, since with decreasing frequency, the fundamental mode transforms continuously into a uniform plane wave, which is no longer guided by the fiber core (Sect. 5.3.4). With increasing wavelength, the spot size of the fundamental mode increases causing large macro- and microbending losses. Thus, although the theoretical cutoff wavelength of the LP_{01} mode in most single-mode fibers is infinite, there is an effective cutoff wavelength for the fundamental mode.

For fibers in which the refractive index is smaller than the cladding index n_2 in part of the core cross section (Figs. 5.3, 4), the LP_{01} mode can have a finite theoretical cutoff wavelength. The condition for a finite cutoff-frequency of the fundamental mode is that the refractive–index difference $n(\varrho) - n_2$ averaged over the core cross section is negative [Hussey and Pask 1982a; Okoshi 1983]:

$$\int_0^\infty [n(\varrho) - n_2] \varrho \, d\varrho < 0 \quad .$$ (6.22)

Simple and accurate formulas for computing the cutoff wavelength of the fundamental mode for depressed inner cladding fibers have been reported [Sammut 1978; Monerie 1982].

6.3 LP$_{11}$ Mode

In the case of single-mode fibers, one is usually interested in the fundamental mode with mode numbers $l = 0$ and $m = 1$ (Chap. 5), and the second-order mode with $l = 1$ and $m = 1$, which is usually the next mode to appear when the operating wavelength is decreased.

6.3.1 Field Distribution

For the LP$_{11}$ mode, using the power-Laguerre-Gaussian approximation (6.7), the field distribution (6.1) is described by [Thyagarajan et al. 1982]:

$$\underline{E}_x(\varrho, \phi, z) = E_0 \frac{\varrho}{w_G} \exp\left[-\left(\frac{\varrho}{w_G}\right)^2\right] \cos\phi \exp(-j\beta z) \tag{6.23}$$

which can also be written in the form

$$\underline{E}_x(x, y, z) = E_0 \frac{x}{w_G} \exp\left[-\left(\frac{\varrho}{w_G}\right)^2\right] \exp(-j\beta z). \tag{6.24}$$

As can be seen from Fig. 6.2, which shows the radial field distribution of the LP$_{11}$ mode along the diameter $y = 0(\phi = 0)$, the field is zero on the fiber axis and has a maximum of $0.429 E_0$ at a distance $x = 0.707\,w_G$ from the fiber axis. In the azimuthal direction, the electric field changes in proportion to $\cos\phi$ (6.23).

For the LP$_{11}$ mode with the other orientation, the factor x in (6.24) must be replaced by

$$y = \varrho \sin\phi \quad . \tag{6.25}$$

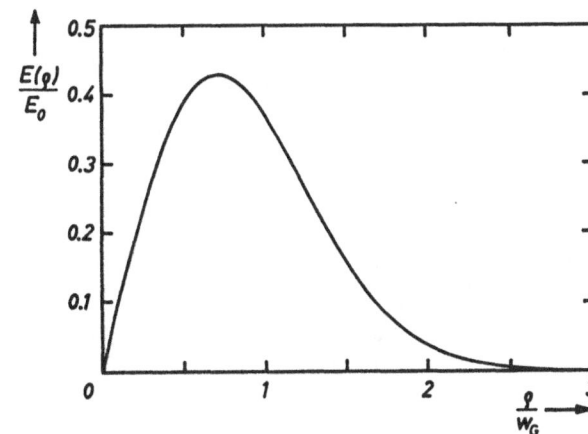

Fig. 6.2. Normalized radial amplitude distribution (6.23) for the LP$_{11}$-mode

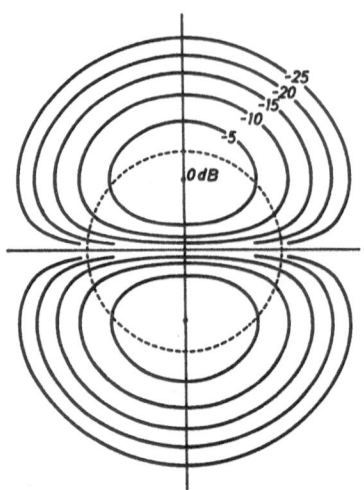

Fig. 6.3. Lines of constant field level $-5\,\mathrm{dB}$, $-10\,\mathrm{dB}\ldots$ of the LP_{11}-mode in a fiber cross section. The level in the maximum of the field is set to $0\,\mathrm{dB}$. The dashed line represents the core-cladding interface. $w_\mathrm{G}/a = 1$

The fundamental mode of a single-mode fiber is very similar to the fundamental Gaussian beam mode. Likewise, the LP_{11} mode of a fiber is similar to the second order Gaussian quasi-TEM_{11} beam mode (Sect. 4.8). Whereas the spot size $w(z)$ of the free beam in a homogeneous medium changes because of diffraction (4.5), the corresponding quantity w_G characterizing the transverse field extent of the second-order fiber mode, does not change along the fiber because the fiber core counteracts diffraction spreading.

In Fig. 6.3, the transverse field distribution of the LP_{11} mode is illustrated by plotting lines of constant field strength (Sect. 3.5.4). The reference point with zero field level is at $x = 0.707\,w_\mathrm{G}$, $y = 0$. The field level changes by $5\,\mathrm{dB}$ between neighboring lines. Whereas there is only one intensity maximum in the fiber cross section for the fundamental mode (Fig. 5.9), there are two for the LP_{11} mode.

Since the field distribution of the LP_{11} mode guided by a fiber is similar to the first higher-order Gaussian beam mode, Fig. 4.14 can also be used to illustrate the different LP_{11} modes. There are four LP_{11} modes with either a $\cos\phi$ (left column) or $\sin\phi$ dependence (right column) of the electric field and either x-polarization (upper row) or y-polarization (lower row).

For the simple step-index fiber, the radial field distributions in the core and in the cladding are described by a Bessel function $\mathrm{J}_1(U\varrho/a)$ and a modified Hankel function $\mathrm{K}_1(W\varrho/a)$ of first order, respectively. The eigenvalues in the core and cladding are obtained by solving the characteristic equation (6.9) for $l = 1$:

$$U\frac{\mathrm{J}_0(U)}{\mathrm{J}_1(U)} = -W\frac{\mathrm{K}_0(W)}{\mathrm{K}_1(W)} \quad , \tag{6.26}$$

combined with the condition (6.10).

For values of V smaller than 2.405, which is the first zero of the Bessel function $\mathrm{J}_0(x)$, there is no solution of the system of equations (6.10, 26). For V values between 2.405 and 3.832, there is only one solution and for values larger than 3.832, which is the first (nonzero) zero of the Bessel function $\mathrm{J}_1(x)$, the characteristic equation has more than one solution.

150

That solution with the smallest value of the unkown quantity U, gives the eigenvalue of the LP_{11} mode.

Using the recurrence relations for the Bessel and modified Hankel functions [Abramowitz and Stegun 1965], the characteristic equation of the LP_{1m} modes can also be written in the form (6.11)

$$U\frac{J_2(U)}{J_1(U)} = W\frac{K_2(W)}{K_1(W)} \quad . \tag{6.27}$$

The power transmitted by one of the four LP_{11} modes can be obtained by integrating the intensity over both the core and cladding cross section:

$$P = \frac{\pi n_e E_0^2 w_G^2}{16 Z_0} \quad . \tag{6.28}$$

In the two-mode wavelength region, the powers propagated by the two fundamental modes and the four LP_{11} modes depend on the launching conditions at the fiber input end (Chap. 7) and on the differential attenuations of the modes.

At the fiber output end, the fundamental fiber mode essentially transforms into a fundamental Gaussian beam mode with its waist located at the fiber end and the width of the waist w_0 being equal to the spot size w_G of the fiber mode (Sect. 8.1). Similarly, the LP_{11} mode is radiated from the fiber end as a second-order beam mode with the beam waist w_0 in (4.56) being equal to the spot size w_G of the LP_{11} mode.

Strictly speaking, an LP_{11} mode is a combination of two modes from a triplet of three exact vector modes (TE_{01}, TM_{01}, and HE_{21}) which are rigorous solutions of the wave equation and which, for small relative refractive-index differences Δ have approximately the same phase constant [Snitzer 1969].

The LP_{01} mode and the LP_{11} mode propagate with slightly different group velocities. For short wavelengths, the second-order mode is the slower one. For a certain "equalization" wavelength λ_{eq}, the group delays of both modes are equal, and for longer wavelengths the LP_{11} mode propagates faster than the fundamental mode.

For a step-index fiber, the equalization wavelength corresponds to a V value of 3 [Sakai and Kimura 1977; Cohen L.G. et al. 1980b].

In a bent fiber, the LP_{11} mode experiences additional losses due to radiation. The attenuation caused by bending is strongly related to the effective cutoff wavelength to be discussed in Sect. 6.3.3.

For step-index fibers, there exists a closed-form expression for the pure bend loss of the LP_{11} mode [Marcuse 1976a; Gambling et al. 1977d; Murakami et al. 1979]. This bend loss formula is very similar to that (5.148) for the bend loss of the fundamental fiber mode. One only has to replace the square $K_1^2(W)$ of the modified Hankel function of zero order in the denominator by the product $0.5\,K_0(W)K_2(W)$, where $K_0(W)$ and $K_2(W)$ are modified

Hankel functions of zero and second order. Of course, U and W are now the eigenvalues of the LP_{11} mode which are different from those of the fundamental mode.

As an example, for a step-index fiber with core diameter $2a = 8\,\mu m$ and relative refractive-index difference $\Delta = 0.003$, Fig. 5.27 also shows the wavelength and curvature dependence of the bend attenuation coefficient α_{b11} of the LP_{11} mode.

The wavelength dependence for the attenuation of the LP_{11} mode can be approximated by the following empirical equation [Lenahan 1983; Duff et al. 1985; Wei et al. 1986]:

$$\alpha_{11}(\lambda, R)L_r = 19.34\,\text{dB}\,X^{0.1[\lambda - \lambda_{ce}(L_r, R)]} \quad , \tag{6.29}$$

where $\lambda_{ce}(L_r, R)$ is the effective cutoff wavelength defined by (6.54) measured for a reference length L_r of fiber. The wavelengths λ and λ_{ce} have to be expressed in nanometers. The quantity X is a constant determined by the index profile and is related to the slope m of the λ_{ce} versus $\log(L/L_r)$ curve (6.51) by [Wei et al. 1986]:

$$X = 10^{10/m} \quad . \tag{6.30}$$

The reported values of m in the range of 15 nm/decade to 100 nm/decade and correspond to X values between 4.6 and 1.3.

A formula has been derived for calculating the bending loss of the LP_{11} mode in depressed-cladding fibers and has been used for analyzing the curvature dependence of the effective cutoff wavelength [Shah 1987].

The pure bend loss of the LP_{11} mode in a polarization-preserving fiber, measured by an optical heterodyne interferometer [Shibata N. et al. 1985], has been found to agree well with that calculated theoretically [Sakai and Kimura 1978a].

A formula for calculating the microbending loss of the LP_{11} mode has been published by Gambling et al. [1977d].

6.3.2 Theoretical Cutoff Wavelength

When the second-order LP_{11} mode propagates in a fiber, intermodal dispersion due to the different group delays of the fundamental and the second-order modes (differences up to several ns/km), will take place, reducing the bandwidth and the maximum possible bit rate. Moreover, for coherent sources with a small linewidth, which will be used for operation at wavelengths other than the wavelength of minimum dispersion, interference between the first- and second-order modes can cause modal noise as in multimode fibers [Heckmann 1981b] (Sect. 11.6.1). Therefore, it is very important to choose an operating wavelength for which the LP_{11} mode cannot propagate. The range of wavelengths for which a fiber supports only a single mode is thus limited at short wavelengths by the cutoff wavelength λ_c of the LP_{11} mode; this is a very important quantity to know.

For the simple step-index fiber, the theoretical normalized cutoff frequency of the LP_{11} mode is (6.18)

$$V_c = 2.405 \quad . \tag{6.31}$$

Inserting this value into (6.21), gives the theoretical cutoff wavelength of the LP_{11} mode:

$$\lambda_c = \frac{2\pi a \sqrt{n_1^2 - n_2^2}}{2.405} \quad , \tag{6.32}$$

where n_1 and n_2 are the refractive indices in the homogeneous core and cladding, respectively. For a typical numerical example, we consider a standard single-mode fiber for an operating wavelength of $1.3\,\mu m$: the core diameter is $2a = 8\,\mu m$ and the relative refractive index difference $\Delta = 0.003$. Inserting these values in (6.32), gives a theoretical cutoff wavelength of $1.18\,\mu m$.

For arbitrary refractive-index profile, the cutoff frequency of the LP_{11} mode is difficult to calculate. It can be determined by using one of the general methods of analysis. These methods, however, are designed to solve much broader problems, e.g. calculation of the field distribution and the propagation constants of the guided modes as functions of wavelength. Therefore, their use appears too time consuming and cumbersome for finding only the single-mode limit. Therefore, several methods have been proposed to calculate the cutoff frequencies directly without having to make a complete analysis.

For arbitrary index profiles, there are direct numerical methods for calculating either the cutoff frequencies of the *vector* modes, i.e. the TE_{01}, TM_{01}, or HE_{21} modes [Clarricoats and Chan 1970; Dil and Blok 1973; Biancardi and Rizzoli 1977; Hashimoto 1980; Sharma E.K. et al. 1981; Meunier J.P. et al. 1982b; Boucouvalas and Papageorgiou 1982] or those of the *scalar* modes, i.e. the LP_{11} mode [Meunier J.P. et al. 1980a, 1984a,b; Sharma A. and Ghatak 1981b; Chen 1982; Alphones 1985]. These direct numerical methods are highly involved and time consuming because of extensive computations.

On the other hand, for arbitrary index profiles, there are more simple, approximate techniques for calculating the theoretical single-mode limit [Hotate and Okoshi 1978b, 1979a; Kokubun and Iga 1980, 1982; Love 1985].

Finally, there are some techniques for special classes of refractive index profiles: step-index fibers [Snitzer 1961; Snyder 1969a], step-index fibers with a central index dip [Gambling et al. 1977c], parabolic-index (square-law truncated index) fibers [Someda and Zoboli 1975; Arnold 1977b; Sammut and Ghatak 1978; Lukowski and Kapron 1977], fibers with power-law refractive-index distributions of arbitrary exponent [Okamoto and Okoshi 1976; Love 1979], W-type fibers [Kawakami et al. 1976; Kumar et al. 1978; Yata and Ikuno 1981], and dielectric tube waveguides [Miyagi and Nishida 1977a,b]. For the step and parabolic profiles, there are analytical expressions for the normalized cutoff frequencies.

Since there are so many methods for calculating mode cutoff, only a few simple formulas will be given here.

Firstly, the normalized cutoff frequency V_c for any form of the index profile is given, to a good approximation, by [Gambling et al. 1978d, 1979b; Gambling and Matsumura 1979]

$$V_c = \frac{2.405}{(2G)^{1/2}} \quad , \tag{6.33}$$

where the guidance factor

$$G = \int_0^1 \frac{\varepsilon(R) - \varepsilon_2}{\varepsilon_1 - \varepsilon_2} R \, dR \tag{6.34}$$

reflects the integrated effect of the refractive index

$$n = \varepsilon^{1/2} \quad , \tag{6.35}$$

over the whole of the core. The symbol ε denotes the relative permittivity and the symbol

$$R = \varrho/a \quad , \tag{6.36}$$

the normalized radial coordinate.

A second simple, explicit, and accurate expression for the theoretical cutoff of the LP_{11} mode is [Love 1985]:

$$V_c = 0.6575 \left\{ \int_0^1 [1 - g(R)] R^3 \exp(-2R^2) dR \right\}^{-1/2} \quad . \tag{6.37}$$

Thus to calculate the normalized cutoff wavelength one only requires knowledge of the normalized refractive-index profile function $g(R)$ (5.6), and not of the transverse field distribution.

For a step-index fiber, the approximation (6.37) gives $V_c = 2.413$ instead of 2.405, and for a fiber with a truncated parabolic index profile 3.575 instead of the exact value of 3.518.

As a consequence of (6.33) or (6.37), both the rounding of the profile at the core-cladding boundary [Gambling and Matsumura 1978a] and the index dip on the fiber axis [Gambling et al. 1978d] of nominally step-index fibers increase the theoretical normalized cutoff frequency V_c to a value higher than 2.405.

The normalized cutoff frequency V_c of fibers with power-law index profiles (5.6, 7) as a function of the profile exponent g has been calculated and plotted by Gambling et al. [1977b]. For a parabolic (truncated square law) index profile ($g = 2$), $V_c = 3.518$. A simple approximation for V_c for power-law profiles is [Okamoto and Okoshi 1976]:

$$V_c = 2.405(1 + 2/g)^{1/2} \quad . \tag{6.38}$$

A fiber with a depressed inner cladding and a higher-index outer cladding never cuts off in the infinite cladding sense; instead, the LP_{11} mode leaks out of the core by evanescent coupling to the outer cladding [Wang C.C. et al. 1983; Tomita and Cohen 1985].

For comparing the properties of fibers with different refractive index profiles, it is useful to relate the normalized frequency V to the normalized cutoff frequency V_c of the second-order mode by defining an effective normalized frequency [Brinkmeyer 1979a,b]

$$V_e = 2.405\,V/V_c \quad . \tag{6.39}$$

By introducing the factor of 2.405 in the definition of V_e, the single-mode limit for any fiber is given by

$$V_{ec} = 2.405 \quad . \tag{6.40}$$

For fibers with an elliptical core, the four LP_{11} modes differing in orientation and polarization have slightly different effective cutoff wavelengths [Klein and Heinlein 1982; Saad 1985].

Although the theoretical cutoff frequency λ_c of the LP_{11} mode is a well-defined quantity, it is not very useful for the system designer because of several deficiencies:

1. Contrary to the implicit assumption of an infinite cladding made in the definition of the theoretical cutoff wavelength, a real fiber has a finite cladding diameter. Near cutoff, the electromagnetic field extends far into the cladding. Therefore, the properties of the LP_{11} mode are influenced by the substrate tube cladding, the primary and secondary coating, and the air around the coating. Absorption in the substrate tube cladding and the coatings will attenuate the LP_{11} mode.

2. Near cutoff, because of the large transverse field extent, the second-order mode will suffer large losses caused by macro- and microbends [Gambling et al. 1977d,1978c,f]. For example, for a step-index fiber with core diameter $2a = 8\,\mu m$ and relative refractive-index difference $\Delta = 0.003$, the theoretical cutoff wavelength calculated from (6.32) is $1.18\,\mu m$. The wavelength and curvature dependence of the bend attenuation coefficient α_{b11} of the LP_{11} mode in this fiber is represented in Fig. 5.27. For a radius of curvature of $R = 10\,cm$, which is typical for a cable, the attenuation coefficient is $10^3\,dB/km$ at $\lambda = 1.03\,\mu m$ and $10^6\,dB/km$ at $1.12\,\mu m$.

    Thus, in a factory length of cable with a length of about $1\,km$, for $\lambda > 1.03\,\mu m$, the LP_{11} mode is attenuated by more than $1000\,dB$. For the shorter fiber lengths of order $1\,m$ encountered in pigtails, cable jumpers, or repair sections, an attenuation of greater than $1000\,dB$ is obtained for $\lambda > 1.12\,\mu m$. Thus, because of the macrobending loss of the LP_{11} mode, the operating wavelength may be significantly shorter than the theoretical cutoff wavelength of $1.18\,\mu m$ without having signal degradation caused by the second-order mode. The longer the fiber, the smaller the allowed operating wavelength. Stated alternatively, the theoretical cutoff wavelength may be higher than the operating wavelength. A high cutoff wavelength in standard fibers, for operation at a wavelength of $1.3\,\mu m$ is advantageous, since bending losses at $1.55\,\mu m$ will remain small, and future use of this fiber at $1.55\,\mu m$ is not precluded.

3. Again due to the large transverse field extent of the LP_{11} mode near cutoff, the laser beam optimized for coupling maximum power in the fundamental mode will launch little power in the second-order mode (Sect. 7.2). Likewise, at a misaligned connector or splice, only a small amount of power will be coupled from the fundamental mode to the LP_{11} mode.

4. There is no method for measuring the theoretical cutoff wavelength directly (Sect. 13.8).

5. Usually, the refractive-index profile is not known very accurately. Thus, it is also difficult to determine the theoretical cutoff wavelength indirectly from the index profile by applying numerical methods.

For avoiding intermodal dispersion and modal noise due to interference between the LP_{01} and LP_{11} modes (Sect. 11.6.1), it is sufficient that at splices, at connectors, and at the fiber end the power P_{11} in the second-order mode is small compared to the power P_{01} in the fundamental mode. An extinction ratio of $10\log(P_{01}/P_{11}) = 40\,dB$ at points such as splice joints, where modal noise can be generated, is considered sufficient to avoid modal noise and intermodal dispersion [Tomita and Cohen 1985]. Because of the small launching efficiency and large attenuation of the LP_{11} mode near cutoff, this condition can also be satisfied for wavelengths shorter than the theoretical cutoff.

6.3.3 Effective Cutoff Wavelength

From the previos discussion we see that the system designer is not interested in the theoretical cutoff wavelength λ_c, but in the so-called effective cutoff wavelength λ_{ce}, which limits the region for which the fiber is "effectively" single mode. The effective cutoff wavelength is usually measured by increasing the signal wavelength in a fixed length of fiber, until the LP_{11} mode is undetectable. For wavelengths shorter than λ_{ce}, the attenuation of the higher-order modes is insufficient and modal noise can be generated which increases the bit error rate (Sect. 11.6.1).

Because of the large attenuation of the LP_{11} mode near cutoff, the experimentally determined effective cutoff wavelength λ_{ce} is always smaller than the theoretical cutoff wavelength λ_c by as much as $100-200\,nm$. Experimentally, for a fiber of length $1\,m$ and for fibers stranded in a cable of $2\,km$ length, the effective cutoff wavelengths were found to correspond to V-values of 2.8 and 3, respectively, as compared to the theoretical value of 2.405 [Gambling et al. 1977d; Kato et al. 1981].

Practical transmission systems are operated close to the effective cutoff wavelength to enhance fundamental mode confinement, but sufficiently far from cutoff such that no power is transmitted in the second-order mode.

Since the attenuation of the LP_{11} mode depends on the fiber length and the radius of curvature, the effective cutoff wavelength is not a very well defined quantity, but depends on the measuring method with which it is defined and on the fiber length and curvature used in the measuring method.

Since there are many criteria for detecting the LP_{11} mode (Sect. 13.8), numerous definitions for the effective cutoff wavelength are possible.

The CCITT [1986] provisionally defines the effective cutoff wavelength as follows:

"The cutoff wavelength is the wavelength greater than which the ratio between the total power, including launched higher-order modes, and the fundamental mode power has decreased to less than 0.1 dB in a quasi straight 2 m fiber length with one single loop of 14 cm radius."

One CCITT Reference Test Method for measuring the effective cutoff wavelength corresponding to this definition is the bending method (Sect. 13.8.1). Launching conditions must be chosen to excite both LP_{01} and LP_{11} modes. In the following analysis of the bending method for measuring λ_{ce}, the powers guided by the two fundamental modes and by the four LP_{11} modes are denoted P_{01} and P_{11}, respectively. The arguments 0 and L are used to denote the fiber input and output, respectively.

When an incoherent source excites each of the two LP_{01} modes and each of the four LP_{11} modes with equal powers (Sect. 7.3), the input power in the second-order modes is twice as large as that in the fundamental modes

$$P_{11}(0) = 2P_{01}(0) \quad , \tag{6.41}$$

and the level difference between the powers is

$$a_1 = 10 \log\left(\frac{P_{01}(0)}{P_{11}(0)}\right) = -3.01 \, \text{dB} \quad . \tag{6.42}$$

In the bending method, one first measures the total transmitted power

$$P_1 = P_{01}(L) + P_{11}(L) \quad , \tag{6.43}$$

including launched higher-order modes, as a function of wavelength at the end of a quasi straight fiber with a length of 2 m containing one single loop of 14 cm radius.

Then, by bending the fiber into at least one additional loop with a radius of 30 mm, the higher-order modes are filtered out due to their large bending loss while the LP_{01} mode attenuation is negligibly small. With the small loop, the output power equals the fundamental mode power alone:

$$P_2 = P_{01}(L) = P_{01}(0) \quad . \tag{6.44}$$

The level difference

$$\Delta a = 10 \log(P_1/P_2)$$
$$= 10 \log \left(\frac{P_{01}(L) + P_{11}(L)}{P_{01}(L)} \right) \quad , \tag{6.45}$$

between the two measured powers is calculated and plotted versus wavelength (Fig. 13.18). The effective cutoff wavelength λ_{ce} as defined by the CCITT is then the longest wavelength for which the level difference (6.45) is

$$\Delta a = 0.1 \, \text{dB} \quad . \tag{6.46}$$

At the effective cutoff wavelength λ_{ce}, the level difference a_2 at the fiber end between the power $P_{01}(L)$ transmitted by the fundamental modes and the power $P_{11}(L)$ transmitted by the higher-order modes can be calculated from (6.45) and (6.46) to be

$$a_2 = 10 \log \left(\frac{P_{01}(L)}{P_{11}(L)} \right) = 16.33 \, \text{dB} \quad . \tag{6.47}$$

Forming the level difference $a_2 - a_1$ and taking (6.44) into account gives the attenuation a_{11} of the LP_{11} mode in the 2 m fiber with a single loop of 14 cm radius:

$$a_{11} = 10 \log \left(\frac{P_{11}(0)}{P_{11}(L)} \right) = 19.34 \, \text{dB} \quad . \tag{6.48}$$

Thus the effective cutoff wavelength λ_{ce} corresponds to that wavelength for which the attenuation of the LP_{11} mode in the test fiber is 19.34 dB [Shah 1987].

The fiber loop with a radius of 14 cm has a length of

$$L_b = 2\pi \, 0.14 \, \text{m} = 0.88 \, \text{m} \quad . \tag{6.49}$$

Since the CCITT requires the remainder of the fiber to be "quasi straight", the LP_{11} mode is attenuated only in the fiber loop with a bend attenuation constant

$$\alpha_{b11} = 19.34 \, \text{dB}/0.88 \, \text{m} = 22 \, \text{dB/m} = 22\,000 \, \text{dB/km} \quad . \tag{6.50}$$

For a given operating wavelength, a compromise must be found for the effective cutoff wavelength. On one hand, it must be small enough to avoid modal noise due to interference with the LP_{11} mode. On the other hand, it should be as large as possible to limit bending losses at long wavelengths, which might be used in the same fiber at a later time. For single-mode fibers optimized for operation in the 1300 nm window, the CCITT provisionally recommends a value of λ_{ce} in the wavelength range from 1100–1280 nm.

Presently, the fiber manufacturers specify the effective cutoff wavelength $\lambda_{ce}(0.88\,\mathrm{m},\ 14\,\mathrm{cm})$ as recommended by the CCITT for a fiber of length 2 m containing a single loop with a radius of curvature of 14 cm. In practical single-mode transmission systems, there will be other values for the fiber length L and the radius of curvature R of the fiber axis. Therefore, for the system designer, it is important to know the length and curvature dependence of the effective cutoff wavelength $\lambda_{ce}(L, R)$ as defined by a level difference of 0.1 dB between the total power and the fundamental mode power.

Several authors have found experimentally that the effective cutoff wavelength decreases as a logarithmic function of length [Murakami et al. 1979; Murayama et al. 1983; Wang C.C. et al. 1983; Nijnuis and van Leeuwen 1984; Kitayama Y. and Tanaka 1984, 1985; Anderson W.T. and Lenahan 1984; Kitayama K. et al. 1984; Franzen 1985b; Müller 1985; McMillan and Robertson 1985]:

$$\lambda_{ce} = \lambda_{cer} - m \log(L/L_r) \quad , \tag{6.51}$$

where λ_{cer} is the effective cutoff wavelength measured at a fiber reference length L_r and m is the slope of the λ_{ce} versus $\log(L/L_r)$ curve.

The length dependence of the effective cutoff wavelength described by (6.51) can be derived from the approximate expression (6.29) for the wavelength dependence of the attenuation of the LP_{11} mode [Wei et al. 1986] with (6.30)

$$m = 10/\log X \quad . \tag{6.52}$$

Depending on the particular index profile type and the particular coating and cable structures, different values for the slope m in the range 15 nm/decade to 100 nm/decade have been found.

Beyond a fiber length of approximately 4 km, the effective cutoff wavelength has been found to become constant [Kitayama K. 1984]. This effect can be explained by mode coupling between the fundamental mode and the LP_{11} mode. Microbends couple the LP_{01} mode to the LP_{11} mode, leading to a power exchange. The evolution of the powers P_{01} of the fundamental and P_{11} of the second-order modes has been investigated for different ratios P_{11}/P_{01} at the fiber input [Cancellierei and Orfei 1985].

As has been stated before, the effective cutoff wavelength λ_{ce} also depends on the radius of curvature R of the fiber axis. The change of λ_{ce} with R has been investigated experimentally by several authors [Lazay 1980; Shah 1984; Nijnuis and van Leeuwen 1984; Kitayama Y. and Tanaka 1984; Franzen 1985b; Müller 1985].

For *matched*-cladding fibers, the effective cutoff wavelength decreases with increasing curvature $\kappa = 1/R$. The experimental data can be approximated by a function

$$\lambda_{ce} = \lambda_{ce0} - A(1/R) + B(1/R)^2 \quad , \tag{6.53}$$

where λ_{ce0} is the effective cutoff wavelength of the straight fiber ($1/R = 0$) and where the coefficients A and B characterize the curvature sensitivity of λ_{ce}. For different fibers, the coefficient A has values in the range between about $300\,\text{nm/cm}^{-1}$ and $1400\,\text{nm/cm}^{-1}$.

For *depressed*-cladding fibers, totally different results are reported. Some authors [Nijnuis and van Leeuwen 1984; Franzen 1985b] find that λ_{ce} decreases with increasing curvature, whereas others [Shah 1984; Müller 1985] first find an increase of λ_{ce}, then a maximum at a certain curvature, and finally a decrease with increasing curvature. Theoretically, in contrast to matched-cladding fibers, the effective cutoff wavelength of depressed-cladding fibers is curvature insensitive for bend radii (as low as 5 cm) of practical interest [Shah 1987].

The different curvature dependence of the effective cutoff wavelength in matched-cladding and depressed-cladding fibers can be explained by the fact that there are different loss mechanisms in the two fiber types: in matched-cladding fibers, the LP_{11} mode is attenuated by radiation from bends, and in depressed-cladding fibers by tunneling of power through the depressed cladding [Nijnuis and van Leeuwen 1984; Anderson W.T. and Lenahan 1984].

Because of the different length and curvature dependence of λ_{ce} in different fibers, it is generally not sufficient to characterize the cutoff behavior of the second-order mode by a single number, namely the value of λ_{ce} for a particular combination of fiber length and radius of curvature, e.g. 0.88 m and 14 cm. The true effective cutoff wavelength may differ considerably from the value λ_{ce} determined by the bending method at a short fiber section. For the system designer, it would be helpful to know the effective cutoff wavelength for a number of combinations of fiber length and radius of curvature.

The effective cutoff wavelength is closely related to the bend loss of the LP_{11} mode. As has been shown before, the effective cutoff wavelength λ_{ce} as defined by the CCITTT corresponds to that wavelength, for which the ratio of total power to fundamental power is 0.1 dB, or, alternatively, for which the attenuation of the LP_{11} mode is 19.34 dB. The loss a_{11} of the LP_{11} mode is the product of its attenuation coefficient $\alpha_{11}(\lambda, R)$, which depends on the wavelength λ and the radius of curvature R, and the fiber length L.

The cutoff condition for an arbitrary fiber length L and radius of curvature of the fiber axis R can thus be written in the form [Kitayama Y. and Tanaka 1985]:

$$a_{11} = \alpha_{11}(\lambda_{ce}, R)L = 19.34\,\text{dB} \quad . \tag{6.54}$$

This represents an implicit equation for the effective cutoff wavelength $\lambda_{ce}(L, R)$ as a function of fiber length and radius of curvature. For the standard test fiber with a length of 2 m with one single loop of 14 cm radius, (6.54) gives the CCITT cutoff wavelength described before.

It can easily be shown that a description of cutoff in terms of the function $\alpha_{11}(\lambda, R)$ is completely equivalent to the description using the function

$\lambda_{ce}(L, R)$ [Nijnuis and van Leeuwen 1984; Kitayama Y. and Tanaka 1985; Wei et al. 1986].

Methods for selectively measuring the attenuation of the LP_{11} mode are described in Sect. 13.3.6. Papers analyzing the bend loss of the LP_{11} mode have been listed at the end of Sect. 6.3.1.

Other proposals have been made for defining the effective system cutoff wavelength in terms of the attenuation of the LP_{11} mode:

$$a_{11}(\lambda_{ce}, L) = 3\,\text{dB} \tag{6.55}$$

[Nijnuis and van Leeuwen 1984],

$$\alpha_{11}(\lambda_{ce}, R) = 5\,\text{dB/m} \tag{6.56}$$

[Tomita and Cohen 1985], or

$$\alpha_{11}(\lambda_{ce}, R) = 10\,\text{dB/m} \tag{6.57}$$

[van Leuwen and Nijnuis 1984]. In the opinion of the present author, these proposals are not as logical as the 19.34 dB definition (6.54) corresponding to the CCITT cutoff wavelength.

For the system designer, it would be very advantageous, if the fiber manufacturer were to measure and specify the attenuation coefficient $\alpha_{11}(\lambda, R)$ of the LP_{11} mode as a function of wavelength and radius of curvature. Then, by inserting the calculated attenuation a_{11} into a formula for the modal noise to signal ratio [Duff et al. 1985], he could check whether there might be problems with modal noise. However, determination of the complete functions $\alpha_{11}(\lambda, R)$ for every reel of fiber to be used is, of course, not realistic. But it would be helpful to know the bend loss of the LP_{11} mode for a number of combinations of wavelengths and radii, which would have to be standardized.

Because of uncertainties in most cables and systems, the creation of a general prescription relating system modal noise to a λ_{ce} requirement on a 2 m fiber is difficult, if not impossible. Instead, a generic description of fiber cutoff has been proposed specifying the LP_{11} attenuation which completely determines the fiber's sensitivity to modal noise [Sears et al. 1986a]. At present, the CCITT is discussing recommendations to specify the LP_{11} attenuation instead of the effective cutoff wavelength.

In a cable, the loss of the LP_{11} mode varies with temperature, and thus the effective cutoff wavelength is unstable.

6.4 Leaky Modes

Waveguide modes which continuously lose power by radiation into the surrounding medium are called leaky modes. Since any fiber mode continuously loses power by Rayleigh scattering (Sect. 5.9.2), which is a special kind of radiation, all fiber modes are, in principle, leaky modes. Usually, however, leakage due to Rayleigh scattering is not considered sufficient to name a mode leaky.

In a leaky mode, the amplitude *decreases* longitudinally, even if the materials used for making the dielectric waveguide are lossless. Since the squares of the complex wave numbers in the longitudinal and transverse directions must sum to the square of the wave number k_n of the material (4.14), the wave amplitude must *increase* in the transverse direction. Since the cross section outside of the fiber is unbounded, this implies that the leaky wave field must increase transversely to infinity and must transmit infinite power, yielding an unphysical result. However, after a certain distance ϱ from the axis, related to the distance z from the location of the source, the field rapidly drops off. A leaky wave, with its peculiar behavior, is thus defined only in a cone shaped region around the fiber axis with the apex of the cone at the source [Oliner 1984].

A first example of a leaky wave is the fundamental mode in a fiber with a depressed inner cladding and higher-index outer cladding: For long wavelengths, the phase constant β becomes smaller than the wavenumber $n_2 k$ in the outer cladding (6.22). When the evanescent field in the inner depressed cladding extends into the outer cladding, power leaks through the depressed cladding into the surrounding outer cladding [Marcuse 1982b; Cohen L.G. et al. 1982c].

One method for manufacturing directional couplers is to polish the fiber cladding down to near the core (Sect. 12.2.11). When a high-index fluid or prism is put onto the polished face, power leaks from the core into the high-index material making the fundamental fiber mode a leaky wave.

Another important class of leaky modes is found in multimode fibers. Waveguide theory states that a fiber mode will not be guided for wavelengths larger than its cutoff wavelength. On the other hand, a mode in a multimode step-index fiber can be decomposed into a family of light rays, a so called ray congruence [CSELT 1980]. This ray congruence in geometric optics corresponds to the mode in wave theory.

The angles determining the direction of propagation of one member of the family of rays can be calculated from the parameters of the corresponding mode [Snyder and Love 1983]. Each ray of the congruence is incident at the same angle on the interface between core and cladding. Interestingly, for wavelengths slightly larger than the theoretical cutoff wavelength, i.e. in the cutoff region, it is found that this angle of incidence (measured against the interface) is smaller than the critical angle necessary for total internal reflection. Thence, arguing in terms of geometric optics, one concludes that the ray congruence must be guided without losses by multiple total internal reflection.

Thus, for wavelengths somewhat larger than the theoretical cutoff wavelength, wave theory and geometric optics disagree [Snyder and Mitchell 1973; Snyder et al. 1974]: the former predicts that the mode is not guided, whereas the latter predicts that the corresponding ray congruence is guided without attenuation. The physical explanation for this discrepancy is that the interface between core and cladding is not a plane, as assumed in the classical theory of total internal reflection, but is in fact curved. For a curved boundary, Fresnel's law for the reflection at a plane interface between two transparent media must be generalized [Snyder 1974; Heyman 1984]. The most important result of the analysis is that the incident wave is only partially reflected at the curved boundary, even if the angle of incidence is smaller than the critical angle for total internal reflection. Therefore, at each reflection of the ray, a portion of the incident ray power is transmitted into the cladding, causing leakage losses.

If the wavelength is only slightly larger than the cutoff wavelength and if the azimuthal mode number l of the corresponding LP_{lm} mode is large, the leakage attenuation is small. Therefore, these high-order leaky modes [Snyder and Love 1983] must be taken into account, e.g. when measuring the near-field distribution at the output end of a multimode fiber in order to determine the refractive-index profile [Petermann 1977b].

For guided modes, the phase constant β is larger than the wave number $n_2 k$ in the cladding and smaller than the maximum wave number $n_1 k$ in the core (5.48). Compared with this, the phase constant of leaky modes is smaller and lies in the interval

$$\sqrt{n_2^2 k^2 - \nu^2/a^2} < \beta < n_2 k \quad , \tag{6.58}$$

where ν is the azimuthal mode number of the exact vector mode (Sect. 6.1) and is either $l+1$ or $l-1$. This relation only holds for fibers with a large value of the normalized frequency V, i.e. for multimode fibers.

For single-mode fibers, the operating wavelength is very large compared to the cutoff wavelengths of most higher-order modes. Therefore, leakage attenuation of these modes is large and the influence of these high-order fiber modes need not be considered. But in a single-mode fiber, the operating wavelength is not much larger than the cutoff wavelength of the LP_{11} mode. This mode then propagates as a leaky mode when the carrier wavelength is longer than its cutoff wavelength. According to Snyder and Mitchell's theory [1974], the transmission loss of the leaky mode, which propagates at a wavelength longer than λ_c by 10 nm, is 1 dB/mm [Murakami et al. 1979]. Because of this very large attenuation, and because of very large bend losses, this leaky LP_{11} mode will not usually be observed.

Yet another example of a leaky mode is the fundamental mode propagating along a bend with constant curvature (Sect. 5.9.7).

Note that leaky modes do not obey the power orthogonality condition (7.7) [Snyder and Love 1983]. However, by truncating the field of the leaky modes, these modes become approximately power orthogonal [Sammut and Snyder 1976a,b].

6.5 Cladding Modes

In general, the cladding is surrounded by a medium with a lower refractive index, either a plastic coating or air. At the interface between the cladding and this lower-index medium, total internal reflection can take place. Thus a single-mode fiber can also guide so-called cladding modes, which correspond to rays traversing the core and the cladding. If the angle between the direction of propagation and the fiber axis is small, they are guided by multiple total internal reflections at the outer boundary of the fiber.

The phase constant of the cladding modes is smaller than that of the guided core modes. If the lower-index medium is air, the interval for β is:

$$k < \beta < n_2 k \quad . \tag{6.59}$$

Strictly speaking, for the upper boundary of the interval (6.59) one has to take the lower boundary for the β-values of the leaky modes (6.58). But, since we are interested in single-mode fibers, for which higher-order leaky modes can be neglected, the lower boundary $n_2 k$ for guided modes has been inserted.

Cladding modes are inevitably exited by the source at the fiber input end, at misaligned joints, at bends and other discontinuities in the fiber. These cladding modes are undesirable for two reasons [Marcuse 1971b]: They couple some of their power back into the core causing problems of delay distortion of the guided mode and in addition, they may reach the end of the fiber and enter the detector giving rise to further unwanted delayed signals.

Cladding modes are usually more strongly attenuated than the fundamental mode because of the losses in the outer cladding (made from the substrate tube for the inside CVD fabrication techniques or the sleeve tube for the VAD technique) and in the coating [Marcuse 1971b; Kuhn 1975a]. For long fibers, cladding modes will thus not be observed at the fiber end, i.e. there will not be any light power in the cladding apart from the evanescent field of the fundamental mode.

On the other hand, with measuring methods using short sections of fibers, cladding mode strippers (Sect. 12.2.8) have to be used in order to avoid errors caused by light guided by the outer fiber boundary instead of the fiber core.

Since the cladding diameter is large compared to the wavelength, the methods of geometric optics may be applied to analyze the propagation of the family of rays that correspond to a cladding mode.

When considering the behavior of cladding modes, most authors assume that the core region may be neglected, i.e. they set $n(\varrho) = n_2$ also for $\varrho < a$. The range of validity of this approximation has been examined by Clarricoats and Chan [1973].

Because of the large step in refractive index at the boundary between the cladding (or primary coating) and the surrounding air, it is not possible, in general, to apply the weakly guiding approximation [Snyder 1969a; Gloge 1971a] for calculating the fields of the cladding modes. However, when the phase constant β of the cladding modes remains close to the cladding wave number $n_2 k$, the concept of scalar LP modes can be extended to cladding modes [Vassallo 1983].

6.6 Radiation Modes

The fiber discontinuities mentioned above will also generate rays at larger angles with respect to the fiber axis. At the interface between the cladding and the lower index medium, these rays will not experience total internal reflection, but partial reflection and refraction. Thus these rays are not guided by the fiber and contribute a negligible amount to the total electromagnetic field except near the discontinuity. The modes corresponding to these rays are called radiation modes. If the lower index medium is air, the radiation modes have phase constants between zero and the vacuum wave number:

$$0 < \beta < k \quad . \tag{6.60}$$

The guided modes and cladding modes have discrete values of the phase constant β, which are determined by the index profile and the operating wavelength. In contrast, the radiation modes can have any β value in the interval (6.60). The electromagnetic radiation field must therefore be described as an integral over a continuous spectrum of radiation or unbound modes. The radiation field includes all the energy that leaves the core and radiates into the surroundings of the fiber.

6.7 Dual Mode Fiber

Initially, the small value of the spot size of the fundamental mode in a single-mode fiber was thought to be a substantial drawback, since splicing and connectorizing were believed to be difficult. In order to relieve splice tolerances, Sakai and Kimura [1977] proposed to use fibers with a larger core radius than that allowed by the single-mode condition (6.14), so that besides the LP_{01} mode also the LP_{11} modes are guided. In order to avoid intermodal dispersion they proposed to use a particular value for the normalized frequency V, for which the time delays of the fundamental and the second-order modes have the same value. Several papers on the design of these dual-mode fibers have been published [Sakai et al. 1978; Kitayama et al. 1981; Kato et al. 1982a; Cvijetic 1984] as well as experimental investigations [Cohen L.G. et al. 1980b; Kitayama K. et al. 1979, 1980, 1981, 1982].

For step-index fibers, zero intermodal dispersion occurs at an equalization V-number of 3 [Sakai and Kimura 1977; Cohen et al. 1980b]. The equalization V-number and, therefore, the core diameter can be increased by grading the index profile.

An obvious drawback is the narrow operating range which limits this dual-mode fiber to a predetermined wavelength of operation [Gloge 1979b]. It is then impossible to later upgrade a dual-mode fiber link simply using the methods of wavelength division multiplexing (WDM), or frequency division multiplexing (FDM). Moreover, for coherent sources, modal noise effects caused by the interference between the LP_{01} and the LP_{11} modes are likely to occur [Heckmann 1981b; Thyagarajan et al. 1982].

7. Launching of Modes

In this chapter, we will discuss methods for launching guided modes in dielectric optical waveguides. We are, of course, mainly interested in exciting the LP_{01} mode, but e.g. for measuring purposes it is also important to know how to excite a LP_{11} mode.

At this point, it is very important to note that the methods of geometric optics may not be applied when discussing launching problems for single-mode fibers.

For instance, the numerical aperture A_N for a multimode fiber is the sine of the maximum angle of incidence in air with respect to the fiber axis, for which an incident ray will produce a guided ray in the fiber. For single-mode fibers, the fundamental mode can be excited by a plane wave even if the angle of incidence is larger than the numerical aperture $(n_1^2 - n_2^2)^{1/2}$ calculated from the core and cladding indices n_1 and n_2 (5.10). Therefore, the concept of a ray cannot be used, and the more complicated methods of wave optics have to be applied to analyze launching problems.

There are several methods to excite a guided mode. The most important is illumination of the front end with a free wave, e.g. with the fundamental Gaussian beam mode. One is interested, of course, in maximizing the power in the guided mode relative to the power transmitted by the incident beam, i.e. in maximizing the launching efficiency. The launching efficiency will be discussed both for coherent (Sect. 7.2) and incoherent illumination (Sect. 7.3) of the fiber input end.

Launching by butt-jointing two fiber ends is similar to launching by a Gaussian beam. Fiber joints will be described in Chap. 9 after we have studied radiation from the fiber end (Sect. 8.1).

Another method for launching a guided wave consists in reducing the distance between two parallel fiber cores, until the evanescent fields in the claddings overlap. Power then becomes transversely coupled from one fiber into the other (Sect. 7.4).

In a uniform optical dielectric waveguide, no power is coupled from one guided mode to another. If e.g. at the beginning of an ideal fiber, which can guide both the LP_{01} and the LP_{11} modes, only the fundamental mode is being excited, no power will be found in the second-order modes at the end of the fiber. In reality, due to fiber discontinuities like microbends etc., there is a slight unwanted coupling between the guided modes. By intentionally increasing the disturbances, it is possible to excite a guided mode by coupling power into it from another guided mode.

In order to excite guided modes, one can also illuminate a fiber from the side and use the light produced in the core by fluorescence. Mode launching by mode coupling and by fluorescence will be discussed in Sect. 7.5.

Finally, Sect. 7.6 describes methods for selectively exciting a particular mode.

7.1 Mode Orthogonality

For calculating the launching efficiency, use is made of the orthogonality of the modes. Thus in this section, mode orthogonality is explained in simple terms.

If a dual-mode fiber transmits only the fundamental mode or only the LP_{11} mode, the transmitted power can be calculated approximately from (5.37) or (6.28), respectively. We now address the question of the total power transmitted when the fiber transmits both LP_{01} and LP_{11} modes simultaneously. We assume initially that both waves are launched by the same coherent light source.

Since the waves are assumed to be coherent, for calculating the transmitted power, one must superpose the fields of the two modes (Sect. 3.7.2). The Poynting vector is then calculated as the cross product (3.16) of the total electric and magnetic fields. Finally, the total power transmitted by the fiber is obtained by integrating the power density, which is one half of the real part of the complex Poynting vector, over the *infinite* fiber cross section (3.17).

If we denote the complex electric and magnetic field vectors (Sect. 3.4) of the LP_{01} and LP_{11} modes by $\underline{E}_0$, $\underline{H}_0$, $\underline{E}_1$, and $\underline{H}_1$, respectively, the total electric and magnetic fields are

$$\underline{E} = \underline{E}_0 + \underline{E}_1 \quad , \tag{7.1}$$

$$\underline{H} = \underline{H}_0 + \underline{H}_1 \quad . \tag{7.2}$$

Inserting the total fields into the formula (3.16) for the time averaged Poynting vector we obtain

$$\begin{aligned} \underline{S} = {} & \tfrac{1}{2}\mathrm{Re}\{\underline{E}_0 \times \underline{H}_0^*\} + \tfrac{1}{2}\mathrm{Re}\{\underline{E}_1 \times \underline{H}_1^*\} \\ & + \tfrac{1}{2}\mathrm{Re}\{\underline{E}_0 \times \underline{H}_1^*\} + \tfrac{1}{2}\mathrm{Re}\{\underline{E}_1 \times \underline{H}_0^*\} \quad . \end{aligned} \tag{7.3}$$

The asterisk denotes the complex conjugate. The first and second terms on the right hand side are the Poynting's vectors $\underline{S}_0$ and $\underline{S}_1$ of the fundamental and the second-order mode, respectively. The third and fourth terms on the right-hand side can be combined to form the interference term known from (3.18), which depends on the phase difference $\Delta\phi$ between the fields of the LP_{01} and LP_{11} modes at the point of observation. Because of the interference term, the intensity distribution over a fiber cross section will change when the phase difference is changed. One observes a "speckle pattern" formed by the LP_{01} and LP_{11} modes.

For convenience, we introduce the symbols S_0, S_1, S_{01}, and S_{10} for the four cross products on the right-hand side of (7.3). First, we consider the power P_A transmitted through an arbitrary part A of the fiber cross section. It can be obtained by integrating the Poynting vector over the area A and is given by

$$P_A = \int_A S_0 dA + \int_A S_1 dA + \int_A S_{01} dA + \int_A S_{10} dA \quad . \tag{7.4}$$

The first two integrals are the powers transmitted through the area A by the LP_{01} or the LP_{11} mode alone. Because of the third and fourth integrals, the power P_A is different from the sum of the powers of the two modes and for a finite area A, it will depend on the relative phase angle of the two modes. Thus, the law of superposition does not hold for the powers transmitted through an arbitrary part of the fiber cross section!

If one takes for the area of integration in (7.4) the infinite cross section A_∞ of the fiber, the power calculated is the total power P transmitted. For this choice of the area A, the third and the fourth integrals are equal to zero

$$\int_{A_\infty} S_{01} dA = 0.5 \int_{A_\infty} \text{Re}(\underline{E}_0 \times \underline{H}_1^*) dA = 0 \quad , \tag{7.5}$$

$$\int_{A_\infty} S_{10} dA = 0.5 \int_{A_\infty} \text{Re}(\underline{E}_1 \times \underline{H}_0^*) dA = 0 \quad . \tag{7.6}$$

Therefore, the total power is simply the sum of the power of the two individual modes. This is not a trivial statement, since the law of superposition is valid for amplitude quantities such as current, voltage, electric and magnetic field, but not in general for quantities like power, power density, or energy.

Since the integrals over the imaginary parts of the cross products are also zero, we have:

$$\int_{A_\infty} (\underline{E}_0 \times \underline{H}_1^*) dA = \int_{A_\infty} (\underline{E}_1 \times \underline{H}_0^*) dA = 0 \quad . \tag{7.7}$$

Because of these relations, the LP_{01} and LP_{11} modes are said to be orthogonal. The equations (7.7) are called orthogonality relations.

Our consideration for the special case of the LP_{01} and LP_{11} modes can be generalized: any guided, cladding, or radiation mode is orthogonal to any other guided, cladding, or radiation mode. This means that if $\underline{E}_0$ is the electric field vector of an arbitrary mode and $\underline{H}_1$ is the magnetic field vector of a different mode, the integral (7.7) over the total cross section is zero.

The derivation of these orthogonality relations for dielectric waveguides can be found in a number of papers and books [Goubau 1952; Bresler et al. 1958; Collin 1960; Snyder 1969b; Marcuse 1974, 1982a; Kogelnik 1975; Sammut et al. 1975; Unger 1977a; Nyquist et al 1981; Vassallo 1981; Snyder and Love 1983; Hardy and Streifer 1985; Qian and Huang 1986a].

Because of orthogonality, the *total* power transmitted by a fiber is simply the sum of the powers transmitted by the individual modes. Note, however, that this applies only for the powers transmitted through the *infinite* cross section A_∞. For a finite area A, the law of superposition does not hold for the powers!

The orthogonality relations are very useful for deriving formulae for the launching efficiency at the fiber input end (Sect. 7.2), or for the coupling efficiency of fiber joints (Chap. 9).

Up to now, we have made the assumption of a sinusoidal time dependence of the electromagnetic fields. The wavelength spectrum of this radiation consists of a single line of zero width. For such a monochromatic wave, there is a fixed phase relationship between the fields of the LP_{01} and LP_{11} modes derived from the same optical source, i.e. the two waves are coherent (Sect. 3.7.2) and will interfere for arbitrary fiber lengths.

Now we consider the case in which the LP_{01} and LP_{11} modes are launched from a quasi-monochromatic source. The two modes have different group velocities and therefore different specific group delays τ_{g01} and τ_{g11} (Sect. 5.7). A fiber of length L will introduce a time delay Δt_g between the modes of

$$\Delta t_g = (\tau_{g01} - \tau_{g11})L \quad . \tag{7.8}$$

Thus the fiber length determines whether the modes behave coherently or incoherently. For short fiber lengths, the delay time Δt_g is very small compared to the coherence time t_c (3.35) of the source. Since t_c is the average duration of a wave train (Sect. 3.7.6), waves derived from the same wave train will superimpose. Because each wave train in itself is coherent, constructive and destructive interference between the LP_{01} and LP_{11} mode fields, so-called speckle, will be observed.

On the other hand, for long fiber lengths, the time delay Δt_g is larger than the coherence time t_c, and waves derived from different wave trains will superimpose at the fiber end. Because the wave trains are emitted randomly, there is no fixed phase relationship between the LP_{01} and the LP_{11} mode, i.e. the phase difference between the two modes fluctuates randomly with time. The time average of the interference term in (3.18) will be zero, i.e. interference effects cannot be observed and the resulting intensity is simply the sum of the intensities of the LP_{01} and LP_{11} modes. The same holds for the powers transmitted through a finite area A, and of course for the powers transmitted through the infinite waveguide cross section A_∞.

7.2 Launching by Coherent Illumination of the Fiber Front End

7.2.1 Overlap Integral

Consider a spatially coherent wave (Sect. 3.7.5) incident on the fiber front face, which is assumed to be situated in the plane $z = 0$. We are interested in launching either the LP_{01} or the LP_{11} mode in the fiber. For simplicity, we first assume the medium in front of the fiber to be an immersion fluid with a refractive index which equals the cladding index of the fiber. The wave reflected from the front face of the fiber is then very weak and can be neglected. Later, the losses caused by reflection can be taken into account (7.15).

The complex electric and magnetic fields of the incident wave will be denoted $\underline{E}_i$ and $\underline{H}_i$. The electromagnetic field produced in the fiber is the superposition of the fields $\underline{E}_{lm}$, $\underline{H}_{lm}$ of the LP_{lm} wave ($l = 0$ or 1, $m = 1$) which we want to launch and a residual field consisting of cladding modes, radiation modes and possibly other guided modes. Usually, one is only interested in the power P_{lm} guided by the desired mode relative to the power P_i transmitted by the incident wave, i.e. the launching efficiency η defined by

$$\eta = P_{lm}/P_i \quad . \tag{7.9}$$

The launching loss a in decibels is

$$a = -10 \log \eta \quad . \tag{7.10}$$

A formula for the launching efficieny can be derived as follows [Goubau 1952; Jones 1965; Snyder 1966, 1969b; Marcuse 1970b, 1982a; Kapany et al. 1970; Ulrich 1971; Clarricoats and Sharpe 1972; Snyder 1974; Snyder and Mitchell 1974; Clarricoats 1976; Barrell and Pask 1979; Saijonmaa et al. 1980]. The total electric field in the fiber is expressed as a sum of the field of the desired mode with its amplitude coefficient still unknown, and a residual field consisting of other guided, cladding, and radiation modes. In other words, the input field is expanded in terms of the guided and unguided modes of the fiber, which form a set of orthogonal functions.

As is well known from electromagnetic field theory [Stratton 1941], the tangential component of the electric field must be continuous at the boundary between two regions, taken here as the infinite transverse plane $z = 0$ through the fiber front face. Since the reflected wave can be neglected, the field for $z = -0$, i.e. just to the left of the boundary plane, can be taken to equal the field of the incident wave. For $z = +0$, i.e. just inside the fiber, we have the superposition of the desired mode and the residual field. Equating the tangential components of the electric fields at $z = -0$ and $z = +0$ gives an equation, which besides the unknown amplitude of the wanted mode, unfortunately includes the unknown amplitudes of the other guided, cladding, and radiation modes.

But at this point, one can make use of the orthogonality relations (7.7) to get rid of the unknown amplitudes of the residual field: we cross multiply the equation derived by the complex conjugate of the magnetic field vector of the desired mode and integrate the products on both sides over the entire cross-sectional plane $z = 0$. Because of the orthogonality between the desired LP_{lm} mode and the other modes, the integral containing the residual field will be zero. Thus we can now determine the amplitude of the LP_{lm} mode.

Mathematically, the method used corresponds to the well-known technique for determining the coefficients in a Fourier series using the orthogonality of the cosine and sine functions.

Having determined the amplitude of the LP_{01} or LP_{11} mode, its power then can be calculated using (5.37) or (6.28). The power P_i transmitted by the incident wave can be obtained by integrating its Poynting vector over the plane $z = -0$ (3.17). Inserting P_{lm} and P_i in the definition (7.9) for the launching efficiency, one finally obtains the very important equation:

$$\eta = \frac{\left| \int\limits_{A_\infty} (\underline{E}_i \times \underline{H}_{lm}^*) dA \right|^2}{\mathrm{Re}\left(\int\limits_{A_\infty} (\underline{E}_i \times \underline{H}_i^*) dA \right) \int\limits_{A_\infty} (\underline{E}_{lm} \times \underline{H}_{lm}^*) dA} . \tag{7.11}$$

The integrals in the denominator are easily seen to represent the powers of the incident wave and the guided LP_{lm} mode for arbitrarily chosen amplitude factors. Because the tangential components of the electric and magnetic fields are in phase for a guided mode, the second integral in the denominator is real and the symbol "Re" denoting "real part of" has been omitted. For both the incident and the launched waves, arbitrary values for the field amplitudes can be taken in (7.11), since the electric and magnetic fields are proportional to the amplitude, and since the second power of the field amplitude is included both in the numerator and in the denominator of (7.11).

The integrand in the numerator is essentially the product of the local fields of the incident wave and the wave to be excited taken in the coupling plane $z = 0$. The product is only different from zero, if at the point considered both fields are different from zero, i.e. if the fields overlap. For this reason, the integral in the numerator is usually called "overlap integral". Correspondingly, this method for calculating the launching efficiency is called the "overlap integral method".

Because of the law of energy conservation, the power in the excited mode cannot be larger than the power transmitted by the incident wave. The maximum value of the launching efficiency is thus $\eta = 1$. This optimum launching efficiency is obtained if in the coupling plane $z = 0$, the incident field equals the field of the LP_{lm} mode to be excited:

$$\underline{E}_i = \underline{E}_{lm} , \tag{7.12}$$

$$\underline{H}_i = \underline{H}_{lm} , \tag{7.13}$$

as can be seen by inserting (7.12, 13) into (7.11). For optimum coupling efficiency, the incident field must be "matched" to the field of the required mode.

The formula (7.11) for the launching efficiency can be somewhat simplified: first, the arbitrary amplitudes of the incident wave and the wave to be launched can be chosen in such a way that both power integrals in the denominator equal unity. The field distributions of the two modes are then said to be normalized. In the overlap integral in the numerator, because of the properties of a scalar triple product, we may replace the total field vectors by the transverse vector components $\underline{E}_{it}$, $\underline{H}_{it}$, $\underline{E}_{lmt}$, and $\underline{H}_{lmt}$ which are parallel to the plane $z = 0$. Moreover, for the guided mode, the transverse vector component of the magnetic field can be expressed by the cross product of the unit vector in the z-direction and the transverse vector component of the electric field divided by the wave impedance (5.29). The resulting simplified expression for the launching efficiency is

$$\eta = \left| \frac{n_2}{Z_0} \int\limits_{A_\infty} \underline{E}_{it} \cdot \underline{E}_{lmt}^* \, dA \right|^2 \; . \tag{7.14}$$

In this formula, the cross product of the total electric field of the incident wave and the magnetic field of the wave to be launched has been replaced by the simpler dot product of the transverse vector components of the electric fields of the two waves.

This expression can be interpreted as follows: each elemental area dA of the fiber input face produces a part of the complex amplitude of the guided wave proportional to the dot product $\underline{E}_{it}\underline{E}_{lmt}^*$ in the area dA. The total amplitude is the sum of the complex contributions of all elemental areas. The launched power is proportional to the squared modulus of the total amplitude.

As has been shown above, the launching efficiency is maximum if the incident field is equal to the field to be launched (7.12, 13). Any deviation from this condition causes the launching efficiency to be smaller than 100%. The power which is not coupled into the required mode will be found in undesired cladding modes, radiation modes, and perhaps in other guided modes. From a practical point of view, the residual part of the total field produced by the incident wave represents a loss of power.

Several simple kinds of field mismatch can be distinguished:

1. "amplitude mismatch", when the amplitude distributions of the incident and the mode field are different in the cross section.
2. "phase" or "wavefront mismatch", when the phase distributions differ. Then the dot products for different elemental areas will not add up in phase, leading to a smaller amplitude for the guided mode.
3. "polarization mismatch", when the directions of the electric field vectors of the launching wave and the launched wave do not coincide at each point of the total waveguide cross section. Then, because of the finite angle between the vectors, the scalar product and the contribution to the amplitude of the launched wave are reduced.

Until now, we have assumed that the incident wave propagates in an immersion fluid and were thus allowed to neglect the reflected wave. However, the medium in front of the fiber end is often air and in this case reflection losses have to be taken into account. In principle it is possible to calculate both the launched wave, the reflected wave, and the radiation both in the forward and in the backward directions. To do this, one requires the additional boundary condition that the tangential component of the magnetic field vector must be continuous in the coupling plane $z = 0$ [Stratton 1941]. Since this method is very complicated [Morishita et al. 1979], in practice, it is sufficient to use the simple formula (5.129) for the power transmission factor for a homogeneous plane wave incident normally on a plane boundary between two media with refractive indices n_1 and n_2. For air, $n_1 = 1$. For the quantity n_2, an average value has to be taken, either the effective index n_e (5.45) or more simply, because of the small index difference, the cladding index. The effective launching efficiency η_e is then the product of the power transmission factor T of the interface and the launching efficiency η as given by (7.11) or (7.14)

$$\eta_e = T\eta \quad . \tag{7.15}$$

The overlap integral method for calculating the launching or coupling efficiency is an approximate one, since the waves radiated from the discontinuity in the backward direction are neglected. Despite its frequent use, the validity of this method is not well known and it is hard to predict the degree of accuracy of the approximation [Marcuse 1970c]. Some information about the accuracy of the method can be found in a paper on the radiation losses caused by discontinuities in a dielectric slab waveguide, which represents a mathematically simple two-dimensional model of a circular glass fiber [Finegan 1985].

7.2.2 Excitation of the LP$_{01}$ Mode by the Fundamental Gaussian Beam

As has been explained in Sect. 2.2, the fundamental mode of a single-mode fiber is very similar to the fundamental Gaussian beam mode (Chap. 4). The main difference is that the field diameter of the LP$_{01}$ mode does not change during propagation, whereas the field diameter of the Gaussian beam changes because of diffraction. Because of this similarity, the fundamental fiber mode can be launched very effectively by illuminating the fiber front end with a fundamental Gaussian beam mode.

The Gaussian beam is assumed to be incident through a homogeneous transparent medium with refractive index n_h. Most often, the medium will be air with $n_h = 1$. For optimum launching efficiency, the following conditions must be met: the beam axis must coincide with the fiber axis, the beam waist must be located in the fiber front face, and the beam spot size must be matched to the spot size of the fundamental fiber mode. Under these conditions, very high values of the launching efficiency can be calculated from (7.11) when inserting the beam field (4.1) for $\boldsymbol{E}_{it}$ and the exact LP$_{01}$ mode field for $\boldsymbol{E}_{01t}$.

For the special case of a step-index fiber, the launching efficiency η has been calculated rigorously by inserting the true field distribution of the HE$_{11}$-mode into the overlap integral (7.11) [Marcuse 1970c]. Figure 7.1 represents the launching efficiency η as a function of the spot size w_0 of the Gaussian beam at the waist relative to the core radius a for different

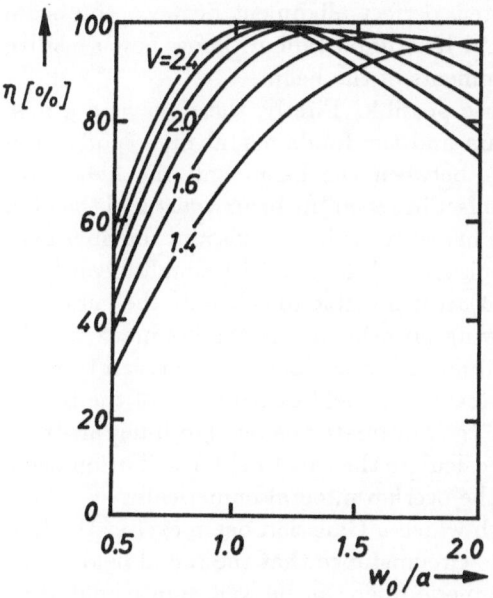

Fig. 7.1 Launching of the fundamental mode of a step-index fiber by a Gaussian beam. Launching efficiency η as a function of the spot size w_0 of the Gaussian beam at the waist relative to the core radius a for different values of the normalized frequency V (5.8)

values of the normalized frequency V (5.8). There is a broad maximum for the launching efficiency. In the maximum, the launching efficiency is near to 100%.

For V values near the cutoff of the LP_{11} mode ($V_c = 2.405$), the optimum spot size w_0 is somewhat larger than the core radius a. For smaller V values, because of the larger transverse field extent of the fundamental mode, a larger spot size w_0 must be chosen to maximize the launching efficiency. Later, in Sect. 10.1.2, the optimum value of w_0 will be taken to define the spot size w_G of the fundamental fiber mode. Thus the optimum value for w_0 can be read from the w_G versus V data in Table 5.1.

The optimum relative spot size w_G/a and the maximum launching efficiency are represented in Fig. 7.2 as functions of the normalized frequency V. For $V = 2.4$, the launching efficiency is as high as 99.7%. With decreasing normalized frequency, η becomes smaller since the Gaussian approximation for the radial field distribution $E(\varrho)$ becomes worse. However, even at $V = 1.5$, η is still larger than 97%.

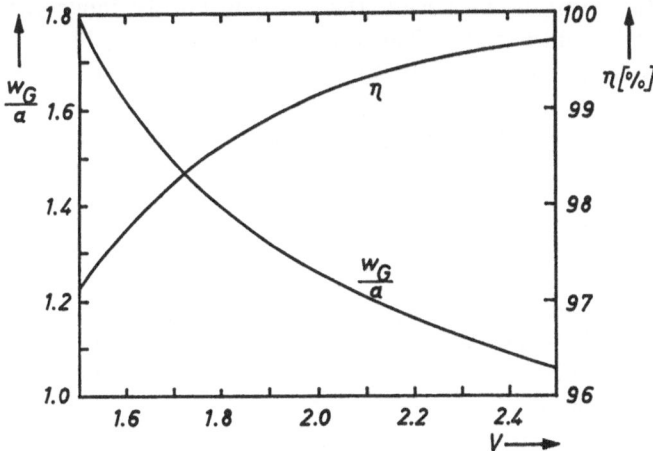

Fig. 7.2. Optimum relative spot size w_G/a of the Gaussian beam, and maximum launching efficiency η as functions of the normalized frequency V (5.8) for a step-index fiber

175

Because of the tiny core diameter, perfect alignment between the beam and the fiber cannot be achieved, and it is important to know how sensitive the launching efficiency is to misalignments of the beam.

Several types of misalignment are possible. Firstly, there may be a mismatch of the field widths of the beam and the fundamental fiber mode. Secondly, a transverse offset may exist between the beam and the fiber axis. Thirdly, there may be a longitudinal offset between the beam waist and the fiber input face. Lastly, there may be a tilt between the beam axis and the fiber axis. Usually, several different sources of mismatch loss will exist simultaneously.

In principle, using (7.11) or (7.14), it is possible to calculate the launching efficiency for a single-mode fiber having an arbitrary refractive index profile. One has to calculate both the field produced by the incident Gaussian beam in the input plane of the fiber and the transverse field distribution of the fundamental mode in the fiber. One must then evaluate the overlap integral. However, in practice, it is very tedious to calculate the exact field distribution using numerical methods and to evaluate the overlap integral numerically.

For estimating the efficiency with which a Gaussian beam excites the fundamental fiber mode, it is a fortunate circumstance that the radial field distribution of the LP_{01} mode in a single-mode fiber can be well approximated by a simple Gaussian function (5.25). For the fiber spot size, the quantity w_G as defined by Marcuse [1978a] should be taken (Sect. 10.1.2): it is that beam waist spot size w_0, which maximizes the launching efficiency. Using this approximation in the overlap integral, gives simple, but of course approximate, equations for the launching efficiency for the fundamental mode.

We first consider a mismatch of the field distributions of the Gaussian beam and the fundamental fiber mode, i.e. $w_0 \neq w_G$. The beam axis is assumed to coincide with the fiber axis and the beam waist to be located in the fiber front face. Since the beam has a plane wavefront in the waist (Sect. 4.2), the wavefront of the beam is matched to that of the LP_{01} mode. Locally, the state of polarization of the total field in the fiber input end equals the state of polarization of the incident Gaussian beam, because of the continuity of the tangential components of the electric field vector. When the incident Gaussian beam is linearly polarized, the polarizations are (approximately) matched, since the excited fundamental fiber mode is (approximately) uniformly linearly polarized (Sect. 5.3.5). For a spot size mismatch only, the launching efficiency is [Kogelnik 1964a; Marcuse 1977b]:

$$\eta = \left(\frac{2w_0 w_G}{w_0^2 + w_G^2} \right)^2 . \tag{7.16}$$

Figure 7.3 represents the launching efficiency as a function of the ratio of the spot sizes. If the spot size of the incident beam matches that of the fiber, then in this approximation, the launching efficiency is 100%. A 15% mismatch of the spot sizes causes a launching loss of only 0.1 dB.

Next, we assume the waist to be located in the fiber front face, the spot sizes to be matched, and the beam and fiber axes to be parallel, but transversely

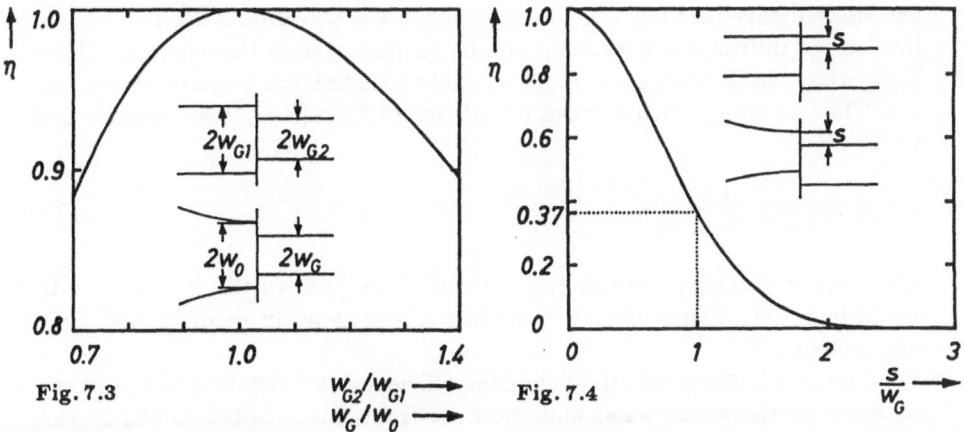

Fig. 7.3. Launching of the fundamental fiber mode by a Gaussian beam. Launching efficiency (7.16) for a mismatch between the spot sizes w_0 and w_G of the beam and the fundamental mode. Also coupling efficiency (9.3) between two single-mode fibers with different spot sizes w_{G1} and w_{G2}. The radial field distribution of the fundamental fiber mode has been approximated by a Gaussian function (5.25)

Fig. 7.4. Launching of the fundamental fiber mode by a Gaussian beam. Launching efficiency (7.17) for a transverse offset s. Also coupling efficiency (9.4) between two single-mode fibers for a transverse offset. The radial field distribution of the fundamental fiber mode has been approximated by a Gaussian function (5.25)

offset by a distance s. Now, the wavefronts and the polarizations are matched and we have another kind of amplitude mismatch. The launching efficiency is [Kogelnik 1964a; Marcuse 1977b]:

$$\eta = \exp[-(s/w_G)^2] \quad , \tag{7.17}$$

and the launching loss in decibels

$$a = -10 \log \eta = 4.34 (s/w_G)^2 \quad . \tag{7.18}$$

Figure 7.4 represents the launching efficiency as a function of the offset s relative to the spot size w_G. For a transverse offset of one spot size, the launching efficiency is 37% and the launching loss 4.34 dB. For the launching loss to be smaller than 0.1 dB, the offset must be smaller than $0.15 w_G$, i.e. for $w_G = 5\,\mu$m, smaller than $0.8\,\mu$m! Thus it is not a simple task to couple laser light efficiently into a single-mode fiber (Sect. 12.2.2).

Using the Gaussian approximation, the launching efficiency is found to be independent of the direction of offset with respect to the electric field vector. But even when using the true field distribution of the fundamental mode, the direction in which the beam is offset with respect to polarization, has been found to be unimportant [Marcuse 1970c].

We next consider the case where the axes of the beam and the fiber are coincident, the spot sizes are matched ($w_0 = w_G$), but the beam waist is located in a plane $z = -z_w$ in front of the fiber input face. The incident beam now has curved wavefronts in the coupling plane; the wavefronts are not matched. Because of diffraction spreading, the beam diameter in the coupling plane is larger than the fundamental mode diameter; there is an amplitude mismatch, too. The launching efficiency can be calculated from [Kogelnik 1964a; Marcuse 1977b]:

$$\eta = \frac{1}{1 + (0.5z_w/z_R)^2} \quad , \tag{7.19}$$

where $z_R = \pi n_h w_G^2/\lambda$ is the Rayleigh distance (4.6) of the beam wave that would be radiated from the fiber end into a homogeneous medium with refractive index n_h.

Figure 7.5 represents the launching efficiency as a function of the distance z_w between the beam waist and the fiber front face relative to the Rayleigh distance. If the distance equals one Rayleigh distance, the launching loss is 1 dB. The launching loss is smaller than 0.1 dB, if the distance z_w is smaller than $0.3z_R$. For $\lambda = 1.3\,\mu m$, $w_G = 5\,\mu m$, and air ($n_h = 1$), the Rayleigh distance is 60 μm and, therefore, a slight defocussing of the laser beam causes only a small loss increase. When the (virtual) waist is located within the fiber in the plane $z = z_w$, the launching efficiency is also given by (7.19).

Finally, we assume that the spot sizes are matched, that there is neither a transverse nor a longitudinal offset, but that the beam axis and the fiber axis are misaligned by an angle θ. The tilt causes a wavefront mismatch and

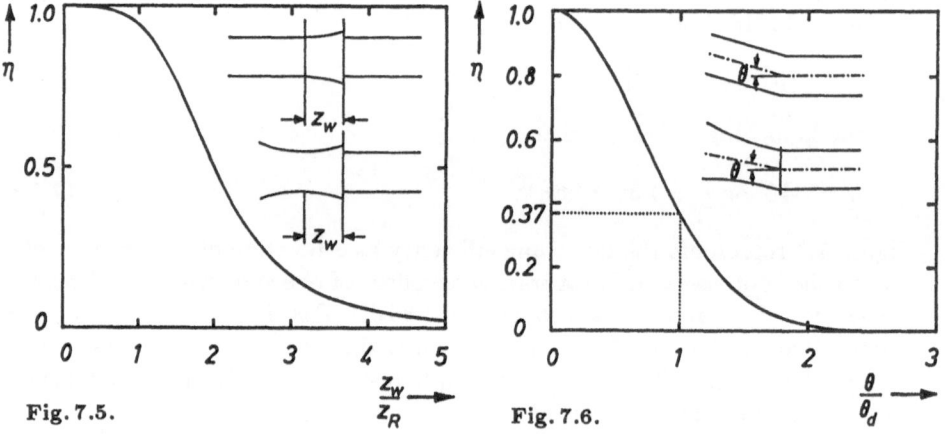

Fig. 7.5.

Fig. 7.6.

Fig. 7.5. Launching of the fundamental fiber mode by a Gaussian beam. Launching efficiency (7.19) for a longitudinal offset z_w between the beam waist and the fiber input face. Also coupling efficiency (9.6) between two single-mode fibers for an air gap of width z_w between the fiber ends. The radial field distribution of the fundamental fiber mode has been approximated by a Gaussian function (5.25)

Fig. 7.6. Launching of the fundamental fiber mode by a Gaussian beam. Launching effiency (7.20) for an angular offset θ. Also coupling efficiency (9.8) between two fibers with matched spot size for a tilt between the fiber axes. The radial field distribution of the fundamental fiber mode has been approximated by a Gaussian function (5.25)

therefore a decrease in the power launched given by the factor [Kogelnik 1964a; Marcuse 1977b]:

$$\eta = \exp[-(\theta/\theta_d)^2] \quad , \tag{7.20}$$

where $\theta_d = \lambda/(\pi n_h w_G)$ is the divergence angle (8.30) of the beam which the fiber end would radiate into a homogenous medium with refractive n_h (Fig. 7.6). The launching loss in decibels is

$$a = -10 \log \eta = 4.34 \, (\theta/\theta_d)^2 \quad . \tag{7.21}$$

To ensure a launching loss smaller than 0.1 dB, the tilt angle must be smaller than $0.15 \, \theta_d$. For $\lambda = 1.3 \, \mu m$, $w_G = 5 \, \mu m$, and air ($n_h = 1$), the divergence angle is $\theta_d = 4.7°$, and the tilt angle must be smaller than 0.7°.

The simple formulas given here only hold when a single cause of launching loss is present. In general, however, there will be a spot size mismatch, a transverse offset, a longitudinal offset, and an angular offset at the same time. Using the Gaussian approximation for the radial field distribution of the fundamental fiber mode, various approximate formulas have been derived for the launching efficiency under such circumstances. Since these formulas also describe the coupling efficiency for fiber joints with a general misalignment, they will be reviewed in Sect. 9.5.

Note that for small V values and for dispersion-flattened fibers, the true field distribution cannot be adequately approximated by a simple Gaussian function. One must then, in general, calculate the radial field distribution and the overlap integral in (7.11) by numerical methods. This more rigorous analysis has been done for various combinations of incident field and fiber type:

1. a fundamental Gaussian beam exciting a step-index fiber [Stern and Dyott 1970, 1971; Stern et al. 1970; Heyke 1970; Marcuse 1970c; Gambling et al. 1973; Imai and Hara 1974, 1975; Barrell et al. 1978];
2. a fundamental Gaussian beam exciting a W-type fiber [Chandra et al. 1978];
3. a fundamental Gaussian beam exciting a parabolic-index fiber [Saijonmaa et al. 1980];
4. a fictitious truncated uniform plane wave (plane wave segment) exciting a step-index fiber [Snyder 1966, 1969b; Barrell and Pask 1979], and
5. a fictitious truncated parabolic plane wave exciting a step-index fiber [Pask and Barrell 1975].

A number of authors have investigated axial power launching experimentally [Stern and Dyott 1970, 1971; Stern et al. 1970; Dakin et al. 1972].

A uniform plane wave can be considered as a Gaussian beam with infinite spot size w_0 at the waist (Sect. 4.5). For some applications, e.g. for probing optical fields it is interesting to know the power of the fundamental mode launched by a uniform plane wave incident on the fiber front face.

The fiber front end acts as a receiving optical dielectric rod antenna. For a uniform plane wave with intensity S incident through a homogeneous medium with refractive index n_h in the direction of the fiber axis, the received power P can be obtained by evaluating the overlap integral (7.11) between the uniform field distribution of the incident wave and the approximately Gaussian field distribution of the fundamental fiber mode (5.25)

$$P = 2\pi w_G^2 TS \quad , \quad \text{where} \tag{7.22}$$

$$T = \frac{4 n_h n_2}{(n_h + n_2)^2} \tag{7.23}$$

is the power transmission factor (5.129) of the interface between the homogeneous medium and the glass. For air, $n_h = 1$, and $n_2 = 1.46$, giving $T = 0.965$.

Because a uniform plane wave transmits infinite power, in this case, it is not possible to define a launching efficiency. But one can resort instead to a quantity used in antenna technology, namely the effective antenna area A_e, which is the ratio of the power P received by an antenna to the intensity S of the incident uniform plane wave [Ramo et al. 1965]

$$A_e = P/S = 2\pi w_G^2 T = 2A_b T \quad . \tag{7.24}$$

Since T is nearly unity, the effective antenna area of the fiber input end is approximately twice the cross section (4.29, 5.38) of the fundamental fiber mode

$$A_b = \pi w_G^2 \quad . \tag{7.25}$$

At the single-mode limit of a step-index fiber $V = 2.4$, the spot size is $w_G = 1.09a$ (Table 5.1) and the effective antenna area 2.4 times the core cross section. For $V = 1.6$, the radial field extent is much larger ($w_G = 1.62a$) and the effective area equals 5.3 core cross sections. Reflection losses have been neglected.

For very small values of V, most of the power is transmitted in the cladding ($P_2 \gg P_1$), and the radial field distribution can be approximated by a modified Hankel-function of zero order $K_0(W\varrho/a)$ for all distances ϱ from the fiber axis. The effective antenna area is then given approximately by [Neumann 1964c, 1967b]:

$$A_e = 4\pi(a/W)^2 T \quad . \tag{7.26}$$

The excitation of guided modes in a step-index fiber by a uniform plane wave has been analyzed more rigorously [Snyder 1966; Clarricoats and Chan 1970; Clarricoats 1976]. According to this theory, the effective antenna area is

$$A_e = \frac{4\pi a^2 V^2}{(UW)^2 T} \quad . \tag{7.26a}$$

Since the directional characteristic of a receiving antenna is the same as that of a radiating antenna [Ramo et al. 1965], for an obliquely incident uniform plane wave, the effective antenna area is smaller by the directional factor $e^{-2(\theta/\theta_d)^2}$ (8.29) than the values given previously.

7.3 Launching by Incoherent Illumination of the Fiber Front End

Incoherent emitters such as tungsten lamps or light emitting diodes (surface emitters) have an emitting surface which is large compared to the cross section of the LP_{01} mode.

Furthermore, they radiate into a wide angular range. Thus only a small amount of power can be launched from incoherent emitters into single-mode fibers. Nevertheless, these sources are used for measuring the wavelength dependence of fiber attenuation (Sect. 13.3.1), the fiber spot size (Sect. 13.7.2), or chromatic dispersion (Sects. 13.9.2, 9.3). Moreover, it has recently been shown that, contrary to previous belief, it is possible to realize short-haul, low-bandwidth single-mode transmission systems with LED's as optical sources [de Bortoli and Moncalvo 1986]. Therefore, in this section, we discuss the axial coupling between an incoherent source and a single-mode fiber.

The radiation properties of an incoherent source are characterized by its radiance (brightness) B, which is defined from the power ΔP radiated from an area element ΔA on the emitting surface into an element $\Delta\Omega$ of solid angle [Born and Wolf 1975; Marcuse 1975b; Friberg 1986]:

$$B = \Delta P/(\Delta A \Delta \Omega \cos\theta) \quad , \tag{7.27}$$

where θ is the angle between the direction of observation and the direction normal to the surface of the emitter. Since the product $\Delta A \cos\theta$ represents the radiation area as seen by an observer in a direction at angle θ, the radiance is the amount of power that the projected unit area of the source radiates into the unit solid angle; its dimensions are $W/(cm^2 sr)$ (sr = steradians).

The data sheets of LED's specify their radiance; typical values are 10–100 $W/(cm^2 sr)$ for surface emitters [Burrus 1972] and up to 1000 $W/(cm^2 sr)$ for edge emitters [Ettenberg et al. 1976].

In principle, the radiance B can depend on the position of the element ΔA on the emitting surface and the direction in space of the solid angle element $\Delta\Omega$. Thus the radiance is a function of two spatial coordinates, e.g. ϱ' and ϕ', which characterize the position of an emitting element on the planar surface, and of two angular coordinates, θ and ϕ, which denote the direction of ray emission (Fig. 8.1). For a uniform emitting surface, B is independent of the position of the emitting element. Often it is also permissible to assume that, to a good approximation, B is independent of the direction of ray emission. The power ΔP radiated into a fixed solid angle is then proportional to $\cos\theta$ Lambert's cosine law) and the source is said to be Lambertian.

In a homogeneous lossless medium, the radiance is constant along the straight trajectory of each ray. In an inhomogeneous lossless medium, the quantity B/n^2 remains constant along the ray trajectory, which is generally curved. According to this so-called radiance theorem or radiance law [Hudson 1974; Born and Wolf 1975; di Vita and Vannucci 1975; MacMahon 1975; Friberg 1986], the product of the cross-sectional area and the square of the numerical aperture of an optical beam must remain constant under any lossless transformation of that beam. Because of the second law of thermodynamics, it is impossible to increase the radiance by means of optical systems consisting of passive elements such as mirrors and lenses.

However, the radiance theorem, which is derived using the principles of geometrical optics, does not hold in the case of single-mode waveguides, for which geometrical optics is not valid [Goodman 1985].

The total power P_r radiated by an incoherent source is obtained by integrating the differential power ΔP over the emitting area A of the source and over the half space corresponding to the solid angle $\Omega = 2\pi$:

$$P_r = \int\limits_{\Omega} \int\limits_{A} B \cos \theta \, dA \, d\Omega \quad . \tag{7.28}$$

For the special case of a uniform Lambertian source, the integration gives [Barnoski 1976]

$$P_r = \pi B A \quad . \tag{7.29}$$

If a single-mode or multimode fiber is butt-jointed to an incoherent source (or to the image of it), only part of the total power P_r radiated into air (7.29) is coupled into guided modes. Two heuristic methods for estimating the power launched into a single-mode fiber will be given.

The first method starts by considering a multimode step-index fiber, for which one is allowed to use the methods of geometric optics to calculate the launched power. Since the angle θ between the fiber axis and rays that are accepted by the fiber is small, the mode power P_m guided by the multimode fiber is approximately obtained by multiplying the radiance B by the core area πa^2 and by the solid acceptance angle $\Omega = \pi A_N^2$ of the fiber [Hudson 1974; Miller and Chynoweth 1979]:

$$P_m = B \pi a^2 \pi A_N^2 \quad . \tag{7.30}$$

Expressing the numerical aperture A_N in terms of the refractive indices (5.10) gives

$$P_m = \pi^2 a^2 (n_1^2 - n_2^2) B \quad . \tag{7.31}$$

The medium between the source and the fiber has been assumed to be air.

If the incoherent source is Lambertian and if the emitting area is large in comparison to the core cross-section, each mode guided by the multimode fiber receives an equal amount of power [Marcuse 1975b; Carpenter and Pask 1976, 1977; Pask 1978]. Thus, the total launched power is uniformly distributed over all fiber modes. In a multimode step-index fiber, the number of guided modes is

$$M = V^2/2 \quad , \tag{7.32}$$

where V is the normalized frequency (5.8). Therefore, each mode, and specifically each of the two orthogonally polarized fundamental modes, carries a power of

$$P_{\mathrm{m}}/M = B\lambda^2/2 \quad . \tag{7.33}$$

Since each of the orthogonal fundamental modes receives the same power, the power launched into *both* fundamental modes from an incoherent source with radiance B butt-jointed to a multimode fiber is estimated to be

$$P = B\lambda^2 \quad . \tag{7.34}$$

Since we have used arguments of geometrical optics, stricly speaking, this consideration applies only to multimode step-index fibers. But (7.34) will presumably also hold for decreasing V values, and in the limit, for single-mode fibers.

In the second approach, we start from a single-mode fiber. Strictly speaking, one is not allowed to use geometric optics for single-mode fibers. But for an estimate, we may set the acceptance area equal to the beam cross section of the fundamental fiber mode (7.25)

$$A_{\mathrm{b}} = \pi w_{\mathrm{G}}^2 \quad , \tag{7.35}$$

and the solid acceptance angle equal to

$$\Omega = \pi \theta_{\mathrm{d}}^2 \quad , \text{ where} \tag{7.36}$$

$$\theta_{\mathrm{d}} = \frac{\lambda}{\pi w_{\mathrm{G}}} \quad , \tag{7.37}$$

is the divergence angle of the beam radiated from the free fiber end into air (8.30). Multiplying the radiance by the acceptance area and by the solid acceptance angle, one obtains for the power launched into the *two* fundamental modes of orthogonal polarization

$$P = B\lambda^2 \quad , \tag{7.38}$$

which agrees with the result (7.34) of the first method for estimating the power.

Thermodynamic reasoning [McMahon 1975] and an analysis based on wave optics [Marcuse 1975b] give the same result as our two heuristic arguments.

A more general formula has been derived and verified experimentally [Christodoulides et al. 1986a,b] from which one can calculate the launched power when the radiance is not constant over the emitting surface or when it depends on the direction of radiation. For the special case of constant radiance, this general formula reduces to our simple relation (7.34).

It is interesting to note, that for a given source, the power launched from an incoherent source does not depend on the fundamental mode spot size w_G. The physical reason for this is that a large spot size corresponds to a large area of acceptance, but a small solid angle of acceptance and vice versa.

For a numerical example, we assume an incoherent source with radiance $B = 100\,\mathrm{W/(cm^2 sr)}$ butt-jointed to a single-mode fiber for $\lambda = 1.3\,\mu\mathrm{m}$. Then, from (7.34) the launched power is estimated to be $P = 0.8\,\mu\mathrm{W}$. In practice, launched powers of this order of magnitude have been reported (Sect. 12.2.2).

When the source is incoherent, large compared to the core cross section, and Lambertian, the LP_{01}^{x} and LP_{01}^{y} modes propagate equal powers in the single-mode region. In the two-mode region, each of the four LP_{11} modes (Sect. 6.3) carries as much power as one of the two orthogonal fundamental modes, i.e. the power transmitted by the four second-order modes is twice the power transmitted by the two fundamental modes.

In deriving (7.34), it has been implicitly assumed that the surface of the emitter is larger than the area over which the mode field has appreciable amplitude and that the source radiation pattern is Lambertian. Since the radiance cannot be increased, (7.34) shows that under these assumptions the launched power cannot be increased by inserting a passive optical system between the source and the fiber input end.

In our simplified analysis, we assumed a large, uniform, and Lambertian source. In the general case, the power launched into a fiber mode from an *incoherent* source is proportional to the overlap of the *intensity* distributions of the source and the mode [Snyder and Pask 1973; Snyder 1974]. Note, that for a *coherent* source, the launched power is proportional to the square of the overlap of the *amplitude* distributions (7.14).

If the emitting area is smaller than the cross section of the fundamental fiber mode or if the radiation is confined in a solid angle smaller than the solid acceptance angle of the fiber, it is possible to increase the launched power by using optical elements.

Recently, a theoretical formalism for calculating the coupling efficiency of LED's to single-mode fibers with arbitrary coupling optics has been published [Hillerich 1986, 1987].

7.4 Launching by Evanescent Field Coupling

When two straight single-mode fibers are placed parallel to one another and when the distance between the axes is reduced until the evanescent fields of the two guides overlap, the fundamental field on one fiber interacts with the neighboring fiber and power is coupled between the two. This type of coupling is called evanescent field coupling or transverse coupling. The word "transverse" is used to distinguish between two dielectric waveguides lying side by side, where the transfer of power takes place in a radial direction, and the axial coupling (Chap. 9) between two waveguides placed end to end, where the transfer of power takes place along the z-axis [Arnaud 1974a].

The strength of the coupling is determined by the separation of the fiber cores, the extent a/W (5.15) to which the field spreads into the cladding, and the length of the coupling region.

This distributed electromagnetic coupling can cause unwanted cross talk between parallel dielectric waveguides. Nonetheless, it is used as the basis of four-port directional couplers, which are very useful components in single-mode fiber transmission systems (Sect. 12.2.11).

The interaction between two coupled parallel dielectric waveguides can be analyzed by three different methods: coupled-mode theory, normal-mode analysis, and by the beam propagation technique.

Using coupled-mode theory, the coupling of parallel fibers has been analyzed by a large number of authors [Miller S.E. 1954; Bracey et al. 1959; Jones 1965; Vanclooster and Phariseau 1970; Marcuse 1971a, 1973b, 1974, 1975a; Kapany and Burke 1972; Snyder 1972; Yariv 1973; McIntyre and Snyder 1973; Taylor and Yariv 1974; Kogelnik 1975; Sporleder and Unger 1979; Tekippe 1981; Eyges and Wintersteiner 1981; Snyder and Love 1983; Hung-chia 1984; Marom et al. 1984; Thyagarajyn and Tewari 1985; Hardy and Streifer 1985, 1986; Hardy et al. 1986b; Qian 1986; Qian and Huang 1986a].

In the coupled-mode theory, one considers, in the individual waveguides a and b, the fundamental individual LP_{01} modes with complex wave amplitudes $\underline{A}$ and $\underline{B}$. First, we only consider coupling between identical, lossless fibers. Because of the power exchange, the amplitudes $\underline{A}(z)$ and $\underline{B}(z)$ become dependent on the longitudinal coordinate z. It is extremely convenient to define the amplitudes $\underline{A}$ and $\underline{B}$ in such a way that $|\underline{A}(z)|^2$ and $|\underline{B}(z)|^2$ correspond to the powers $P_a(z)$ and $P_b(z)$ carried by the waveguides a and b, respectively.

The differential equations describing the evolution of the complex amplitudes of the coupled modes are

$$d\underline{A}/dz = j\kappa\underline{B} \quad , \tag{7.39}$$

$$d\underline{B}/dz = j\kappa\underline{A} \quad . \tag{7.40}$$

For coupling between identical waveguides, the coupling coefficient κ from guide a to guide b is identical to that from b to a [Hardy and Streifer 1985] and is also a real quantity [Miller S.E. 1954].

The coupling coefficient is proportional to the overlap integral of the field of guide a with the field of guide b taken over the core cross-section of guide b [Marcuse 1971a]. The coupling coefficient κ decreases approximately exponentially with increasing distance d between the fiber axes, since the cladding field decreases approximately exponentially with increasing distance ϱ from the fiber axis (5.15, 5.16). Formulas for the coupling coefficient between two circular single-mode fibers have been published by several authors [Bracey et al. 1959; Jones 1965; Marcuse 1971a; McIntyre and Snyder 1973, 1974].

In the wavelength region of interest for single-mode operation, the coupling coefficient κ is proportional to the third power of the wavelength [Digonnet and Shaw 1982b]:

$$\kappa = c\lambda^3 \quad , \tag{7.41}$$

where c is a constant. The coupling increases with wavelength because the transverse field extent a/W in the cladding and therefore the field overlap increase.

For the special case of coupled step-index fibers, the coupling coefficient between the fundamental modes is [McIntyre and Snyder 1973, 1974]:

$$\kappa = \frac{1}{n_1 k} \left[\frac{U}{aVK_1(W)} \right]^2 K_0 \left(\frac{Wd}{a} \right) \quad , \tag{7.42}$$

where n_1 is the core index, k the wavenumber in vacuum, a the core radius, d the distance between the core axes, V the normalized frequency (5.8), U the radial phase parameter in the core (5.21), W the cladding decay parameter (5.15), and K_0 and K_1 modified Hankel functions of zero and first order, respectively.

The coupled differential equations (7.39, 40) for the amplitudes of the coupled waves can easily be solved [Miller S.E. 1954; Marcuse 1971a; Yariv 1973; Taylor and Yariv 1974; Marom et al. 1984]. Assuming that at the beginning ($z = 0$) of the coupled waveguides, only guide a is excited with zero phase angle, i.e.

$$\underline{A}(0) = A_0 \quad , \tag{7.43}$$

$$\underline{B}(0) = 0 \quad , \tag{7.44}$$

the z-dependence of the wave amplitudes is

$$\underline{A}(z) = A_0 \cos(\kappa z) \quad , \tag{7.45}$$

$$\underline{B}(z) = j\underline{A}_0 \sin(\kappa z) \quad . \tag{7.46}$$

Note the factor j which implies that the fields in waveguides a and b are in phase quadrature.

The powers guided by the coupled waveguides are

$$P_a(z) = P_a(0) \cos^2(\kappa z) \quad , \tag{7.47}$$

$$P_b(z) = P_a(0) \sin^2(\kappa z) \quad , \tag{7.48}$$

where $P_a(0)$ is the power in guide a at the beginning of the coupled section.

Coupling of two identical lossless parallel single-mode fibers causes a complete periodic interchange of power. The total power is preserved, since

$$P_a(z) + P_b(z) = P_a(0) \quad . \tag{7.49}$$

After a length z, for which

$$\kappa z = \tfrac{\pi}{2} \ , \tag{7.50}$$

all the power is in the guide b; and after a length L_{p}, for which $\kappa L_{\mathrm{p}} = \pi$, the total power is back in guide a again. The period L_{p} of power exchange is related to the coupling coefficient κ by

$$\kappa = \pi/L_{\mathrm{p}} \ . \tag{7.51}$$

Sometimes a coupling length [Ragdale et al. 1984] or beat length [Tekippe 1981] is defined as the length $L_{\mathrm{p}}/2$ over which there is total power transfer between the fibers.

A second method for analyzing wave interaction in parallel dielectric waveguides is normal-mode analysis. In this method, one does not consider the individual modes of the coupled waveguides, but considers the eigenmodes of the total structure consisting of the two guides and their surrounding media [Louisell 1955; Marcatili 1969a; Snitzer 1969; Yariv 1973; Wijngaard 1973; Matsuhara and Kumagai 1974; Cullen et al. 1975; Suematsu and Kishino 1977; Snyder and Young 1978; Eyges and Wintersteiner 1981; Snyder and Love 1983; Marom et al. 1984; Mao et al. 1986; Ankiewicz et al. 1986].

There are two modes of the total structure that propagate with transverse field distributions that remain constant along the length: an even (symmetric) and an odd (antisymmetric) mode, of higher order. The field patterns of these so-called supermodes [Hardy and Streifer 1986] are invariant with the longitudinal coordinate z except for a very slow exponential decay caused by losses. The normal modes are well approximated by linear combinations of the modes of the individual waveguides [Yariv 1973; Marom et al. 1984; Ankiewicz et al. 1986]. This is illustrated in Fig. 7.7, which shows a cross section of the coupled region with the two fiber cores, and the electric field distributions of the individual modes in waveguide a and b and of the even and odd supermodes. The even eigenmode is essentially the sum of the individual modes, and the odd eigenmode their difference [Kotrotsios et al. 1986].

Whereas the fundamental mode of one isolated, uncoupled fiber has a phase constant β, the phase constant is $\beta + \kappa$ for the even eigenmode, and $\beta - \kappa$ for the odd eigenmode, respectively [Marom et al. 1984].

A single fiber mode incident in guide a at $z = 0$ excites the two supermodes with equal amplitudes and equal phase. At a distance z from the input, for which the differential phase shift of the supermodes is π, i.e. for

$$[(\beta + \kappa) - (\beta - \kappa)]z = 2\kappa z = \pi \ , \tag{7.52}$$

the normal modes are in antiphase. This corresponds to total power in guide b and no power in guide a. At $z = L_{\mathrm{p}}$ with

$$2\kappa L_{\mathrm{p}} = 2\pi \ , \tag{7.53}$$

the normal modes are in phase again, and the power is back in guide a. Coupled-

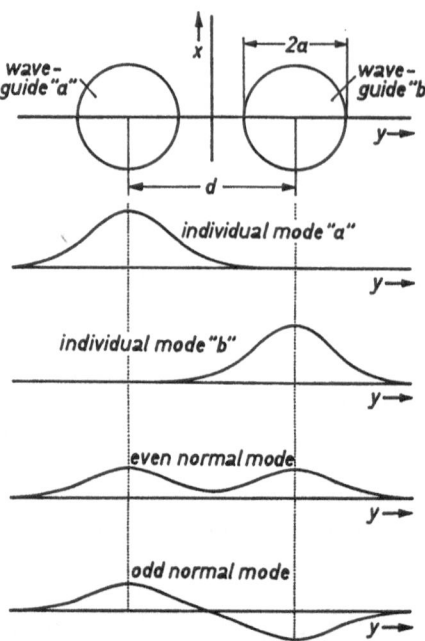

Fig. 7.7. Cross-section of a directional coupler and transverse field distributions of the individual and normal modes

mode theory and normal-mode theory give the same expression for the period L_{p} (7.51, 53). Thus, the periodic power exchange can also be explained with the normal-mode approach in terms of interference between the even and odd supermodes, which causes a beat phenomenon.

The even supermode in the coupled region has the same normalized cutoff frequency V_{c} as the fundamental mode in a single fiber [Love and Ankiewicz 1984b]. Usually, $V_{\mathrm{c}} = 0$; but in a depressed-inner cladding or W-fiber, the fundamental mode is cut off at a finite value of V if the first-order moment of the refractive-index difference is negative (6.22) [Hussey and Pask 1982a].

The odd supermode in a directional coupler always has a finite cutoff frequency which depends on the separation of the coupled fibers [Love and Ankiewicz 1984a, 1985; Snyder and Rühl 1985]. For frequencies below this cutoff frequency, the odd supermode becomes a leaky mode and is attenuated as its power radiates away into the cladding. Consequently, the fraction of coupler input power which excites the odd supermode is lost and low-loss couplers cannot be made.

The finite cutoff frequency of the odd supermode can cause a directional coupler to fail at long wavelengths. Moreover, the odd supermode has a slightly larger attenuation than the even supermode. This phenomenon has detrimental effects on the performance of directional couplers: the power that can be coupled from one fiber to the other is less than 100% and the phase shift in coupled light is less than the expected value of 90° [Youngquist et al. 1983].

The relations between the coupled-individual-mode and the uncoupled-normal-mode analyses are discussed in a paper by Marom et al. [1984]. The normal-mode formalism is more exact [Yeh et al. 1978]. When the distance between the cores is too small or when the two cores are touching (the "strong" coupling case), the coupled-mode theory can be seriously in error [Eyges and Wintersteiner 1981]. The analytical methods based on coupled-mode theory are difficult to apply to strongly coupled waveguides. Even in the case of identical fibers, the power transfer is not complete if the fibers are strongly coupled [Suematsu and Kishino 1977].

Besides the coupled-mode and normal-mode analyses, there is a third method for analyzing coupled dielectric waveguides: the scalar-wave fast-Fourier-transform technique [Yeh et al. 1979a] which is also called the propagating beam method [Feit and Fleck 1981].

In parallel fibers, power is coupled only in the forward direction. Thus it is possible to fabricate four-port directional couplers by transversely coupling two fiber waveguides (Sect. 12.2.11).

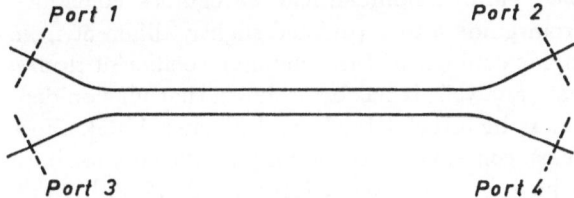

Fig. 7.8. Scheme of an optical four-port directional coupler showing the ports

In a practical optical directional coupler, the guides can only be coupled over a finite distance. Thus, they must therefore be bent at the beginning and the end of the coupled region (Fig. 7.8). The power exchange between two dielectric optical waveguides with varying spacing has been studied by solving the coupled-mode equations with continuously variable coupling coefficient [McIntyre and Snyder 1974; Findakly and Chen 1978; Alferness 1979] and by using an improved coupled wave theory [Hardy et al. 1986a]. The coupling of curved dielectric waveguides has been analyzed by Matsuhara and Watanabe [1975]. Directional couplers consisting of two circularly curved, single-mode *slab* waveguides have been analyzed by McHenry and Chang [1984]. However, coupling between nonparallel *fibers* is still not clearly understood, even though any physically realizable directional coupler must have nonparallel connecting fiber sections which will also contribute to the mode coupling.

Up to now, we have only considered directional couplers consisting of two identical dielectric waveguides. When waveguides with different phase constants are coupled, the coupled power is

$$P_b(z) = P_a(0)(\kappa z)^2 \, \text{sinc}^2[(\kappa z)^2 + (\Delta\beta z)^2]^{1/2} \quad \text{where} \tag{7.54}$$

$$\text{sinc}(x) = \frac{\sin(x)}{x} \tag{7.55}$$

and where $\Delta\beta$ is the difference of the phase constants of the coupled dielectric waveguides.

The exchange of energy between the two different dielectric waveguides is no longer complete but remains periodic as a function of the longitudinal coordinate [Miller S.E. 1954]. The difference in the phase constants $\Delta\beta$ relative to the coupling constant κ determines the fraction of the total energy which is exchanged. When this ratio is very large compared to unity, appreciable energy exchange between the coupled guides is prevented. In other words, for a mismatch of the phase velocities, the power exchange between the coupled waveguides is negligible. Appreciable coupling of power from one guide to the other parallel guide is only possible if the modes have nearly identical phase velocities.

The coupling of nonidentical dielectric waveguides has been analyzed by Kuester and Chang [1975a].

Former analyses of coupled parallel nonidentical waveguides concluded that the coupling coefficient from guide b to a (defined slightly different from our κ) equals the negative complex conjugate of the coupling coefficient from a to b. Recently, in a more accurate analysis, it has been shown that for nonidentical waveguides this statement is incorrect [Hardy and Streifer 1985]. More accurate expressions for the phase constants and coupling coefficients for both identical and dissimilar fibers have been derived [Hardy et al. 1986b]. Note, however, that this "new theory" of Hardy and Streifer has been claimed to contain "a slight discrepancy" [Chuang 1987] and "fundamental errors" [Snyder and Ankiewicz 1986]. On the other hand, new coupled-mode theories have been derived recently [Chuang 1987; Haus et al. 1987]; these give numerical results almost indistinguishable from the Hardy and Streifer values on the plots of propagation constants and coupling coefficients, although slight differences exist between the three theories.

In coupled-mode theory, there are four parameters: the two propagation constants of the individual modes and the two coupling coefficients. These four parameters can be expressed in terms of just two quantities, the asynchronism δ and the butt coupling c [Marcatili 1986, Marcatili et al. 1987]. The quantity δ is essentially the difference between the phase constants of the individual modes, and the butt coupling c is the overlap integral of the fields of these modes.

It is unfortunate, but inevitable, that the coupled mode equations are usually difficult to analyze. The exact analysis of devices based on directional couplers often reqires a numerical solution of the coupled-mode equations. It is therefore difficult to formulate a mental picture of the evolution of both the magnitudes and phases of the individual modes in the two waveguides.

To overcome this problem, three graphical methods have been proposed for studying mode coupling. The first graphical representation uses a generalization of the Smith chart which is well known to microwave engineers [Marcatili 1980]. The second [Ulrich 1977; Frigo 1986; Korotky 1986] utilizes the Poincaré sphere which graphically describes the transformation of the state of polarization by birefringent optical elements. The third technique [Kotrotsios et al. 1986], represents the complex wave amplitudes as vectors in the complex plane.

If the coupled fibers are bent into two concentric circles, this changes the difference of the phase constants of the fundamental modes and, therefore, the coupling between the fibers. This effect can be used [Murakami 1980; Murakami and Sudo 1981] to change the power transfer between identical fibers or, alternatively, to obtain complete power transfer between two different fibers.

The theory of coupled dielectric waveguides can be generalized to the case of more than two coupled waveguides [Snyder 1972; Eyges and Wintersteiner 1981; Hardy and Streifer 1986].

A directional coupler composed of a single-mode fiber and a multimode fiber may be capable of serving as a drop/insert device of a node in multimode fiber local area networks. Thus the process of transverse power transfer between a single-mode and a multimode fiber has been analyzed using the beam propagation method [Berthou et al. 1984] and a coupled-mode theory [Chang H. et al. 1986].

In a directional coupler, two guided modes in parallel fibers are coupled through their evanescent fields. Evanescent field coupling is also possible between a fiber mode and a free beam wave by using a prism coupler. A prism coupler consists of a high-index prism with its base extending into the evanescent field of the fiber mode. For making a prism coupler, it is necessary to expose the guided mode field. This can be achieved either by polishing the fiber after pulling, by polishing the preform before pulling the fiber from it, or by etching the fiber.

The first technique is equivalent to that of manufacturing polished directional couplers (Sect. 12.2.11): a curved single-mode fiber is cemented into a fused silica block. A flat surface is polished to expose a small region close to the core. The base of the prism with a large refractive index $n_p > n_1$ is placed in contact with the flat surface.

For explaining the operation of a prism coupler, we first consider the inverse case in which a guided mode with phase constant β is propagated by the fiber. Since its evanescent field extends through the thin cladding into the homogeneous prism, a wave is radiated into the prism. The direction of this beam, characterized by the angle θ between its axis and the fiber axis, can be obtained by matching the z-component of the phase velocity of the beam to the phase velocity of the guided mode:

$$\frac{\omega}{n_p k \cos \theta} = \frac{\omega}{\beta} \quad , \text{ i.e.} \tag{7.56}$$

$$\cos \theta = \frac{\beta}{n_p k} \quad . \tag{7.57}$$

If a fiber guides several modes, each mode generates a radiated beam. Relation (7.57) can be used to determine the phase constant β_m of the mth mode from the measured radiation angle θ_m. This method corresponds to the so-called m-line technique in integrated optics [Tien et al. 1969].

Because of reciprocity, a wave incident on the fiber through the prism at the synchronous coupling angle θ given by (7.57) optimally launches the guided mode with phase constant β. This method of mode launching by a prism coupler is well known in integrated optics [Tien and Ulrich 1970; Ulrich 1970; Tien 1971; Tamir 1975a].

In the second technique for making a fiber-prism coupler, a flat is polished on the fiber preform before fiber pulling [Dyott and Schrank 1982; Millar et al. 1986]. This technique results in fibers with a D-shaped cross section, where the distance between the core and the flat side of the D can be so small, that the evanescent fields of the guided modes extend into the surrounding air or into the base of a prism pressed against the flat surface.

In a dual-mode D-fiber, as long as the core is not exposed, the beam couples more efficiently to the LP_{11} mode than to the LP_{01} mode, since the transverse field of the LP_{11} mode extends further towards the flat and beyond it.

In the third technique [Szcepanek and Berthold 1978], the fiber is prepared by etching such that the diameter is linearly tapered. Coupling of a laser beam to the fiber is achieved through a prism and a thin glycerine film on the prism surface.

7.5 Miscellaneous Methods of Mode Launching

Besides the methods so far described for launching light power into a guided fiber mode, i.e. coherent or incoherent axial launching and distributed transverse launching, there are also some other techniques for exciting guided modes and these will be summarized in this section.

In an ideal straight uniform dual-mode fiber, the two LP_{01} modes and four LP_{11} modes propagate independently of one another, i.e. they do not exchange power. However, deviations of the fiber axis from a perfectly straight line, caused perhaps by the environment, will couple the modes together. Power coupled from the desired fundamental mode into the unwanted LP_{11} mode causes excess loss, and when the attenuation of the LP_{11} mode is insufficient it causes bandwidth reduction due to intermodal delay differences or modal noise. Equations describing the coupling of guided modes by fiber imperfections have been derived [Marcuse 1973b; Grau 1978]. Mode coupling can be caused by localized, periodic, or random perturbations. An important example for a local perturbation causing mode conversion is a splice or connector with lateral, longitudinal, or angular misalignment, or field profile mismatch. In general, each mode guided in the first fiber couples to several modes in the second fiber (Sect. 9.6). An example for a periodic perturbation of the fiber geometry is a sinusoidal variation of the core diameter along the fiber. In the single-mode regime, this perturbation causes coupling between the LP_{01} mode and radiation modes, i.e. radiation losses (Sect. 5.9.10).

In the dual-mode regime, power is exchanged between the LP_{01} and the LP_{11} modes by periodic bends [Youngquist et al. 1984]. The coupling is maximum if the difference of the phase constants of the coupled modes equals the spatial frequency of the perturbation.

In a real fiber, there are also random perturbations along the fiber, e.g. random fluctuations of the core diameter, which cause mode coupling. The small differences in propagation constants between the guided modes allow relatively long spatial fluctuations, which are also relatively likely, to couple light from one mode to the other.

A fundamental mode fiber can guide two orthogonally polarized LP_{01} modes. These modes can also become coupled by perturbations to the fiber geometry. Polarization coupling is particularly disadvantageous in polarization-maintaining fibers.

At macrobends or microbends, power is radiated from a single-mode fiber guiding the fundamental fiber mode. Vice versa, by illuminating a macrobend or microbend with a laser beam, part of the power transmitted by the beam can be launched into the fundamental fiber mode.

Fiber taps for coupling power out of or into a fiber are described in Sect. 12.2.13. The technique of "local power injection and detection" is used for aligning the fiber cores prior to fusion splicing (Sect. 12.2.4).

In fibers with a Germanium-doped core, guided modes can also be excited by fluorescence [Presby 1981b]. The fiber is illuminated from the side by ultraviolet light with a wavelength below 350 nm. The wavelength of the fluorescent light is near 420 nm. Ultraviolet-excited fluorescence has been used for local power injection in splicing machines (Sect. 12.2.4) and for measuring the refractive-index profile of preforms (Sect. 13.10.1).

7.6 Selective Mode Excitation

Usually in a fiber that can guide several modes, all modes will be simultaneously launched by an optical source coupled to the fiber input end. The distribution of power in the different modes is described by the so-called mode-power spectrum, or mode spectrum.

Sometimes, one is interested in launching only one of several guided modes, e.g. when experimentally studying the properties of the LP_{11} mode, which can cause modal noise and intermodal dispersion. In this section, methods for selectively launching particular modes are reviewed.

Selective mode excitation can be obtained either by axial or by transverse (side) illumination. For axial illumination, the general rule for selectively exciting a certain guided mode is to provide an incident electric field whose tangential field components match those of the desired mode (Sect. 7.2.1). The launching efficiency is then unity, and no unwanted modes are launched.

A spatial-filtering technique has been used to generate an incident field closely matched to that of the desired mode [Kapany et al. 1970; Kapany and Burke 1972]. Appropriate amplitude and phase filters are inserted into the pupil of a launching lens which focuses collimated light onto the end of the fiber. Though no combination of polarizers, spatial filters, and lenses can match the incident field perfectly to that of the desired mode, it has been found that even quite crude approximations can provide adequate mode selection for some purposes.

Other simple experimental methods for selectively launching low-order modes are to focus the exit slit of a monochromator on the end of the fiber and to vary the angle between the axis of the cone of light and the fiber, or to focus a pinhole on the fiber end and adjust it laterally [Snitzer and Osterberg 1961].

A technique of axial mode launching providing good efficiency, mode selectivity, and easy mode to mode switching is based on the fact that the Gaussian modes of open laser cavities are very similar to the fiber modes [Facq et al. 1984]. A gas or dye laser can easily be made to oscillate in a single Gaussian beam mode by inserting thin wires into the optical resonator. In order to obtain optimum launching efficiency, the exit laser beam must be focused by an adapted microscope objective onto the fiber core. Note, that the spot size is independent of the mode numbers. By moving a wire into and out of the laser beam, one can e.g. switch easily between the LP_{11} and LP_{01} modes of the fiber.

A method for selectively launching high-order leaky modes in multimode fibers consists in irradiating the fiber obliquely through a high-index prism coupler with a laser beam [Stewart W.J. 1975; Midwinter 1975; Zemon and Fellows 1976]. Different leaky modes can be excited by changing the angle between the fiber axis and the beam axis. The excitation efficiency is fairly low.

A similar technique for selective mode launching in few-mode fibers uses the fiber-prism coupler described in Sect. 7.4. There is a simple relation (7.57) between the angle θ of the incident laser beam and the phase constant β of the excited guided modes. In a dual-mode fiber, or in a single-mode fiber for wavelengths smaller than the cutoff wavelength of the LP_{11} mode, the LP_{01} or the LP_{11} mode can be launched selectively by choosing the angle θ corresponding to (7.57).

Medium and low-order modes can be side-launched through a prism coupler pressed against the etched fiber [Szczepanek and Berthold 1978].

8. Radiation from the Fiber End

As has been explained in Sect. 2.4, most of the power of a guided wave is radiated from the fiber end in the form of a free diverging wave beam. Essentially, the fundamental fiber mode is transformed into the fundamental Gaussian beam mode and the second order fiber mode into the second order Gaussian beam mode (Chap. 4). The fiber end acts as as transmitting dielectric antenna, transforming a guided electromagnetic wave into a freely propagating wave [Neumann 1967a].

The structure of the wave beam radiated from the fiber end must be known, e.g. for estimating the loss of a butt-joint connector with a gap (Sect. 9.3), or for designing expanded beam connectors (Sect. 12.2.5). In Sect. 8.1, the end-fire radiation from a single-mode fiber will thus be discussed more rigorously.

A number of interesting mathematical relations exist between the functions describing the radial field distribution of the fundamental fiber mode and the radiation pattern of the beam radiated from the fiber end (Sect. 8.2). These are very useful for determining the mode field diameter from the measured radiation pattern (Sect. 13.7).

8.1 Far-Field Radiation Pattern

We assume the fiber end to be planar and normal to the fiber axis. The fiber endface is assumed to lie in the plane $z = 0$ (Fig. 8.1). The half-space $z > 0$ is filled with air. An LP_{lm} mode is assumed to be incident from $z = -\infty$. We are interested in the radiation field in a point of observation P with Cartesian coordinates x, y, z and spherical polar coordinates r, θ, ϕ.

The fiber endface can be considered as an aperture radiator and the radiation field can be calculated from the aperture illumination using methods known from antenna theory [Ramo et al. 1965]. Alternatively, the fiber endface can be considered as a diffracting aperture and the diffracted field can be obtained using methods known from optical diffraction theory [Born and Wolf 1975]. In essence, the two methods are, of course, identical.

First, one has to determine the electromagnetic field in the plane $z = +0$. Because of the reflected guided wave and a back-radiated unguided wave, the field will deviate from the field of the incident wave. Therefore, one proceeds as for the case of mode launching at the fiber front end (Sect. 7.2.1): at first, the reflected wave is neglected and the aperture field is assumed to be equal to the field of the incident mode field. The radiation field is then calculated as

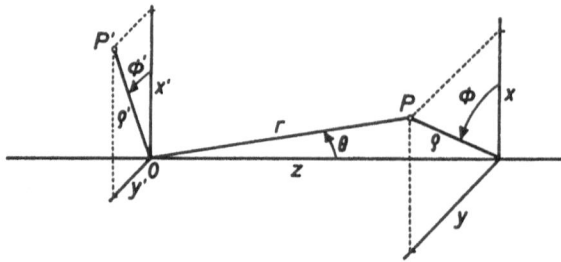

Fig. 8.1. Cylindrical polar cooordinates ϱ', ϕ', 0 of the source point in the fiber end face and cylindrical polar coordinates ϱ, ϕ, z and spherical coordinates r, θ, ϕ of the observation point in the field radiated from the fiber end

will shortly be described. Finally, the reflection losses are taken into account by multiplying the radiation field by the amplitude transmission factor for a normally incident homogeneous plane wave traversing the plane interface between glass and air (5.127).

Since an electromagnetic field is a vector field, in principle one has to apply the complicated Kirchhoff-Huygens *vector*-diffraction formula for calculating the radiation field [Ramo et al. 1965]. But for the LP_{lm} fiber modes (Chaps. 5, 6), one of the three cartesian components of the electric field vector, e.g. $\underline{E}_x$, is much larger than the other two, and the field distribution in the radiating fiber endface can be described approximately by a single scalar function $E(x', y')$ or $E(\varrho', \phi')$, which, by suitably choosing the origin of the time scale, can be made real. The variables x', y' and ϱ', and ϕ' are the Cartesian and cylindrical coordinates, respectively, of a Huygens source point P' in the radiating plane $z = 0$.

Moreover, the transverse field extent in the radiating aperture is assumed to be large as compared to the wavelength. There is then no radiation for polar angles θ much larger than about 10°. In order to get a simple formula for the radiation field, the distance r of the observation point P from the center of the radiating fiber endface is assumed to be much larger than the Rayleigh distance z_R (4.6) of the radiated beam:

$$r \gg z_R = \pi w_G^2 / \lambda \quad , \tag{8.1}$$

where w_G is the spot size of the fundamental fiber mode. The observation point is then said to be located in the "far field".

With these assumptions, one may apply the simpler *scalar* diffraction theory for calculating the electric field $E_f(x, y, z)$ at the observation point; this is the so-called far-field radiation pattern [Goodman 1968]:

$$E_f(x, y, z) = \frac{j}{\lambda z} \exp(-jkr) \int\!\!\int\limits_{-\infty}^{+\infty} E(x', y')$$

$$\times \exp\left[j2\pi\left(\frac{xx'}{\lambda z} + \frac{yy'}{\lambda z}\right)\right] dx' dy', \quad \text{where} \tag{8.2}$$

$$r = \sqrt{x^2 + y^2 + z^2} \tag{8.3}$$

is the distance of the observation point from the center of the radiating fiber endface.

Introducing normalized transverse coordinates for the observation point P:

$$f_x = x/(\lambda z) \quad , \tag{8.4}$$

$$f_y = y/(\lambda z) \quad , \tag{8.5}$$

the far-field radiation pattern can be written in the form

$$E_f(x,y,z) = j(\lambda z)^{-1} \exp(-jkr) E_F(f_x, f_y) \quad , \tag{8.6}$$

where

$$E_F(f_x, f_y) = \int\!\!\!\int\limits_{-\infty}^{+\infty} E(x', y') \exp[j2\pi(f_x x' + f_y y')] dx' dy' \tag{8.7}$$

is the two-dimensional inverse Fourier transform [Goodman 1968] of the near-field amplitude distribution $E(x', y')$. The variables f_x and f_y are called spatial frequencies.

As a result of (8.6), the far-field distribution $E_f(x, y, z)$ is apart from multiplicative factors, the inverse two-dimensional Fourier transform of the near-field distribution $E(x', y')$. Because of the properties of the Fourier transform, a wide aperture field produces a narrow wave beam and vice versa.

The near-field amplitude distribution can either be measured (Sect. 13.5) or calculated from the refractive-index profile. For a given near field, (8.6, 7) can be used for determining the radiation pattern of the fiber end.

A new technique for calculating the Fourier transform of the near field [Narasimhan and Karthikeyan 1984] combines the accuracy of the trapezoidal rule in numerical integration and the efficiency of the fast Fourier transform (FFT) algorithm and minimizes the aliasing error.

Conversely, the near field $E(x', y')$ can be calculated from the far field $E_f(x, y, z)$ by solving (8.6) for $E_F(f_x, f_y)$:

$$E_F(f_x, f_y) = -j\lambda z \exp(jkr) E_f(x, y, z) \quad , \quad \text{with} \tag{8.8}$$

$$x = \lambda z f_x \quad , \tag{8.9}$$

$$y = \lambda z f_y \quad , \tag{8.10}$$

and by inserting $E_F(f_x, f_y)$ into the formula for the two-dimensional Fourier transform [Goodman 1968]:

$$E(x', y') = \int\!\!\!\int\limits_{-\infty}^{+\infty} E_F(f_x, f_y) \exp[-j2\pi(f_x x' + f_y y')] df_x df_y \quad . \tag{8.11}$$

Equations (8.8, 11) can be used to determine the near-field amplitude distribution from the far-field radiation pattern, which can be measured more easily (Sect. 13.6) than the near-field intensity distribution (Sect. 13.5).

The relation between the near field and the far field can also be expressed using polar coordinates. Replacing the Cartesian coordinates x', y' of the source point P' by cylindrical polar coordinates ϱ', ϕ', and the Cartesian coordinates x, y, z of the observation point by spherical coordinates r, θ, ϕ (Fig. 8.1), gives for the far field [Gambling et al. 1976d; Anderson W.T. and Philen 1983a; Anderson W.T. 1984]:

$$E_f(r, \theta, \phi) = \frac{j}{\lambda r} \exp(-jkr) \int_0^{2\pi} \int_0^\infty E(\varrho', \phi')$$

$$\times \exp[jk\varrho' \sin \theta \cos(\phi - \phi')]\varrho' d\varrho' d\phi' \quad . \tag{8.12}$$

This is the so called Fraunhofer diffraction equation, which can be applied to calculate the radiation field caused by an arbitrary fiber mode. One has only to insert for $E(\varrho', \phi')$ the transverse field distribution (6.1) of the mode considered.

For the special case of the fundamental mode, the aperture illumination does not depend on the azimuthal angle ϕ' and the integral over ϕ' can be obtained in closed form by using the integral representation [Gradshteyn and Ryzhik 1965] of the zero-order Bessel function J_0:

$$E_f(r, \theta) = \frac{j}{\lambda r} \exp(-jkr)2\pi \int_0^\infty E(\varrho')J_0(k \sin \theta \varrho')\varrho' d\varrho' \quad . \tag{8.13}$$

By introducing a normalized angular coordinate, the spatial frequency

$$q = \frac{\sin \theta}{\lambda} \quad , \tag{8.14}$$

the far-field radiation pattern of the fundamental fiber mode can be written in the form [Williams C.S. 1973; Hotate and Okoshi 1979b; Anderson W.T. and Philen 1983a]

$$E_f(r, \theta) = \frac{j}{\lambda r} \exp(-jkr)E_H(q) \quad , \quad \text{where} \tag{8.15}$$

$$E_H(q) = 2\pi \int_0^\infty E(\varrho')J_0(2\pi q\varrho')\varrho' d\varrho' \tag{8.16}$$

is the Hankel transform of zero order (Fourier-Bessel transform) of the radial field distribution $E(\varrho')$ [Goodman 1968]. Note that the radiation pattern (8.15) is independent of the azimuth ϕ.

Some authors use the spatial angular frequency u with

$$u = 2\pi q \tag{8.17}$$

instead of the spatial frequency q.

In short, the radiation pattern (8.15) of the fundamental fiber mode is the Hankel transform of its radial field distribution. Equations (8.15, 16) can be used to calculate the far-field pattern $E_f(r, \theta)$ from the radial near-field distribution $E(\varrho')$. Conversely, the radial amplitude distribution $E(\varrho')$ of the fundamental fiber mode can be determined from the far-field radiation pattern $E_f(r, \theta)$ by solving (8.15) for $E_H(q)$:

$$E_H(q) = -j\lambda r \exp(jkr) E_f(r, \theta) \quad , \quad \text{with} \tag{8.18}$$

$$\theta = \sin^{-1}(q\lambda) \quad , \tag{8.19}$$

and by calculating the inverse Hankel transform of $E_H(q)$ [Hotate and Okoshi 1979b]:

$$E(\varrho') = 2\pi \int\limits_0^\infty E_H(q) J_0(2\pi q \varrho') q \, dq \quad . \tag{8.20}$$

Note that the inverse Hankel transform operator (8.20) has the same form as the direct Hankel transform operator (8.16).

Equations (8.18, 20) can be used to determine the near-field amplitude distribution of the LP_{01} mode from the far-field radiation pattern, which can be measured more easily (Sect. 13.6) than the near-field intensity distribution (Sect. 13.5).

Like the far field of any antenna, since $z \approx r$, the radiation field amplitude decays as $1/r$, and the intensity decays with to $1/r^2$. Since the amplitude distribution $E(\varrho')$ of the fundamental fiber mode in the radiating endface is positive real, the Hankel transform (8.16) is also real, and is positive on the beam axis. From (8.15), the phase angle of the electric field amplitude on the beam-axis is $-kz + \frac{\pi}{2}$, as is to be expected for a free Gaussian beam for points far from the beam waist [(4.12) with (4.16)]. From the factor $\exp(-jkr)$ in (8.15), the wavefronts are seen to be spheres. Equation (8.15) also shows that the phase of the radiation field is constant on a sphere with radius r as long as the sign of the real Hankel transform $E_H(q)$ is constant. At the angles θ for which the sign of the Hankel transform changes, i.e. at the zeros of the radiation pattern, the phase of the radiation field changes by $180°$.

Summarizing, the near-field and the far-field patterns of the fundamental fiber mode essentially form a pair of Hankel transforms.

Since the radial field distribution is not usually known in closed form, one has to use numerical methods for calculating its Hankel transform, i.e. the far-field radiation pattern. However, the radial field distribution of the fundamental mode can be approximated remarkably precisely by a Gaussian function (5.25) [Marcuse 1970c]. Under optimal conditions, in a step-index fiber with $V = 2.4$, 99.7% of the fundamental mode power is found in a Gaussian distribution at the fiber endface and propagates as a Gaussian beam mode into free space.

A simple Gaussian function approximates the radial field distribution not only for step-index fibers, but also for many fibers with other refractive-index profiles. It is thus interesting to know the Hankel transform of a Gaussian function.

For a Gaussian function

$$E_G(\varrho') = (2/w_G)\exp[-(\varrho'/w_G)^2] \quad , \tag{8.21}$$

the Hankel transform

$$E_{GH}(q) = (2/W_G)\exp[-(q/W_G)^2] \quad , \tag{8.22}$$

is also Gaussian [Anderson W.T. 1984]. The width W_G of the Hankel transform (8.22) is related to the width w_G in (8.21) by (10.30)

$$W_G = 1/(\pi w_G) \quad . \tag{8.23}$$

Both the functions $E_G(\varrho')$ and $E_{GH}(q)$ are normalized, i.e. the power integrals are unity:

$$\int_0^\infty E_G^2(\varrho')\varrho' d\varrho' = 1 \quad , \tag{8.24}$$

$$\int_0^\infty E_{GH}^2(q)q \, dq = 1 \quad . \tag{8.25}$$

In order to appoximately calculate the radiation field of the fiber end including the field amplitude and taking into account the reflection losses at the fiber endface, the radial field distribution of the fundamental mode is approximated by a Gaussian function (5.25)

$$E(\varrho) = E_0 \exp[-(\varrho/w_G)^2] \quad . \tag{8.26}$$

The aperture field distribution in the radiating endface is then

$$E(\varrho') = E_A \exp[-\varrho'^2/w_G^2] \quad , \tag{8.27}$$

where (5.127)

$$\frac{E_A}{E_0} = \frac{2n_2}{n_2 + n_h} \quad , \tag{8.28}$$

is the amplitude transmission factor of the fiber endface between glass (n_2) and the homogeneous medium (n_h) into which the fiber end radiates.

When we set $w_G = w_0$, (8.27) represents the field distribution of a Gaussian beam in the plane of the waist (4.3). The far-field $(z \gg z_R)$ radiation pattern can be obtained from the formula (4.1) for the field of a Gaussian beam by approximating $\varrho/z = \theta$

$$E_f(r,\theta) = j(\lambda r)^{-1}\exp(-jkr)E_A\pi w_G^2 n_h \exp[-(\theta/\theta_d)^2] \quad , \tag{8.29}$$

where

$$\theta_d = \frac{\lambda}{\pi n_h w_G} \quad , \tag{8.30}$$

is the beam divergence angle (4.8).

Alternatively, the radiation pattern (8.29) of the fiber end can be derived from (8.15) and (8.22) by approximating the spatial frequency q by θ/λ and by expressing W_G by w_G (8.23).

Thus, in the Gaussian approximation, the fiber end can be considered as an optical dielectric aerial, radiating a beam of $1/e$-half-angle θ_d (8.30) in the direction of the fiber axis.

For step-index fibers, in the typical range of operating wavelengths, w_G is approximately proportional to λ (Sect. 5.3.2). Equation (8.30) then indicates that the width of the far-field radiation pattern hardly changes with wavelength [Geckeler 1986a].

The intensity on the beam axis ($\theta = 0$) at a distance r from the fiber end is

$$S = \frac{n_h}{2Z_0}\frac{4n_2^2}{(n_h + n_2)^2}E_0^2\frac{\pi^2 w_G^4 n_h^2}{\lambda^2 r^2} \quad . \tag{8.31}$$

The power guided by the fundamental mode available for radiation is (5.37)

$$P = \frac{\pi n_2 E_0^2 w_G^2}{4Z_0} \quad . \tag{8.32}$$

A hypothetical antenna radiating this available power P isotropically, would produce a spherical wave with intensity S_i at a distance r

$$S_i = P/(4\pi r^2) \quad , \tag{8.33}$$

since the power is uniformly distributed over the surface $4\pi r^2$ of a sphere with radius r.

Since the fiber end acts as a directional antenna, the intensity S in the maximum of the radiation lobe ($\theta = 0$) is larger than the intensity S_i produced by the hypothetical isotropic radiator by a factor gi known as the antenna gain [Ramo et al. 1965]

$$g_i = S/S_i \quad . \tag{8.34}$$

Inserting (8.31) for S, (8.33) for S_i, and (8.32) for P gives [Unger 1984]

$$g_i = \frac{4n_h n_2}{(n_h + n_2)^2}8\pi^2 w_G^2\left(\frac{n_h}{\lambda}\right)^2 = 8\pi^2 w_G^2(n_h/\lambda)^2 T \quad , \tag{8.35}$$

where the symbol T denotes the power transmission factor (5.129) of the interface between glass and the homogenous medium, which is approximately unity.

For a single-mode fiber with $w_G = 5\,\mu m$ radiating at $\lambda = 1.3\,\mu m$ into air ($n_h = 1$), the antenna gain is $g_i = 1127$ or $30.5\,dB$.

Comparing the equations for the effective antenna area A_e (7.24) and the antenna gain g_i (8.35), one sees that these quantities are proportional

$$A_e = g_i \frac{1}{4\pi} \left(\frac{\lambda}{n_h} \right)^2 \quad . \tag{8.36}$$

This relation between the receiving and transmitting properties of an electromagnetic antenna is well known from lower frequencies [Ramo et al. 1965].

Since the Gaussian function decreases monotonically to zero, the approximate radiation pattern (8.29) has no side lobes and no zeros.

For real fibers, the near-field distribution deviates from the Gaussian function, and the field radiated from the fiber end has sidelobes. However, the side lobes are very difficult to detect experimentally since they are very small.

As an example we consider the radiation of the fundamental mode from the end of a step-index fiber. The normalized far-field intensity distribution is given by [Gambling et al. 1976b,d; Hotate and Okoshi 1979b]:

$$|E_f(\alpha)|^2 = \left[\frac{U^2 W^2}{(U^2 - \alpha^2)(W^2 + \alpha^2)} \left\{ J_0(\alpha) - \alpha J_1(\alpha) \frac{J_0(U)}{U J_1(U)} \right\} \right]^2 \quad , \tag{8.37}$$

for $\alpha \neq U$ and

$$|E_f(\alpha)|^2 = \left[\frac{U^2 W^2}{2V^2} \frac{1}{U J_1(U)} (J_0^2(\alpha) + J_1^2(\alpha)) \right]^2 \tag{8.38}$$

for $\alpha = U$.

The normalized angular variable α is

$$\alpha = 2\pi a q = ka \sin\theta \quad . \tag{8.39}$$

A simple approximate expression for the far-field distribution has been derived by Timmermann [1977].

Equations (8.37, 38) have been evaluated for normalized frequencies of $V = 1.6$ and 2.4. Figures 8.2, 3 show the calculated far-field radiation pattern which displays zeros and side lobes. However, the secondary maximum is more than $40\,dB$ below the central maximum.

For other refractive-index profiles, the secondary maxima were experimentally found to have intensity levels $28\,dB$ to more than $68\,dB$ below the central maximum [Tynes et al. 1979].

When analyzing the insertion loss of a fiber connector with an air gap, it is also necessary to know the radiation field of the fundamental fiber mode for small distances r from center of the radiating endface. The Fraunhofer diffraction equation (8.12) or (8.13) may not be applied, because it assumes the observation point to be far from the fiber end. Usually, it is extremely difficult

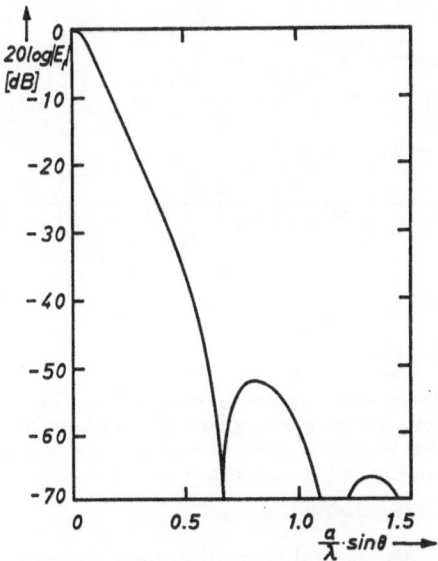

Fig. 8.2. Normalized far-field radiation pattern of a step-index fiber with a normalized frequency of $V = 1.6$. Independent variable: normalized polar angle $aq = \alpha/(2\pi) = a \sin \theta / \lambda$

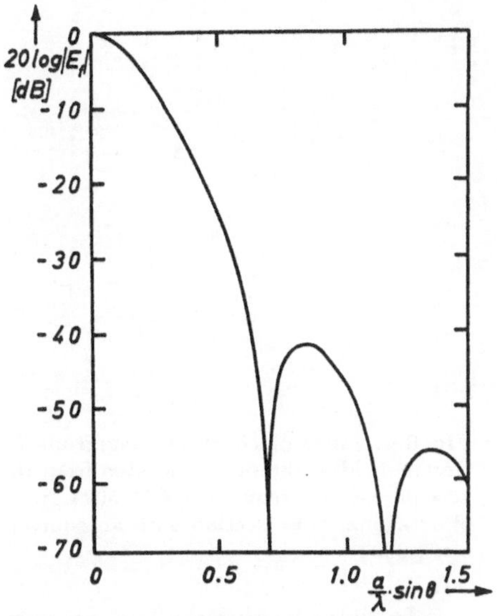

Fig. 8.3. Normalized far-field radiation pattern of a step-index fiber with a normalized frequency of $V = 2.4$. Independent variable: normalized polar angle $aq = \alpha/(2\pi) = a \sin \theta / \lambda$

to calculate radiation fields close to the source. But fortunately, the transverse field distribution of the LP_{01} mode can be approximated both in the core and in the cladding by a simple Gaussian function. The Fourier integral then can be evaluated in closed form and the radiation field both for small and large values of r is given by the well-known formula (4.1) for the field distribution in the fundamental Gaussian beam in a homogeneous medium.

When the radial field distribution of the fundamental fiber mode cannot adequately be described by a Gaussian function, the field strength at each observation point can be determined by numerically evaluating the Kirchhoff-Huygens vector diffraction formula.

A more efficient method is to expand the near-field amplitude distribution $E(\varrho', \phi')$ as a truncated series of orthogonal Hermite-Gaussian or Laguerre-Gaussian functions [Bogush and Elkins 1986]. Physically, this corresponds to decomposing the radiated beam into Gaussian beam modes (Sect. 4.8). Apart from a continuous increase in spot size and a change in the wavefront curvature, each Gaussian beam mode maintains its functional form during propagation for all distances from the fiber end. The beam field is approximated by the sum of the simple fields of the constituent Gaussian beam modes both for small distances z from the radiating fiber end as well as for large distances.

This method is a generalization of the fundamental Gaussian beam mode approximation for the radiated beam described before. It is more accurate than the fundamental Gaussian beam approximation, and computationally more convenient than the numerical evaluation of diffraction integrals.

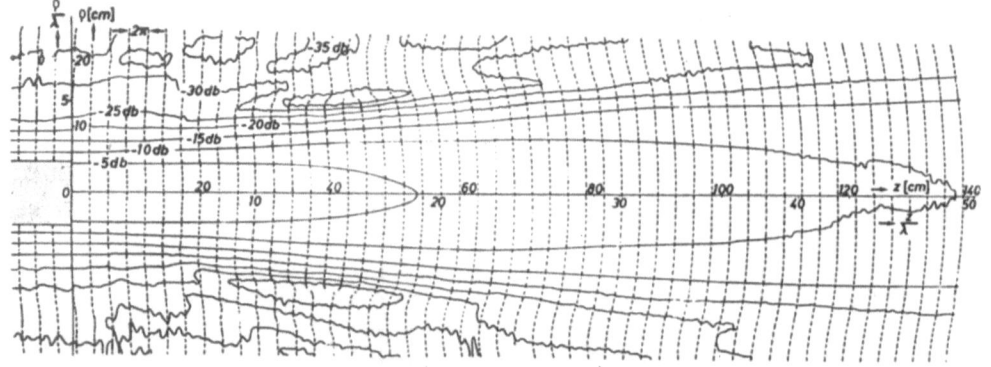

Fig. 8.4. Lines of constant magnitude (*solid lines*) and wavefronts (*dashed lines*) of the electric field in the beam radiated from the end of a dielectric waveguide. Model experiment at a microwave frequency of 10.692 GHz. The fiber model consisted of a foamed plastic rod of octagonal cross section with an equivalent diameter of 9.8 cm and a relative permittivity $\varepsilon = 1.033$

In order to give a visual impression of the field distribution near the radiating free end of a single-mode fiber, a model experiment has been performed [Albrecht and Neumann 1979]. The fiber core was modeled by a rod made of polystyrol-foam with a diameter of 10 cm. The cladding of the model was air. With the apparatus described in Sect. 5.9.7, the distribution of the electric field was measured at a frequency of 10.7 GHz. In Fig. 8.4, solid lines represent curves of constant field strength and dashed lines wavefronts in a plane containing the beam axis. Behind the fiber end and near the beam axis, one recognizes the radiation lobe with increasing beam diameter and spherical wavefronts. In the region of high field strength, the field distribution is very similar to that calculated for a fundamental Gaussian beam with a spot size $w_G = 3\lambda$ (Fig. 4.9).

There is also some interest in the radiation patterns of higher order fiber modes:

> For step-index fibers, closed form expressions for the radiation pattern of the LP_{11} mode [Pocholle 1979] and for the exact vector $TE_{0\mu}$, $TM_{0\mu}$, $EH_{\nu\mu}$, and $HE_{\nu\mu}$ modes [Kapany et al. 1965; Kapany and Burke 1972] have been derived.

When a fiber transmits both the LP_{01} and the LP_{11} mode, the coherence properties of the source and the fiber properties have to be taken into account when considering the radiation field. For a coherent source the *fields* superimpose and interference effects will occur both in the near field and in the far field (Sect. 3.7.1). For a partially coherent source, when the difference in time delay (13.99) of the two modes in the fiber is larger than the coherence time, the *intensities* of the modes add up in both the near and far field and the patterns will not change if the phase difference between the modes changes, e.g. due to drift of laser wavelength.

8.2 Relations Between the Near-Field and Far-Field Functions

In Sect. 8.1, we have learned that the near- and far-field amplitude distributions of the fundamental fiber mode essentially form a pair of Hankel transforms. There are two other relations between the near-field and the far-field pattern which can be obtained from the two-dimensional autocorrelation (Wiener-Khintchine) theorem [Goodman 1968]. This theorem states that the autocorrelation function and the power spectrum of a complex function of two independent variables form a pair of two-dimensional Fourier transforms. In the special case of functions with circular symmetry, they form a pair of Hankel transforms.

We discuss the three connections between the near field (NF) and the far field (FF) of the fundamental mode using the scheme plotted in Fig. 8.5 and start with the NF amplitude distribution $E(\varrho')$, which, in order to emphazise the main points, is assumed to be positive real and to be normalized, i.e.

$$\int_0^\infty E^2(\varrho)\varrho\, d\varrho = 1 \quad . \tag{8.40}$$

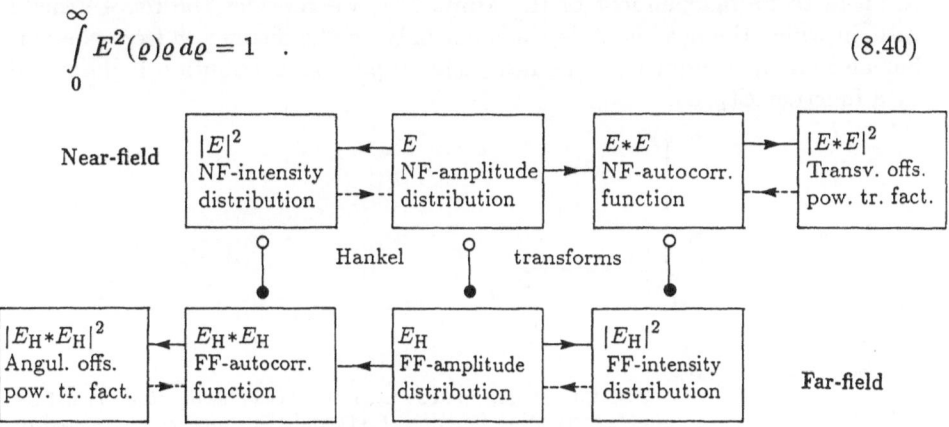

Fig. 8.5. Relations between the amplitude distributions, intensity distributions, and autocorrelation functions of the near-field (NF) and the far-field (FF) of a single mode fiber

Instead of the actual FF-amplitude distribution $E_f(r, \theta)$, the function $E_H(q)$ is considered, which differs from $E_f(r, \theta)$ only by the factor $-j\lambda r \exp(jkr)$ (8.18). The functions $E(\varrho)$ and $E_H(q)$ form a pair of Hankel transforms (8.16, 20). In Fig. 8.5, this first connection between near field E and far field E_H is represented graphically by a vertical line with an open and a closed circle at its ends. The reduced FF amplitude distribution $E_H(q)$ is also real and normalized [Anderson W.T. 1984]:

$$\int_0^{1/\lambda} E_H^2(q)q\, dq = 1 \quad . \tag{8.41}$$

For obtaining the two other relations between the near field and the far field, the NF amplitude distribution E is manipulated twice.

Firstly, we form the square of the NF amplitude distribution to get the NF intensity distribution $|E|^2$ of the fundamental mode. In general, when the squared modulus of a complex quantity is formed, the information about its phase angle is lost. But for the fundamental mode of a fiber, it is known that the phase fronts are planes normal to the fiber axis. Therefore, the phase of the NF distribution is known to be independent of the radial coordinate, and one can get the NF amplitude distribution in the fiber endface by taking the square root of the measurable NF intensity distribution, neglecting an unimportant constant phase angle. This is indicated by a dashed arrow pointing from the NF intensity $|E|^2$ to the NF amplitude distribution E.

Secondly, we calculate the two-dimensional autocorrelation function $E * E$ of the transverse field distribution of the fundamental fiber mode by forming the overlap integral between the transverse NF amplitude distribution with a transversely offset version of the same field distribution. Before assuming the NF field to be independent of the azimuth ϕ, we consider the more general case, in which the near field depends not only on the distance ϱ from the axis but also on the azimuth ϕ. The near-field amplitude distribution is described by a function $E(\varrho, \phi)$.

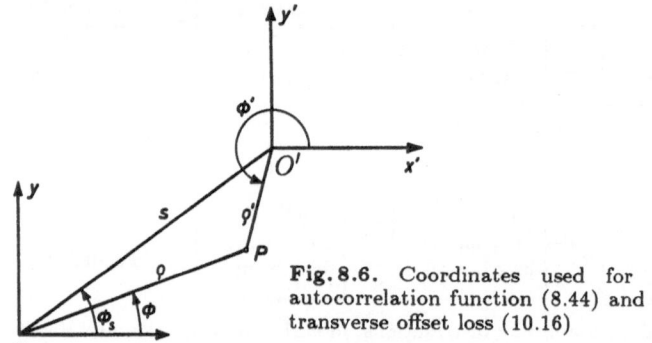

Fig. 8.6. Coordinates used for calculating the near-field autocorrelation function (8.44) and the spot size w_a related to transverse offset loss (10.16)

Figure 8.6 shows the coordinates used for calculating the two-dimensional autocorrelation function $E_a(s, \phi_s)$. The center of the unshifted near field is at the origin O of the xy coordinate system, and the center of the shifted field at the origin O' of the $x'y'$ coordinate system. The offset between the centers is s, and the direction of offset is characterized by the angle ϕ_s. (Note that in this connection, the coordinates ϱ, ϕ, ϱ', and ϕ' have a meaning different to that in Fig. 8.1). An arbitrary point P in the cross-section has the cylindrical polar coordinates ϱ, ϕ in the unshifted, unprimed coordinate system, and the coordinates ϱ', ϕ' in the shifted, primed coordinate system. The coordinate transformation is described by the equations

$$\varrho'^2 = \varrho^2 + s^2 - 2\varrho s \, \cos(\phi - \phi_s) \quad , \quad \text{and} \tag{8.42}$$

$$\tan \phi' = (\varrho \sin \phi - s \sin \phi_s)/(\varrho \cos \phi - s \cos \phi_s) \quad . \tag{8.43}$$

At the point P, the unshifted field is $E(\varrho, \phi)$ and the shifted field $E(\varrho', \phi')$, where E denotes the *same* function of two variables. For calculating the two-dimensional autocorrelation function, one first has to multiply the unshifted and shifted fields at P, and then to integrate their product $E(\varrho, \phi)E(\varrho', \phi')$ over the entire fiber cross-section:

$$E_a(s, \phi_s) = \int_0^{2\pi} \int_0^\infty E(\varrho, \phi)E(\varrho', \phi')\varrho \, d\varrho \, d\phi \quad . \tag{8.44}$$

In general, the two-dimensional autocorrelation function of the near-field amplitude distribution depends both on the magnitude (s) and the direction (ϕ_s) of the offset.

The field of the fundamental mode is approximately circularly symmetric, i.e. independent of the azimuth ϕ. The autocorrelation function of the near field is thus independent of the direction of the shift, i.e. independent of the angle ϕ_s. For simplicity, one can choose $\phi_s = 0$, and obtains for the coordinate transformation (8.42)

$$\varrho'^2 = \varrho^2 + s^2 - 2\varrho s \cos \phi \quad . \tag{8.45}$$

and for the autocorrelation function of the near-field distribution

$$E * E = E_a(s) = \int_0^{2\pi} \int_0^\infty E(\varrho)E(\varrho')\varrho \, d\varrho \, d\phi \quad , \tag{8.46}$$

Since it is not possible to reconstruct the NF amplitude distribution directly from the autocorrelation function, in Fig. 8.5, there is only one arrow pointing from the radial NF field distribution E to the NF autocorrelation function $E * E$.

The power transmission factor for a joint with a transverse offset s equals the squared modulus $|E * E|^2$ of the normalized autocorrelation function (Sect. 10.1.5). Since we have assumed a positive NF amplitude distribution, the autocorrelation function is real and can be obtained (dashed arrow) by taking the square root of the power transmission factor, which can easily be measured using the transverse offset method (Sect. 13.7.2).

In a manner analogous to the near field, one can manipulate the far-field distribution. Firstly, the FF intensity pattern $|E_H|^2$ is calculated by taking the squared modulus of the FF amplitude distribution. In the FF radiation pattern produced by the fundamental fiber mode, the phase of the electric field was shown (Sect. 8.1) to be constant on a sphere except for sudden changes by π (a reversal in sign of the electric field strength) at the zeros of the radiation pattern. Therefore, in this special case, the FF amplitude pattern can be obtained by taking the square root of the measured intensity pattern and applying the appropriate sign (dashed arrow pointing to the left).

Secondly, in formal analogy to the autocorrelation function of the NF-amplitude distribution (8.46), we define the autocorrelation function $E_H * E_H$ of the FF amplitude distribution:

$$E_H * E_H = E_{Ha}(q_s) = \int_0^{2\pi} \int_0^{1/\lambda} E_H(q)E_H(q')q\,dq\,d\phi \quad , \quad \text{where} \tag{8.47}$$

$$q'^2 = q^2 + q_s^2 - 2qq_s \cos\phi \quad , \tag{8.48}$$

with q given by (8.14) and

$$q_s = (1/\lambda)\sin\theta_s \quad . \tag{8.49}$$

The FF autocorrelation function depends on the normalized angular misalignment q_s which corresponds to the radial offset s in the NF autocorrelation function. For the reason mentioned above, in Fig. 8.5, there is only one arrow pointing from the FF amplitude distribution E_H to the FF autocorrelation function $E_H * E_H$.

We now speculate as to whether the FF autocorrelation function $E_{Ha}(q_s)$ could be determined experimentally. The integrand in (8.47) is the product of the function $E_H(q)$ and an angle-shifted version $E_H(q')$ of the same function. The function $E_H(q)$ is related to the amplitude distribution in the far field $E_f(\varrho, \theta)$ by (8.18). Note that the phase angles of $E_H(q)$ and E_f differ by $kr - \frac{\pi}{2}$. Neglecting this angle for the moment, the integral can be interpreted as an overlap integral between an unshifted and a shifted FF-amplitude distribution.

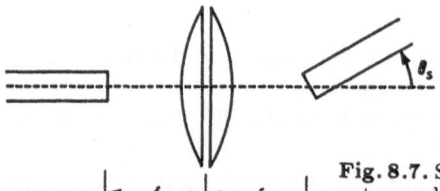

Fig. 8.7. Scheme of an experiment to measure the far-field autocorrelation function

Experimentally, the overlap integral could be formed by the set-up of Fig. 8.7. An ideal lens with focal length $f \gg z_R$ is placed in the field radiated from the end of the first fiber. The focus of the lens coincides with the center of the fiber endface. The lens collimates the radiated beam. At the input face of the lens, the field distribution is $E_f(r, \theta)$ and at the output face it is $E_H(q)$.

Similarly, a second tilted fiber produces a shifted version of the function $E_H(q)$ in the back plane of a second ideal lens. The overlap integral of the two fields in a plane between the two lenses gives the far-field autocorrelation function, and its square the power transmission factor between the two fibers.

Alternatively, one can calculate the coupling efficiency between the first and the second fiber by determining the incident field at the input face of the second fiber, and by calculating the square of the overlap integral between the incident field and the tilted fundamental mode field of the second fiber. The two lenses form a real image with unit magnification of the near field of the first fiber on the input face of the second fiber. The power transmission factor thus equals that of a tilt in a fiber.

This reasoning shows that, in principle, the squared modulus of the FF-autocorrelation function can be determined by measuring the power transmission factor of a fiber tilt.

Having defined the squared moduli and the autocorrelation functions of both the NF- and FF-amplitude distributions, the Wiener-Khintchine theorem can be applied to show that the NF-autocorrelation-function $E_a(s)$ (8.46) and the FF-intensity distribution $|E_H(q)|^2$ form a second pair of Hankel transforms [Pask 1984, Anderson W.T. 1984]:

$$E_a(s) = H\{|E_H(q)|^2\} \quad , \tag{8.50}$$

$$|E_H(q)|^2 = H\{E_a(s)\} \quad , \tag{8.51}$$

and that the FF-autocorrelation function $E_{Ha}(q_s)$ (8.47) and the NF-intensity distribution $|E(\varrho)|^2$ form a third pair of Hankel transforms

$$E_{Ha}(q_s) = H\{|E(\varrho)|^2\} \quad , \tag{8.52}$$

$$|E(\varrho)|^2 = H\{E_{Ha}(q_s)\} \quad . \tag{8.53}$$

In these equations, $H\{\ldots\}$ denotes the Hankel transform operator (8.16, 20).

The relations between near-field and far-field functions illustrated in Fig. 8.5 relate four measurable functions: near-field intensity distribution $|E|^2$, far-field intensity distribution $|E_H|^2$, transverse offset power transmission function $|E * E|^2$, and tilt power transmission function $|E_H * E_H|^2$ to one another and to the near- and far-field amplitude distributions (E, E_H) and the autocorrelation functions $(E * E, E_H * E_H)$ which cannot be measured directly. The relations can be utilized to determine one of these functions, which itself may be difficult to measure directly, from another that is easier to measure. These relations are also very useful for determining the values of the various types of spot size of the fundamental fiber mode (Sect. 13.7).

There is an alternative formulation of fiber-optics theory, which emphasizes the far-field properties instead of the near-field properties of the guided mode field [Pask 1984]. In this, the scalar wave equation is Hankel transformed to give an integral equation for the the FF-amplitude distribution. This integral equation contains the Hankel transform of the refractive index distribution, which can be measured directly by illuminating the fiber transversely with a laser beam and observing the resulting diffraction pattern [Okoshi and Hotate 1976b; Brinkmeyer 1979b].

9. Joints Between Fibers

The method most often used to connect two single-mode fibers is the simple butt joint: the fiber endfaces are prepared such that they are flat and normal to the fiber axis (Sect. 12.1.1) and then butt-joined, either permanently to form a splice (Sect. 12.2.4), or detachably to form a connector (Sect. 12.2.5). If two identical fibers are jointed with no geometrical misalignment, the two fibers behave like a single continuous fiber and there is no reason for power loss.

In reality, the fibers to be connected will be slightly different, and because of an amplitude distribution mismatch, so called "intrinsic" losses will result. Moreover, because of the small size of the fiber core, it is difficult to avoid misalignments such as a transverse offset of the fiber axes, an air gap between the fiber endfaces, or a finite tilt angle between the fiber axes. These misalignments will cause so called "extrinsic" losses.

In this chapter, we will first summarize simple approximate formulas for the losses caused by a spot size mismatch only (Sect. 9.1), by a transverse offset only (Sect. 9.2), by a gap only (Sect. 9.3), and by a tilt only (Sect. 9.4). Section 9.5 reviews more general formulas for the joint loss for the case of several simultaneous misalignments. In the two-mode region, joints not only cause radiation losses, but also power conversion between the fundamental and the LP_{11} modes (Sect. 9.6). Finally, it is interesting to know the losses of a butt joint between a single-mode fiber and a multimode fiber for both directions of transmission (Sect. 9.7).

The power transmission factor η of a joint is defined by the ratio of the power P_2 propagated by the fundamental mode in the secondary fiber and the power P_1 propagated by the fundamental mode in the primary fiber:

$$\eta = P_2/P_1 \quad . \tag{9.1}$$

The corresponding joint loss a in optical decibels is

$$a = -10 \log \eta \quad . \tag{9.2}$$

The power transmission factor can be calculated using the overlap integral method (Sect. 7.2.1): in the plane input face of the outgoing fiber, the field distribution of the fundamental mode in this fiber has to be multiplied locally by the field of the incident wave. This product then has to be integrated over the infinite input face of the outgoing fiber [Cook et al. 1973]. The squared modulus of the overlap integral gives the power transmission factor (7.11).

The classical theoretical approach to light-energy coupling between single-mode optical waveguides placed end to end has been reviewed by Vatoux et al. [1983].

For practical purposes, it will usually suffice to approximate the fundamental mode field distributions by Gaussian functions of widths w_{G1} and w_{G2} (5.25). The overlap integral and the two power integrals in the denominator of (7.11) can then be evaluated in closed form. The resulting approximations for the transmission factor are formally equal to the expressions for the efficiency of launching the fundamental mode by a Gaussian beam known from Sect. 7.2.2. Notice, that for single-mode fibers, it is the mode field diameter rather than the core diameter that is significant for fiber joining. In contrast to multimode fibers, where the insertion loss of a joint sensibly depends on the mode spectrum [Dalgleish 1980], the insertion loss of a single-mode fiber joint is a well defined quantity and is independent of the direction of power transmission.

9.1 Loss Caused by Spot Size Mismatch

We consider two different fibers to be butt joined. The losses related to the properties of the two fibers to be joined are referred to as fiber-mismatch or intrinsic losses. Even if it were possible to avoid transverse offsets, gaps, and tilts, radiation losses would still occur because of the different field distributions in the fibers to be connected. Using the Gaussian approximation for the radial field distribution, the coupling efficiency is [Kogelnik 1964a; Marcuse 1977b]

$$\eta = \left(\frac{2w_{G1}w_{G2}}{w_{G1}^2 + w_{G2}^2} \right)^2 \tag{9.3}$$

which is formally equal to the equation (7.16) for the efficiency with which the fundamental mode is launched by a Gaussian beam illuminating the fiber front face. Thus we can also use Fig. 7.3 to represent the change in power transmission factor caused by a mismatch in spot size.

In the Gaussian approximation, the loss is zero if the spot sizes of the two fibers are the same. Since the same spot size can be obtained for different combinations of the core radius and the relative refractive-index difference, we note the interesting result that a core radius mismatch or a mismatch in the radial index distribution do not necessarily cause losses.

Moreover, fiber joints are relatively insensitive to differences between two similar fibers provided that they are otherwise perfectly aligned; a 15% mismatch in spot size causes a coupling loss of only 0.1 dB.

However, if a standard fiber with a spot size of $w_{G1} = 5\,\mu\text{m}$ at $\lambda = 1300\,\text{nm}$ is butt joined to a dispersion-shifted fiber with a markedly smaller spot size of $w_{G2} = 3.5\,\mu\text{m}$ [Shah et al. 1986; Anderson W.T. et al. 1986], there is a considerable mismatch in the amplitude distributions which causes the coupling efficiency estimated from (9.3) to be only 88.3% corresponding to an intrinsic loss of 0.54 dB.

When a fiber with a small spot size w_{G1} is to be field-matched to another one with a larger spot size w_{G2}, one has to transform a thin collimated Gaussian beam into a wider collimated Gaussian beam. Since the total power transmitted by the beam is constant, one has to redistribute the intensity within the beam cross-section: power has to be taken from the region near the beam axis and to be transferred to the region far from the axis. To this end, the time averaged Poynting vector, which in a uniform single-mode fiber has only an axial component, must obtain a positive radial component. Since the Poynting vector is normal to the wavefronts, the plane wavefronts of the fundamental mode in a uniform fiber must be deformed into spherical wave-fronts with positive radius of curvature.

The beam transformation can be accomplished by means of a taper (Sect. 12.2.6). At the beginning of the taper, the core radius equals the core radius of the first fiber with the smaller spot size, and at the end it equals the core radius of the second fiber with the larger spot size. At the beginning of the taper, the focusing power of the fiber is reduced a little, and therefore, the diffraction spreading tendency of the beam predominates (Sect. 2.2). In other words, the beam gradually begins to widen. However, the core of the tapered fiber only allows the beam to widen very slowly. Tapers have to be extremely gradually and are, in fact, only slightly more efficient than the abrupt transition between the fibers to be joined [Baets and Lagasse 1982].

Sometimes one is interested in beam transformers of minimum length. Then one must choose a scheme with the fastest increase of spot size, and a fiber taper is not the optimum solution. A better method is to remove a section of fiber core completely. The beam radiated from the end of the first fiber into the intermediate homogeneous medium diverges because of diffraction. The absence of a focussing core causes the beam to broaden faster than in the taper method. Of course, in order to match the radiated beam to the field of the fundamental mode in the second fiber with respect to both amplitude and phase distribution, one has to insert a positive lens into the beam at a certain position in order to make the curved wavefronts plane. Beam transformers of this kind are also used for transforming the fundamental fiber mode into a wide collimated free beam and vice versa (12.2.1).

The best method, however, to obtain a very short beam transformer, is to joint the first fiber to a short section of defocusing graded-index medium in which the core index is smaller than the cladding index. The defocusing medium increases the natural tendency of the beam to diffract, i.e. the beam widens faster than in the homogeneous medium and much faster than in the tapered fiber. The necessary wavefront match can be obtained by subsequently passing the beam through a focusing medium which acts as a positive lens.

The length and defocusing power of the defocusing medium and the length and focusing power of the focusing medium necessary for obtaining field matching can be calculated by applying the methods (Sect. 4.9) used for transforming Gaussian beams [Neumann 1987].

Another method for reducing losses caused by spot size mismatch consists in splicing a section of intermediate fiber between the two fibers to be connected. This method has a certain formal similarity to the quarter-wavelength transformer of transmission line techniques, since the length of the intermediate fiber must be equal to one quarter of the pitch (12.9) of an equivalent focusing graded-index medium and since the spot size of the intermediate fiber must equal the geometric mean of the spot sizes of the two fibers to be joined [Neumann 1987].

Still another device for reducing the loss of joints between fibers with different spot sizes is a biconical fiber taper. By tapering both fiber ends to reduce the diameter of the splice junction, the splice loss between a standard fiber ($2w_G = 9.86\,\mu$m) and a dispersion-shifted fiber ($2w_G = 4.34\,\mu$m) has been reduced from 1.5 dB to 0.56 dB [Mortimore and Wright 1986]. The residual loss is probably due to wavefront mismatch.

9.2. Loss Caused by Transverse Offset

The losses related to the characteristics of the joining technique are called extrinsic losses. These include power losses caused by a transverse offset between the fiber cores, end separation, and axial tilt.

When discussing extrinsic losses, we assume that two identical fibers are to be joined. First, the fiber axes are assumed to be offset transversely by a distance s with no tilt or gap. Since the incident field and the field to be launched are offset, there is an amplitude distribution mismatch which, using the Gaussian approximation (5.25) for the radial field distribution of the fundamental mode, gives a coupling efficiency [Kogelnik 1964a; Marcuse 1977b]

$$\eta = \exp[-(s/w_G)^2] \quad . \tag{9.4}$$

This equation corresponds to (7.17), which describes the launching efficiency when an offset free Gaussian beam with matched spot size excites the fundamental mode at the fiber input end. Thus Fig. 7.4 can also be used to represent the coupling efficiency.

The insertion loss

$$a = 4.34(s/w_G)^2 \quad , \tag{9.5}$$

caused by a transverse offset, is proportional to the square of the offset s. Since an offset of only $0.15\,w_G$ causes a loss of 0.1 dB, one can see that off-axis misalignment is most critical for practical single-mode fiber joints.

For a small transverse offset, the best approximation for the true joint loss is obtained by inserting the Petermann II spot size w_d (10.10) instead of w_G into (9.4, 5).

213

For step-index fibers, a more accurate expression for the factor $(4.34/w_G^2)$ relating the square of the transverse offset to the splice loss has been reported [Murakami et al. 1978].

The functional dependence (9.4) of the coupling efficiency on the offset can be used both to define an autocorrelation spot size w_a (Sect. 10.1.5) and to measure it (Sect. 13.7.2).

9.3 Loss Caused by a Gap

In a fiber connector, a gap of length z_w may exist between the endfaces. The gap is filled with a homogeneous intermediate medium with refractive index n_i, this can be air $(n_i = 1)$ or an immersion liquid $n_i \approx n_2$. Since the beam radiated from the end of the incoming fiber diverges, there is now both an amplitude and a wavefront mismatch in the input face of the outgoing fiber. The coupling efficiency can be calculated by inserting into the general formula (7.11) for the launching efficiency the field radiated from the first fiber and the field of the fundamental mode in the second fiber. Neglecting reflection losses and using the Gaussian approximation for the fundamental mode fields, the radiated field can be approximated by that of the fundamental Gaussian beam wave (4.1) and the coupling efficiency is simply [Kogelnik 1964a; Marcuse 1977b]

$$\eta = \frac{1}{1 + (0.5z_w/z_R)^2} \quad , \text{ where} \tag{9.6}$$

$$z_R = \pi n_i w_G^2 / \lambda \quad , \tag{9.7}$$

is the Rayleigh distance (4.6) of the beam radiated into the gap medium. The same equation applies to launching of the fundamental fiber mode by a defocused Gaussian beam (7.19); thus Fig. 7.5 also represents the power losses caused by a gap in a connector.

For the gap loss to be smaller than 0.1 dB, the longitudinal separation must be smaller than $0.3\,z_R$. For $\lambda = 1.3\,\mu m$ and $w_G = 5\,\mu m$, the Rayleigh distance in air is $60\,\mu m$ and the gap width must therefore be smaller than $18\,\mu m$. Single-mode fiber joints are thus relatively insensitive to longitudinal endface separation.

Note, however, that with air in the gap, multiple reflections between the two fiber endfaces cause the insertion loss to fluctuate as a function of the width of the gap. The reflection losses resulting from the two fiber endfaces have been discussed in Sect. 5.9.5. Because of the multiple reflections, the coupling efficiency of the gap (9.6) has to be multiplied by the factor $T = 1 - R$ (5.130) with R given by (5.132) to give the true power transmission factor.

Replacing the air between the fiber ends by an immersion fluid not only reduces the reflection losses, but also reduces the gap losses, since the Rayleigh distance (9.7) becomes larger. Other methods for reducing the reflection losses will be reviewed in Sect. 12.2.5.

9.4 Loss Caused by a Tilt

The last single geometrical misalignment to be considered is a tilt with angle θ between the axes of the incoming and outgoing fibers. The tilt causes a wavefront mismatch, resulting in a coupling efficiency [Kogelnik 1964a; Marcuse 1977b]

$$\eta = \exp[-(\theta/\theta_d)^2] \quad , \tag{9.8}$$

where the divergence angle (4.8)

$$\theta_d = \lambda/(\pi n_2 w_G) \quad , \tag{9.9}$$

is the angular $1/e$-width of the beam radiated from the fiber end into the homogenous cladding medium. A corresponding equation (7.20) holds for the launching efficiency of the fundamental mode launched by an obliquely incident Gaussian beam (Fig. 7.6).

For small angular misalignment, the best approximation for the true joint loss is obtained by inserting the effective spot size w_e (10.8) instead of the launching-related spot size w_G into (9.9).

Since the coupling loss of a tilted joint is

$$a = 4.34(\theta/\theta_d)^2 \quad , \tag{9.10}$$

the angular offset must be smaller than $0.15\theta_d$ if the loss is to be less than 0.1 dB. For typical values of $\lambda = 1.3\,\mu$m, $w_G = 5\,\mu$m, and $n_2 = 1.5$, the beam divergence angle is $\theta_d = 3.2°$ and for a loss of 0.1 dB, the tilt angle must remain smaller than $0.5°$.

For step-index fibers, a more accurate expression for the factor $(4.34/\theta_d^2)$ relating the square of the angular displacement to the splice loss has been reported [Murakami et al. 1978].

Comparing the equations for the coupling loss of a transverse offset (9.5) and a tilted joint (9.10), and taking into account that the divergence angle is inversely proportional to the spot size w_G (9.9), one recognizes that for given radial and angular tolerances of a fiber joint, a large spot size gives small transverse offset losses but larger losses due to any angular misalignment.

The product of the offset $0.15\,w_G$ and the tilt angle $0.15\,\theta_d$ which, each by themselves, cause a splice loss of 0.1 dB, equals $7.2 \times 10^{-3}\,\lambda/n_2$; it is a constant for a given operating wavelength. This "uncertainty principle for single-mode fibers" [Marcuse 1977a,b] tells us that tolerances with respect to offsets and tilts are mutually exclusive, i.e. any attempt to increase tolerance to lateral displacement by increasing the spot size will automatically lead to a decrease in tolerance to angular misalignment.

Since one usually has no problem with the angular tolerances, it might seem desirable to design fibers with a large spot size. However, a fiber with large spot size has large macro- and microbending losses (Sects. 5.9.7 and 5.9.9). Thus a compromise for the fiber spot size must be found. For a wavelength of $1.3\,\mu$m, the optimum value of the spot size value is between $4.5\,\mu$m and $5\,\mu$m (Sect. 10.1.7).

In connectors of the expanded-beam type (Sect. 12.2.5), the fiber end is allowed to radiate freely. A lens placed in the radiated diverging beam makes the spherical wavefronts plane and thus produces a wide collimated beam. Because of the increased spot size of the expanded beam, a larger radial offset can be tolerated, but because of the "uncertainty principle", the angular tolerance is then smaller.

9.5 General Loss Formulas

Until now we have only considered individual sources of joint losses. In general, two butt-joined fibers will have transverse offset, gap, and tilt simultaneously. For small losses, the total loss in decibels is approximately equal to the sum of the individual losses caused by spot-size mismatch (9.3 with 9.2), lateral (9.5), longitudinal (9.6 with 9.2), and angular offsets (9.10) [Cook et al. 1973; Gambling et al. 1978e].

The coupling efficiency between two different fibers with a gap but without transverse or angular misalignment is [Kogelnik 1964a]

$$\eta = \frac{4w_{G1}^2 w_{G2}^2}{(w_{G1}^2 + w_{G2}^2)^2 + [\lambda z_w/(\pi n_i)]^2} \quad . \tag{9.11}$$

A formula for the coupling efficiency of joints with spot size mismatch, transverse offset, and angular offset, but excluding longitudinal offset has been published by Grau et al. [1980].

A simple approximate formula for the coupling efficiency in the general case excluding the gap influence is

$$\eta = \left(\frac{2w_{G1}w_{G2}}{w_{G1}^2 + w_{G2}^2}\right)^2 \exp\left[-\frac{2w_{G1}w_{G2}}{w_{G1}^2 + w_{G2}^2}\left(\frac{s^2}{w_{G1}w_{G2}} + \frac{\theta^2}{\theta_{d1}\theta_{d2}}\right)\right], \tag{9.12}$$

where θ_{d1} and θ_{d2} are the divergence angles (9.9) of fiber 1 and 2, respectively.

Based on Marcuse's [1977b] Gaussian analysis, a general equation for the coupling efficiency between nonidentical single-mode fibers with combined transverse, longitudinal, and angular misalignments has been derived [Nemoto and Makimoto 1979; Miller C.M. 1984]:

$$\eta = (4\sigma/q)\exp(-pu/q) \quad . \tag{9.13}$$

The quantities σ, q, p, and u appearing in this equation can be calculated from the measurable quantities w_{G1}, w_{G2}, s, z_w, and θ as follows:

The quantity σ simply is the squared ratio of the spot sizes:

$$\sigma = (w_{G2}/w_{G1})^2 \quad . \tag{9.14}$$

The parameter p is defined by the equation

$$p = 2(\pi n_i w_{G1}/\lambda)^2 \quad , \tag{9.15}$$

and is two times the inverse of the squared divergence angle of the first fiber (9.9).

The radial offset s and the longitudinal offset z_w are normalized to the Rayleigh distance (4.6)

$$z_{R1} = \pi n_i w_{G1}^2/\lambda \tag{9.16}$$

of the first fiber and are written

$$F = s/z_{R1} \quad , \tag{9.17}$$

$$G = z_w/z_{R1} \quad . \tag{9.18}$$

Finally, the symbols q and u denote abbreviations for

$$q = G^2 + (\sigma + 1)^2 \quad , \text{ and} \tag{9.19}$$

$$u = (\sigma + 1)F^2 + 2\sigma FG \sin\theta + \sigma(G^2 + \sigma + 1)\sin^2\theta \quad . \tag{9.20}$$

When only one of the four mismatch parameters $\sigma - 1$, s, z_w, or θ is different from zero, the general formula (9.13) reduces to the equations (9.3, 4, 6, 8), respectively. For $s = 0$ and $\theta = 0$, one obtains (9.11) and for $z_w = 0$, one obtains (9.12). The general expression (9.13) has been verified experimentally [Kummer and Fleming 1982].

The formulas for the splice losses given previously only hold when the radial field distribution of the fundamental fiber mode can be approximated by a Gaussian function. For longer wavelengths and dispersion-shifted or dispersion-flattened fibers, the Gaussian approximation becomes worse.

A better approximation for the radial field distribution of the fundamental fiber mode is a Gaussian function near the axis, an exponential function in the core near the core-cladding boundary, and a modified Hankel function of order zero in the cladding [Sharma A. et al. 1982]. Using this Gaussian-exponential-Hankel approximation, it is possible to express the losses for angular and transverse offsets in analytical form [Hosain et al. 1982]. For a very accurate determination of the splice losses, one has to calculate the field overlap (7.11) by numerical integration [Guttmann and Krumpholz 1973; Cook et al. 1973; Anderson W.T. et al. 1986].

In elliptical core fibers, the field distribution is also slightly elliptical. Formulas for joint losses caused by transverse offset, longitudinal separation, angular misalignment, and azimuthal mismatch have been published by Sarkar et al. [1984]. Because the noncircularity of the field is much smaller than the noncircularity of the core, losses caused by azimuthal mismatch are very small and can generally be neglected.

Joint losses have been investigated experimentally by a number of authors [Bisbee 1971a; Guttmann and Krumpholz 1973; Murakami et al. 1978; Gambling et al. 1978a; Kitayama K. et al. 1981; Kummer and Fleming 1982; Shah et al. 1986; Anderson W.T. et al. 1986].

9.6 Mode Conversion at Joints

A single-mode fiber operated at short wavelengths becomes a dual-mode fiber. At a joint in a dual-mode fiber with either transverse or angular misalignment, power is transfered from the fundamental LP_{01} mode into the second order LP_{11} mode and vice versa. The mode conversion efficiency can be calculated by multiplying the transverse field distribution of the incident mode with the offset or tilted field distribution of the mode to be launched and by integrating the product over the infinite fiber cross-section. The coupling efficiency is proportional to the square of this overlap integral (7.11).

Mode conversion at butt joints between fibers with square-law index profiles has been studied both theoretically and experimentally by Krivoshlykov et al. [1983]. Mode conversion at a splice in a two-mode fiber link has been analyzed by Thyagarajan et al. [1982]. Curves showing the coupling efficiencies $LP_{01} \rightarrow LP_{01}$, $LP_{11} \rightarrow LP_{11}$, and $LP_{01} \rightarrow LP_{11}$ as a function of the radial or angular offset have been calculated numerically by Sakai and Kimura [1978b]. Axial coupling between a Hermite-Gaussian or Laguerre-Gaussian beam (Sect. 4.8) of arbitrary azimuthal and radial order and another Hermite-Gaussian or Laguerre-Gaussian beam of arbitrary azimuthal and radial order has been analyzed for arbitrary mismatch and misalignment between the two beams [Bayer-Helms 1984]. These results can also be applied to determine the axial coupling between arbitrary modes in each of the two fibres.

9.7 Joints Between Multimode and Single-Mode Fibers

Laser diodes are sometimes pigtailed to a multimode fiber. It is then useful to know the losses introduced by butt joining a single-mode fiber to the multimode pigtail. For simplicity, we will assume the fibers to be aligned without transverse or angular axis offset and without a gap.

We start our consideration with the simpler inverse problem in which a single-mode fiber is joined to a multimode fiber. The fundamental mode incident along the single-mode fiber will excite mainly guided modes in the multimode fiber. The radiation losses will be very small, and the insertion loss of the joint also small. Therefore, splicing the end of a single mode fiber link to the multimode fiber pigtail of a photoreceiver does not introduce significant power losses. Moreover, due to the short length of the pigtail, pulse broadening caused by intermodal dispersion is also negligibly small.

Quantitatively, the power launched in a particular LP_{lm} mode of the multimode fiber can be calculated by evaluating the overlap integral (7.11) between the LP_{01} mode of the single-mode fiber and the LP_{lm} mode of the multimode fiber. Because of mode orthogonality (Sect. 7.1), the total power transmitted by the multimode fiber equals the sum of the powers of the individual modes.

The number M of modes guided by a graded-index multimode fiber is $V^2/4$, where V is the normalized frequency (5.8). For standard multimode fibers, the core diameter is $2a = 50\,\mu$m, and the numerical aperture $(n_1^2 - n_2^2)^{1/2} = 0.2$, giving, for a wavelength of $1.3\,\mu$m a normalized frequency of $V = 24$. The number of guided modes is 146. Thus for calculating the insertion loss of a joint between a single-mode and a multimode fiber, one would have to evaluate a large number of overlap integrals.

In contrast, when a multimode fiber is connected to a single-mode fiber, the insertion loss is very large. In order to calculate the loss, one first has to determine the electric field at the output face of the multimode fiber. This consists of the superposition of the fields of the more than 100 modes, and, for a coherent source, appears as a complicated "speckle pattern" (Sect. 3.7.5). The speckle pattern fluctuates in time because of changes in the relative phases of the different modes e.g. due to a slight wavelength drift.

The power transmission factor of the joint between the multimode and single-mode fiber can, in principle be calculated by forming the overlap integral between the field distribution in the endface of the multimode fiber and the field distribution of the fundamental mode in the single-mode fiber. Because of the fluctuations of the speckle pattern, the power coupled into the LP_{01} mode will also fluctuate. This power fluctuation is called modal noise (Sect. 11.6.1).

Because of the fluctuations of the insertion loss, it is not worthwhile to do a rigorous analysis. For a rough estimate, one can assume that there is effective coupling only between the two fundamental modes of the multimode and the two fundamental modes of the single-mode fiber. If each guided mode of the multimode fiber transmits the same fraction of the total power, one expects a coupling efficiency of 2/146, corresponding to a coupling loss of 19 dB.

The experimental coupling loss is reported to be e.g. 14.5 dB [Shumate et al. 1985]. The difference can be explained by assuming that the power is not uniformly distributed over all guided modes and that the fundamental modes in the multimode fiber carry a relatively large amount of power. Because of the high insertion loss of the joint, it is not advisable to splice a single-mode fiber to the multimode fiber pigtail of an optical source.

A joint between a single-mode and a multimode fiber exhibits optical reciprocity since no nonreciprocal media are present. Thus, in principle, the transmission loss should be independent of the direction of wave propagation. However, for obtaining the same small loss in the direction from the multimode to the single-mode fiber, it would be necessary to reverse not only the direction of propagation of the many guided modes but also the direction of propagation of the radiation field while maintaining their relative amplitudes and phases. Since it is practically impossible to replicate the total field radiated from the single-mode fiber into the multimode fiber with reversed direction of propagation, efficient coupling back into the single-mode fiber cannot be achieved.

Although a joint between a multimode and a single-mode fiber has an insertion loss which depends on the direction of wave propagation, it cannot be termed a nonreciprocal device.

10. Spot Size and Width of the Radiation Pattern

Many properties of the fundamental mode are determined by the radial extent of its electromagnetic field; examples are losses caused by launching, jointing, microbends, waveguide dispersion, the capture fraction for backscattered light, and the width of the radiation pattern. Thus, the radial field extent is an important parameter for characterizing single-mode fiber properties [Alard et al. 1982b] and it is necessary to define a spot size as a quantitive measure for the width of the radial field distribution.

Several different definitions have been proposed for the near-field spot size and the most important of these are summarized in Sect. 10.1.

There is an inverse relationship between the spot size and the angular width of the beam radiated from the fiber end. Several different definitions for the angular width of the far-field radiation pattern have been proposed; the most important of these are reviewed in Sect. 10.2. In Sect. 10.3, quantitative relations between three of the different spot sizes and three of the different far-field widths will be given. These interrelations are very useful for determining the spot size from the far-field radiation pattern (Sect. 13.7).

10.1 Definitions for the Width of the Near-Field

In Sect. 5.3.2, the radial field extent of the fundamental fiber mode field was characterized by the spot size w_G, which is the spot size w_0 at the waist of that incident Gaussian beam which couples maximum power into the fundamental fiber mode. If the radial field distribution of the fundamental mode is assumed to be exactly Gaussian, the spot size w_G is the distance from the fiber axis at which the field amplitude is $1/e = 0.37$ and the intensity $1/e^2 = 0.135$ of the corresponding values on the axis. The mode field diameter is twice the spot size.

For real fibers, the radial field distribution $E(\varrho)$ is not strictly Gaussian and in general can only be calculated numerically from the refractive index profile or determined experimentally (Sect. 13.5). For example, the radial field distribution in a step-index fiber is described in the core by the Bessel function $J_0(U\varrho/a)$ (5.21) and in the cladding by the modified Hankel function $K_0(W\varrho/a)$ (5.15); a Gaussian function (5.25) only represents a useful approximation to the true radial field distribution (Fig. 5.6).

The problem of defining a spot size for non-Gaussian field distributions is a difficult one, as can be judged from the fact that at least eight definitions seem to exist, not counting, say $1/2$, $1/e$ or $1/e^2$ points as different. Depending on the problem in question, it turns out that one is led to different definitions for the field width.

Up to now the standardization committees have not reached agreement upon an acceptable definition for the spot size of the fundamental mode and the matter is presently being strongly debated [Caponi et al. 1984]. Several definitions have been proposed during the past few years. Each has its virtues, problems, and supporters and the standards organizations are struggling to develop a consensus [Dick and Shaar 1986].

Since at present there is no final recommendation for a particular definition of the mode field diameter, in this book, the different definitions proposed in the literature will be summarized. We introduce only the five most important ones and just mention other possible definitions (Sect. 10.1.6).

At first sight, the equations defining the various measures for the near-field width will look completely different. But we will see that the differently defined spot sizes agree, if the near-field is assumed to be strictly Gaussian. For the near-field distributions of actual standard fibers, the numerical values of the spot size calculated according to the various definitions usually differ only by some percent. Near cutoff of the second order mode, the differences are small, but they become larger for longer wavelengths.

Since it is a wave property, namely the transverse field extent (characterized by the spot size), which determines the launching, microbending and jointing losses as well as the waveguide dispersion, for single-mode fibers, it is not necessary to recommend a particular type of refractive-index profile or special values for the core diameter $2a$ or the relative refractive-index difference Δ. This allows different fiber designs and manufacturing methods to be used for obtaining the same spot size. Thus, the most important dimensional characteristics of single-mode fibers are specified in terms of field properties and not in terms of core dimensions and material properties as for multimode graded-index fibers.

Even if the core of a single-mode fiber is slightly elliptical, the field distribution of the fundamental mode is approximately circularly symmetric [Sarkar et al. 1984]. In the following, the electric field is thus assumed to be independent of the azimuth angle ϕ.

10.1.1 Simple Definition of the Spot Size

The simplest method for defining the width of the near-field is the following: the spot size w_s is that distance from the axis, for which the *amplitude* of the true electric field is $1/e = 0.37$ of its value on the fiber axis. At this distance, the intensity is $1/e^2 = 0.135$ of its maximum. Note that some authors use the $1/e$-width of the *intensity* distribution instead, which for a Gaussian field distribution is $0.707\,w_s$.

It has been recognized that the simple $1/e$-radius w_s of the electromagnetic field supplies too rough an approximation to the actual mode field radius and is not satisfactory for definition and specification purposes [Anderson and Philen 1983a; Bonaventura and Rossi 1984; Caponi et al. 1984].

10.1.2 Spot Size Related to Launching

The launching problem (Sect. 7.2.2) was first analyzed by Marcuse [1977b]: A fundamental Gaussian beam mode in air is incident on the fiber input end. In order to launch as much power as possible into the fundamental mode of the fiber, it is first necessary to match the wavefronts. Since the wavefronts of the LP_{01}-mode are planes normal to the axis, the axis of the Gaussian beam must coincide with the fiber axis. Moreover, since the Gaussian beam has a plane wavefront only at its waist, the waist must be located on the fiber input face. Under these conditions, the launching efficiency is given by the launch efficiency integral (7.14) which is essentially the square of the overlap integral [Marcuse 1978a]

$$\eta = \frac{\left[\int\limits_0^\infty E(\varrho)E_G(\varrho)\varrho\,d\varrho\right]^2}{\int\limits_0^\infty E^2(\varrho)\varrho\,d\varrho \int\limits_0^\infty E_G^2(\varrho)\varrho\,d\varrho} \quad . \tag{10.1}$$

In the integrals, the function $E(\varrho)$ is the true radial field amplitude distribution of the fundamental mode to be launched and

$$E_G(\varrho) = E_0 \exp[-(\varrho/w_0)^2] \quad , \tag{10.2}$$

is the radial field distribution (4.1) of the Gaussian beam in the waist at $z = 0$. The parameter w_0 is the spot size at the waist of the Gaussian beam.

The launching efficiency clearly depends on the spot size w_0. For a certain value of w_0, the launching efficiency becomes a maximum, i.e.

$$\frac{d\eta}{dw_0} = 0 \quad , \quad \text{for} \quad w_0 = w_G \quad . \tag{10.3}$$

This optimum spot size w_G of the incident Gaussian beam is the second measure proposed to define the spot size of the fundamental fiber mode [Gambling and Matsumura 1977, 1978a; Marcuse 1978a]. The subscript "G" is intended to remind us that launching is by a Gaussian beam. The spot size w_G of the Gaussian beam which would optimally excite the fundamental fiber mode is *not* equal to the simple $1/e$-width w_s (Sect. 10.1.1) of the fundamental mode!

Taking the derivative of (10.1) with respect to w_0 and equating this to zero, gives a transcendental equation for w_G [Matsumura and Suganuma 1980; Matsumura et al. 1980b].

For a single-mode fiber with an arbitrary refractive-index profile, the curve representing the launching efficiency η as a function of the spot size w_0 of the incident Gaussian beam is very similar to the coupling curve shown in Fig. 7.3, and has a broad maximum.

For the special case of a step-index fiber, the η versus w_0/a curve is shown in Fig. 7.1 for several values of the normalized frequency V.

The maximum launching efficiency obtained for $w_0 = w_G$ is larger than 95% for the fibers and operating wavelengths used in practice, showing the strong similarity between the free Gaussian beam and the LP_{01} mode of a fiber. This has been shown for step-index fibers [Marcuse 1977b], graded index fibers with a power-law index profile [Marcuse 1978a], step-index fibers with a central index dip [Gambling et al. 1978d], and fibers with power law profile, central dip, and step at the core-cladding interface [Streckert and Brinkmeyer 1982].

The optimum launching efficiency depends on the wavelength used. It is maximum for wavelengths close to the cut-off wavelength of the second-order mode (Sect. 6.3.2), since there the radial field distribution can be approximated quite well by a simple Gaussian function. With increasing wavelength, the optimum launching efficiency becomes smaller, since a larger portion of the transmitted power is in the cladding, where the field distribution is described by the modified Hankel function $K_0(W\varrho/a)$ (5.15, 16) instead of a Gaussian function (5.25).

The value w_G of w_0 that maximizes the launching efficiency also minimizes the area-weighted rms-deviation

$$I_2 = \int\limits_0^\infty [E(\varrho) - E_G(\varrho)]^2 \varrho \, d\varrho \qquad (10.4)$$

between the amplitude distributions $E(\varrho)$ and $E_G(\varrho)$ (10.2) of the fundamental fiber and Gaussian beam modes, respectively [Anderson W.T. and Philen 1983; Anderson W.T. 1984].

For the simple step-index fiber, a useful approximate formula for the spot size is [Marcuse 1977b]:

$$w_G = a(0.65 + 1.619/V^{1.5} + 2.879/V^6) \quad . \qquad (10.5)$$

Numerical values for w_G are listed in Table 5.1. The spot size w_G depends on the relative index difference Δ, but changes only slightly with wavelength for V varying between 2.0 and 2.4 [Gambling and Matsumura 1977].

For a wide class of fibers including those with power-law index profiles or a central dip, the spot size w_G depends to a good approximation only on the effective normalized frequency V_e (6.39) and an effective core radius a_e [Streckert and Brinkmeyer 1982]:

$$w_G = a_e(0.6043 + 1.755 \, V_e^{-1.5} + 2.78 \, V_e^{-6}) \quad . \qquad (10.6)$$

The effective core radius a_e is derived from a diffraction experiment (Sect. 13.8.11) which, for step-index fibers, yields the actual core radius [Brinkmeyer 1979b]:

$$a_e = \frac{3.832}{k \sin \theta_{\min}} \quad , \tag{10.7}$$

where $\theta_{\min}$ is the first minimum in the diffraction pattern obtained by transmitting a laser beam transversely through the fiber.

The spot size w_G is inversely proportional to the far-field width W_G (10.30).

10.1.3 Spot Size Related to Microbending Losses

For analyzing microbending losses in single-mode fibers, Petermann [1976a; 1976b; 1977a] introduced a third kind of spot size w_e by the definition

$$w_e^2 = 2 \frac{\int\limits_0^\infty E^2(\varrho)\varrho^2 \varrho \, d\varrho}{\int\limits_0^\infty E^2(\varrho)\varrho \, d\varrho} \quad . \tag{10.8}$$

Since Petermann also defined a further spot size (Sect. 10.1.4), w_e sometimes is called the Petermann I spot size.

The integrals in this definition have simple meanings: the integral in the denominator is proportional to the power transmitted by the fundamental mode. In the integral in the numerator, the square of the distance ϱ from the fiber axis is weighted by the square of the electric field strength, i.e. by the mode intensity. Thus, w_e represents the root-mean-square (rms) or effective value of ϱ, as indicated by the subscript "e".

Alternatively, the spot size w_e can be regarded as representing the second moment (3.28) of the near-field intensity distribution [Stewart W.J. 1980b].

In the original definition of Petermann, the factor of 2 is missing. I have introduced it in order to facilitate comparison of the different definitions.

Not only microbending losses, but also losses due to small tilts, i.e. angular misalignment, depend on the effective spot size [Caponi et al. 1984]. Low tilt losses thus require small values of w_e [Petermann 1987].

For the special case of a step-index fiber, the effective spot size can be calculated from [Gambling and Matsumura 1977]

$$\left(\frac{w_e}{a}\right)^2 = \frac{2}{3}\left(\frac{J_0(U)}{U J_1(U)} + \frac{1}{2} + \frac{1}{W^2} - \frac{1}{U^2}\right) \quad . \tag{10.9}$$

Numerical values for w_e are listed in Table 5.1.

The spot size w_e is inversely proportional to the far-field width W_d (10.29).

10.1.4 Spot Size Related to Waveguide Dispersion

When analyzing waveguide dispersion D_w (Sect. 5.8), Petermann [1983] defined a fourth, "strange" [Pask 1984] kind of spot size:

$$w_\mathrm{d}^2 = 2 \frac{\displaystyle\int_0^\infty E^2(\varrho)\varrho\, d\varrho}{\displaystyle\int_0^\infty (dE/d\varrho)^2 \varrho\, d\varrho} \quad . \tag{10.10}$$

The subscript "d" indicates "dispersion" or the "derivative" of the radial field distribution in the denominator. This spot size is the inverse of the rms-value of the relative slope $(dE/d\varrho)/E$ of the radial near-field distribution $E(\varrho)$. Often, the quantity w_d is called the Petermann II spot size.

On carrying out the integral in the denominator by parts the square of the first derivative is replaced by the transverse Laplace operator [Andreasen 1986a]. Some authors [Povlsen 1985a] thus call w_d the "Laplace spot size".

The spot size w_d must be used when deriving the waveguide dispersion D_w from the wavelength dependence of the spot size (5.89) [Petermann 1983; Coppa et al. 1983b; Pask 1984].

It can be shown that the spot size w_d is related to the coupling loss for a fiber joint with a very small ($s \ll w_\mathrm{d}$) transverse offset [Petermann 1983; Anderson W.T. 1984; Pask 1984]. Denoting the power transmission coefficient by $\eta(s)$ (Sect. 10.1.5), the spot size w_d is related to the second derivative of $\eta(s)$ taken at $s = 0$, i.e. to the initial curvature of the power transmission function [Calzavara et al. 1986a]

$$w_\mathrm{d}^2 = -\frac{2}{d^2\eta/ds^2|_{s=0}} \quad . \tag{10.11}$$

Inserting w_d into the simple formula (9.5) for the insertion loss caused by a small transverse offset between equal fibers approximates the true losses very well, even for fibers whose radial field cannot be described by a simple Gaussian function. Thus low transverse offset losses demand large values of w_d [Petermann 1987].

The "strange" spot size w_d is also related to the difference between the phase and group velocities, if material dispersion can be ignored [Pask 1984].

For step-index fibers, the spot size w_d can be obtained either from the eigenvalues U and W in the core and cladding (5.23, 5.24) [Hussey and Martinez 1985]

$$\frac{w_\mathrm{d}}{a} = \frac{\sqrt{2}}{W} \frac{J_1(U)}{J_0(U)} \quad , \tag{10.12}$$

via the the quantities b (5.51) and $d(Vb)/dV$ (5.74) characterizing the phase and the group velocity, respectively [Hussey 1984]

$$\frac{w_\mathrm{d}}{a} = \frac{2}{V\sqrt{d(Vb)/dV - b}} \quad , \tag{10.13}$$

or from Marcuses approximation (10.5) for the spot size w_G related to launching [Hussey and Martinez 1985]

$$\frac{w_d}{a} = \frac{w_G}{a} - (0.016 + 1.561\,V^{-7}) \quad .$$ (10.14)

This last formula is accurate to within 1% in the range

$$1.5 \le V \le 2.5 \quad ,$$ (10.15)

which is the range of most practical interest in single-mode fiber transmission.

Numerical values for w_d are listed in Table 5.1.

The spot size w_d is inversely proportional to the effective far-field width W_e (10.31).

10.1.5 Spot Size Related to Transverse Offset Loss

In 1980, Streckert [1980] introduced the transverse offset method for measuring the spot size; this will be described in detail in Sect. 13.7.2. The theory underlying this measuring method can also be used to define a fifth kind of spot size, w_a [Streckert 1985].

The power transmission function $\eta(s, \phi_s)$ for a butt joint with a transverse axes offset is (7.11):

$$\eta(s, \phi_d) = \left[\frac{\displaystyle\int_0^{2\pi}\!\!\int_0^\infty E(\varrho, \phi) E(\varrho', \phi')\varrho\, d\varrho\, d\phi}{\displaystyle\int_0^{2\pi}\!\!\int_0^\infty E^2(\varrho, \phi)\varrho\, d\varrho\, d\phi} \right]^2$$ (10.16)

where s is the transverse offset with azimuth ϕ_s (Fig. 8.6). The variables ϱ' and ϕ' are related to the variables of integration ϱ and γ by the two equations (8.42, 43)

$$\varrho'^2 = \varrho^2 + s^2 - 2\varrho s\, \cos(\phi - \phi_s) \quad , \text{ and}$$ (10.17)

$$\tan \phi' = \frac{\varrho \sin \phi - s \sin \phi_s}{\varrho \cos \phi - s \cos \phi_s} \quad .$$ (10.18)

The overlap integral in the numerator of (10.16) is proportional to the amplitude transmission coefficient of the joint with transverse offset. In its integrand, the near-field amplitude distribution of the fundamental fiber mode is multiplied by a shifted version of the same field distribution. Thus the integral in the numerator represents the two-dimensional autocorrelation function of the near-field amplitude distribution (8.44) [Goodman 1968]. The normalized power transmission function $\eta(s, \phi_s)$ is proportional to the squared modulus of the autocorrelation function.

Now, from the normalized power transmission function $\eta(s, \phi_s)$, the autocorrelation or mode field convolution spot size w_a is defined as that value of the transverse offset s, for which the power transmission is $1/e = 0.37$ of its maximum value of unity at $s = 0$. For a circular near-field distribution, any direction ϕ_s of transverse offset can be taken.

There are several advantages of this definition for the mode field radius [Streckert 1985]:

1. The spot size w_a (like w_e and w_d) is an intrinsic property of the fundamental mode field distribution. No comparison with an assumed Gaussian distribution need be made.

2. The spot size w_a differs only slightly from the widely used w_G. For fibers with power-law profiles or a central index dip, the spot sizes w_a and w_G differ by less than 2.3% for $1.5 < V_e < 3$, where V_e is the effective normalized frequency (6.39) [Streckert 1985]. Thus, in practice, all formulas containing w_G an be applied using w_a instead.

3. The autocorrelation spot size w_a can be measured in strict correspondence with its definition using the well-known, simple, and popular transverse offset method (Sect. 13.7.2).

4. For noncircular fundamental mode fields, the definition can be generalized by introducing two spot sizes $w_{a\,\text{max}}$ and $w_{a\,\text{min}}$ for offsets along the principal axes of the field distribution. The mode field noncircularity then can be defined by the ratio $(w_{a\,\text{max}} - w_{a\,\text{min}})/w_{a\,\text{max}}$.

For the special case of step-index fibers, numerical values for w_a are listed in Table 5.1.

10.1.6 Other Definitions for the Spot Size

We conclude this section on the different possibilities for defining the spot size of the fundamental mode, by briefly mentioning some other possible definitions:

Instead of minimizing the rms-deviation I_2 (10.4) between the true near-field *amplitude* distribution and the *amplitude* distribution of the Gaussian model one can also minimize the rms-deviation of the near-field *intensity* distributions [Anderson W.T. 1984] to obtain a sixth kind of spot size.

Another method for defining the spot size [Snyder and Sammut 1979; Snyder 1981] is to insert a Gaussian function into a stationary expression for the phase constant β and to maximize β by varying the width of the Gaussian approximation. Formulas for this spot size for Gaussian, step-index, and smoothed out step-index profiles can be found in the second paper by Snyder [1981].

Still other definitions can be formulated with respect to the measuring techniques denoted far-field filtering and variable-aperture-lens (VAL) methods [Povlsen 1984] (Sect. 13.7.3) or by fitting an error function to the transmitted power curve in the scanning knife-edge method [Dick and Shaar 1986] (Sect. 13.7.3).

Thus there are many ways of defining a measure for the transverse field extent. As long as there is no general consensus as to the best definition, it is very important to provide information not only about the definition but also about the measuring method being used, whenever one specifies a value for the spot size.

10.1.7 Relations Between the Different Spot Sizes

If a hypothetical, circular, and strictly Gaussian near-field distribution

$$E(\varrho) = E_0 \exp[-(\varrho/w_s)^2] \quad , \tag{10.19}$$

is substituted into the definitions [Sect. 10.1.1; 10.3, 8, 10; Sect. 10.1.5] for the spot sizes, all five definitions give the same value, i.e.

$$w_s = w_G = w_e = w_d = w_a \quad . \tag{10.20}$$

Since the near-field distribution of real fibers is never strictly Gaussian in practice, the different definitions give slightly different values for the spot size.

The spot sizes related to waveguide dispersion (w_d), launching (w_G), microbending (w_e), and the transverse field extent a/W in the cladding (5.15, 5.16) are related by [di Vita et al. 1984; Coppa et al. 1984a; Caponi et al. 1984; Povlsen 1985b; Petermann and Kühne 1986; Petermann 1987]:

$$w_d \leq w_G \leq w_e < 2a/W \quad . \tag{10.21}$$

Low offset and tilt losses demand large w_d and low w_e. Because of the inequality (10.21), the optimum fiber design for low splice loss would yield $w_d = w_e$ which occurs only for a Gaussian fundamental mode field [Petermann 1987].

For power-law profiles with a dip, it has been found empirically that w_e can be calculated from w_d and $dw_d/d\lambda$ and that the wavelength dependence of the quantities w_d, w_e and D (5.85) can be calculated from the values of w_d and D measured at a fixed wavelength [Povlsen 1985b].

It is not possible to make further general statements about the relations between the different kinds of spot size. The differences between the definitions can be studied by assuming a typical refractive index profile, solving the wave equation (5.20) for the radial distribution $E(\varrho)$, and calculating the different spot sizes corresponding to the definitions.

As examples, Figs. 10.1–10.3 show the calculated values of the spot sizes as functions of the effective normalized frequency V_e (6.39) for three typical examples: a simple step-index fiber (Fig. 10.1), a fiber with a truncated parabolic-index profile (Fig. 10.2), and a quadruple-clad (QC) fiber (Fig. 10.3) [Wilczewski 1985]. The curves for the truncated parabolic-index fiber have also been published by Coppa et al. [1985b].

From these figures and the literature, the following trends can be observed:

1. The simple spot size w_s usually deviates appreciably from the other four spot sizes.

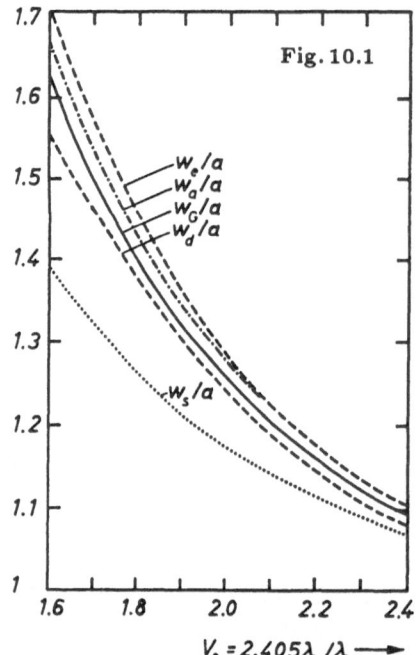

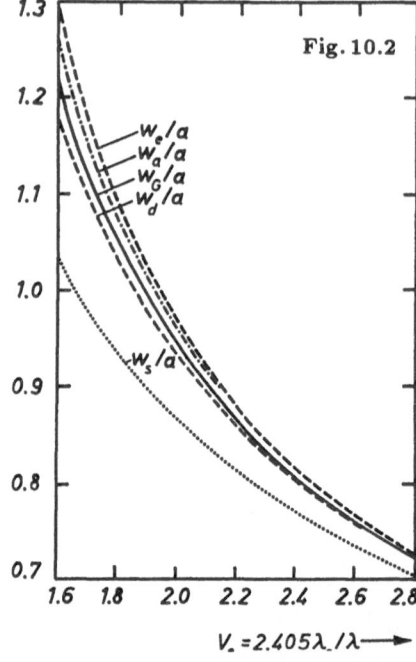

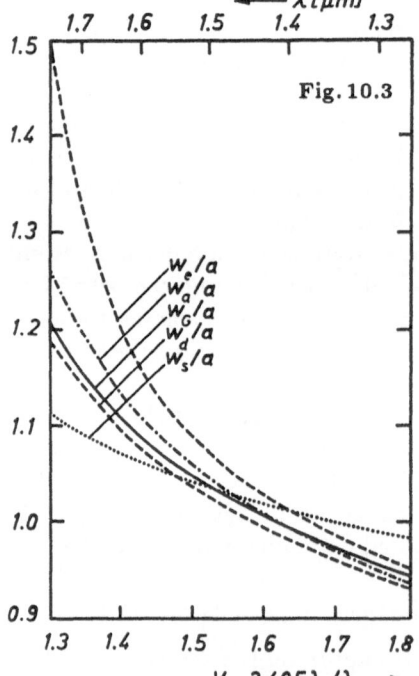

Fig. 10.1. Spot sizes w_s, w_d, w_G, w_a, and w_e as functions of the normalized frequency V (5.8) for a step-index fiber

Fig. 10.2. Spot sizes w_s, w_d, w_G, w_a, and w_e as functions of the effective normalized frequency $V_e = 2.405\,V/V_c$ (6.39) for a fiber with a truncated parabolic index profile (5.6, 5.7, $g = 2$)

Fig. 10.3. Spot sizes w_s, w_d, w_G, w_a, and w_e as functions of the effective normalized frequency $V_e = 2.405\,V/V_c$ (6.39) for a quadruple-clad (QC-) fiber with the refractive index profile shown in Fig. 5.4:
$a_1 = 2.95\,\mu$m, $a_2 = 6.22\,\mu$m, $a_3 = 9.5\,\mu$m, $a_4 = 13.1\,\mu$m,
$\Delta_1 = 0.55\%$, $\Delta_2 = -0.56\%$, $\Delta_3 = 0.08\%$, $\Delta_4 = -0.10\%$.
The cutoff-wavelength of the fundamental mode is $1.84\,\mu$m and that of the second order LP_{11} mode $0.95\,\mu$m

2. for values of the effective normalized frequency near 2.4, i.e. close to the cutoff frequency of the second-order mode, w_G, w_e, w_d and w_a have similar values. This can be explained by the fact that the power of the fundamental mode is concentrated in the core, where the radial field distribution can be well approximated by a Gaussian function.

3. With increasing wavelength, i.e. decreasing V_e, the transverse field extent increases, as has already explained in Sect. 5.3.2. Since more power is transmitted in the cladding, where the radial field distribution is described by the modified Hankel function (5.15), the Gaussian approximation gets worse and the differences between the spot sizes w_G, w_e, w_d, and w_a become larger.

4. for the QC-fiber (Fig. 10.3), at long wavelengths the effective spot size is much larger than the other spot sizes. This can be explained as follows: For the parameters chosen, the fundamental mode of this QC-fiber has a finite cutoff wavelength (6.22), which is 1.84 μm. When the wavelength approaches the cutoff wavelength, the field distribution consists of a central peak in the core (Fig. 5.4: $\varrho < a_1$) resting on a flat pedestal which extends far into the claddings. Since w_e is the average of ϱ^2 weighted by the field intensity, w_e becomes larger than the other spot sizes, which mainly characterize the width of the central intensity maximum.

5. The autocorrelation spot size w_a gives a surprisingly good guide to the effective spot size w_e, particularly around the cutoff wavelength of the LP_{11} mode [di Vita et al. 1984; Caponi et al. 1984]. This coincidence of w_a and w_e is better for monotonic profiles and W-fibers than for fibers with an index dip on the axis or for segmented core fibers (Fig. 10.3).

6. For single-mode fibers with a power-law index profile with or without a central dip, the autocorrelation spot size w_a also deviates only slightly from the spot size w_G related to launching [Streckert 1985]. Note that the corresponding plot, Fig. 6 of Streckert and Brinkmeyer [1982] is in error.

The different definitions of the spot size have also been compared in the technical literature [Anderson W.T. 1984; Coppa et al. 1984a; Auge et al. 1985].

Since the radial field distribution cannot be completely characterized by a single quantity that simultaneously describes all the fiber properties, the choice of spot size definition is, in principle, only a matter of agreement. But because the standardization committees have yet to agree upon a particular definition, the user will ask which definition to take.

There are several problems where only the ratio of two spot sizes must be known, e.g. when jointing two fibers with different spot sizes (Sect. 9.1) or when correcting the splice losses measured by the backscattering method for different fibers (Sect. 13.4.3). Since the slopes of the spot size versus normalized frequency curves for the different definitions are very similar, in such cases, any of the definitions can be used. Of course, the same definition must be used for both fibers!

There are other problems for which the quantity to be determined depends only slightly on the spot size, e.g. the launching of the LP_{01} mode with a Gaussian beam (Sect. 7.2.2). Since the launching efficiency as a function of the spot size has a broad maximum (Fig. 7.3), it does not matter which definition is taken.

In other cases, a quantity may be sought which depends not only on the spot size, but also on other parameter that are not well known or difficult to measure. For instance, when one is interested in microbending losses (Sect. 5.9.9), the statistics of the random bends are largely unknown and it is impossible to estimate the microbending loss from the spot size only.

Finally, it has been proposed that one might obtain e.g. the effective cutoff wavelength of the second-order mode [Millar 1981b] (Sect. 6.3.2) or the waveguide dispersion (5.89) indirectly from the wavelength dependence of the spot size. The values of the cutoff wavelength determined in such a way are very uncertain (Sect. 13.8.3) and the pure waveguide dispersion determined from the spot size w_d is of interest mainly from the standpoint of a fibers designer. The user is more interested in the total dispersion, which can be determined more directly by several methods (Sect. 13.9).

In the Comité Consultatif International Telephonique et Telegraphique (CCITT), most of the delegations at the 1984 meeting were in favor of measuring the autocorrelation spot size w_a using the transverse offset technique (Sect. 13.7.2). A few others were in favor of calculating the spot size w_G related to launching either from the measured near-field distribution or from the measured far-field radiation pattern (Sect. 13.7.1). As has been stated before, both definitions usually give similar numerical values for the spot size that lie in the middle of the range of values obtained from the five definitions (Figs. 10.1–10.3).

Although the CCITT has not yet agreed on a particular definition for the mode field diameter, its Recommendation G.652 [CCITT 1984] for an operating wavelength of 1.3 μm recommends a mode field diameter between 9 and 10 μm. This recommendation also points out that a value of 10 μm is commonly employed for matched cladding designs and a value of 9 μm for depressed cladding designs. The recommended range of mode field diameter is a compromise: for smaller values, the higher GeO_2 doping level in the core of matched cladding fibers causes an increase of the attenuation by Rayleigh scattering, and for larger values, the micro-bending losses increase.

At the 1986 and 1987 meetings of the CCITT, most of the representatives were in favor of accepting the so-called Petermann II definition for the spot size, i.e. the spot size w_d related to waveguide dispersion. The reasons for preferring w_d are:

1. There are three equivalent definitions of this quantity: firstly the original definition (10.10) which determines w_d from near-field data, secondly the definition of the effective far-field width W_e (10.28) which using $w_d = 1/(\pi W_e)$ (10.31) gives w_d in terms of far-field data, and thirdly the

relation (10.11) which relates w_d to the second derivative of the transverse offset data.

2. Inserting the spot size w_d into the simple approximation (9.5) for the joint loss caused by transverse offset gives values which, for equal fibers and small losses, agree very well with the exact ones even for dispersion-shifted and dispersion-flattened fibers which have non-Gaussian radial field distributions.

3. It is possible to measure the spot size w_d by at least five different methods, the near-field or the far-field scan (Sect. 13.7.1), the transverse offset (Sect. 13.7.2), the knife-edge, and the variable-aperture far-field method (Sect. 13.7.3).

10.2 Definitions for the Angular Width of the Far-Field Radiation Pattern

In Sect. 8.1, it has been shown that there are simple relationships (8.13, 15) between the radial field distribution $E(\varrho)$ of the fundamental fiber mode and the directional characteristic $E_f(r, \theta)$ of the light radiated from the fiber end: the near-field and far-field distributions form a pair of Hankel transforms (8.16, 20). The wider the near-field distribution, the narrower the far-field radiation pattern and vice versa.

This relationship can be used e.g. for determining the transverse field distribution of the fundamental mode from the measured far-field radiation pattern (Sect. 13.6). Mode field diameter and angular width of the far-field are inversely proportional. It is thus possible to calculate the fundamental mode spot size from the width of the far-field radiation pattern.

There are as many possibilities for defining the angular width of the far-field radiation pattern, as there are methods for defining the near-field width. In this section, we will summarize the most important definitions for the far-field width.

In order to get simple relations between the near- and far-field widths (Sect. 10.3), it is useful to describe the far-field radiation pattern not by the electric field $E_f(r, \theta)$ but by the function $E_H(q)$ which is the Hankel transform (8.16) of the near-field amplitude distribution $E(\varrho)$. The function $E_H(q)$ is related to the far-field amplitude distribution $E_f(r, \theta)$ by (8.18)

$$E_H(q) = -j\lambda r \exp(jkr) E_f(r, \theta) \quad , \tag{10.22}$$

where the variable q, the so-called spatial frequency, is related to the polar angle θ by (8.14)

$$q = (1/\lambda) \sin \theta \quad . \tag{10.23}$$

Some authors use the spatial angular frequency (8.17)

$$u = 2\pi q \quad , \tag{10.24}$$

instead of q.

Whereas lower case w's have been chosen for the symbols of the near-field width, upper case W's will be used for the far-field widths.

10.2.1 Simple Definition of the Far-Field Width

A simple definition of the far-field width is that value W_s of the spatial frequency q, for which the electric field strength is $1/e = 0.37$ of its value on the beam axis, i.e. for $q = 0$.

10.2.2 Far-Field Width Related to Launching

A second measure W_G for the far-field width is obtained by analyzing the launching problem [Anderson W.T. 1984].

First we recall some properties of the radiation from the fiber end (Sect. 8.1): the guided LP_{01} mode transforms in a free wave beam whose far-field directional characteristics $E_H(q)$ can be calculated from (8.16). If we imagine reversing the direction of propagation of this beam, it would launch the fundamental mode with an efficiency of 100% (reflection losses must be considered separately).

We now replace the incident wave by a fundamental Gaussian beam with variable spot size w_0 at the waist. The normalized near- and far-field amplitude distributions of this Gaussian beam are described by (8.21) and (8.22), respectively. The $1/e$-width of the Gaussian far-field is $W_0 = 1/(\pi w_0)$ (8.23).

In order to obtain maximum launching efficiency, we assume that the axes of the Gaussian beam and the fiber coincide and that the waist of the Gaussian beam is located at the fiber input face. Because of a mismatch of the far-field distributions $E_H(q)$ and $E_{GH}(q)$, the launching efficiency η is smaller than 100%. It can be calculated from the overlap integral of the far-fields instead of from the overlap integral (10.1) of the near-fields [Anderson W.T. 1984]:

$$\eta = \frac{\left| \int\limits_0^{1/\lambda} E_H(q) E_{GH}(q) q \, dq \right|^2}{\int\limits_0^{1/\lambda} E_H^2(q) q \, dq \int\limits_0^{1/\lambda} E_{GH}^2(q) q \, dq} . \tag{10.25}$$

Note that in the far-field overlap integral (10.25), the variable of integration is the spatial frequency q and not the angular variable θ.

The launching efficiency depends on the far-field width W_0 of the Gaussian beam. For a certain value of W_0, the launching efficiency becomes a maximum. This optimum far-field width of the incident Gaussian beam is chosen to define the far-field width W_G of the fundamental fiber mode. The subscript "G" refers to launching by a Gaussian beam.

That value W_G of W_0, which maximizes the launching efficiency, also minimizes the area-weighted rms deviation

$$I_3 = \int\limits_0^{1/\lambda} [E_{\mathrm{H}}(q) - E_{\mathrm{GH}}(q)]^2 q \, dq \qquad (10.26)$$

between the far-field amplitude distributions [Anderson W.T. 1984].

The far-field width W_{G} is inversely proportional to the spot size w_{G} (10.30).

10.2.3 "Strange" Definition for the Far-Field Width

A third measure for the far-field width can be defined by a formula, which is analogous to the "strange" definition (10.10) of the near-field width w_{d} [Nishimura and Suzuki 1984]:

$$W_{\mathrm{d}}^2 = 2 \frac{\displaystyle\int_0^{1/\lambda} E_{\mathrm{H}}^2(q) q \, dq}{\displaystyle\int_0^{1/\lambda} (dE_{\mathrm{H}}(q)/dq)^2 q \, dq} \, . \qquad (10.27)$$

The subscript "d" indicates the derivative of the radial field distribution in the denominator. This far-field width is the inverse of the rms value of the relative slope $(dE_{\mathrm{H}}/dq)/E_{\mathrm{H}}$ of the far-field radiation pattern.

Note that in the paper of Nishimura, the spatial angular frequency $u = 2\pi q$ is used (10.24).

The far-field width W_{d} is inversely proportional to the spot size w_{e} (10.29).

10.2.4 Effective Far-Field Width

A fourth measure for the far-field width has been defined in analogy to the effective spot size w_{e} (10.8) [Anderson W.T. 1984]:

$$W_{\mathrm{e}}^2 = 2 \frac{\displaystyle\int_0^{1/\lambda} E_{\mathrm{H}}^2(q) q^2 q \, dq}{\displaystyle\int_0^{1/\lambda} E_{\mathrm{H}}^2(q) q \, dq} \, . \qquad (10.28)$$

The integrals in this definition have simple meanings: the integral in the denominator is proportional to the power transmitted by the radiation field. In the integral in the numerator, the square of the transform variable q (10.23) is weighted by the far-field intensity distribution. Thus, W_{e} represents the rms or effective far-field width. The subscript "e" indicates effective.

Pask [1984] essentially introduced the same definition, but used the spatial angular frequency u (10.24) instead of q and also omitted the factor of 2.

The far-field width W_{e} is inversely proportional to the spot size w_{d} (10.31).

10.2.5 Other Definitions for the Far-Field Width

In Sect. 8.2, the autocorrelation function $E_H * E_H$ of the far-field amplitude distribution $E_H(q)$ was introduced (8.47). For the sake of completeness, one could also define a fifth measure for the far-field width in analogy to the definition of the autocorrelation spot size w_a (Sect. 10.1.5): W_a is that value of the variable q_s, for which the squared modulus of the autocorrelation function (8.47) is $1/e = 0.37$ of its maximum value.

Besides the far-field widths introduced so far, there are, of course, many other possibilities to define the width of the radiation lobe. In antenna techniques, one often specifies the full width at half maximum intensity (FWHM width).

10.3 Relations Between the Near-Field and Far-Field Widths

The near-field and far-field amplitude distributions form a first pair of Hankel transforms. The near-field autocorrelation function and the far-field intensity distribution form a second pair, and the near-field intensity distribution and the far-field autocorrelation function form a third pair of Hankel transforms (Sect. 8.2). These three relations between near-field and far-field functions are represented graphically in Fig. 8.5.

Since there are mathematical relations between the near- and far-field functions, there are corresponding equations relating the quantities characterizing the widths of these curves, i.e. the spot-size and the far-field width. There is an essentially inverse relationship between the near- and far-field widths.

It can be shown that for arbitrary refractive-index profiles, i.e. for arbitrary near- and far-field distributions, the following relations hold exactly: between the effective spot size w_e and the strange far-field width W_d [Nishimura and Suzuki 1984]:

$$W_d = 1/(\pi w_e) \quad , \tag{10.29}$$

between the Gaussian spot size w_G and the Gaussian far-field width W_G [Anderson W.T. 1984]:

$$W_G = 1/(\pi w_G) \quad , \tag{10.30}$$

and between the strange spot size w_d and the effective far-field width W_e [Pask 1984]:

$$W_e = 1/(\pi w_d) \quad . \tag{10.31}$$

These relations are very useful for saving computer time. For instance, determining w_e by measuring the far-field pattern (Sect. 13.6), calculating from it the near-field pattern by a Hankel transform (8.18, 20), and using the definition (10.8) takes ten to twenty times the computer time necessary to insert the far-field pattern in the definition for W_d (10.27) and use (10.29) to calculate w_e [Nishimura and Suzuki 1984].

The five measures for the width of the far-field are defined in such a way that they have the same value when the far-field distribution is strictly Gaussian (8.22):

$$W_s = W_G = W_d = W_e = W_a \quad . \tag{10.32}$$

For the far-field distributions of real fibers, the various far-field widths differ typically by up to ten percent. From the relation (10.21) and the relations between the far-field and near-field widths (10.29–31), one can deduce, that

$$W_d \leq W_G \leq W_e \quad . \tag{10.33}$$

These relations can also be obtained by observing that the definitions of the far-field widths W_d, W_G, and W_e are formally analogous to the definitions of the spot sizes w_d, w_G, and w_e.

11. Signal Transmission Through Single-Mode Fibers

The main application of single-mode fibers is in signal transmission. The electrical signal to be transmitted is modulated onto an optical carrier wave which is guided by the fiber in the form of the LP_{01} mode. Different modulation techniques can be used: intensity, amplitude, phase, or frequency-modulation.

In this chapter, the fundamental aspects of signal transmission by single-mode fibers will be considered.

First, in Sect. 11.1, an optical transfer function is introduced which describes the change in amplitude and phase of a sinusoidal optical carrier wave transmitted along the fiber as functions of the optical frequency.

When the optical wave is modulated by a signal, chromatic dispersion (Sect. 5.8) leads to a detected output signal that does not represent an exact replica of the input signal, even if one does not consider the time delay and the change in amplitude caused by the optical transmission line.

In Sect. 11.2, we will discuss the conditions under which a single-mode fiber link behaves as a linear time-invariant transmission system. In these cases, one can define a baseband transfer function which describes the change in amplitude and phase of a sinusoidal signal transmitted by the fiber as functions of the modulation frequency.

In Sect. 11.3, the impulse response of a single-mode fiber is introduced and related to the baseband transfer function.

The transmission of analogue signals through single-mode fibers behaving as linear time-invariant systems will be discussed in Sect. 11.4.

As a result of chromatic dispersion, a pulse transmitted through a single-mode fiber broadens. The relation between the input pulse duration and the output pulse duration will be discussed in Sect. 11.5 for sources with arbitrary linewidth and for arbitrary operating wavelength.

Besides chromatic dispersion, there are several other effects which cause signal distortions (Sect. 11.6). One important effect which can occur at short wavelengths is modal noise due to interference between the fundamental and second order modes (Sect. 11.6.1). Other effects causing signal distortions are unwanted frequency modulation of the source (Sects. 11.6.2–6), reflections (Sects. 11.6.7, 8), conversion of phase modulation to amplitude modulation (Sect. 11.6.9), and polarization effects (Sects. 11.6.10 to 11.6.12).

Thus, this chapter emphasizes the fundamentals of signal transmission through single-mode fibers.

11.1 Optical Transfer Function

As long as the transmitted power is so small that nonlinear effects (Sect. 5.4) can be neglected, in propagating along a fiber of length L with attenuation coefficient α and phase constant β, the amplitude of the fundamental mode decreases by a factor $\exp(-\alpha' L)$ and the zero phase angle changes by $-\beta L$. The phasor of the electric field of the fundamental fiber mode depends on the spatial coordinates ϱ and z as follows (5.105)

$$\underline{E}_x(\varrho, z) = E(\varrho)\exp(-\alpha' z)\exp(-j\beta z) \quad . \tag{11.1}$$

The attenuation coefficient α' in Nepers per kilometer is related to the fiber loss α in dB/km by (5.110)

$$\alpha = 8.69\alpha' \quad . \tag{11.2}$$

The optical transfer function $H_o(\omega)$ of the fiber is defined as the complex ratio of the phasors of the electric field strength on the fiber axis ($\varrho = 0$) taken at the far ($z = L$) and near ($z = 0$) ends of the fiber as a function of the optical angular frequency ω:

$$H_o(\omega) = \frac{\underline{E}_x(0, L)}{\underline{E}_x(0, 0)} = \exp(-\alpha' L)\exp(-j\beta L) \quad . \tag{11.3}$$

Introducing the complex propagation constant of the LP_{01} mode (5.106)

$$\gamma = \alpha' + j\beta \quad , \tag{11.4}$$

the optical transfer function can be written

$$H_o(\omega) = \exp(-\gamma L) \quad . \tag{11.5}$$

The optical transfer function should not be confused with the baseband transfer function to be discussed in the next section.

11.2 Baseband Transfer Function

Optical fibers cannot be used for direct baseband transmission. Signals can only be transmitted by modulating them onto an optical carrier wave.

An optical fiber link consisting of optical source, modulator, fiber, photoreceiver, and demodulator can be considered as a special case of a general transmission system in which a signal $s_i(t)$ at the input produces an output signal $s_o(t)$. Of course, one wants to transmit the signal without distortions. Since a change in amplitude can be compensated for by an amplifier, and since a time shift is usually of no concern, a transmission system is called distortionless if the output function $s_o(t)$ is a replica of the input function $s_i(t)$ apart from a possible alteration in magnitude and a delay in time [Kapron and Keck 1971].

In practice, the signal will inevitably be distorted to some extent by the transmission system. Both linear and nonlinear distortions will be present.

Linear distortions can be caused only by linear transmission systems. By definition, a transmission system is called linear if the law of superposition holds, i.e. if the response to a sum of excitations is equal to the sum of the responses to the excitations acting separately [Schwartz 1980]. Usually, the relation between the input and output does not change in time: we have a time-invariant system. For a system which is both linear and time invariant, a purely sinusoidal input signal causes a sinusoidal output signal of constant amplitude at the same frequency as the input, differing only in amplitude and phase [Carlson 1975; Föllinger 1982]. For a linear time-invariant system, a transfer function can be defined from the change in amplitude and the phase shift between output and input. For an arbitrary signal, the output spectrum of a linear and time-invariant system contains only those frequency components which were already contained in the input spectrum. A linear time-invariant system does not create new frequency components.

There are two types of linear distortions: amplitude and phase distortions. For amplitude distortions, the relative amplitudes of the different frequency components in the output signal differ from those in the input signal. For phase distortions, which are also called delay distortions, the relative phases of the different frequency components at the output differ from the relative phases at the input. Since the signal is the superposition of the different frequency components, the output signal will be different from the input signal in both cases.

In order to avoid amplitude distortions, the amplitude response must be constant, and in order to avoid phase distortions, the phase shift caused by the transmission system must increase in proportion to the modulation frequency over those frequencies for which the signal spectrum is nonzero [Carlson 1975].

The linear distortions caused by a linear time-invariant transmission system − i.e. amplitude and delay distortions − are theoretically curable through the use of equalization networks [Carlson 1975]. After transmission through the distorting system, the signal is transmitted through an equalizer. By tailoring the transfer function of the equalizer, the overall system can be made to have negligible linear distortions.

Nonlinear distortions are caused by nonlinear systems. In these systems, the law of superposition does not apply. The nonlinearities create output frequency components that were not present in the input, e.g. harmonics or intermodulation products. Since the product of the input spectrum and a transfer function cannot describe the creation of new frequency components, a nonlinear system cannot be described by a transfer function [Carlson 1975]. Nonlinear distortions cannot be cured by an equalizer.

Whether a single-mode fiber link behaves as a linear or nonlinear transmission system depends on the kind of modulation used and the linewidth of the optical source in relation to the modulation bandwidth.

At present, most optical transmission systems use the simple intensity modulation (IM) and direct detection scheme. The question of baseband linearity in these IM systems will be discussed in Sect. 11.2.1.

Future optical transmission systems will use amplitude, phase, or frequency modulation of monochromatic (coherent) optical carriers in combination with homodyne or heterodyne detection. These coherent systems have higher receiver sensitivities and smaller channel bandwidths than IM systems. Therefore system performance can be improved toward longer repeater spacing and larger information capacity. Baseband linearity in coherent systems is discussed in Sect. 11.2.2.

11.2.1 Systems with Intensity Modulation

Because of the (approximately) linear relationship between the optical output power of a light emitting diode (LED) or a semiconductor laser (LD) and the injection current, it is very simple and popular to use optical power modulation, which is also called intensity modulation.

A semiconductor light source emits only for positive values of the injection current. Therefore, for a bipolar input signal $s(t)$, the injection current $i_i(t)$ must contain a dc bias current i_{bi}:

$$i_i(t) = i_{bi}[1 + s(t)] \quad , \quad \text{with} \tag{11.6}$$

$$|s(t)| < 1 \quad . \tag{11.7}$$

Assuming an ideal LED or LD, for which the optical power P coupled into the fundamental fiber mode depends linearly on the injection current, the current (11.6) generates an optical power

$$P_i(t) = P_{bi}[1 + s_i(t)] \quad , \quad \text{where} \tag{11.8}$$

$$s_i(t) = [i_{bi}/(i_{bi} - i_{th})]s(t) \quad . \tag{11.9}$$

The symbol i_{th} denotes the threshold current of the laser diode. For an LED, $i_{th} = 0$, and $s_i(t) = s(t)$.

Optical power refers to the power averaged over a time interval which is very long as compared to the optical period but very short as compared to the period of the highest modulation frequency to be transmitted. The optical power is always positive.

When the optical carrier is monochromatic with a sinusoidal time variation with angular frequency ω_c, intensity modulation is related to the more common amplitude modulation. Since the transmitted power is proportional to the square of the electric field, the time variation of the electric field in a intensity-modulated wave is [Meslener 1982]:

$$E_i(t) = E_{0i}[1 + s_i(t)]^{1/2} \cos(\omega_c t) \quad , \tag{11.10}$$

where E_{0i} is the electric field of the unmodulated carrier. Thus, intensity modulation with the unipolar signal $1 + s_i(t)$ is equivalent to amplitude modulation with the unipolar signal $[1 + s_i(t)]^{1/2}$ [Grau 1981].

At the end of the fiber, the time variation of the optical power will be

$$P_o(t) = P_{bo}[1 + s_o(t)] \quad , \tag{11.11}$$

where the bias power P_{bo} will be smaller than P_{bi} due to fiber attenuation (5.111)

$$P_{bo} = P_{bi} \exp(-2\alpha' L) \quad , \tag{11.12}$$

and where the output signal $s_o(t)$ will deviate from $s_i(t)$ due to linear and nonlinear distortions.

For the PIN and avalanche photodiodes used today, the photocurrent is proportional to the number of incident photons, i.e. to the optical power (5.113)

$$i = RP \quad , \tag{11.13}$$

where R is the responsivity (5.114) of the photodetector. Therefore, the power-modulated carrier can simply be demodulated using a photodiode. The optical output power (11.11) produces a photocurrent

$$i_o(t) = RP_{bo}[1 + s_o(t)] \quad . \tag{11.14}$$

Direct detection by a photodiode is equivalent to linear rectification followed by squaring.

In order to show that intensity modulation and direct photodetection do not necessarily produce linear and nonlinear distortions, we first omit the fiber and place a semiconductor photodiode directly in front of the emitting face of the LED or LD. The time variation of the optical power at the photodetector is then equal to that at the optical transmitter

$$P_o(t) = cP_i(t) \quad , \tag{11.15}$$

where c is a constant of proportionality. From (11.8, 11, 15), one obtains

$$s_o(t) = s_i(t) \quad . \tag{11.16}$$

Thus the output signal is identical to the input signal, i.e., there are neither linear nor nonlinear distortions. The same holds if between optical source and detector, the wave propagates through a medium without chromatic dispersion, e.g. a short air path [Meslener 1984].

Thus, distortions in single-mode IM transmission systems are not caused by the square-root dependence of the input field on the input power and the square dependence of the output power on the output field. When the envelop of the optical wave is not distorted by the transmission medium, the square-root input and the square output characteristics combine to give an overall linear characteristic.

The real reason for distortions in single-mode fibers is the wavelength dependence of the group delay, i.e. chromatic dispersion, which for longer fiber lengths distorts the wave envelope and makes the input-output power relationship nonlinear. Assume for instance, that a monochromatic optical wave is sinusoidally modulated in power. This corresponds to amplitude modulation with the square root of the sum of a constant plus a sine function (11.10), i.e. amplitude modulation with a spectrum of modulation frequencies. If the relative amplitudes or the relative phase angles of the components in the optical spectrum are changed during transmission, the output envelope will be different from the input envelope and the square of the output envelope will not correspond to a pure sine oscillation. The sine will be distorted, i.e. new modulation frequencies have been produced. Thus, an optical transmission system consisting of a power-modulated semiconductor source, a single-mode fiber, and a direct photodetector is intrinsically nonlinear.

In order to describe the signal transmission characteristics of the fiber alone, i.e., excluding those of the source and the photodiode, one does not compare the received photocurrent (11.14) with the current (11.6) injected into the semiconductor source, but discusses the relationship between the time variations of the optical output and input powers $P_o(t)$ (11.11) and $P_i(t)$ (11.8). For linear characteristics of the source and the photodiode, the optical input and output powers are linearly related to the electrical input and output signals, respectively.

Now, for both multimode and single-mode fibers, the relation between input and output power is intrinsically nonlinear, i.e. the law of superposition does not hold, and for a sinusoidal input signal the output signal will contain harmonics. Therefore, strictly speaking, a baseband transfer function cannot be defined for optical fiber.

For *multimode* fibers, the question of baseband linearity and the problem of equalization in fiber optic digital communication systems have been investigated [Personick 1971, 1973a,b; Personick et al. 1974; Vassallo 1977]. In most practical cases, a multimode fiber is approximately linear in its input-output power relationship. However, it is possible to construct examples where the use of highly coherent sources would lead to results which do not follow from the power linearity assumption [Miller and Chynoweth 1979].

Most light sources currently used for optical communications, i.e. semiconductor LED's or LD's, produce optical spectra with widths which, though narrow as a fraction of their central frequency, are very large as compared with the expected signal bandwidth (at most a few gigahertz). For these sources with large linewidths, *single-mode* fiber links are also substantially linear [Bennett 1983].

However, for single frequency lasers with very small linewidths, when the signal bandwidth is comparable with, or larger than, the source bandwidth, the relationship between output pulse and input pulse for the source-fiber system is a nonlinear one and, therefore, a baseband transfer function cannot be defined for coherent sources [Kapron 1980, 1984b; Bennett 1983].

In the following, we discuss only the case of sources with large linewidths, in which a fiber link is approximately linear and time-invariant, so that a baseband transfer function can be defined. One assumes a sinusoidal time variation of the optical input power around a bias power P_{bi}:

$$P_i(t) = P_{bi}[1 + m_i \cos(\omega_m t)], \tag{11.17}$$

where the quantities m_i and $\omega_m = 2\pi f_m$ denote the input intensity modulation index and the modulation angular frequency, respectively. The intensity modulation index is the ratio of the peak change in optical power to the average optical power. The subscript m is used to differentiate between modulation frequencies (nowadays at most some gigahertz) and optical frequencies (of the order of some hundred terahertz).

The modulation index is assumed to be smaller than unity

$$m_i \leq 1 \quad . \tag{11.18}$$

For the partially coherent source assumed, there will be negligible nonlinear distortions and, thus, the output power will consist of a constant part and a sinusoidal component:

$$P_o(t) = P_{bo}[1 + m_o \cos(\omega_m + \Phi)] \quad . \tag{11.19}$$

For low modulation frequencies and short fibers, the output modulation index m_o is equal to the input modulation index m_i, whereas for higher modulation frequencies and long fibers it will be smaller because of chromatic dispersion, i.e. there will be amplitude distortions.

The output power oscillation is phase shifted by the angle Φ relative to the input power oscillation. The value for this phase shift can be estimated by remembering that the wave envelope propagates with the group velocity v_g (5.65). In propagating through the fiber, the modulation is therefore delayed by a time L/v_g. Since a time delay of one modulation period $T_m = 1/f_m$ corresponds to a phase change of 2π, the phase shift of the modulation is found to be approximately

$$\Phi(f_m) = -[L/(v_g T_m)]2\pi = -\omega_m L \tau_g \quad , \tag{11.20}$$

where $\tau_g = 1/v_g$ is the group delay per unit length (5.66).

This linear relationship between the phase delay of the modulation and the modulation frequency only holds for low modulation frequencies. Because of chromatic dispersion, the relation between phase delay Φ and modulation frequency f_m becomes nonlinear for higher modulation frequencies, i.e. there will also be phase distortions.

Just as for sinusoidal voltages or currents, phasors for the sinusoidal part of the optical power can be defined: the modulus of the phasor equals the modulation index and its phase angle equals the zero phase angle of the power oscillation. Thus, from (11.17), the power phasor at the fiber input is

$$\underline{P}(0, f_m) = m_i \quad , \tag{11.21}$$

and that at the output

$$\underline{P}(L, f_m) = m_o \exp(j\Phi) \quad , \tag{11.22}$$

respectively.

Now, the baseband transfer function $H_m(f_m)$ of the fiber link is defined by the ratio of the power phasors at the output and input ends:

$$H_m(f_m) = \underline{P}(L, f_m)/\underline{P}(0, f_m) = (m_o/m_i)\exp(j\Phi). \tag{11.23}$$

The baseband transfer function may also be called the power transfer function or modulation transfer function.

By definition, the magnitude $|H_m(f_m)|$ of the baseband transfer function, the so-called amplitude response of the fiber, is the ratio of the output and input modulation indices, whereas the phase angle of the baseband transfer function

$$\arg H_m(f_m) = \Phi(f_m), \tag{11.24}$$

the phase response, is the phase shift of the modulation.

For lasers with a linear current/power characteristic and photodetectors with a linear power/photocurrent characteristic, the baseband transfer function as defined by (11.23) differs only by a constant factor from the electrical transfer function defined by the ratio of the phasors of the photocurrent received and the current injected into the laser.

Often, a slightly different definition of the baseband transfer function is used. Since the large phase shift (11.20) caused by the time delay does not cause phase distortions, in the definition of the phase angle of the transfer function one takes the difference, $\Phi - (-\omega_m L \tau_g)$ instead of Φ.

The baseband transfer function is determined by the optical transfer function (11.3) of the fiber, i.e. by the frequency dependencies of the attenuation coefficient α' and of the phase constant β, and by the linewidth of the source.

Since the spectrum of the modulated optical wave is relatively narrow, the fiber attenuation α' can usually be considered to be constant over that narrow interval of wavelengths [Kapron 1980].

Because of the complicated dependence of the phase constant β on the optical frequency (Sect. 5.5), it is useful to expand the phase constant $\beta(\omega)$ in a power series of the optical frequency difference $\omega - \omega_c$ at the optical carrier frequency ω_c:

$$\beta = \beta_c + \frac{d\beta}{d\omega}(\omega - \omega_c) + \frac{1}{2}\frac{d^2\beta}{d\omega^2}(\omega - \omega_c)^2 + \frac{1}{6}\frac{d^3\beta}{d\omega^3}(\omega - \omega_c)^3 + \ldots \quad .$$

(11.25)

In this expansion, β_c is the phase constant at the frequency ω_c of the optical carrier. The first three derivatives of the phase constant with respect to angular frequency, $d\beta/d\omega$, $d^2\beta/d\omega^2$, and $d^3\beta/d\omega^3$, are related (5.86, 94) to quantities that can be measured (Sect. 13.9) [Marcuse 1980]:

$$\frac{d\beta}{d\omega} = \tau_g \quad , \tag{11.26}$$

$$\frac{d^2\beta}{d\omega^2} = -\frac{\lambda^2}{2\pi c}D \quad , \tag{11.27}$$

$$\frac{d^3\beta}{d\omega^3} = \frac{\lambda^2}{(2\pi c)^2}[\lambda^2 S + 2\lambda D] \quad , \quad \text{where} \tag{11.28}$$

$$\tau_g = 1/v_g \tag{11.29}$$

is the specific group delay time (5.66),

$$D = d\tau_g/d\lambda \tag{11.30}$$

the chromatic dispersion (5.85), and

$$S = dD/d\lambda \tag{11.31}$$

the dispersion slope (5.93).

If the operating wavelength λ is different from a wavelength λ_0 of minimum dispersion for which $D = 0$, one can truncate the expansion (11.25) with the term proportional to the square of the frequency deviation. If $\lambda = \lambda_0$, this term vanishes and it is necessary to take the next term containing the dispersion slope S into account.

Present-day semiconductor lasers usually have linewidths which are very large compared to the bandwidth of the signal. For these sources, the fiber is a linear transmission system and the complex baseband transfer function (11.23) can be calculated. Assuming a Gaussian source spectrum, it is [Gloge et al. 1980; Cohen L.G. et al. 1982a; Bennett 1983]:

$$H_m(f_m) = \frac{1}{\sqrt{1 + jf_m/f_2}}\exp\left[-\frac{1}{2}\frac{(f_m/f_1)^2}{1 + jf_m/f_2}\right] \quad , \tag{11.32}$$

where the quantities f_1 and f_2 have been introduced as abbreviations for

246

$$f_1 = \left(2\pi\sigma_\omega \frac{d^2\beta}{d\omega^2}L\right)^{-1} \quad \text{and} \tag{11.33}$$

$$f_2 = \left(2\pi\sigma_\omega^2 \frac{d^3\beta}{d\omega^3}L\right)^{-1} \quad . \tag{11.34}$$

The quantity σ_ω is the rms angular frequency spectral width of the source (3.45) and L is the fiber length. The factor $\exp(-j\tau_g\omega_m L)$, which constitutes a uniform time delay, has been suppressed in the formula (11.32) for the baseband transfer function.

In (11.33, 34), the theoretical quantities σ_ω, $d^2\beta/d\omega^2$, and $d^3\beta/d\omega^3$ can be expressed by the measurable parameters σ_λ, D, and S by using (3.45, 11.27, 28) to give

$$f_1 = (-2\pi\sigma_\lambda DL)^{-1} \quad \text{and} \tag{11.35}$$

$$f_2 = [2\pi\sigma_\lambda^2(S + 2D/\lambda)L]^{-1} \quad . \tag{11.36}$$

Remember that for a Gaussian form of the spectrum, the rms linewidth σ_λ in the wavelength domain is related to the full width at half maximum, $\Delta\lambda_{\mathrm{FWHM}}$ by (3.43)

$$\sigma_\lambda = 0.42\Delta\lambda_{\mathrm{FWHM}} \quad . \tag{11.37}$$

Note that when comparing the dispersion properties of e.g. ordinary and dispersion-shifted fibers on the basis of the dispersion parameters, one should compare the derivatives $d^2\beta/d\omega^2$ and $d^3\beta/d\omega^3$ of the phase constant with respect to the angular frequency instead of the derivatives D and S of the group delay with respect to wavelength [Marcuse and Stone 1985].

Most often, one is only interested in the modulus of the complex baseband transfer function (11.32) [Cohen L.G. et al. 1982a]:

$$|H_m(f_m)| = \frac{1}{[1 + (f_m/f_2)^2]^{1/4}} \exp\left[-\frac{1}{2}\frac{(f_m/f_1)^2}{1 + (f_m/f_2)^2}\right] \quad . \tag{11.38}$$

In electrical network theory and communication techniques, the bandwidth of a low pass filter is usually defined by that frequency for which the magnitude of the transfer function is $2^{-1/2} = 0.707$ ($-3\,\mathrm{dB}$) of its value for small frequencies. For ideal source and detector, the baseband transfer function defined by (11.32), differs only by a constant factor from the electrical transfer function. Therefore, it seems logical to define the bandwidth B of a single-mode fiber link as that modulation frequency f_m, for which the modulus of the baseband transfer function is 0.707 of its value at small frequencies:

$$|H_m(B)| = 0.707 \quad . \tag{11.39}$$

For $f_m = B$, the modulation index is reduced by 3 dB:

$$20\log(m_o/m_i) = -3\,\text{dB} \quad . \tag{11.40}$$

Note however, that in optical communications, another definition of the bandwidth is sometimes used [Gloge et al. 1980; Cohen L.G. et al. 1982a]. This other modulation bandwidth is defined by that modulation frequency for which the amplitude response is 0.5 (-6 dB) of its value at low frequencies.

The baseband transfer function (11.32) depends both on fiber properties (L, D, S) and on source properties (σ_λ). Therefore, we see that the terms "transfer function of a single-mode fiber" and "bandwidth of a single-mode fiber" are inappropriate. Strictly speaking, we have the baseband transfer function or bandwidth of the source-fiber system. This means, that a single-mode fiber transfer function does not exist in the classical sense, since it is dependent on source properties [Kapron and Lazay 1983].

This is in contrast to the situation with multimode fibers where, although the source-dependent dispersion (chromatic or intramodal dispersion) is present in each propagating mode, the group delay variation over the various modes (intermodal dispersion) is usually dominant. Source-fiber bandwidth is then nearly independent of source spectral width, and one can usefully speak of "fiber bandwidth" at a given operating wavelength [Bennett 1983].

The general formula (11.38) for the magnitude of the baseband transfer function can be simplified if the operating wavelength is either close to a wavelength λ_0 of minimum dispersion or is far from a wavelength of minimum dispersion.

When the operating wavelength λ is close to a wavelength λ_0 of minimum dispersion, the dispersion parameter D is very small and thus from (11.33, 34)

$$f_1 \gg f_2 \quad . \tag{11.41}$$

For wavelengths for which (11.41) is fulfilled, the amplitude response (11.38) can be approximated by the simple formula

$$|H_m(f_m)| = [1 + (f_m/f_2)^2]^{-1/4} \quad , \tag{11.42}$$

from which the 3-dB bandwidth defined by (11.39) is calculated to be

$$B = \sqrt{3}f_2 = 1.73f_2 \quad . \tag{11.43}$$

Thus, in essence, the quantity f_2 defined by (11.34) is the bandwidth at the wavelength of minimum dispersion.

Since the natural linewidth has been assumed to be very large in comparison with the signal bandwidth, (11.43) in combination with (11.34) erroneously predicts infinite bandwidth for a monochromatic source ($\sigma_\lambda = 0$).

For a numerical example, we consider a matched cladding standard fiber with germanium doped core with diameter $2a = 8.6\,\mu$m, a pure silica cladding,

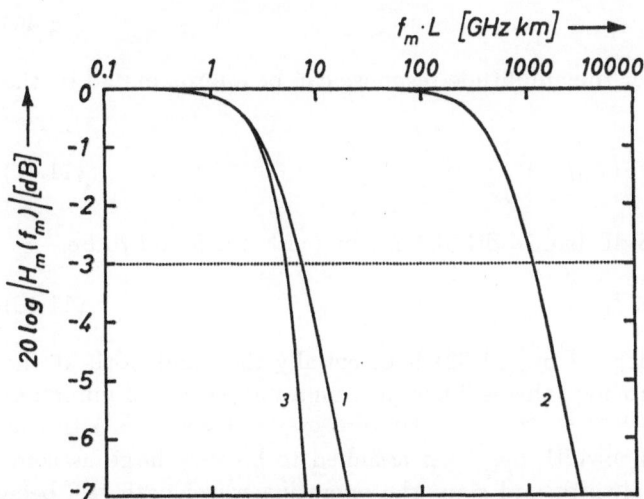

Fig. 11.1. Modulus of the baseband transfer function (11.38) for matched cladding standard fiber with $2a = 8.6\mu$m, $\Delta = 0.003$, $\lambda_0 = 1325$ nm, $S_0 = 0.085$ ps/(km nm^2) as a function of the product of modulation frequency f_m and fiber length L for
1. ELED with $\Delta\lambda_{\mathrm{FWHM}} = 50$ nm, $\lambda = \lambda_0$,
2. LD with $\Delta\lambda_{\mathrm{FWHM}} = 4$ nm, $\lambda = \lambda_0$,
3. LD with $\Delta\lambda_{\mathrm{FWHM}} = 4$ nm, $\lambda = 1550$ nm, $D = 15.4$ ps/(km nm), $S = 0.058$ ps/(km nm^2)

and a relative refractive-index difference of $\Delta = 0.003$. For this fiber, the wavelength of minimum dispersion is calculated to be at $\lambda_0 = 1325$ nm. The dispersion slope at λ_0 is $S_0 = 0.085$ ps/(km nm^2).

First, the source is assumed to be an edge-emitting light emitting diode (ELED) with a FWHM linewidth of 50 nm operating at λ_0. Curve 1 in Fig. 11.1 represents the modulus of the baseband transfer function as a function of the product of modulation frequency and fiber length. Inserting the values for S_0 and σ_λ into (11.34) and (11.43), gives for the bandwidth × length product

$$BL = 7.2 \,\mathrm{GHz\,km} \quad . \tag{11.44}$$

For an LED link with a typical length of 10 km, the bandwidth is 720 MHz.

Next, the source is taken to be a semiconductor laser with a FWHM linewidth of 4 nm operating at λ_0. Curve 2 in Fig. 11.1 represents the modulus of the baseband transfer function. Inserting the values for S_0 and σ_λ into (11.34) and (11.43), gives for the bandwidth × length product

$$BL = 1128 \,\mathrm{GHz\,km} \quad . \tag{11.45}$$

For a LD link at 1.3 μm with a typical length of 30 km, the theoretical bandwidth would be 37.6 GHz. Note however, that it is difficult to match the laser wavelength exactly to the minimum dispersion wavelength λ_0.

For operating wavelengths far from a wavelength λ_0 of minimum dispersion, f_1 (11.33) is much smaller than f_2 (11.34) and for

$$f_1 \ll f_2 \quad , \tag{11.46}$$

the expression (11.38) for the amplitude response can be approximated by the simple function

$$|H_{\mathrm{m}}(f_{\mathrm{m}})| = \exp\left[-\frac{1}{2}(f_{\mathrm{m}}/f_1)^2\right] \quad . \tag{11.47}$$

From (11.47), the 3-dB bandwidth defined by (11.39) is found to be

$$B = \sqrt{\ln 2}\, f_1 = 0.833 f_1 \quad . \tag{11.48}$$

Thus the quantity f_1 defined by (11.33) is essentially the bandwidth at the operating wavelength provided this is different from a wavelength of minimum dispersion.

Since the natural linewidth has been assumed to be very large as compared to the signal bandwidth and since the operating wavelength has been assumed to be different from a wavelength of minimum dispersion, (11.48) in combination with (11.33) erroneously predicts infinite bandwidth both for a monochromatic source ($\sigma_\lambda = 0$) and at a wavelength of minimum dispersion ($D = 0$).

For the matched cladding fiber considered before, at the wavelength of minimum loss $\lambda = 1550\,\mathrm{nm}$, the chromatic dispersion is $D = 15.4\,\mathrm{ps/(km\,nm)}$ and the dispersion slope is $S = 0.058\,\mathrm{ps/(km\,nm^2)}$. Assuming a laser source with a FWHM linewidth of $4\,\mathrm{nm}$, one finds $f_1 L = 6.1\,\mathrm{GHz\,km}$ and $f_2 L = 711\,\mathrm{GHz\,km}$. The amplitude response is represented by curve 3 in Fig. 11.1.

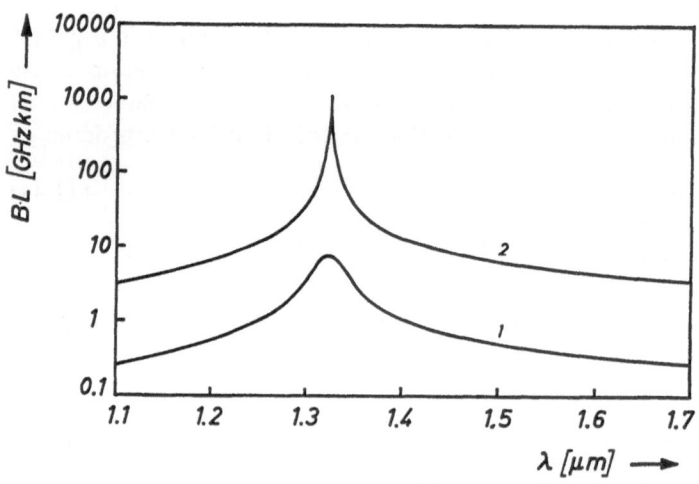

Fig.11.2. Product of the 3-dB bandwidth B and fiber length L of a matched cladding standard fiber with $2a = 8.6\,\mu\mathrm{m}$, $\Delta = 0.003$, $\lambda_0 = 1325\,\mathrm{nm}$, $S_0 = 0.085\,\mathrm{ps/(km\,nm^2)}$ as a function of wavelength λ for
1. ELED with $\Delta\lambda_{\mathrm{FWHM}} = 50\,\mathrm{nm}$,
2. LD with $\Delta\lambda_{\mathrm{FWHM}} = 4\,\mathrm{nm}$.
The wavelength dependence of D was assumed to be described by (11.50)

The bandwidth × length product is

$$BL = 5.1\,\text{GHz km} \quad . \tag{11.49}$$

For a fiber length of 100 km, which, with regard to the power budget is possible due to the very low fiber losses at 1.55 μm, the bandwidth is only 51 MHz. This bandwidth is disappointingly small considering the very high carrier frequency used.

In order to get larger bandwidths at 1.55 μm, one has either to use dispersion-shifted or dispersion-flattened fibers with smaller chromatic dispersion (Sect. 5.8) or to use a source with a smaller linewidth.

The wavelength dependence of the 3-dB bandwidth times length product is represented in Fig. 11.2 both for an ELED (Curve 1) and a laser diode (Curve 2) as sources. For calculating these curves, it has been assumed that the wavelength dependence of the chromatic dispersion can be described by the function (13.109)

$$D(\lambda) = S_0(\lambda - \lambda_0^4/\lambda^3)/4 \quad , \tag{11.50}$$

with zero dispersion slope $S_0 = 0.085\,\text{ps}/(\text{km nm}^2)$ at $\lambda_0 = 1325\,\text{nm}$. For simplicity, constant linewidths of $\Delta\lambda_{\text{FWHM}} = 50\,\text{nm}$ for the ELED and $\Delta\lambda_{\text{FWHM}} = 4\,\text{nm}$ for the LD have been used.

Taking another special functional form for the wavelength dependence of the chromatic dispersion $D(\lambda)$, simple asymptotic formulas for the chromatically limited bandwidth × length product have been derived [Engelmann 1981].

In this section, we assumed the linewidth of the optical source to be very large in comparison to the modulation bandwidth. A single-mode fiber link using intensity modulation and direct photodetection then behaves similarly to a linear time-invariant system. However, for coherent sources, if the linewidth of the source is much smaller than the modulation bandwidth, the chromatic dispersion of the fiber causes nonlinear distortions in IM direct-photodetection transmission systems. For a purely sinusoidal input signal with frequency f_m, the output signal contains harmonics with frequencies $2f_m$, $3f_m$, $4f_m$, ... as well as the fundamental with frequency f_m. Remember that the behaviour of this system cannot be described by a baseband transfer function.

Nonlinear distortions in IM systems caused by chromatic dispersion will be reviewed in Sect. 11.4 on analogue signal transmission along single-mode fibers.

11.2.2 Coherent Systems

Future coherent optical transmission systems will use amplitude modulation of a sinusoidal carrier wave and homodyne or heterodyne detection instead of intensity modulation and direct detection. Chromatic dispersion will then cause linear, but no nonlinear distortions. One can define a baseband transfer function analogous to (11.23) where the m's are now the input and output *amplitude* modulation indices instead of the *intensity* modulation indices, and where Φ is the phase shift experienced by the modulation during propagation through the fiber.

The baseband transfer function of AM systems has been calculated [Kimura et al. 1978; Furuya et al. 1979a]. For operating wavelengths different from a wavelength λ_0 of minimum chromatic dispersion, the modulus of the baseband transfer function is

$$|H_m(f_m)| = \cos\left(2\pi^2 \frac{d^2\beta}{d\omega^2} f_m^2 L\right) \quad . \tag{11.51}$$

Using (11.27), the second derivative $d^2\beta/d\omega^2$ of the phase constant can be replaced by the chromatic dispersion D

$$|H_m(f_m)| = \cos(\pi\lambda^2 DL f_m^2/c) \quad . \tag{11.52}$$

The 3-dB bandwidth B defined as that value of the modulation frequency f_m which gives (11.39)

$$|H_m(f_m)| = 0.707 \quad , \tag{11.53}$$

is calculated from (11.52) to be

$$B = (4\lambda^2 DL/c)^{-1/2} \quad . \tag{11.54}$$

In AM systems, the fiber bandwidth is inversely proportional to the square root of the fiber length, in contrast to IM systems where the bandwidth (11.43, 48) is inversely proportional to fiber length.

For a numerical example, we consider a standard fiber with $D = 15.4$ ps/(km nm) at $\lambda = 1550$ nm. The product of the bandwidth and the square root of fiber length is

$$B\sqrt{L} = 45\,\text{GHz}\,\sqrt{\text{km}} \quad , \tag{11.55}$$

giving a very large bandwidth of 4.5 GHz for a fiber length of 100 km despite the large value of chromatic dispersion.

The relations for the baseband transfer function (11.52) and the bandwidth (11.54) of AM systems only hold for $\lambda \neq \lambda_0$. For $\lambda = \lambda_0$, $d^2\beta/d\omega^2 = 0$, and the third derivative $d^3\beta/d\omega^3$ becomes dominant in the transfer function causing phase distortions. However, these phase distortions can, in principle, be compensated for electrically, so that the ultimate upper limit of the bandwidth at zero dispersion wavelength is determined by the fourth derivative $d^4\beta/d\omega^4$ [Furuya et al. 1979a].

11.3 Impulse Response

If a very short light pulse is injected into the fiber (in the mathematical idealization an "impulse" of zero duration, infinite amplitude and unit area) the output power pulse is called the impulse response $P_\delta(t)$ of the fiber.

Practically, any pulse will have a finite duration. When the spectrum of a pulse is still narrow as compared to the spectral width of the optical carrier wave, the source-fiber system is approximately linear, and the impulse response in the time domain can be obtained from the inverse Fourier transform of the baseband transfer function (11.32) [Geckeler 1986a]:

$$P_\delta(t) = \frac{1}{2\pi} \int\limits_{-\infty}^{+\infty} H_{\mathrm{m}}(\omega_{\mathrm{m}}) \exp(\mathrm{j}\omega_{\mathrm{m}}t)d\omega_{\mathrm{m}} \quad . \tag{11.56}$$

Conversely, the baseband transfer function can be obtained by Fourier transforming the impulse response:

$$H_{\mathrm{m}}(\omega_{\mathrm{m}}) = \int\limits_{-\infty}^{+\infty} P_\delta(t) \exp(-\mathrm{j}\omega_{\mathrm{m}}t)dt \quad . \tag{11.57}$$

These relations can be used to measure the baseband transmission characteristics of the fiber in the frequency or in the time domain.

11.4 Transmission of an Analogue Signal

Although optical fiber transmission systems are best suited for digital signals (Sect. 11.5), they can also be used to transmit analogue signals. The advantage of analogue signal transmission lies in the fact that it is not necessary to use expensive AD converters, coders, decoders, and DA converters.

As has been discussed in Sect. 11.2, an analogue signal transmitted via a single-mode fiber will, in general, be affected both by linear and nonlinear distortions. We start by considering linear distortions.

For intensity modulation, the analogue signal is defined as the time variation of the optical power, i.e. $P_{\mathrm{i}}(t)$ (11.8) and $P_{\mathrm{o}}(t)$ (11.11) at the fiber input and output, respectively. For optical sources with large linewidths, the single-mode fiber can be considered as a linear, time-invariant tranmission system (Sect. 11.2.1). There are then two methods for calculating the output signal waveform from the input signal and the transfer properties of the fiber: one in the frequency domain and one in the time domain.

In order to calculate the output signal in the frequency domain, the input signal is first decomposed into its spectral components by a Fourier transform:

$$F_i(\omega_m) = \int\limits_{-\infty}^{+\infty} P_i(t)\exp(-j\omega_m t)dt \quad . \tag{11.58}$$

During propagation along the fiber, each Fourier component of the signal is attenuated by the factor $|H_m(\omega_m)|$ and phase shifted by the angle $\arg[H_m(\omega_m)]$, i.e. the output signal spectrum is obtained by multiplying the input spectrum by the complex baseband transfer function (11.23):

$$F_o(\omega_m) = F_i(\omega_m)H_m(\omega_m) \quad . \tag{11.59}$$

Finally, the output signal waveform is obtained by taking the inverse Fourier transform of the output spectrum:

$$P_o(t) = \frac{1}{2\pi}\int\limits_{-\infty}^{+\infty} F_o(\omega_m)\exp(j\omega_m t)d\omega_m \quad . \tag{11.60}$$

In the time domain, the output signal is the convolution of the input signal and the impulse response (11.56) [Marcuse 1981a]:

$$P_o(t) = \int\limits_{-\infty}^{+\infty} P_i(t-\tau)P_\delta(\tau)d\tau \quad . \tag{11.61}$$

In Sect. 11.2, it has been emphasized that a single-mode fiber link is an intrinsically nonlinear transmission system that can only be considered approximately linear if the natural linewidth of the source is large compared to the maximum modulation frequency. Since a linear time-invariant transmission system has been assumed up to now, evaluating (11.60) or (11.61) gives the linear distortions of the signal only.

In single-mode fiber systems using coherent optical carriers, intensity modulation, and direct photodetection, chromatic dispersion causes nonlinear distortions. In order to characterize these nonlinear distortions, we assume a sinusoidal input signal as in (11.17) and specify the relative nth order harmonic distortion at the output by the ratio m_n/m_0 of the intensity modulation indices m_n and m_0 of the nth order order harmonic [with frequency $(n+1)f_m$] and of the fundamental oscillation, respectively.

Nonlinear distortions caused by chromatic dispersion have been analyzed by several authors [Unger 1977b; Kapron 1979, 1980; Furuya et al. 1979a,b; Arnold G. and Petermann 1980; Marcuse 1980; Yamamoto A. and Kimura 1981; Meslener 1982, 1984; Heckmann 1983].

For a monochromatic optical carrier without residual frequency modulation and for operation far from a wavelength of minimum dispersion, the relative second-order harmonic distortion is given by [Meslener 1984]:

$$m_2/m_0 = \pi^2 c^2 m_i L^2 D^2 (f_m/f_c)^4 \quad . \tag{11.62}$$

The second-order harmonic distortion factor m_2/m_0 (11.62) is proportional to the intensity modulation index m_i. Therefore, even for a monochromatic source, the fiber appears linear in intensity if the modulation depth is not too large [Kapron 1979].

The distortion factor (11.62) is proportional to the fourth power of the ratio of the modulation frequency f_m and the optical carrier frequency f_c. For small values of f_m, the amplitudes of the harmonics are very small; but they increase very rapidly with increasing modulation frequency. Before there is a noticeable decrease in the amplitude of the fundamental, i.e. linear distortions, there will be harmonic distortions. Therefore, the maximum permissible modulation frequency is not determined by the frequency response of the fundamental, but by the onset of nonlinear distortions. The harmonic distortions reduce the effective bandwidth by the factor 0.5 to 0.1 compared to the bandwidth (11.54) of the same fiber when using amplitude modulation and heterodyne detection [Furuya et al. 1979b].

For a numerical example, we assume a 100 km single-mode fiber for $\lambda = 1.55\,\mu m$ with $D = 15.4\,ps/(km\,nm)$, a maximum input modulation index $m_i = 1$, and a modulation frequency of 1 GHz. From (11.62), the distortion factor is calculated to be

$$m_2/m_0 = 1.5 \times 10^{-3} \quad . \tag{11.63}$$

Nonlinear distortions are thus very small for the fiber lengths and the modulation frequencies used today. A more complete study [Meslener 1984] has shown, that for repeater-free optical links of less than 100 km length and modulating frequencies below about 400 MHz, dispersion induced harmonic distortion in IM systems will not be a consideration when source frequency modulation (FM) is zero.

However, the addition of relatively low levels of source FM (wavelength chirping, Sect. 11.6.2) can produce substantial harmonic distortion if chromatic dispersion is large or if the modulation index is high [Meslener 1982, 1984]. High levels of FM will induce high levels of harmonic distortion and will restrict the use of fibers in analogue applications, such as cable TV.

The waveform distortion of a sinusoidal signal, amplitude or phase modulated onto a coherent optical carrier far from a wavelength of minimum dispersion has also been studied by Yamamoto A. and Kimura [1981]. Linear chromatic dispersion (D) degrades the modulation depth and causes AM to PM as well as PM to AM conversion. Examples of originally sinusoidal waveforms distorted by linear chromatic dispersion can be found in the paper mentioned.

Nonlinear distortions can also be specified by the intermodulation factor instead of the second-order harmonic distortion. This measure is used when several TV signals amplitude modulate auxiliary electrical carriers which power modulate a laser.

11.5 Pulse Transmission

A digital signal is transmitted through an optical fiber as a time sequence of optical pulses. The duration of the output pulses determines the maximum bit rate which can be transmitted by the fiber (11.70). Therefore, one is interested in the minimum output pulse width possible. The width of the output pulse is determined by the linewidth of the optical source, the input pulse width, and by the dispersion properties of the fiber.

There are several measures for the pulse duration, e.g. the rms width, the $1/e$ halfwidth, the 50% halfwidth, or the corresponding full widths.

For an arbitrary form of the optical pulse $P(t)$, the rms width σ_t represents a universal measure of the pulse duration. It is defined in analogy to the rms linewidth (3.38) using the moments of the pulse form (3.28):

$$M_n = \int_{-\infty}^{+\infty} t^n P(t)dt \quad . \tag{11.64}$$

The moment of zero order is simply the optical power integrated over time, i.e. the energy W in a single pulse

$$M_0 = \int_{-\infty}^{+\infty} P(t)dt = W \quad . \tag{11.65}$$

The first order moment relative to the zero order moment

$$\frac{M_1}{M_0} = \frac{\int\limits_{-\infty}^{+\infty} tP(t)dt}{\int\limits_{-\infty}^{+\infty} P(t)dt} = t_{\rm c} \quad , \tag{11.66}$$

is the center of gravity $t_{\rm c}$ of the function $P(t)$, or loosely the time of maximum power.

The rms width of the pulse is defined by the equation (3.30)

$$\sigma_t^2 = \frac{\int\limits_{-\infty}^{+\infty} (t - t_{\rm c})^2 P(t)dt}{\int\limits_{-\infty}^{+\infty} P(t)dt} = \frac{M_2}{M_0} - \left(\frac{M_1}{M_0}\right)^2 \quad . \tag{11.67}$$

This equation applies for any pulse form, but in order to determine the rms width, it is necessary to measure the pulse form $P(t)$ with a photoreceiver and an oscilloscope and to calculate the three moments by evaluating the integrals in (11.67).

Thus in order to simplify the characterization of the pulse duration, the practitioner often assumes a Gaussian pulse form:

$$P(t) = P(t_c) \exp[-0.5(t - t_c)^2 / \sigma_t^2] \quad . \tag{11.68}$$

The rms width σ_t is the time interval between the maximum and the 60.7% point. Other measures for the pulse duration are the 50% halfwidth Δt_{50} and the $1/e$ halfwidth Δt_{37}. The relations between these three measures for the pulse width are the same as for the corresponding three measures for the spectral linewidth (3.42):

$$\sigma_t = 0.849 \Delta t_{50} = 0.707 \Delta t_{37} \quad . \tag{11.69}$$

Note that some authors prefer to cite the full widths, which are twice the corresponding half widths defined above.

In order to avoid intersymbol interference due to overlapping of consecutive output pulses, the pulse separation T is commonly chosen to be 4 rms widths ($4\sigma_{to}$) at the fiber output end resulting in the following expression for the maximum permissible bit rate R_b [Lin and Marcuse 1981; Marcuse and Lin 1981; Kapron 1984b]

$$R_b = \frac{1}{T} = \frac{1}{4\sigma_{to}} \quad . \tag{11.70}$$

Due to chromatic dispersion, i.e. the wavelength dependence of the group delay, the output pulse will generally be longer than the input pulse. This effect is called pulse broadening or pulse dispersion.

Pulse broadening for optical carriers of arbitrary coherence, for a general form of the input pulse, and for arbitrary dispersion characteristics of the fiber has been analyzed by Gloge [1979a].

In order to obtain a simple formula for the pulse broadening, we only consider the case where light of a Gaussian source spectrum with rms width σ_ω (in angular frequency) is intensity modulated by a Gaussian input pulse with rms width σ_{ti}. When the pulse has propagated a distance L along the fiber, its rms width has increased to [Marcuse 1980; Marcuse and Stone 1985]

$$\sigma_{to} = \sigma_{ti} \left[1 + \left(\frac{d^2\beta}{d\omega^2} \frac{L}{2\sigma_{ti}^2} \right)^2 (1 + 4\sigma_\omega^2 \sigma_{ti}^2) \right.$$
$$\left. + \left(\frac{d^3\beta}{d\omega^3} \frac{L}{4\sqrt{2}\sigma_{ti}^3} \right)^2 (1 + 4\sigma_\omega^2 \sigma_{ti}^2)^2 \right]^{1/2} \quad . \tag{11.71}$$

Since the general expression (11.71) for the pulse broadening is complicated, we discuss four important special cases and replace the quantities σ_ω, $d^2\beta/d\omega^2$, and $d^3\beta/d\omega^3$, by the measurable quantities σ_λ, D, and S, by using (3.45, 11.27), and (11.28). The cases are arranged in the order of increasing pulse broadening:

Case 1. Source with small spectral width operating at a wavelength λ_0 of minimum dispersion,

Case 2. Source with small spectral width operating far from a wavelength of minimum dispersion,

Case 3. Source with large spectral width operating at a wavelength λ_0 of minimum dispersion,

Case 4. Source with large spectral width operating far from a wavelength of minimum dispersion.

Case 1. The linewidth is very small, so that the product $2\sigma_\omega\sigma_{ti}$ in (11.71) is small compared to unity:

$$2\sigma_\omega\sigma_{ti} \ll 1 \quad . \tag{11.72}$$

Moreover, the operating wavelength λ is equal to a wavelength λ_0 of minimum dispersion,

$$\lambda = \lambda_0 \quad , \tag{11.73}$$

so that

$$d^2\beta/d\omega^2 = 0 \quad . \tag{11.74}$$

Under these assumptions, the output pulse duration is

$$\sigma_{to}^2 = \sigma_{ti}^2 + \left(\frac{S_0 L\lambda^2}{4\sqrt{2}2\omega_c^2\sigma_{ti}^2}\right)^2 \quad . \tag{11.75}$$

At a wavelength λ_0 of minimum dispersion, the pulse broadening depends on the zero dispersion slope S_0 instead of the first-order dispersion parameter D. For small fiber lengths, there is no pulse broadening, whereas for large fiber lengths, the second term on the right hand side of will prevail.

For a numerical example, we take $\lambda = \lambda_0 = 1.3\,\mu\text{m}$, $\sigma_{ti} = 1\,\text{ns}$, $L = 100\,\text{km}$, $S = 0.1\,\text{ps}/(\text{km nm}^2)$ and find for the increase in pulse duration

$$\sigma_{to} - \sigma_{ti} = 10^{-27}\,\text{s} \quad . \tag{11.76}$$

Thus, at a wavelength of minimum dispersion and for a coherent source, there is practically no pulse broadening.

Pulse propagation through single-mode fibers with third-order dispersion $(d^3\beta/d\omega^3 \neq 0)$ has been studied in more detail with respect to width and velocity of a Gaussian-shaped pulse when the optical source oscillates with a single frequency [Miyagi and Nishida 1979a].

Case 2. The linewidth is very small, so that the product $2\sigma_\omega\sigma_{ti}$ is small as compared to unity:

$$2\sigma_\omega\sigma_{ti} \ll 1 \quad . \tag{11.77}$$

The operating wavelength λ is far from a wavelength λ_0 of minimum dispersion:

$$\lambda \neq \lambda_0 \quad , \tag{11.78}$$

so that the term in (11.71) containing $d^3\beta/d\omega^3$ is negligibly small. Thus the relation reduces to

$$\sigma_{to}^2 = \sigma_{ti}^2 + \left(\frac{DL\lambda}{2\omega_c\sigma_{ti}}\right)^2 \quad . \tag{11.79}$$

This equation can be interpreted physically by considering the spectrum of the input pulse:

When a coherent carrier is power modulated by a Gaussian pulse of rms width σ_{ti} as described by (11.68), the power spectrum $P_\omega(\omega)$ of the modulated optical wave is also Gaussian [Unger 1977a]:

$$P_\omega(\omega) = \frac{dP}{d\omega} = P_\omega(\omega_c) \exp\left[\frac{0.5(\omega - \omega_c)^2}{\sigma_{\omega m}^2}\right] \quad . \tag{11.80}$$

The quantity $\sigma_{\omega m}$ is the rms width of the spectrum caused by modulation. It is inversely proportional to the rms pulse width σ_{ti}:

$$\sigma_{\omega m} = 1/(2\sigma_{ti}) \quad . \tag{11.81}$$

Replacing the rms width $\sigma_{\omega m}$ in the angular frequency domain by the rms width $\sigma_{\lambda m}$ caused by modulation in the wavelength domain (3.45)

$$\sigma_{\omega m} = \omega_c \sigma_{\lambda m}/\lambda \quad , \tag{11.82}$$

(11.79) gives

$$\sigma_{to}^2 = \sigma_{ti}^2 + (DL\sigma_{\lambda m})^2 \quad . \tag{11.83}$$

Since the chromatic dispersion D is defined (5.85) as the change in group delay time per unit length per unit change in wavelength, the product $DL\sigma_{\lambda m}$ can be interpreted as the spread in delay times of the different spectral components contained in the light pulse.

Applying (11.79) for calculating the output pulse width for wavelengths near a dispersion minimum will result in values which are smaller than the real values.

As a numerical example, we assume a standard single-mode fiber with a length of $L = 100\,\text{km}$ operated in the third low-loss transmission window ($\lambda = 1.55\,\mu\text{m}$) where the first-order dispersion parameter is $D = 15.4\,\text{ps}/(\text{km}\,\text{nm})$. For an input pulse with rms width $\sigma_{ti} = 1\,\text{ns}$, one calculates for the pulse spreading

$$\sigma_{\text{to}} - \sigma_{\text{ti}} = \frac{\sigma_{\text{to}}^2 - \sigma_{\text{ti}}^2}{\sigma_{\text{to}} + \sigma_{\text{ti}}} \approx \frac{1}{2\sigma_{\text{ti}}} \left[\frac{DL\lambda}{2\omega_c\sigma_{\text{ti}}} \right]^2 = 5 \times 10^{-15}\text{s} \qquad (11.84)$$

From this example, one concludes that for a monochromatic source, pulse broadening is negligibly small even far from a wavelength of minimum dispersion. Thus, by using single-frequency lasers with very small linewidth, very high bit rates can be transmitted through very long conventional single-mode fibers even if the operating wavelength (e.g. 1.55 μm) is far from the wavelength of minimum dispersion (e.g. 1.3 μm).

An alternative approach for obtaining a large bit rate times length product at 1.55 μm, is to use conventional semiconductor lasers with a typical linewidth of $\Delta\lambda_{\text{FWHM}} = 4$ nm with dispersion-shifted or dispersion-flattened fibers (Sect. 5.8).

Using a coherent carrier wave, it is possible to transmit pulses with durations much shorter than 1 ns. Since the output rms pulse width determines the maximum posssible bit rate (11.70), one is interested in the smallest output pulse width possible. Considering the dependence of σ_{to} on σ_{ti} in (11.75) or (11.79), one recognizes that for given chromatic dispersion $D(\lambda)$ and length L of the fiber, there is an optimum value of the input pulse width which gives the shortest output pulses. The optimum input pulse width can be obtained by setting the derivative of σ_{to} with respect to σ_{ti} equal to zero [Kapron and Keck 1971; Lin C. and Marcuse 1981].

Note that the results (11.75, 79) are only valid in the case of a single pulse. If a sequence of pulses is transmitted along the fiber, they are not transmitted independently and additional distortion occurs [Jürgensen 1977, 1978]. In the region where two adjacent pulses overlap, an oscillating distortion is superimposed.

Case 3. The linewidth is very large, so that the product $2\sigma_\omega\sigma_{\text{ti}}$ in the general formula (11.71) is large as compared to unity:

$$2\sigma_\omega\sigma_{\text{ti}} \gg 1 \quad . \qquad (11.85)$$

Moreover, the operating wavelength λ is equal to a wavelength λ_0 of minimum dispersion:

$$\lambda = \lambda_0 \quad , \qquad (11.86)$$

so that $d^2\beta/d\omega^2 = 0$. One then has

$$\sigma_{\text{to}}^2 = \sigma_{\text{ti}}^2 + \tfrac{1}{2}(S_0 L\sigma_\lambda^2)^2 \quad . \qquad (11.87)$$

Note the quadratic, rather than linear, dependence on spectral width of the output pulse duration for very large fiber lengths.

Introducing the parameter f_2 (11.34, 36), which is essentially the fiber bandwidth at λ_0 (11.43), one gets

$$\sigma_{to}^2 = \sigma_{ti}^2 + \tfrac{1}{2}(2\pi f_2)^{-2} \quad . \tag{11.88}$$

For a laser with a linewidth $\Delta\lambda_{\mathrm{FWHM}} = 4\,\mathrm{nm}$ ($\sigma_\lambda = 1.7\,\mathrm{nm}$), a fiber length of $L = 100\,\mathrm{km}$, and a zero dispersion slope $S_0 = 0.1\,\mathrm{ps/(km\,nm^2)}$, the minimum output pulse duration for $\sigma_{ti} \to 0$ is

$$0.707\sigma_\lambda^2 L S_0 = 20\,\mathrm{ps} \quad . \tag{11.89}$$

Thus, even for a partially coherent source, pulse broadening is no problem provided if the operating wavelength exactly equals a wavelength of minimum dispersion.

Pulse propagation through single-mode fibers with third-order dispersion ($d^3\beta/d\omega^3 \neq 0$) has been studied in more detail with respect to width and velocity of a Gaussian-shaped pulse for a source of finite spectral width [Miyagi and Nishida 1979b].

Case 4. In this case, the linewidth is again assumed to be very large, so that the product $2\sigma_\omega\sigma_{ti}$ in (11.71) is large compared to unity:

$$2\sigma_\omega\sigma_{ti} \gg 1 \quad . \tag{11.90}$$

The operating wavelength λ is far from a wavelength λ_0 of minimum dispersion:

$$\lambda \neq \lambda_0 \quad , \tag{11.91}$$

so that the term in (11.71) containing $d^3\beta/d\omega^3$ is negligibly small. In this case:

$$\sigma_{to}^2 = \sigma_{ti}^2 + (DL\sigma_\lambda)^2 \quad . \tag{11.92}$$

In this expression, σ_λ is the natural rms linewidth of the source.

Introducing the parameter f_1 defined by (11.33, 35) – effectively the fiber bandwidth at $\lambda \neq \lambda_0$ (11.48) – one gets

$$\sigma_{to}^2 = \sigma_{ti}^2 + (2\pi f_1)^{-2} \quad . \tag{11.93}$$

For very short input pulses, (11.92) can be simplified to give the duration of the impulse response [Marcuse 1979b]

$$\sigma_{to} = DL\sigma_\lambda \quad . \tag{11.94}$$

This relation is often used to estimate the duration of the output pulse [Midwinter 1985]: one multiplies the first order dispersion parameter D in $\mathrm{ps/(km\,nm)}$ by the fiber length L in kilometers and by the laser rms linewidth σ_λ in nanometers.

As a numerical example, we assume a laser with a linewidth of $\Delta\lambda_{\mathrm{FWHM}} = 4\,\mathrm{nm}$ ($\sigma_\lambda = 1.7\,\mathrm{nm}$) with an emission wavelength of $1.55\,\mu\mathrm{m}$. For a standard single-mode fiber of length $100\,\text{km}$ with a first order dispersion parameter of $D = 15.4\,\text{ps}/(\text{km}\,\text{nm})$, the product $DL\sigma_\lambda$ equals $2.6\,\text{ns}$. The maximum bit rate (11.70) is only $95\,\text{Mb/s}$!

For Gaussian pulses, a relation linking the width of the impulse response $(\sigma_{\text{ti}} \ll \sigma_{\text{to}})$ with the $-3\,\text{dB}$ bandwidth B can be obtained by multiplying (11.48) by (11.94) and using (11.35):

$$B\sigma_{\text{to}} = 0.136 \quad .\tag{11.95}$$

Applying (11.93) for calculating the output pulse width for wavelengths near a dispersion minimum will result in values which are smaller than the real values.

Practical approximate formulas for the pulse dispersion and the baseband bandwidth have been published for the case where the operating wavelength λ is only slightly different from the wavelength λ_0 of minimum dispersion [Kapron 1984b].

In multimode fibers, pulse broadening depends on launching, coating, cabling, and cable deployment conditions. In contrast to this, dispersion in single-mode fibers is essentially unaffected by these conditions and meaningful data can be acquired with fibers wound on reels [Kaiser 1983].

In a uniform fiber, the pulse spreading depends on the products DL and SL of chromatic dispersion D or dispersion slope S and the fiber length L. For a link consisting of n cascaded fiber sections with different parameters L_i, τ_{gi}, D_i, S_i, the group delay time t_{g} at any wavelength is the sum of the individual delays at that wavelength

$$t_{\text{g}} = \tau_{\text{g}} L = \sum_{i=1}^{n} \tau_{\text{gi}} L_i \quad ,\tag{11.96}$$

where the total fiber length L is:

$$L = \sum_{i=1}^{n} L_i \quad .\tag{11.97}$$

The quantity τ_{g} in (11.96) is the effective or average group delay time per unit length of the concatenated fiber link.

The average chromatic dispersion coefficient of the link is given by differentiation as

$$D(\lambda) = \frac{d\tau_{\text{g}}}{d\lambda} = \frac{1}{L} \sum_{i=1}^{n} D_i L_i \quad ,\tag{11.98}$$

and the average dispersion slope as

$$S(\lambda) = \frac{d^2\tau_{\text{g}}}{d\lambda^2} = \frac{1}{L} \sum_{i=1}^{n} S_i L_i \quad .\tag{11.99}$$

Equations (11.98, 99) for the average dispersion and dispersion slope have been verified experimentally [Olson et al. 1984].

For calculating pulse broadening in a fiber link consisting of different fibers, the average values of dispersion (11.98) and dispersion slope (11.99) have to be used in (11.71, 75, 79, 87, 92), i.e. the products DL and SL in the formulas must be replaced by the sums $\sum D_i L_i$ and $\sum S_i L_i$, respectively.

Because of (11.98), it is possible to minimize the first-order dispersion of a fiber link, e.g. by concatenating a first fiber with D_1, L_1 and a second fiber with D_2, L_2, where

$$D_2 L_2 = -D_1 L_1 \quad . \tag{11.100}$$

Thus, if the operating wavelength λ of the laser is not matched to the wavelength λ_0 of minimum dispersion of fiber 1, chromatic dispersion can be compensated for by transmitting the light through a second fiber (an "optical pulse equalizer") with a large chromatic dispersion D_2 which has the opposite sign to D_1. The optimum length L_2 of the equalizer fiber is obtained from the condition

$$L_2 = -D_1 L_1 / D_2 \quad . \tag{11.101}$$

This method of optical pulse-delay equalization has been demonstrated experimentally [Lin et al. 1980b,c; Larner and Bhagavatula 1985].

Another possibility for reducing the first order chromatic dispersion is to insert into the fiber link with D and L a compact dispersion equalizer or compensator (Sect. 12.3.2). This is an optical two-port with large chromatic dispersion $dt_g/d\lambda$ and small insertion loss. For

$$dt_g/d\lambda = -DL \quad , \tag{11.102}$$

the chromatic dispersion of the fiber link is compensated for by that of the dispersion equalizer.

The formula (11.71) for the output pulse width has been derived under the assumption that the spectral distribution of the light source is Gaussian. Whereas this approximation is adequate for describing a light emitting diode or a laser oscillating at a single frequency, it is insufficient for a typical Fabry-Perot injection laser oscillating simultaneously in several longitudinal modes. For multi-longitudinal lasers, a formula for the output rms pulse duration has been derived by Marcuse in part 2 of a series of papers on pulse distortions in single-mode fibers [1981b].

Chromatic dispersion causes a pulse broadening and a degradation of the pulse height at the sampling time, which can cause intersymbol interference at high bit rates. Assuming the fiber to be a linear channel, intersymbol interference due to pulse overlap can be reduced by using an equalizer amplifier. Since the amplification of the equalizer is in general an increasing function of the baseband frequency (for large initial pulse overlap), noise enhancement occurs [Personick 1976]. Therefore, in order to achieve a prescribed bit error rate, e.g. $p_e = 10^{-9}$, additional optical power is required to compensate for the increased noise. A "spectral penalty" ("dispersion penalty", "power penalty") is defined as the increase in received power in dBm, over that sufficient in the absence of dispersion, which is required for the maintainance of a set error ratio [Anslow et al. 1984]. The receiving power penalty due to linear and quadratic chromatic dispersion is given by Yamamoto A. and Kimura [1981] for various modulation-demodulation schemes.

11.6 Sources of Signal Quality Impairment

In addition to chromatic dispersion, there are also other effects that distort the transmitted signal or degrade the signal-to-noise ratio. In this section, an overview is given of these other sources of signal quality impairment. Reviews of performance limitations and degradations in single-mode transmission systems have also been published in the literature [Ogawa 1982b; Heckmann 1983, 1984].

11.6.1 Interference with the LP$_{11}$ Mode

In multimode fibers and for coherent sources, the superposition of the fields of the more than one hundred modes gives a complicated interference pattern in a fiber cross section. This so called speckle pattern changes its form if the relative phases of the modes vary. These phase variations can be caused by a thermal drift in the wavelength of the laser source or because of thermal or mechanical changes of the phase constants of the modes. Although the speckle pattern fluctuates, mode orthogonality (Sect. 7.1) ensures that the *total* power transmitted by all modes will be constant. On the other hand, if an aperture or a misaligned splice is present, the transmitted power will fluctuate. This problem of "modal noise" in multimode fibers was first discussed by Epworth [1978].

Some authors have claimed that modal noise can be avoided by using single-mode fibers instead of multimode fibers [e.g. Baack et al. 1980; Wenke and Elze 1981; Großkopf et al. 1982]. This is not quite true, since a "single-mode" fiber can propagate *two* fundamental modes with orthogonal polarization and, at shorter wavelengths, an additional *four* LP$_{11}$ modes. The interference between the orthogonally polarized fundamental modes or between the fundamental and the second-order modes give rise to two different kinds of modal noise in single-mode fibers.

We start by describing the modal noise that occurs in single-mode fibers when the second-order mode is not sufficiently attenuated. Polarization modal noise caused by interference between the two orthogonally polarized fundamental modes will be discussed in Sect. 11.6.10.

For small wavelengths, a nominally single-mode fiber can also guide four second-order LP$_{11}$ modes (Sect. 6.3). Thus modal noise can be introduced into single-mode fiber systems by the time varying interference between the LP_{01} and LP_{11} modes, when the fiber is operated at a wavelength smaller than the cutoff wavelength of the second order modes (Sect. 6.3.2). The existence of modal noise in overmoded single-mode fibers was predicted in 1981 [Heckmann 1981b], analyzed theoretically [Stone F.T. 1984], and observed experimentally for different fiber interconnection schemes containing intentionally overmoded fiber sections [Cheung and Kaiser 1984a,b; Cheung 1984; Heckmann 1984; Tomita and Cohen 1984; Cheung et al. 1985].

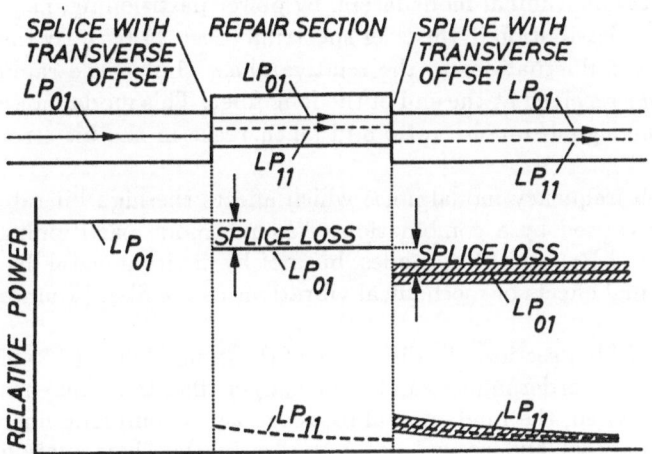

Fig. 11.3. Scheme for explaining the generation of modal noise at a repair section or a cable jumper

The power versus length graph Fig. 11.3 [Sears et al. 1986b] explains the conditions that produce this kind of modal noise. A short cable jumper or repair section is inserted into a long fiber. A transverse offset at the first junction produces a loss in the incident fundamental LP_{01} mode and excites the second-order LP_{11} mode (Sect. 9.6). The majority of the power lost from the LP_{01} mode is transferred into the LP_{11} mode [Sears et al. 1985]. In the short fiber section, the LP_{01} mode travels essentially unattenuated whereas the LP_{11} mode suffers considerable attenuation since the operating wavelength is near the cutoff wavelength of this mode.

The fundamental mode incident on the second misaligned joint launches a fundamental and a second order mode in the second long fiber section. When the attenuation of the LP_{11} mode in the short section is small, this mode still has finite power at the second misaligned joint and will also launch a fundamental and a second-order mode. At the distant receiver, the second-order mode cannot be observed because of its relatively large attenuation. But the fundamental modes launched in the output fiber at the second misaligned joint by the fundamental and second order modes propagating in the short fiber section interfere. Depending on the difference of the phase shifts of these modes in the repair section, the power transmitted by the fundamental mode in the second long fiber will be greater or less larger than a median value.

In practice, the relative phase shift varies in time. Slow changes are caused by a drift in frequency for single-line lasers or by the temperature dependence of the phase constants β_{01} and β_{11} of the two waves propagating in the short fiber section. These slow variations of the phase difference cause variations of the reveived power which are much slower than the bit rate and can be filtered out by the receiver automatic gain control (AGC) circuit.

Rapid phase fluctations are caused by a residual frequency modulation of the laser or, for multi-longitudinal-mode lasers, by power partitioning, i.e. a fluctuation of the power distribution in the laser spectrum (Sect. 11.6.5). By the mechanism described, the fluctuations in the relative phase shift cause rapid fluctuations in the power received at the end of the long fiber. This modal noise causes a decrease in the signal-to-noise ratio and an increase in the bit error rate.

Therefore, the high frequency modal noise which affects the high bit rate system performance is caused by a combination of source mode partitioning and interference between LP_{01} and LP_{11} modes, but not by environmental disturbances, such as thermal effects or mechanical vibrations of the fiber [Tomita and Cohen 1984].

For a longer inserted fiber section, the difference of the group delays (7.8) of the fundamental and second-order modes can become larger than the coherence time of the light (3.35). Then, the fundamental modes in the second long fiber launched by the fundamental and second order modes in the short section are incoherent, and their powers superimpose and the fluctuations caused by interference will vanish [Stone F.T. 1984].

The conditions that produce this kind of modal noise can be summarized as follows:

1. a short repair section or cable jumper with a single-mode fiber which is operated at a wavelength shorter than the theoretical cutoff wavelength;
2. two joints at the ends of the short fiber with transverse offsets which cause $LP_{01} \leftrightarrow LP_{11}$ conversion,
3. a relatively small attenuation of the second-order mode in the short section,
4. a difference of the group delay times of the LP_{01} and LP_{11} modes in the short section which is small compared to the coherence time;
5. a time fluctuation of the relative phase shifts of the two modes propagating through the short fiber section.

From this information, one can derive the following rules for reducing modal noise:

1. for repair sections and cable jumpers use special fibers with a lower value of the effective cutoff wavlength λ_{ce} than in the long cable, so that the operating wavelength is larger than the cutoff wavelength of the second-order mode,

2. minimize the transverse offsets at the junctions,
3. use a short fiber which is sufficiently long to quench the second-order mode [Duff et al. 1985; Cheung et al. 1985] or to make the two modes incoherent. The recommended mimimum splicing length is some meters [Sears et al. 1985, 1986b],
4. insert a suitable mode filter (several loops of fiber, Sect. 12.2.14) which effectively attenuates the LP_{11} mode without introducing substantial loss for the LP_{01} mode [Cheung and Kaiser 1984a].

A formula for the rms signal-to-noise ratio (SNR) as a function of fiber length has been derived [Duff et al. 1985; Sears et al. 1986a] and verified experimentally by measuring the SNR as a function of the length of the repair section. This "modal noise design equation" allows one to compute the minimum required loss $\alpha_{11}(\lambda, R)$ of the LP_{11} mode for the shortest unspliced fiber length L (e.g. 10 m) at the minimum operating laser wavelength λ_{min}. The approximate formula (6.29) for the wavelength dependence of the attenuation $\alpha_{11}(\lambda, R)$ of the LP_{11} mode can then be used to determine the maximum allowable cutoff wavelength [Wei et al. 1986].

For an intermediate fiber of length 10 m, the factory cutoff wavelength can be as much as 95 nm above the minimum system operating wavelength without noticeable signal degradation due to modal noise [Wei et al. 1987].

Modal noise can also be introduced at the detector if the speckle pattern formed by the LP_{01} and LP_{11} modes passes through a restricting aperture, or falls on a detector with spatially varying sensitivity [Tanifuji 1985]. Even if the total optical power is constant, the photodiode current will fluctuate because of the change of the speckle pattern.

11.6.2 Wavelength Chirping

Most semiconductor lasers used at present are far from being ideal sources which emit a single frequency with zero linewidth [Abdula and Saleh 1986]. Different frequency modulation (FM) effects are observed in single-frequency lasers (wavelength chirping, mode hopping, and mode switching) and in multi-longitudinal mode lasers (mode partitioning and drift in mean wavelength). The degradation of system performance by these FM effects is reviewed in Sects. 11.6.2–6.

In single-frequency (single-line) lasers, e.g. distributed-feedback (DFB) lasers, the current pulse changes the carrier density in the semiconductor. This results in a change in the refractive index which, in turn, causes a continuous variation in the instantaneous optical frequency. The wavelength shift during the pulse can be as large as 1 nm [Linke 1984].

This unwanted "wavelength chirping" affects pulse propagation in a single-mode fiber in two different ways [Marcuse 1981c]: Since the change in wavelength causes the laser spectrum to cover a wider band, it augments first- and second-order dispersion effects that tend to broaden the light pulse (Sect. 11.5).

On the other hand, the leading and trailing edges of the chirped pulse are supported by slightly different carrier frequencies. When the leading edge due to chromatic dispersion, moves slower than the trailing edge, pulse broadening by dispersion can be reduced or even reversed so that the pulse may become shorter than it was at the fiber input [Buus 1985; Mozer 1986]. This effect corresponds to the pulse compression technique known from radar.

The maximum transmission capacity of IM systems is limited by the amount of frequency chirp which is introduced by the current injection into the laser. The effect of wavelength chirping on pulse propagation in dispersive single-mode fibers has been analyzed theoretically by a number of authors [Koch and Bowers 1984; Buus 1985; Koyama and Suematsu 1985; Tajima and Washio 1985; Koch and Corvini 1986; Mozer 1986; Petermann and Krüger 1986; Agrawal and Potasek 1986; Anderson and Lisak 1986; Abdula and Saleh 1986].

The time variation of wavelength during the light pulse generated by a DFB laser and the effect of the chirp on a 1 Gb/s-80 km transmission link with a chromatic dispersion of 15 ps/(km nm) has been investigated experimentally [Frisch and Henning 1984]. In this study, negligible changes in the receiver pulse shape and no significant effect on receiver sensitivity were found.

In contrast to these findings, Linke [1984] found a dramatic degradation of system performance at bit rates above 1 Gb/s for a fiber length of 100 km.

Optical pulse compression of the chirped pulses emitted by distributed feedback laser diodes has been demonstrated experimentally in highly dispersive single-mode fibers. The pulse duration has been compressed from 1.7 ns to 0.35 ns [Iwashita et al. 1982] and from 26 ps to 8.3 ps [Takada et al. 1985].

The elimination of chirp using external modulation can improve the bit rate distance product by an order of magnitude [Agrawal and Potasek 1986].

11.6.3 Mode Hopping

When a laser oscillates in a dominant longitudinal mode, it is possible that the mode order changes abruptly with ambient temperature or with drive current. When the mode suddenly hops from one order to an adjacent order, the wavelength changes abruptly. This effect is called mode hopping [Yamada M. and Suematsu 1979; Garrett and Todd 1982; Heckmann 1984] or mode jumping [Ogawa 1982b]. If the mode changes wavelength by 1 nm, and if the chromatic dispersion is 17 ps/(km nm), the change in propagation time over 60 km of fiber is 1 ns, sufficient to cause loss of frame synchronization in systems operating at 500 Mb/s and above. This results in a burst of errors which lasts until the retiming circuit locks into the shifted phase. Such bursts of errors will happen randomly at any repeater.

11.6.4 Mode Switching

An effect similar to mode hopping occurs when a nominally single-frequency laser comes to operate in two modes. The laser frequently switches between a main mode and a side mode. Mode switching was observed in a distributed feedback (DFB) laser when the laser was modulated at a bias current around the threshold. This was found to significantly affect the bit error rate performance in a single-mode fiber transmission system with $R_b = 140\,\mathrm{Mb/s}$, $\lambda = 1550\,\mathrm{nm}$, $D = 17\,\mathrm{ps/(km\,nm)}$, and $L = 113\,\mathrm{km}$ [Yamamoto S. et al. 1986]. Mode switching in single-frequency DFB lasers is similar to mode partitioning in multi-longitudinal-mode lasers, which will be described next.

11.6.5 Mode Partitioning

For multi-longitudinal mode laser diodes of the Fabry-Perot type (FP lasers), the distribution of power among the different longitudinal modes fluctuates randomly, even when the total emitted power is constant. The phenomenon of a time-varying laser spectrum is called "mode partitioning". The power fluctuation of each longitudinal mode in FP laser diodes has been investigated both theoretically [Ogawa and Vodhanel 1982] and experimentally [Okano et al. 1980; Ogawa and Vodhanel 1982; Liu and Ogawa 1984].

In a dispersive fiber, because of the different time delays of the longitudinal modes, mode partitioning causes an amplitude fluctuation of the signal at the decision circuit in the receiver. This fluctuation is called "mode partition noise" [Okano et al. 1980] or "competition noise" [Arnold and Petermann 1980].

Mode partition noise increases the bit error rate as compared to its value without mode partitioning or without chromatic dispersion. In order to achieve a prescribed bit error rate, e.g. $p_e = 10^{-9}$, the received power must be larger than without dispersion. The degradation of system performance due to mode partition noise is characterized by the power penalty (Sect. 11.5) which is the increase in received power in dBm, over that sufficient in the absence of mode partition noise.

Mode partitioning combined with chromatic dispersion result, essentially, in a pulse-delay fluctuation. The signal-to-noise ratio due to mode partition noise is independent of the signal power. Therefore, the bit error rate as a function of the received power in dBm does not show the usual steep decrease, but approaches a floor value (asymptotic error rate, error-ratio plateau) as the optical power is increased [Larner 1985]. The power penalty increases with decreasing bit error rate and becomes infinite at the asymptotic error rate. Error rates smaller than the floor value cannot be obtained by increasing the received signal power.

Mode partition noise and the corresponding power penalty have been studied both theoretically [Arnold and Petermann 1980; Okano et al. 1980; Cohen L.G. and Lumish 1981; Yamamoto S. et al. 1982, Petermann and Arnold 1982; Ogawa 1982a,b; Anslow et al. 1984; Geckeler 1986b] and experimentally [Okano et al. 1980; Yamamoto S. et al. 1982; Ogawa 1982a; Ogawa and Vodhanel 1982; Cheung 1983]. Dispersion penalties become significant for high bit rates, for wavelengths far from a minimum-dispersion wavelength, and for large fiber lengths. Mode partition noise can be the dominant limitation on bit rate or on repeater spacing.

In an analogue transmission system, the carrier to noise ratio has been found to be predominantly determined by mode partition noise [Großkopf et al. 1982].

Power penalties due to mode partition noise can be reduced by [Geckeler 1986a]:

1. using a fiber with smaller chromatic dispersion;
2. using a laser diode with a smaller spectral width;
3. using a shorter fiber;
4. introducing a low pass filter in the photoreceiver (if necessary, reducing the bit rate).

The mode partition noise described so far is due to mode partitioning in the laser combined with chromatic dispersion in the fiber. Another kind of mode partition noise is generated, when the different longitudinal laser modes experience differential mode attenuation due to wavelength-dependent fiber attenuation. Mode partition noise of this kind is an especially important consideration in the design of wavelength-domain-multiplexing (WDM) systems where transmission characteristics near cutoff wavelengths of multiplexers and demultiplexers are highly wavelength dependent [Tomita et al. 1983].

11.6.6 Mean Wavelength Shift

In addition to random mode partitioning, there is another cause of spectral penalty in Fabry-Perot-type lasers [Anslow et al. 1984, Anslow and Goddard 1986]. This effect takes the form of a shift in the mean wavelength of the laser within the first few nanoseconds after the laser is turned on, which combined with chromatic dispersion, gives a distortion of the received waveform. During the pulse, the mean wavelength shifts to larger values. It has been predicted that, in some cases, these wavelengths shifts in conjunction with positive chromatic dispersion could cause catastrophic intersymbol interference [Anslow and Goddard 1986].

11.6.7 Optical Feedback

The adverse effects described so far, i.e. chirping, mode hopping, mode switching, and mode partitioning are intrinsic to the laser, in so far as they arise without a fiber being coupled to the laser [Petermann and Arnold 1982]. Additional impairment occurs if light is reflected from the fiber back into the laser. This is a further source of noise and distortions in single-mode communication systems.

In single-mode fiber optic systems, reflections can be introduced by lens surfaces, fiber endfaces, or connectors with an air gap. The wave reflected from a nearby discontinuity is coherent with the emitted wave, whereas the reflection from a far discontinuity is incoherent [Kuwahara et al. 1983]. The reflecting discontinuity and the laser facet form an optical resonator that acts as an external cavity for the laser [Goldberg et al. 1982; Petermann and Arnold 1982].

Optical feedback distorts laser characteristics in a very complicated manner. The reflection changes the laser's output power, emission wavelength and optical spectrum; it enhances low and high frequency noise, and introduces nonlinearities into the power versus current characteristics. These effects degrade the signal-to-noise ratio at the receiver. Index-guiding lasers exhibit a higher sensitivity to optical feedback. A review of the noise and distortion characteristics of semiconductor lasers in optical fiber communication systems has been published by Petermann and Arnold [1982]. We mention only the most important effects caused by optical feedback.

Reflections from a discontinuity, e.g. the fiber input endface, cause the power coupled into the fiber to change periodically as a function of the separation between the laser and the discontinuity [Weidel 1976; Wenke and Zhu 1983]. Similarly, the emission wavelength changes periodically with separation [Lang and Kobayashi 1980; Wenke and Zhu 1983].

For single-frequency lasers with very small linewidth, a change of the phase angle of the wave reflected back into the laser causes a noticeable change in emission frequency. For a phase change of π in the reflection from a microlens tip on the endface of a single-mode fiber, frequency shifts of 6 GHz have been observed [Vodhanel and Ko 1984].

Optical feedback also changes the form of the spectrum of the emitted light [Lang and Kobayashi 1980; Wenke and Elze 1981; Dandridge and Miles 1981; Petermann and Arnold 1982; Goldberg et al. 1982]. Both linewidth narrowing and broadening can occur. Reflections can split a single longitudinal mode into several submodes [Petermann and Arnold 1982] or satellite modes [Dandridge and Miles 1981]. Optical feedback also causes random fluctuations of the laser output power, so called intensity noise [Lang and Kobayashi 1980, Hirota et al. 1981; Bludau and Rossberg 1982; Kuwahara et al. 1983; Bludau and Rossberg 1985]. Intensity noise increases the bit error rate [Farrington 1981].

Reflection induced intensity noise has been investigated experimentally for light transmitted both through an air path [Mazurczyk 1984] and through single-mode fibers [Cheung 1984; Wang S.J. 1986]. Reflection noise introduces only an insignificant power penalty for buried heterostructure lasers, but could make transmission impossible in certain types of stripe geometry lasers [Cheung 1984]. The degradation of system performance becomes more serious at higher bit rates [Wang S.J. 1986].

Optical feedback also leads to nonlinearities in the output power versus injection current characteristics of the laser [Lang and Kobayashi 1980], and nonlinear signal distortions [Davis and Allsop 1981; Wenke and Elze 1981].

Because of these detrimental effects of optical feedback, coupling arrangements between the laser and the fiber with very small reflections (power reflection factor smaller 10^{-6}) are mandatory.

Methods for calculating and measuring the reflection factor as seen by the laser and techniques for reducing optical feedback are reviewed in Sect. 12.2.2.

11.6.8 Multiple Reflections

The power reflection factor R of a connector with air gap (5.132) depends on the wavenumber k in vacuum, i.e. on the operating wavelength $\lambda = 2\pi/k$. Correspondingly, the power transmission factor (5.130)

$$T = 1 - R \quad , \tag{11.103}$$

varies periodically with wavelength due to multiple reflections between the fiber endfaces. Direct intensity modulation of a laser diode is inevitably accompanied by frequency modulation (Sects. 11.6.2, 6). The transmission factor of the connector changes in accordance with the wavelength changes. Thus the Fabry-Perot resonator formed by the fiber connector generates harmonic distortions which may be harmful in analogue transmission systems [Kuwahara and Goto 1981].

In digital systems utilizing lasers with low relaxation oscillation frequency, connector reflections jeopardize the system operation, and endface contact or index matching of the connectors is mandatory [Cheung and Sandahl 1983; Kaiser 1983].

11.6.9 Self-Amplitude Modulation

In coherent optical transmission systems using phase-shift keying (PSK), chromatic dispersion ($D \neq 0$) causes an unwanted amplitude modulation [Tajima 1985]. This effect is called "self-amplitude modulation". Unlike self-phase modulation, it is a linear effect, occurring also at very small powers. Significant amplitude modulation is predicted to occur after 50 km of transmission. Self-amplitude modulation may thus place a serious limitation on system performance, when the PSK demodulator also responds to AM.

The waveform distortion of a sinusoidal signal, amplitude or phase modulated onto a coherent optical carrier far from a wavelength of minimum dispersion has also been studied by Yamamoto A. and Kimura [1981]. Linear chromatic dispersion (D) degrades the modulation depth and causes AM to PM as well as PM to AM conversion. Examples of originally sinusoidal waveform distorted by linear chromatic dispersion can be found in the paper mentioned.

11.6.10 Polarization Selective Loss

Interference between the orthogonally polarized fundamental modes can cause "polarization modal noise" if the fiber is slightly birefringent and if there is a component with polarization-dependent loss.

If spectrally narrow, polarized light is launched into a real single-mode fiber, the depolarizing length is large compared to the fiber length, and a high *degree* of polarization is observed at the far end of the fiber. The output *state* of polarization (Sect. 5.3.5) however, can vary considerably due to variations of the laser wavelength, or to fluctuations of the fiber birefringence, or to fluctuations in the random polarization coupling. For direct photodetection, these fluctuations of the state of polarization will not cause corresponding fluctuations of the received photocurrent, since, mode orthogonality means that the total power is simply the sum of the powers transmitted by both fundamental modes (Sect. 7.1).

However, if the fiber link contains an element whose insertion loss depends on the state of polarization, the transmitted power will depend on the phase difference of the normal modes, and will fluctuate if the wavelength or the fiber birefringence changes. Thus any polarization-sensitive loss will cause modal noise also in single-mode fibers [Heckmann 1981a; Epworth and Pettitt 1981; Enning and Wenke 1983; Hillerich and Weidel 1982, 1983; Heckmann 1984]. This type of polarization modal noise may degrade the SNR to such a degree that high quality analogue transmission is impossible. For digital transmission, one must, at least increase the loss margin.

In comparison with modal noise in multimode fibers [Epworth 1978], the polarization modal noise has a larger amplitude. Therefore it is not advisable to insert a linear polarizer (analyzer) into a single-mode fiber optic link. But there are other elements with polarization-dependent insertion loss, e.g. the blazed reflection gratings used as demultiplexers in wavelength division multiplex systems [Loewen et al. 1977; Hillerich and Weidel 1982, 1983], beam splitters used as monitors, couplers to single polarization (integrated) optical components, polarization selective power dividers, and bends in high-birefringence fibers.

Other forms of fiber perturbations like bends in low-birefringence fibers, splices, and directional coupler taps are not expected to introduce substantial polarization-dependent losses [Kaminow 1981; Suematsu et al. 1982]. Polarization modal noise due to a lossy joint has been investigated experimentally and has been found to be significantly smaller than modal noise in multimode fibers [Suematsu et al. 1982].

If it is not possible to avoid polarization dependent elements, the polarization noise can be reduced by inserting a depolarizer in front of the photodetector. For a highly coherent source, a birefringent wedge or a section of multimode fiber can be used as depolarizer [Hillerich and Weidel 1982, 1983].

If the wavelength fluctuation is synchroneous with the intensity modulation, the power fluctuation is correlated to the signal. Thus the disturbance is not a random noise. Instead, the signal becomes distorted. The effect is named "polarization modal distortion" [Epworth 1981]. Extreme waveform distortions can be observed when laser mode hopping (Sect. 11.6.3) is combined with polarization-dependent loss.

11.6.11 Polarization Mode Dispersion

In a uniformly birefringent fiber, the orthogonal fundamental modes have different phase constants β_x and β_y and different specific group delays (5.66) τ_{gx} and τ_{gy}. The delay difference per kilometer $\Delta\tau_g$ between the two orthogonally polarized waves,

$$\Delta\tau_g = \tau_{gx} - \tau_{gy} \quad , \tag{11.104}$$

is called the polarization mode dispersion [Rashleigh and Ulrich 1978].

Measured values of the polarization mode dispersion are less than $1\,\text{ps/km}$ in conventional telecommunications-grade fibers [Mochizuki et al. 1981a; Imoto and Ikeda 1981] and more than $1\,\text{ns/km}$ in high birefringence polarization-maintaining fibers [Burns and Moeller 1983; Sasaki Y. et al. 1984]. In spun fibers with very small birefringence, the polarization mode dispersion has an immeasurably small value [Payne D.N. et al. 1982].

Injecting a short light pulse into the fiber input end, will generally launch both fundamental modes. Since the two modes have different group velocities, the output after a fiber length L will consist of two pulses separated by a time interval $\Delta\tau_g L$. Thus, polarization mode dispersion in analogue transmission can cause a signal distortion, known as polarization mode distortion, and in digital transmission, intersymbol interference.

In high birefringence fibers, the product $\Delta\tau_g L$ provides a good estimate for the pulse spreading that is measured on long fiber lengths L. The 3-dB bandwidth B is given by [Kitayama K. et al. 1981; Suzuki K. et al. 1983]

$$B = 0.9/(\Delta\tau_g L) \quad , \tag{11.105}$$

and the maximum bit rate R_b for a digital transmission system is

$$R_b = B/0.55 \quad . \tag{11.106}$$

Since the polarization mode dispersion $\Delta\tau_g$ of polarization-preserving fibers can be larger than $1\,\text{ns/km}$, it can severely limit the bandwidth unless all the energy is propagated in one normal mode.

A method for reducing signal distortion caused by polarization mode dispersion is to introduce a splice half-way along the link and to interchange the fast and the slow linearly polarized fundamental modes at the splice point by aligning the fast axis of the first fiber with the slow axis of the second section [Shibata N. et al. 1982].

In theory, it seemed that quite low levels of stress birefringence would yield significant pulse broadening (e.g. $10-30\,\text{ps/km}$) [Rashleigh and Ulrich 1978; Adams et al. 1979]. However, in low birefringence fibers, there is a significant coupling between the orthogonal eigenstates caused by random perturbations, which serves to equalize the propagation times, so that the product $\Delta\tau_g L$ usually overestimates the observed pulse spreading substantially [Day 1983]. The polarization coupling can be quite significant when the axis of linear birefringence is twisted [Monerie and Jeunhomme 1980].

For fiber lengths z larger than a characteristic coupling length L_c, the pulse spreading is proportional to $(LL_c)^{1/2}$ instead of L, as for short fibers [Rashleigh and Ulrich 1978] and the bandwidth is given by [Suzuki K. et al. 1983]

$$B = \frac{0.9}{\Delta\tau_g(LL_c)^{1/2}} \quad . \tag{11.107}$$

It would thus seem that the same random perturbations that degrade polarization preservation in high-birefringence fibers, may act in a beneficial way in "normal" fibers to substantially reduce polarization mode dispersion.

Experimental results seem to indicate that polarization mode dispersion in long conventional fibers will not be a serious bandwidth limitation as initially thought [Bloom et al. 1979; Kawana et al. 1980; Mochizuki et al. 1981a; White K.I. et al. 1981; Rashleigh 1983]. For instance, polarization-mode dispersion after cabling and jointing has been measured for an installed 30-km link to be less than 0.5 ps [Kapron and Lazay 1983]. Thus, polarization mode dispersion is not a practical limitation even for systems with 1 Gb/s pulse rate and 100 km repeater spacing [Kaminow 1984].

The measurements reported in the literature demonstrate that fibers with a variety of relative refractive-index differences (and spliced links) can be readily fabricated into system lengths which (with direct photodetection) behave, to a good approximation, as though only a single LP_{01} mode is propagating, which disperses predominantly via chromatic processes [Garrett and Todd 1982].

Equations (11.105, 107) give the fiber bandwidth for $L \ll L_c$ and for $L \gg L_c$, respectively. The formula for the bandwidth in the intermediate regime can be found in the literature [Suzuki K. et al. 1983].

In birefringent fibers, distortions can also be caused by uniform mode coupling [Nielsen 1982]: The nonlinear distortion of a sinusoidal signal depends in a complicated way on the modulation frequency, the unwanted source frequency modulation, the input state of polarization, the polarization mode dispersion, and the mode coupling coefficient. However, Nielsen [1982] found that this kind of distortion is insignificant in practical applications.

An analysis of the pulse response of birefringent fibers with uniform polarization coupling [Nielsen 1983] has shown that the coupling will introduce an oscillating tail on the excited mode with a duration given by polarization mode dispersion times fiber length, and a magnitude determined by the ratio of the fiber birefringence to mode-coupling coefficient. For polarization-maintaining fibers, this tail may be several nanoseconds long, but its magnitude is likely to be small. For fibers with low birefringence but with a high degree of mode coupling, the theory predicts a strong reduction of the intrinsic polarisation mode dispersion but a rather poor polarization stability.

Random polarization coupling fluctuates with time introducing power fluctuations at the fiber output. The output power fluctuations have been measured both as a function of time [Garlichs 1981; Rashleigh and Marrone 1982; Harmon 1982] and as a function of frequency [Sakai and Shimizu 1984].

For analogue signals, polarization mode dispersion causes signal distortions. As long as the product $\Delta\tau_g \omega_m L$ of the polarization mode dispersion $\Delta\tau_g$, the modulation radian frequency ω_m, and the fiber length L is much smaller than one, the signal distortion is negligible, but it becomes intolerably large if the product is of the order of unity. Examples of originally sinusoidal PM (FM) signals distorted by polarization mode dispersion are presented in [Yamamoto A. and Kimura 1981].

11.6.12 Polarization Mismatch

If a fiber link is composed of two linearly birefringent sections connected by a joint with misaligned principal axes, nonlinear distortions are introduced by the wavelength dependence of the state of polarization at the joint [Petermann 1981a,b]. These distortions can be avoided either by launching only one of the normal modes in the primary fiber or by aligning the polarization axis at the joint. In general, however, both modes will be excited and there will be polarization coupling at the joint.

The distortions remain negligibly small provided the product of the modulation frequency, the polarization mode dispersion (11.104), and the fiber length is small compared to unity. If the product approaches one, however, different spectral components have different states of polarizations, and the distortions may become significant. A numerical evaluation of the analysis [Petermann 1981b] showed that without residual laser frequency modulation, the second- and third-order harmonic distortions increase in proportion to the third power of the above-mentioned product.

An additional unwanted modulation of the laser frequency [Nakamura et al. 1978] further increases the distortions substantially and in proportion to the product.

The practical consequence of unwanted frequency modulation and polarization-dependent loss is that the useful bandwidth is significantly reduced when nonlinear distortions are required to remain small. The distortions can be reduced by using fibers with small polarization mode dispersion.

If frequency modulation and polarization-maintaining fibers are used, interference effects caused by small polarization misalignments produce a significant nonlinear distortion [Nielsen 1984]. To overcome this problem, the input and output polarizers must be aligned very carefully relative to the principal fiber axes. At splices and connectors, the principal axes must be aligned very carefully. Fibers with small polarization mode dispersion are helpful in reducing the nonlinear distortions.

In single-mode fiber links using linearly birefringent polarization-maintaining fibers, joints with mismatched principal axes cause multiple delayed versions of the original signal and thus signal distortions [Cancellieri et al. 1985a].

12. Components for Single-Mode Fibers

Light sources, optical fibers, and optical receivers are fundamental devices needed to construct optical fiber transmission systems. In addition to these devices, such optical components as light source to fiber couplers, splices, connectors, directional couplers, attenuators, multiplexers, isolators, switches, etc. are essential to realize practical transmission systems or are required for measuring purposes.

A study was conducted at NTT [Minowa et al. 1982] into whether existing components for multimode fiber systems would have an adequate performance when applied to single-mode systems. It was found that all the components, except for photodetector module, had to be newly developed to satisfy the requirements of long-haul single-mode transmission systems.

This chapter therefore gives an overview of optical components for single-mode fiber systems. Section 12.1 considers the fiber end as a simple component. In Section 12.2, couplers such as beam to fiber coupler, laser to fiber couplers, couplers between fibers and integrated optic channel waveguides, splices and connectors between fibers, power splitters, directional couplers, star couplers, etc. will be described. Section 12.3 reviews wavelength selective components such as filters, WDM multiplexers and demultiplexers. Active components like switches and modulators, as well as nonreciprocal, birefringent, and dichroic fiber-optic components will be reviewed in a forthcoming volume. Sources and photodetectors will not be described in this book.

A few reviews of the development, fabrication, and performance of optical components for single-mode fibers have been published [Botez and Herskowitz 1980; Minowa et al. 1982; Mossman 1984; Aoyama T. 1985].

Components for single-mode systems can be classified into four groups:

1. standard optical components like lenses, mirrors, beam-splitters, attenuators and polarizers, which modify freely propagating light beams,
2. micro-optical components like ball lenses or graded-index lenses; these also modify freely propagating light beams,
3. in-line fiber-optic components like fused biconically tapered directional couplers, which directly modify the wave beams guided in single-mode fibers,
4. integrated-optical components such as modulators, which directly modify the wave beams guided in dielectric channel waveguides.

Standard optical components used in optical fiber communication, measurement, or sensor systems, e.g. attenuators, polarizers, and isolators, operate with free wave beams. Thus, for using such bulk components, the beam guided by the single-mode fiber has to be transformed into a collimated free beam. This can be done by terminating the fiber and inserting a focusing lens in the free beam radiated from the fiber end (Sect. 12.2.1). Behind the standard optical component, the free wave beam often has to be coupled back into a second single-mode fiber. This is done by focusing the collimated beam onto the input face of the output fiber.

Since the diameter of the free wave beam is usually very small, the usual standard optical components like lenses or waveplates are unnecessarily large in dimensions. Therefore, so called "micro-optic" components have been developed, which with the exception of the graded-index lens (Sect. 12.2.1) are miniaturized versions of known components. Since the micro-optic components still operate with free wave beams, they can be regarded as an intermediate step towards later systems that use only fiber-optic components.

For components operating on free wave beams it is difficult to align the fiber ends and the various bulk components. Losses are introduced by transforming a guided beam in a free beam and back again. Reflection losses are caused by air-glass interfaces. There is thus a tendency to design fiber-optic analogues of standard bulk components which operate directly on the guided wave beam. These so-called in-line fiber-optic components have lower insertion losses, better stability, and more suitable size than their bulk counterparts. Since these components directly influence the guided beam, it is not necessary to change between the guided and free beam mode.

Although several fiber-like components, e.g. directional couplers, filters, polarizers, polarization compensators, Faraday rotators, and isolators, have been developed, it is not yet possible to design communication, measurement, or sensor systems using fiber-components only. In this chapter on components for single-mode systems, we thus need to describe both bulk and fiber-type components.

The fourth group of components are integrated-optic devices operating on the beams guided by dielectric channel waveguides. Because of space limitations, it is not possible to describe these components in this book. However, in Sect. 12.2.3 methods are reviewed for making low-loss couplers between single-mode fibers and integrated-optic channel waveguides.

12.1 Fiber Ends

12.1.1 Preparation of High Quality Fiber Endfaces

Flat fiber endfaces which are normal to the fiber axis are required for splicing two fiber ends, for coupling light into the fiber, for radiating the guided light from the fiber end, and also for some measurement procedures. There are several methods of preparing fiber endfaces.

Flat and perpendicular fiber ends can be made by the conventional grinding and polishing technique [Miller C.M. 1975; Runge and Cheng 1978]. The fiber end is embedded in epoxy and is ground and polished using abrasives. Commercial equipment gives good quality and repeatability. However, the grinding method is both expensive and time consuming.

The most common method for preparing endfaces is fiber cleaving by bending, tensioning, and scribing with a carbide, sapphire, or diamond head [Gloge et al. 1973; Domergue et al. 1981]. A wide range of cleaving tools are available from several manufacturers; they cost from just a few dollars to many thousands of dollars for automated versions [Murata et al. 1975; Hensel 1975; Millar and Gooch 1983; Hoshikawa and Toda 1983; Haibara et al. 1986]. A cutting tool capable of cutting 144 fibers simultaneously has also been reported [Cherin and Rich 1976].

Fiber ends must not have protruding lips or hackle (rough) regions. High quality cleaves have angles measured against the fiber axis which do not depart from 90° by more than 1°. Ends which are not perpendicular to the fiber axis are caused by torsion maintained in the fiber while fracture takes place [Saunders 1979]. A modern version of fiber cleaver uses an ultrasonic vibrating diamond blade to give hackle-free endfaces with end-angles well below 1°. A simple and effective method for producing a flat normal end is to hold the fiber between the fingers, draw it across a scribe, and to pull it, causing it to break [Chesler and Dabby 1976]. In a variant of the scribing technique, a diamond makes a smooth scratch around the complete circumference of the fiber prior to breaking [Khoe et al. 1980, 1981a]. Another method for obtaining good endfaces is to expose the fiber to a focused high-power CO_2 laser beam before fracturing it by applying tensile stress [Kinoshita and Egishira 1978]. Good fiber ends can also be prepared by spark-eroding the fiber in an arc discharge and then breaking it by bending and pulling [Bisbee 1971b; Sklyarov 1975; Caspers and Neumann 1976].

The quality of the fiber end can be checked by viewing it axially in a standard microscope, or in an interference microscope [Gordon K.S. et al. 1977]. The break angle can be measured by focusing the light of a HeNe laser on the endface either from the glass side [Millar 1981a, 1984b] or from the air side [Reitz 1979, 1980a,b] and to observe the direction of the reflected light beam. Both the break angle and the surface quality can be evaluated by directing a collimated light beam parallel to the fiber axis onto the cut where it is reflected and diffracted [Wolffer 1985]. The reflected beam is observed on a far screen. The position of the spot gives the break angle with an accuracy of $10'$ and the size of the diffraction pattern is proportional to the amplitude of the surface defects.

In practical cutting procedures, rough endfaces are sometimes obtained. In such cases, recutting of the fiber is required. Instead of checking the fiber endface with a microscope, it is also possible to check the quality of the endface by observing the acoustic pulses which are emitted at the moment of the fracture [Miyajima et al. 1986a]. The acoustic spectra of good endfaces have smaller amplitudes and maximum components at lower frequencies than the spectra of bad endfaces.

12.1.2 Low Reflectance Fiber Ends

At the plane endface of the fiber, a reflected wave is produced. The power reflection factor of an ideal endface in air is 3.5% (Sect. 5.9.5). The loss in power can usually be tolerated. But the reflected wave can influence the output power and the frequency of the laser source (Sect. 11.6.7) or impair the accuracy of interferometric sensors. In backscattering measurements (Sect. 13.4), Fresnel reflections from fiber ends can saturate the receiver.

For certain applications it is thus important to reduce the reflection factor of the fiber endface. There are four methods for doing this:

1. The end face can be coated by an antireflective dielectric multilayer coating.
2. A wide-spectrum antireflective coating can be made on fused silica by depositing a specific polymer solution which is converted to a porous graded-index SiO_2 film [Yoldas and Partlow 1984].
3. The fiber end can be immersed in a liquid whose refractive index closely matches the index on the axis of the fiber.
4. The simplest and most efficient method is to tilt the end-face of the fiber, e.g. by 10° or 12°, relative to the usual orientation normal to the fiber axis by grinding and polishing the end of the fiber at this angle. By this method, the reflected light beam is separated in angle from the incident beam [Ulrich and Rashleigh 1980; Danielson B.L. 1985]. A disadvantage of this method is that, because of refraction, the fiber axis includes a small angle with the axis of the free beam radiated from the fiber end.

Methods for reducing reflections from fiber connectors will be reviewed in Sect. 12.2.5.

12.1.3 Reflecting Fiber Ends

Reflecting fiber ends are needed e.g., for fiber resonators (Sect. 12.3.4), or in the folded path OTDR technique (Sect. 13.4.3).

A reflecting fiber end can be made by holding the fiber end in a precision-bore glass capillary butted against a multilayer dielectric mirror with a drop of index-matching oil [Franzen and Kim 1981]. Since these mirrors have a finite optical penetration depth, there are losses caused by beam spreading. These losses and those caused by fiber end curvature, mirror tilt, and end separation have been investigated by Marcuse and Stone [1986].

A simple method for fabricating an optical fiber mirror consists in plating the fiber end with an alloyed metal of low melting point [Ishihara et al. 1986]. Power reflectivities of 50–60% have been obtained at $\lambda = 633\,\text{nm}$. In comparison, the reflectivity of an aluminium mirror fabricated using vacuum evaporation was 70%.

12.2 Couplers

12.2.1 Beam to Fiber Couplers

At present, it is not possible to realize all functions in all-fiber components. Many components only work, or work better, with free beams. Therefore, it is often necessary to transform the fundamental fiber mode into a free collimated beam with a spot size much greater than the fiber spot size, or vice versa.

The necesary transformation can be made by terminating the guiding fiber. At the fiber end, the fundamental mode transforms into a diverging beam wave (Sect. 8.1). At that distance from the fiber end where the beam spot size has increased to the desired value, a positive lens transforms the spherical wavefronts into plane wavefronts.

Such devices can be called lensed fiber ends or expanded beam fiber terminations. These components are used e.g., in expanded beam connectors, in fixed or variable attenuators, in switches, in optical isolators, etc.

We start with the problem of transforming the fundamental fiber mode with spot size w_G into a free collimated beam in air with spot size w_0. As has been explained in Sect. 8.1, the beam radiated from the broken or polished fiber end diverges by diffraction and is very similar to the fundamental Gaussian beam described in Chap. 4. The spot size $w(z)$ of the radiated beam at a distance z from the fiber end is given approximately by (4.5)

$$w(z) = w_G[1 + (z/z_R)^2]^{1/2} \quad , \tag{12.1}$$

with the Rayleigh distance z_R (4.6) given by

$$z_R = \pi w_G^2/\lambda \quad . \tag{12.2}$$

Since the Rayleigh distance z_R in air is of the order of $60\,\mu\text{m}$, one can usually assume

$$z \gg z_R \quad , \tag{12.3}$$

so that (4.7)

$$w(z) = w_G z/z_R = z\lambda/(\pi w_G) \quad . \tag{12.4}$$

Under the same assumption, the radius of curvature $R(z)$ of the wavefronts (4.17) equals z. At the distance

$$z = \pi w_G w_0/\lambda \quad , \tag{12.5}$$

where the spot size $w(z)$ equals the spot size w_0 of the collimated beam to be produced, a positive lens is placed with a focal length f equal to z:

$$f = \pi w_G w_0 / \lambda \quad . \tag{12.6}$$

Thus, the center of the fiber end face is to be located at the focal point of the lens. The lens transforms the spherical wavefronts of the diverging beam into plane wavefronts without changing the beam diameter.

For the inverse problem of coupling a wide free Gaussian beam with spot size w_0 at the waist into a single-mode fiber with spot size w_G, a lens with focal length

$$f = \pi w_G w_0 / \lambda \quad , \tag{12.7}$$

has to be inserted into the beam. It transforms the wide collimated beam into a converging Gaussian beam, the waist of which is located in the back focal plane of the lens. When the fiber input end is brought into the focal plane, the spot size of the converging beam matches that of the fundamental fiber mode, giving optimum launching efficiency.

The lenses required for collimation and focusing can be conventional lenses, but these are not practical with respect to size, weight, reliability, and cost. Therefore, various microoptic lenses have been developed [Uchida T. and Sugimoto 1978; Uchida T. and Kobayashi 1982], e.g. small high-index ball lenses [Nicia 1978; Tamura et al. 1985], plano-convex rod lenses, or graded-refractive-index lenses (GRIN-rod lenses, SELFOC lenses) [Uchida T. et al. 1970; Tomlinson 1980; Yamagishi et al. 1983; Khoe et al. 1984a].

In a beam to fiber coupler using conventional lenses or ball lenses, the center of the beam waist must be carefully aligned to the core center in order to reduce coupling losses. By inserting a glass flat into the air path of the beam and rotating it, the beam spot can be shifted transversely by $\pm 15\,\mu$m [Krüger 1983].

Graded-index lenses may be designed in such a way that between the fiber end and the output face of the lens, the beam propagates only in glass, so that the two reflecting glass-air interfaces at the fiber end and the input of a conventional lens can be avoided.

A graded-index lens is a cylindrical glass rod in which the refractive index depends quadratically on the distance ϱ from the rod axis:

$$n(\varrho) = n_0(1 - a\varrho^2/2) \quad , \tag{12.8}$$

where the so called focusing parameter a describes the index grading.

Due to continuous refraction in this inhomogeneous medium, a Gaussian beam launched into a graded-index lens with a radial offset or angular deviation, propagates down the rod with the beam axis undulating sinusoidally around the rod axis. The period of the undulation, the so called pitch p, is related to the focusing parameter [Kogelnik 1965b; Uchida T. and Kobayashi 1982] by

$$p = 2\pi / a^{1/2} \quad . \tag{12.9}$$

When the beam is launched without radial and angular offset, the beam axis is straight and coincides with the rod axis.

In general, the spot size of the Gaussian beam propagating in the rod is not constant, but oscillates periodically with a period that is *one-half* of the pitch. In the cross sections where the beam diameter is minimum or maximum, the wavefronts are plane.

For a certain spot size

$$w_{\mathrm{r}} = (\lambda/\pi)^{1/2} a^{-1/4} \quad , \tag{12.10}$$

the tendency of the Gaussian beam to diverge due to diffraction is compensated for by the focusing power of the graded-index rod [Neumann 1987], and the Gaussian beam propagates in the rod as a normal mode with constant beam diameter $2w_{\mathrm{r}}$ (the index r denotes rod).

The relation between the minimum (w_{min}) and the maximum (w_{max}) spot sizes of the beam is

$$w_{\mathrm{min}} w_{\mathrm{max}} = w_{\mathrm{r}}^2 = (\lambda/\pi) a^{-1/2} \quad . \tag{12.11}$$

A graded-index rod or graded-index multimode fiber with a length of a quarter pitch

$$L = p/4 \quad , \tag{12.12}$$

can be used as a collimator or beam expander [Mathyssek et al. 1984].

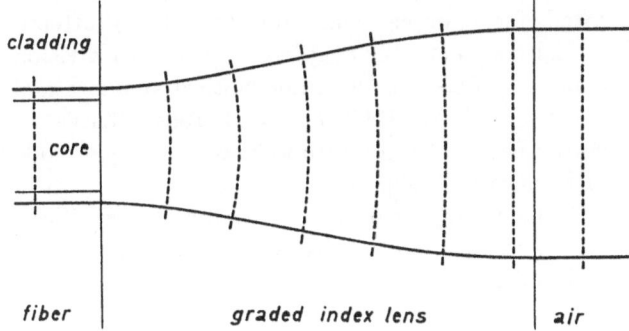

Fig. 12.1. Beam expansion caused by a graded-index lens attached to the end of a single-mode fiber. $w_{\mathrm{G2}} = 2w_{\mathrm{G1}}$, $w_{\mathrm{G1}} = 0.8\lambda$

Assume a single-mode fiber with spot size $w_{\mathrm{G}} < w_{\mathrm{r}}$ to be attached to the input face of the quarter-pitch lens. In Fig. 12.1, the full lines represent the beam contours, and the dotted lines the wavefronts. Since the input wavefront is plane, the beam diameter has a minimum $w_{\mathrm{min}} = w_{\mathrm{G}}$ at the rod input. Since the spot size w_{r} of the normal mode of the graded-index rod has been assumed to be larger than the input spot size w_{G}, diffraction prevails, and the spot size increases along the rod. A quarter pitch down the rod, i.e. at its output end, the wavefront is plane again but the beam width is larger than at the lens input face (12.11)

$$w_{\max} = w_r^2/w_G \quad .\tag{12.13}$$

Because of its larger beam width, the output beam diffracts more slowly than the beam which the fiber end would radiate directly into air. The output beam behaves approximately like a parallel beam.

In the example of Fig. 12.1, the beam diameter is expanded by a factor of two. In order to get a short lens, an unrealistically small value of $w_G = 0.8\,\lambda$ has been assumed. In reality, the beam diverges much more slowly and the lens is much longer.

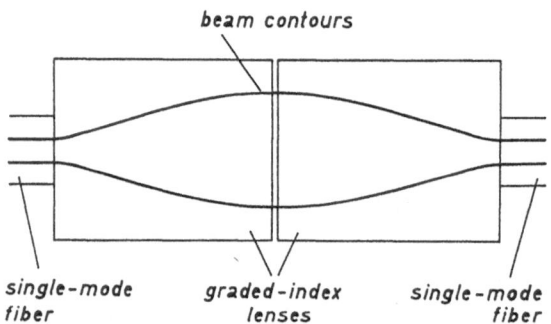

beam contours

single-mode fiber   *graded-index lenses*   *single-mode fiber*

Fig. 12.2. Expanded beam connector with graded-index lenses

With the graded-index lens beam expander as a basic building block, many devices can be fabricated, for example, connectors (Fig. 12.2), attenuators, power dividers, directional couplers, star couplers, wavelength division multiplexers/demultiplexers, laser to fiber couplers, and optical switches and isolators [Kobayashi et al. 1979; Tomlinson 1980; Iga et al. 1984]. These so-called micro-optic devices can be used both in multimode and in single-mode fiber systems [Uchida T. and Kobayashi 1982].

An alternative method for beam expansion is to use fiber tapers (Sect. 12.2.6). For a given operating wavelength and a fiber with a given index difference Δ, the spot size w_G is minimum for a certain core radius a, which for step-index fibers corresponds to a normalized frequency of about $V = 1.9$ (Table 5.1). Single-mode fibers are usually operated at a V-value near to that which gives the minimum spot size. It is then possible to increase the spot size by either increasing or decreasing the core radius, i.e. by using expanding or contracting fiber tapers.

In the first scheme, a single-mode fiber is tapered gradually so as to enlarge its cross section by about one order of magnitude. When the taper expands gradually, the power remains in the fundamental mode even if the taper is a multimode waveguide. At the wide output end of the taper, the beam has a much larger field diameter than at the input end. For a sufficiently slow taper, the wavefronts remain nearly plane. Expanding-taper beam expanders have been found to be essentially lossless [Amitay et al. 1986]. The insertion loss of an expanded beam connector using two expanding fiber tapers can be smaller than 0.1 dB [Presby et al. 1987; Amitay et al. 1987].

In the second scheme [Jedrzejewski et al. 1986], a contracting taper is used. Along the taper, the core radius and the normalized frequency decrease, while the spot size increases. Biconic tapers are produced by overjacketing the fiber with a capillary, collapsing the capillary about the fiber, and pulling the combined fiber and capillary in a miniature graphite furnace. Breaking a biconic taper at the taper waist gives two contracting fiber tapers. Low-loss contracting tapers must be extremely long (Sect. 12.2.6).

12.2.2 Source to Fiber Couplers

Both lasers and LED's are used as sources for single-mode communication systems. We first consider launching by semiconductor lasers.

For efficient launching into single-mode fibers, the semiconductor laser should oscillate in a single stable transverse mode, with a single-lobed circular far-field pattern, and the near field should not move with changing drive current, temperature, or ageing. Index-guiding lasers can fulfill these requirements better than gain-guiding lasers.

A laser diode emits a diverging beam with curved wavefronts. Because the field of this beam is not matched to that of the fundamental fiber mode, the power P coupled into the fiber will be smaller than the power P_r radiated by the laser into air with the fiber removed. The laser to fiber coupling efficiency η is defined by

$$\eta = P/P_r \quad , \tag{12.14}$$

and the coupling loss in decibels by

$$a = -10\log\eta \quad . \tag{12.15}$$

If one succeeds in decreasing the coupling loss by 1 dB, for a fiber loss of 0.5 dB/km, the link length can be increased by 2 km. This numerical example shows that it is very important to maximize the coupling efficiency, and thus the power injected into the fiber.

Additional requirements for a practical source-fiber coupler are as follows:

1. easy fabrication,
2. high environmental reliability,
3. negligible source performance deterioration.

Because of the small spot size of a single-mode fiber, the efficient coupling of a semiconductor laser to a single-mode fiber is more difficult than coupling to a multimode graded-index fiber.

The problem of laser to fiber coupling has been analyzed theoretically by several authors [Cohen L.G. 1972; Saruwatari and Nawata 1979; Sakai and

Kimura 1980; Wagner and Tomlinson 1982; Nicia 1982; Bludau and Rossberg 1985; Joyce and DeLoach 1985].

Since semiconductor lasers are free running oscillators, the output power and emission frequency depend on the load. In optical communications, the load of a generator is not described by the impedance of the load but by its reflection factor. Reflected waves are caused by air-glass interfaces, e.g. by lens surfaces, by the fiber input and output faces, or by connectors with an air gap. A distributed reflection is caused by Rayleigh backscattering (Sect. 5.9.2).

Optical feedback from the fiber has very detrimental effects on the operation of semiconductor lasers, resulting e.g., in variations in output power and emission frequency, changes in the optical spectrum, nonlinearities in the power output versus injection current characteristics, and intensity noise. These adverse effects caused by optical feedback have been reviewed in Sect. 11.6.7.

In order to reduce impairment of system performance, it is necessary to minimize optical feedback from the fiber to the laser by using coupling arrangements with decreased reflections.

A power reflection factor smaller than 10^{-6}, corresponding to a return loss of $-60\,\mathrm{dB}$ is required for single-longitudinal-mode (index-guiding) lasers to maintain high quality signal transmission [Petermann and Arnold 1982]. The reflection factor as seen by the laser has been estimated by several authors [Bludau and Rossberg 1982, 1985; Kuwahara et al. 1983].

Optical feedback can be studied by measuring, as a function of the distance between the laser and the reflecting discontinuity, either the variations of the output power [Bludau and Rossberg 1982] or the baseband phase shift at the end of a long dispersive fiber due to a change in wavelength [Wenke and Elze 1981; Wenke and Zhu 1983]. Feedback due to connectors, far-end reflections, and backscattering can be reduced by inserting an optical isolator between the source and the fiber. Feedback from near-end reflections can be reduced by antireflection coating the coupling lens or the fiber facet [Ulrich and Rashleigh 1980], by a slight tilt of the fiber endface and a corresponding inclination of the fiber axis [Ulrich and Rashleigh 1980], by lateral displacement of the input lens [Kuwahara et al. 1983], or by using a coupling lens with a large radius of curvature [Bludau and Rossberg 1985].

When a single-mode fiber is butt-jointed to the plane radiating face of a laser diode, the maximum coupling efficiency under favorable conditions is only 15% [Keil 1984b]. This is largely due to the very different spot sizes of the laser mode and the fundamental fiber mode. The fiber spot size is typically about five times the laser spot size. The coupling efficiency can be increased by improving the field match. Numerous schemes for implementing this have been described in the literature and some overviews of the methods have been published [Garrett and Todd 1982; Minowa et al. 1982; Wenke and Zhu 1983; Ku 1984; Khoe et al. 1984a; Ghafoori-shiraz and Asano 1986; Kawano 1986].

There are two main approaches for improving the field match, the "microlens methods" and the "large-lens methods". The former method consists in widening the laser beam in an air gap between the laser facet and the fiber front end and transforming the spherical incident wavefront into a plane wave-front by a microlens placed on the fiber input end. This arrangement has the following advantages: the fabrication is simple, the design is compact, there are only a few assembly steps, the coupling efficiency is high (up to 70% [Bludau and Rossberg 1985]), and there is only a single reflecting surface to impair the power and frequency stability of the laser. However, the distance between laser and fiber is very short and the alignment tolerances for the fiber are small.

In the microlens couplers, the curved wavefront in the radiated beam can be transformed into the plane wavefront of the LP_{01}-mode by several means:

1. tapering and rounding the fiber end by heating [Kuwahara et al. 1980a,b, 1983; Minowa 1982; Vatoux et al. 1983; Mathyssek et al. 1984, 1985; Keil et al. 1984b]
2. forming a conical fiber end by etching [Eisenstein and Vitello 1982],
3. forming a hemispherical microlens on the fiber endface by a photolitho-graphic technique [Cohen and Schneider 1974; Lee and Barnes 1985], by heating the fiber end with a CO_2 laser [Paek and Weaver 1975], or by attaching a glass droplet [Murakami et al. 1980],
4. polishing the fiber endface into a special quadrangular pyramid-shaped hemiellipsoidal lens [Sakaguchi et al. 1981],
5. attaching a high-index microlens on the cleaved end of a tapered single-mode fiber [Khoe et al. 1983b, 1984a; Mathyssek et al. 1985; Khoe and Kock 1985],
6. forming a lens on the fiber end by etching [Kayoun et al. 1981; Ku and Dufft 1982; Kawachi et al. 1982a; Vatoux et al. 1983; Ghafoori-shiraz and Asano 1986],
7. forming an integrated microlens on the fiber input face [Yamada et al. 1980b],
8. forming a cylindrical lens on the fiber end by etching and fusing [Weidel 1974],
9. squeezing core material out of a heated fiber with low melting tempera-ture, thus forming an elliptical microlens on the fiber end [Khoe 1979a],
10. splicing a multimode graded-index fiber (GRIN-lens) with quarter pitch length to the single-mode fiber, etching a taper and forming a lens on the end by fusion [Mathyssek et al. 1984],
11. modeling an aspherically shaped lens from the end of a silica thread of appropriate length. The silica thread is fused directly onto the fiber end [Bludau and Rossberg 1985],
12. creating an ellipsoidal lens by depositing a small glass drop on a very thin quartz fiber and sticking it on the single-mode fiber end [Lecoy and Richter 1982].

The focusing characteristics of hemispherical microlenses formed on the end of single-mode fibers have been investigated theoretically by Lee [1986].

The light power that is reflected back from the surface of a hemispherical fiber endface into the laser has been analyzed and found to be 50 dB smaller than the emitted light power [Kuwahara et al. 1983]. This reflected power is 25 dB smaller than that reflected from a flat fiber end.

A simple automatic method of obtaining tapered ends for single-mode fibers is to use a microcomputer-controlled fabrication process [Schwander and Schwaderer 1985; Schwander et al. 1985].

In one laser diode package, the laser and the lensed fiber end are mounted on two separate substrates [Enochs 1986]. Physical alignment of fiber to laser is achieved with computer-controlled piezo-driven micromanipulators. After alignment, an onboard resistive heating element is used as a source of local heating to solder the fiber substrate into the package.

A method for coupling two single-mode fibers simultaneously to the same laser diode has been described by Khoe et al. [1984b].

In the second scheme for improving the field match between the laser and the fiber, an enlarged image of the laser near field is formed on the plane fiber input face by a discrete lens or a lens combination.

The enlarged image of the laser near field can be produced either by a single lens, e.g.:

1. a glass fiber acting as a cylindrical lens [Weidel 1975; Saruwatari and Nawata 1979],
2. a separate cylindrical lens [Minowa et al. 1982],
3. a high-index microbead attached on the inside of the plane cap window [Kuyt et al. 1981],
4. a separate plano-convex graded-index lens [Minowa et al. 1982; Ishikawa et al. 1985; Honmou et al. 1986; Kitano et al. 1986],
5. a single ball lens [Sumida and Takemodo 1982, 1984],
6. a truncated spherical lens [Yamashita et al. 1985],
    or by a system consisting of two or more lenses such as:
7. a system of standard plano-convex lenses [Krumpholz 1975],
8. a cylindrical and a graded-index lens [Odagiri et al. 1977, 1980],
9. a ball lens and a confocal graded-index lens [Saruwatari and Sugie 1980, 1981; Minowa et al. 1982, Saruwatari et al. 1983; Saruwatari 1984; Kawano et al. 1985a,b, 1986; Kawano 1986],
10. a system of one cylindrical and two spherical lenses [Krumpholz and Westermann 1981],
11. two graded-index lenses in a nearly confocal scheme [Reith et al. 1986].

The coupling characteristics and misalignment tolerances of various large lens couplers have been studied and compared by Kawano [1986].

In the large lens methods, the distance between the laser diode and the fiber is large and astigmatic and elliptical laser beams can be matched to the fiber mode using cylindrical lenses.

On the other hand, with large lens couplers, numerous assembly and alignment steps are necessary.

In the two-lens scheme, one can manufacture separately a laser-to-first-lens coupling submodule and a fiber-to-second-lens coupling submodule. Subsequently, the two submodules are coupled to give the laser to fiber coupling module. Because of the increased spot size, the lateral and axial alignment of the two submodules is relatively easy, and alignment errors made in producing the submodules can be compensated for when jointing the submodules.

Another advantage of the large-lens coupling scheme is that it allows one to insert an optical isolator in the path of the approximately collimated beam between two lenses in order to minimize power and frequency fluctuations due to reflected waves [Sugie and Saruwatari 1986]. Alternatively, a single ball lens can be used simultaneously as a spot-size transformer and as the Faraday rotator for an isolator [Hirota et al. 1981; Sugie and Saruwatari 1982, 1983].

Three different criteria can be used for maximizing the launching efficiency during manufacture of laser to fiber couplers [Wittmann et al. 1984]: maximum power coupled into the fiber, maximum reflection from the laser front facet back into the fiber, and maximum detected photocurrent when the laser is operated as a photodetector. In the second and third methods, the laser is inactive and an external laser is connected to the far end of the fiber.

For most coupling schemes, the coupling losses have been limited to around 3 dB. Mimimum coupling losses of 1 dB [Ishikawa et al. 1985; Honmou et al. 1986] and 2 dB [Kitano et al. 1986] have been reported for couplers using a plano-convex gradient-index lens. A coupling loss of 1 dB is even smaller than the coupling loss between laser diodes and multimode graded-index fibers.

A common problem with contemporary laser diode modules is the temperature dependence of coupling efficiency [Minowa et al. 1982]. The temperature influences the coupling efficiency in the following three ways:

1. thermal expansion offsets the coupling circuit,
2. the LD transverse mode field varies with temperature,
3. the factor relating the front face output power to the rear face monitoring power depends on temperature.

Besides laser diodes, light emitting diodes are also used as optical sources. LED's are coupled to single-mode fibers for producing cheap single-mode transmission systems or for measuring e.g. mode field diameter (Sect. 13.7.2) or chromatic dispersion (Sects. 13.9.2, 3). The power that can be coupled from an LED into a single-mode fiber is much smaller than that obtainable from a laser diode. The theoretical fundamentals of launching incoherent light into single-mode fibers have been discussed in Sect. 7.3.

For *surface emitters* (SE LED's), which make use only of spontaneous emission, launched powers between −38 dBm [Pophillat 1984; Gimlett et al. 1985a,b; Shumate et al. 1985] and −26.7 dBm [Uji et al. 1985; Uji and Hayashi 1985] have been reported. Since a surface emitting LED radiates a power of about 0 dBm into air, the launching loss is of the order of 30 dB [Shumate et al. 1985].

For *edge emitting* LED's (ELED's), which additionally amplify the spontaneous emission by stimulated emission, higher powers can be coupled into the fiber, e.g. −28.1 dBm [Shumate et al. 1985], −26.4 dBm [Gimlett et al. 1985a],

−22.1 dBm [Arnold and Krumpholz 1985a; Gimlett et al. 1985b], −21 dBm [Plastow et al. 1985], −20 dBm [Burns et al. 1983; Arnold and Krumpholz 1985b], −18.5 dBm [Ulbricht et al. 1985a,b], −15.2 dBm [Saul et al. 1985], or −12.2 dBm [Olshansky et al. 1985a,b]. An edge-emitting LED coupled to a single-mode fiber shows increased coupling sensitivity to lateral fiber displacement compared with multimode fibers [Reith and Shumate 1987].

The highest power of $250\,\mu$W (−6 dBm) is obtained by a *superradiant* LED (SR LED) which in construction is very similar to a semiconductor laser diode. However, the frontface is antireflection-coated in order to avoid self-oscillations, i.e. lasing, while the backface is highly reflective [Arnold et al. 1985]. A disadvantage of this source is that for high output powers it tends to become unstable. The instability is generated by small external reflections, which can be sufficient to cause lasing of the device.

Although the power launched by LED's is very small, it is possible to make low-bandwidth, short-haul single-mode transmission systems using LED's as sources. The power level which can be launched from a 100 W-tungsten lamp through a monochromator with linewidth 10 nm into a single-mode fiber is of the order of −50 dBm.

12.2.3 Couplers to Integrated Optical Waveguides

The technique of integrated optics uses dielectric waveguides embedded in a transparent substrate chip. Light waves are guided either in thin films (slabs) or strips (channels) of a dielectric material surrounded by other material with a smaller refractive index [Marcuse 1973a; Tamir 1975a; Ostrowski 1979, Alferness 1981, 1987; Hunsberger 1984, Nolting and Ulrich 1985]. Due to the dimensions of the substrate chip, the waveguides are at most a few centimeters long.

By using appropriate materials and waveguide structures, light can be generated, amplitude modulated, phase shifted, frequency shifted, switched, deflected, splitted, combined, filtered, polarization-transformed, amplified, and detected. A large variety of integrated optic components can be made. In the future, one hopes to integrate a large number of functional devices on a single integrated optic chip.

For optical communications, it would be very advantageous to produce optical transmitters, repeaters, and receivers using integrated optics [Stallard et al. 1986]. However, one faces the problem of obtaining a simple, rugged, low-loss connection between a single-mode fiber and a channel waveguide. Jointing of fibers to channel waveguides is a more difficult problem than that of fiber splicing. Channel waveguides are usually single-moded, but in order to accommodate sharp bends, the mode field diameter of the fundamental mode is smaller than that of the LP_{01}-mode in a typical single-mode fiber. Moreover, the intensity distribution is not circularly symmetric. The mode field diameters parallel and normal to the surface of the chip are different, and the intensity distribution in the direction normal to the surface is generally asymmetrical.

Since the wavefronts of the beam guided by a channel waveguide and a single-mode fiber are planar, the condition for wavefront matching (Sect. 7.2.1) can be fulfilled by a simple butt joint [Guttmann et al. 1975c]. However, there will be losses due to the mismatch of the transverse amplitude distributions. The losses caused by field mismatch can be calculated by inserting into the overlap integral (7.11) the transverse field distributions of the fundamental modes of the channel waveguide and the fiber [Bulmer et al. 1980; Fukuma and Noda 1980; Ramer 1981].

Several methods for reducing the losses caused by field mismatch have been explored: tapering the channel waveguide [Campbell 1979], tapering the fiber [Shigihara et al. 1986], and optimizing the refractive-index profile of the channel waveguide [Fukuma and Noda 1980; Ramaswamy et al. 1982; Alferness et al. 1982; Alferness and Divino 1984; Baba et al. 1984; Lin S.H. et al. 1985; Takato et al. 1986]. Using the last method, a fiber to fiber total insertion loss of only 0.5 dB for a 1 cm long channel waveguide has been obtained [Komatsu et al. 1986]. Square channel waveguides made from silica surrounded by glass can be designed to have a cross-section and a refractive-index difference compatible with those of conventional single-mode fibers [Takato et al. 1986] to give coupling losses of less than 0.05 dB. However, with silica, only passive components can be made.

The refractive indices of typical substrate materials like $LiNbO_3$, $LiTaO_3$, GaAs, or InP are different from that of silica glass. There are thus additional losses caused by Fresnel reflection, which for one air − lithium-niobate interface is 0.65 dB [Alferness 1981]. Reflection losses can be reduced by antireflection-coating the endfaces of both the channel waveguide and the fiber or by applying an index matching liquid between the endfaces.

Because of the small mode field diameters, it is difficult to avoid slight geometrical misalignments in a channel waveguide to fiber joint. The losses caused by transverse, longitudinal, or angular displacement have been investigated in a number of papers [Burns and Hocker 1977; Noda et al. 1978; Ramer 1981; Zolotov et al. 1982].

In laboratory setups, the fiber is usually adjusted by precision micro-manipulators for endfire coupling to the channel waveguide [Noda et al. 1978]. Of course, this method is not convenient for field use.

A simple and practical technique for attaching a single-mode fiber to a channel waveguide consists in aligning the fiber with micro-manipulator stages and then fixing it to the waveguide end using a light-cured adhesive [Cameron 1984].

Several authors use V-grooves preferentially etched into a Si-wafer to align and permanently attach single-mode fibers to channel waveguides [Bulmer et al. 1980, 1981; Kaufmann et al. 1986]. This method can also be used to simultaneously couple more than one fiber to an equal number of channel waveguides [Hsu and Milton 1976; Murphy and Rice 1983, 1986; Murphy et al. 1985].

A microlens applied to the fiber endface causes the beam radiated into air first to become narrower. In the beam waist, the spot size is smaller than that of the fundamental fiber mode. Locating the input of an integrated optic film waveguide in the beam waist gives a better launching efficiency than butt jointing [Bear et al. 1980].

12.2.4 Splices

Cables cannot be produced and installed in lengths of more than a few kilometers and it is therefore necessary to splice fiber links together every 1–2 km for inland cables. For a variety of reasons, every kilometer of a typical transmission line will be broken two or three times (by such things as runaway bulldozers) over its 30 year lifetime [Keck 1983a]. Thus, the main application for splices is in the field. However, splices are also used within buildings for attaching factory-installed pig-tailed connectors, sources, detectors, or couplers, and for connection office to field cables.

Since there will be many splices in a practical transmission system, splice losses are an important factor to be taken into account when a system is designed. One book solely devoted to optical fiber splices and connectors has already been published [Miller C.M. 1986a]. Reviews of splicing techniques have been given by several authors [Miller C.M. 1977; Dalgleish 1980; Garrett and Todd 1982; Hoshikawa and Toda 1983; Woods 1984b; Inada 1984; Khoe et al. 1984a; Boscher 1985; Hooper et al. 1985].

In this section on fiber splices, we first discuss splice losses and fiber alignment prior to splicing. Methods for fusion, mechanical, and mass splicing will then be described. Finally, methods for estimating the splice loss are summarized.

Splice losses can be caused by mismatch of mode field diameter, transverse core offset and angular misalignment of axes. Using the Gaussian approximation for the radial field distribution (5.25), these losses can be estimated using the simple formulas (9.3, 5 and 10), respectively, or the general formula (9.12) which describes the splice loss when several sources of loss are combined.

It has been experimentally verified [Kummer and Fleming 1982] that the Gaussian field theory adequately describes the splice loss between single-mode fibers for realistic (small) misalignments and/or spot size mismatches.

The main problem in single-mode fiber splicing is transverse core offset. For a mode field diameter of $10\,\mu$m and an offset of $1\,\mu$m, the loss is (9.5)

$$a = 4.34(1/5)^2 = 0.17\,\text{dB} \quad . \tag{12.16}$$

Therefore, in single-mode splices (and connectors), the core axes must be very carefully aligned in order to obtain tolerable joint losses.

In a practical fiber, the axes of the cylindrical core and the cylindrical cladding do not coincide in general. The distance between the core and the cladding axes is specified by the "concentricity error" or "core eccentricity". When the cladding axes of the fibers to be jointed are aligned, the core eccentricity causes a radial offset between the two core axes, which depends on the relative azimuthal orientation of the two fibers. In the worst case, the radial offset is the sum of the core eccentricities of the two jointed fibers. However, taking into account the spread of fiber parameters and the statistical aspects of splicing, an average loss much less than the worst-case value would be expected.

Assuming typical parameters for the mode field diameter ($10\,\mu m \mp 1\mu m$) and a core eccentricity ($\leq 1\,\mu m$), a theoretical study has been made [Hevey et al. 1984] of the losses due to mode field diameter mismatch and core eccentricity. Utilizing a Monte Carlo simulation method, the splice losses of 10 000 randomly paired fibers have been calculated. The average loss due to spot size mismatch alone was found to be only 0.008 dB. Assuming the worst case azimuthal orientation of the fibers, the loss due to core eccentricity has been reported to be 0.14 dB.

Before splicing, flat and perpendicular endfaces have to be prepared as described in Sect. 12.1.1. In the next step, the fiber ends to be spliced must be aligned with minimum transverse core offset and minimum tilt angle.

For high-quality single-mode fibers [Ainslie et al. 1981], in which the core axis is concentric with the cladding axis to within about $0.5\,\mu m$, it is possible to use the outside fiber surface as a guide in aligning [Payne D.B. et al. 1982; Leach et al. 1982; Tardy et al. 1983]. The claddings can be aligned in a V-groove, for example [Someda 1973].

In another method for coaxial alignment of the claddings [Millar 1983; Millar et al. 1985], the light beam of a He-Ne laser illuminates both fiber ends simultaneously at 90° to their axes, causing two diffraction patterns to appear on a cylindrical ground glass screen around the fibers. Changes in the radial offset and in the tilt between the fiber axes are indicated by the interference of the diffraction patterns. For coaxial alignment of the claddings, a distinct single fringe along the entire length of the diffraction pattern is observed. This technique for cladding axis alignment can also be used to measure the mode spot eccentricity. After having aligned the claddings, the fibers are offset transversely until the transmitted power is maximum. The measured offset gives the mode-spot concentricity error.

From a practical viewpoint, it is appropriate to allow some degree of core eccentricity in order to produce single-mode fibers economically. In current fiber fabrication techniques, core eccentricity and diameter variations are inevitable [Inada 1984]. For these fibers with a larger core eccentricity, it is necessary to align the mode fields on the basis of maximum transmitted power (power monitoring) [Minowa et al. 1982; Kato et al. 1982b,c; Rivoallan et al. 1983]. Systems have therefore been developed using power injection at the beginning of the newly added length of cable and detecting the power at the far end of the fiber link [Tanifuji et al. 1983; Kato et al. 1983]. Such systems work well for new installations but are not suitable for repairing a damaged cable. Power monitoring is almost impossible in laying and repairing work in a long-haul repeatered optical submarine cable system. Even in land-based systems, the necessity of the far-end disposability and the problem of transmitting the received signal back to the jointing crew is a severe drawback.

Local power injection and local detection provide a means to monitor alignment prior to splicing the fiber ends without remote sources and power meters [Kato et al. 1982d; De Blok and Matthijsse 1984; Kato et al. 1984b; Becker and Zell 1985; Hughes et al. 1985]. The light is launched locally by bending the primary fiber and injecting tangentially at the bend into the fiber core through the coating and the cladding. Macrobend and microbend injectors are described in Sect. 12.2.13. Local detection is also accomplished by mechanically bending the secondary fiber and detecting the power radiated from the bend. Fiber taps for local detection are described in Sect. 12.2.13.

It is not necessary to strip the coating from the fiber; both injection and detection can be accomplished through a range of colored coatings. When the refractive index of the coating is smaller than that of the cladding, cladding modes must be suppressed by immersing the stripped fiber ends in a high-index liquid (Sect. 12.2.8). For a high-index coating, the cladding-mode-stripping is performed rigorously by the coating. The fiber ends have to be pre-aligned by eye in order that a signal can be detected. The final alignment is then made by maximizing the detected signal.

Instead of observing and maximizing the light transmitted through the joint, one can also observe and minimize the light scattered from the joint into the cladding of the receiving fiber [Miller C.M. 1982a,b; Kaiser 1983; Khoe et al. 1983a; Fujise and Iwamoto 1984; Fujise et al. 1986a]. The output of the scattering detector can be used as a direct measure of the splice loss. The detector measures minimum signal when the cores are aligned. Residual core misalignment can be detected much more accurately by observing the scattered power than by observing the transmitted power [Aberson and White 1986].

More than 90% of the power lost from a splice with small amounts of offset and tilt is coupled into the LP_{11}-mode. This fact has been utilized to predict the splice loss from the scattered powers measured for zero and for a known transverse offset [White I.A. and Ludington 1986].

For fibers with a GeO_2-doped core, one can also locally inject power by illuminating the fiber with ultraviolet light (Sect. 7.5). Only the GeO_2-doped core fluoresces with an emission wavelength of 420 nm [Presby 1981a]. Part of the fluorescent radiation excites the fundamental fiber mode. The launched power is sufficient for aligning the fiber ends [Khoe et al. 1983a, 1986].

When light is locally injected into both fiber ends, the core endfaces can be observed simultaneously with a microscope by placing a very small prism-cube beam splitter with one silvered face between the two fiber ends [Imon and Tokuda 1983]. For a microscope resolution of $0.9\,\mu m$, average splice losses of 0.09 dB have been reported.

Sometimes it is not possible to use local injection and detection. For instance, in a fiber section between repeaters in a submarine optical-fiber cable, it is impossible to locally launch light into the fiber and to monitor it. Therefore, methods for direct monitoring of the cores at the splice point have been proposed. Direct core monitoring can be made by a normal microscope combined with a high-resolution TV camera [Kawata et al. 1983, 1984; Katagiri et al. 1984; Uenoya et al. 1985; Yamada T. et al. 1986], a phase microscope [Furukawa et al. 1982], a differential interference contrast microscope [Haibara et al. 1983], or the technique of exciting fluorescence in GeO_2-doped silica cores with ultraviolet light [Tatekura et al. 1982].

In a basic study of splicing using direct core monitoring [Kawata et al. 1984], it was found that if the detection accuracy is within $0.25 \mu m$, a splice loss standard of 0.2 dB can be guaranteed with 99% reliability. With the differential interference contrast microscope [Haibara et al. 1983], a core-axis alignment with a transverse offset below $0.3 \mu m$ can be achieved.

For linearly birefringent polarization-maintaining fibers [Rashleigh and Stolen 1983], the principal axes of the fibers to be connected also have to be aligned [Sasaki et al. 1982; Noda et al. 1983; Suzuki M. et al. 1983; Cancellieri et al. 1985; Kato 1985; Ishikura et al. 1986a] in order to avoid polarization crosstalk, i.e. a reduction in fiber bandwidth (Sect. 11.6.12).

After core alignment, the fiber ends must be connected permanently either by fusion, or mechanically. The fusion process is similar for multimode and single-mode fibers. A localized heating applied at the interface between the two butted, prealigned fiber ends causes the fiber ends to soften and to fuse together. Fusion splices can be made by heating the fiber ends in an electric arc, in a gas flame, or in a CO_2 laser beam.

The first and currently most important splicing technique is electric-arc fusion, both for multimode fibers [Bisbee 1971a, 1976; Kohandzadeh 1976; Khoe 1979b] and for single-mode fibers [Dyott et al. 1972].

The principle of single-mode fiber fusion splicing, the problems which may be encountered, and different possible types of equipment have been reviewed [Richin and Moronvalle 1983; Inada et al. 1986].

Refinements of the basic arc-fusion process include [Dalgleish 1980]: rounding of the fiber ends with a low-energy arc discharge prior to fusion to avoid bubble formation [Hirai and Uchida 1977]; controlled inward movement of the fibers during fusion to prevent necking at the fusion joint [Hatekayama and Tsuchiya 1978b]; automation of the prefusion and fiber movement during fusion [Sakamoto et al. 1978]; optical viewing arrangements to simplify fiber prealignment [Franken et al. 1979], and the incorporation of a protective package after fusing [Pacey and Dalgleish 1979].

The surface tension of the melted silica glass provides a high degree of self-alignment of the *cladding* axes during the splicing process. For single-mode fibers with very small core eccentricity, this effect can be used for core alignment [Hatekayama and Tsuchiya 1978a, 1979]. However, for single-mode fibers with larger values of the core eccentricity, the surface tension effect during the fusion process is a serious problem [Inada 1984]. If the fiber cores are aligned by power monitoring or direct core monitoring prior to fusion, the surface tension will cause the cores to be misaligned. Misalignment due to surface tension can be avoided by using the so-called prefusion method [Kato et al. 1982c; Hooper et al. 1985] in which the fibers to be spliced are adjusted to a small gap using micropositioners. The fiber endfaces are prefused, pressed together gently, and then fused with a continual arc discharge. By a narrow, quick fusion method the surface tension effect must be made as small as possible to avoid core metamorphosis (deformation) and core axis misalignment.

In a fusion splice, in addition to the losses mentioned before, there are also losses caused by core deformation during fusion. Core deformations occur to some degree in all fusion-splicing processes and arise from three major sources [Hooper et al. 1985]:

1. Surface-tension forces act to align the claddings of the fiber at the splice point. Thus the effect of any initial cladding misalignment is to introduce a shear force across the faces of the fibers during fusion.

2. An endface angle deviating from the ideal 90° also causes core deformation. When a single-mode fusion splicing machine is set up and operated correctly, end-angle effects are probably the dominant cause of core deformation.

3. Contamination of the fibers by foreign material in the splice region, for example dirt or primary coating material not being completely removed, is an often underestimated contribution to poor-quality fusion splicing.

Splice losses caused by mutual axis movement due to surface tension in a direction perpendicular to the fiber axes have been analyzed statistically [Kawata 1984].

The loss at fusion splices depends on several variables: width of the electrode gap, prefusion time, operating temperature, fusion time, applied pressure stroke, etc. The conditions for minimum splice losses have been studied by several authors [Kato et al. 1982c, 1983; Das A.K. and Bhattacharyya 1984; 1985]. There is an optimum fusion time which depends on the fusion temperature [Krause J.T. et al. 1985, 1986]. When the fusion time is too large, the insertion loss increases, because the dopants begin to diffuse. The index profile is changed (profile blurring) and the waveguiding ceases in the hot zone.

The melting temperatures are different in GeO_2-doped core fibers (about 1.530°C) and pure silica core fibers (about 1.730°C). Nevertheless, the splice loss of fibers joined by the fusion method is negligibly affected by differences in melting temperatures. Average splice losses smaller than 0.1 dB can be obtained when splicing pure silica core fibers to GeO_2-doped core fibers [Asano et al. 1986].

Several automatic single-mode splicing machines have been developed that can splice single-mode fibers with an average insertion of about 0.1 dB or less [Kato et al. 1982b,c, 1984a,b; Tardy et al. 1983; Suzuki M. et al. 1983; Tanifuji et al. 1983; Uenoya 1984; Uenoya et al. 1985; Yamada T. et al. 1986]. Splicing machines are now commerically available from several manufacturers.

Surprisingly, average splices between single-mode fibers are no more lossy than multimode fiber interconnections, despite the much smaller core diameter. With fusion splicing, the median splice loss is less than 0.1 dB, about half the value typical for multimode fibers [Kapron 1984a].

Fusion splices can also be made in a gas flame [Jocteur and Tardy 1976]. When a fiber is heated during fusion splicing, degradation in mechanical strength occurs unless the fiber surface is protected from ambient moisture or absorbed OH. Through the use of a H_2-Cl_2-O_2 flame, it is possible to retain the original high strength of the as-drawn fiber in fusion splices [Krause J.T. and Kurkjian 1985; Krause J.T. 1986].

A third method for making fusion splices is to heat the fiber ends in a CO_2 laser beam [Fujita et al. 1976; Egashira and Kobayashi 1977]. CO_2 laser heating assures splicing with no danger of contamination. Moreover, it enables array splices to be made [Kinoshita and Kobayashi 1979]. Single-mode fusion splices made with a CO_2 laser resulted in average splice losses of 0.18 dB [Rivoallan et al. 1983].

By biconically tapering the splice region after fusion of the fiber ends, splices between fibers with large core eccentricities can be made without having to align the fiber cores [Ishikura et al. 1986b]. These biconically tapered fusion splices have larger losses than splices with core alignment, but are very simple and reliable and therefore suitable for short subscriber single-mode fiber optical lines.

There are a number of techniques for making mechanical splices in which the fibers are held in alignment by some mechanical means. Mechanical optical fiber splices have been reviewed by Miller C.M. [1986b].

The first low-loss splices between single-mode fibers were made by sandwiching the fiber butt-joint between two methyl methacrylate plates with one of them having an embossed groove [Someda 1973] or between a glass V-groove block and a flat glass plate [Tynes and Derosier 1977].

A simple technique for mechanically splicing two single-mode fibers [Nawata et al. 1979; Nawata 1980] utilizes a high-precision ceramic capillary with bore diameter slightly larger than the fiber diameter. Ceramic capillaries can be made with inner diameters increasing in steps of 0.5 μm. One can thus select a capillary with an inner diameter slightly larger than the fiber cladding diameter. The clearance between the inner diameter of the capillary and the fiber cladding must be below one micrometer. The ends of the bore in the capillary are conically tapered for easy installation of the fiber ends. A pinhole is machined half-way into the capillary in order to remove any air and residue index-matching liquid or adhesive. The fiber ends are bonded in the capillary with an adhesive. Since the fiber claddings are aligned by this technique, it works only for precision fibers with small core eccentricity. Average losses below 0.1 dB have been achieved. In another mechanical splice [Woods 1984a], the fiber ends are aligned and bonded between four slightly bent glass rods. The reported average loss is 0.8 dB.

An epoxy bonding technique has been developed [Deveau et al. 1983; Kaiser 1983] which yields average splice losses of 0.03 dB even for fibers with standard tolerances. The technique involves the active alignment of precisely cleaved fibers and the subsequent bonding of the fiber ends with an index-matched, UV-curable cement in a glass reinforcing tube.

The double-eccentric core technique originally proposed for single-mode connectors [Guttmann et al. 1975a] has also be used for making mechanical splices [Miller C.M. and DeVeau 1985]. The method relies on precision glass ferrules installed on the fibers to be jointed and a triangular alignment sleeve. Built-in eccentricities of the bore within the glass ferrules and an offset built into the alignment sleeve allow relatively large rotational movements to provide small movements of the fiber cores relative to one another. The cores are aligned for minimum splice loss using a local detector that provides a signal proportional to splice loss. In a field test with 15 849 mechanical splices of this type, the mean splice loss was only 0.035 dB [Miller C.M. 1986b]. The time required to complete one splice is about 10 minutes.

In a similar technique, the fiber ends are secured in two ferrules with cyanoacrylate adhesive [Stewart J.H. and Hensel 1976]. The ferrules are mounted in an alignment jig, and the fiber cores aligned for maximum transmission. Thereafter, the ferrules are bonded with cyanoacrylate adhesive or rapid-setting epoxy resin. In a more recent version of this technique [Hardwick and Reynolds 1985], the uncoated fiber end is attached with an adhesive into a precise capillary hole in a cylindrical glass plug, a so-called glass terminus. After UV-curing of the adhesive, the end of the terminus is polished flat. Two termini are butt-jointed in a splicing machine. Using local detection, the microprocessor-controlled splicing machine aligns the fiber cores automatically to 0.1 μm. In the final aligned position, the glass termini are bonded together. Using these so-called bonded splices, average losses of less than 0.05 dB have been achieved in the field.

Cables used for communication between offices in large urban areas often contain many fibers. For this type of application, it is very desirable to have techniques for splicing many fibers simultaneously in order to save time.

The simultaneous mechanical splicing of an array of multimode fibers in a ribbon cable has been accomplished using a series of stacked, etched, silicon chips [Miller C.M. 1978]. This principle has recently also been applied to single-mode fibers [Gartside and Baden 1985; Baden et al. 1986; Miller C.M. 1986b]. Two silicon chips encase and align twelve single-mode fibers. Alignment of the fibers within an array is achieved by etched grooves. Each array face is ground and polished to a mirror-like finish. To make a splice, two arrays are sandwiched between two steel-reinforced silicon chips which have a profile inverse to that of the grooves in the array. An index-matching gel is applied between the two abutting arrays. With over 30 000 single-mode fibers array-spliced in the field, a mean splice loss of < 0.4 dB has been reported.

In another mechanical large-scale splicing technique, ten [Le Noane et al. 1985], or twelve [Hardwick and Davies 1986] single-mode fibers are simultaneously butt-jointed in continuous, parallel grooves engraved in the plane surface of a plastic substrate. Mean splice losses ≤ 0.5 dB have been reported [Hardwick and Davies 1986].

Fusion splicing of single fibers has advantages over other methods in terms of the mechanical reliability and loss stability of the spliced fibers. However, with an arc, it is difficult to uniformly heat several fibers at the same time. Therefore, only recently [Tachikura and Kashima 1984], mass splices of multi-mode fibers have been performed using a high-frequency discharge. Using this mass-splice machine designed for graded-index multimode fibers, array-splices of five single-mode fibers have been made [Katsuyama et al. 1985a,b]. Without any power monitoring to align the fiber axes, a mean splice loss of 0.19 dB has been achieved. For a fused mass splice of ten single-mode fibers, an average insertion loss of 0.24 dB has been reported [Osaka et al. 1986].

After completion of a single splice, a mass splice, or a spliced link, it is important to measure the splice losses obtained. For obvious reasons, the destructive cut-back method cannot generally be used.

However, there are several nondestructive methods for determining the splice loss. Some methods for splice loss evaluation have been reviewed by Engel and McNair [1986]. In the "presplicing method" [Kato et al. 1984a], the splice loss is obtained with an accuracy of ± 0.02 dB from four optical power measurements. An LED is used as a light source at the input end of the primary fiber. One first measures the optical power P_1 leaving the end of the primary fiber. A provisional splice is then made, and the power P_2 received at the far end of the second fiber is measured. Third, the second fiber is cut back at a point 1 m behind the presplice and the output power P_3 is measured. Finally, a permanent fusion splice is made and the output power P_4 at the far end is measured. By taking power ratios, both the unknown loss of the presplice and the unknown loss of the second fiber can be eliminated to obtain an equation for the loss of the permanent splice in decibels:

$$a = 10 \log\left(\frac{P_1 P_2}{P_3 P_4}\right) \quad . \tag{12.17}$$

P_1/P_3 is the ratio of the power readings taken near the splice and P_2/P_4 is the ratio of the readings taken at the far end. Thus, the power meters used in the manhole and at the far end need not be calibrated absolutely.

The splice loss can also be determined by placing the splice in an integrating cube scattering detector and by measuring both the power P_r radiated from the splice and the power P transmitted through the splice [Tynes and Derosier 1977]. For small losses, the splice loss a is given by

$$a = 4.34 P_r/P \quad . \tag{12.18}$$

The losses of the splices in a fiber link can be measured with an OTDR (Sect. 13.4.3). The use of OTDR for field measurements of splice loss is invaluable since such measurements are single-ended and do not require accurate measurements of the fiber loss in order to determine splice loss [Payne D.B. et al. 1982].

The splice loss can also be accurately estimated from the difference in the locally monitored powers before and after splicing if an index matching fluid is used to avoid power fluctuations due to multiple reflections [Kato et al. 1984b]. When splicing with an electric arc discharge, the fluid is completely evaporated. The fluorescence of GeO$_2$-doped fiber cores can be used to inspect a spliced joint with a microscope [Khoe et al. 1983a].

The splice losses caused by one splice defect only, e.g., core offset, angular misalignment, core bending, or profile blurring, have a characteristic wavelength dependence. This fact can be utilized to diagnose the cause of residual splice loss [Engel 1986]. When the splice loss is mainly due to transverse offset it decreases with increasing wavelength since the spot size increases with wavelength [Dreyer et al. 1986].

12.2.5 Connectors

Connectors are demountable fiber connections whereas splices are permanent joints. Connectors are used where it is necessary or convenient to easily disconnect and reconnect fibers, e.g. at optical transmitters and receivers, or in fiber test systems. Practical single-mode fiber systems require connectors for terminating components, testing, and maintenance.

Connectors must exhibit a low insertion loss. Since the fiber attenuation may be as low as 0.2 dB/km it is obvious that the connectors' insertion loss can severely limit the distance between repeaters.

Connectors for optical fibers are more difficult to make than splices. The tolerances described for splices must still be maintained, but it is now necessary to accomplish the alignment repeatedly and reproducibly.

In order to consistently achieve adequate alignment accuracy most single-mode fiber connectors are installed in a factory or work center rather than on site.

Compared to multimode connectors, single-mode connectors have the following difficulties [Minowa et al. 1982]:

1. Due to the considerably smaller diameter of the core in comparison to multimode fibers, the excess loss caused by connector alignment errors becomes quite noticeable;
2. Center alignment accuracy of 1 μm or less is necessary to obtain a connection loss below 1 dB. This requires a centering technique with submicron precision;
3. With a high-precision center alignment, deterioration of connector performance may occur due to environmental factors such as temperature or due to mechanical wear after many mating and disconnect cycles;
4. When a laser diode is used as a light source, pulse waveform distortion may be caused by the reflected wave generated at the connecting point;
5. Relative to multimode fibers, the deviation of single-mode fiber structure parameters has a considerable effect on connection loss.

The loss of a butt joint between a multimode and a single-mode fiber is of the order of 15 dB (Sect. 9.7). Thus, in general, multimode connectors cannot be used in single-mode systems. However, since the insertion loss of a joint between a single-mode fiber and a multimode fiber is in fact small (Sect. 9.7), the simpler multimode connectors can also be used advantageously at the receiving end of a single-mode system [Kaiser 1983]. Pulse broadening in the short multimode fiber connecting the end of the single-mode link to the photodetector is negligibly small.

Several reviews of single-mode fiber connectors have been published [Young W.C. 1984; Khoe et al. 1984a; Roßberg 1985; Cannon 1986].

Connector designs can be grouped into butt-jointed and expanded-beam types. In butt-joint connectors, the fiber ends are fixed in cylindrical ferrules or conical plugs which are aligned by inserting into a sleeve. A ferrule is a precisely machined cylindrical part which has a small hole along its axis that is only slightly larger than the fiber itself. For small insertion loss, the axis of the fiber core must coincide with the axis of the ferrule or plug to within a few tenths of a micrometer. In the case of butt-jointed connectors, the radial offset must be smaller than 0.15 times the spot size (Sect. 10.1) in order to obtain insertion losses smaller than 0.1 dB, i.e. $0.8\,\mu$m for $w_G = 5\,\mu$m. Butt-jointing two fibers with $0.4\,\mu$m core eccentricity can, in the worst case, cause a core offset of $0.8\,\mu$m. Because of the relatively wide divergence angle of the beam radiated from the fiber end, angular alignment of butt-jointed connectors is not difficult (Sect. 9.4).

For precision fibers with a very small core eccentricity and an accurate cladding diameter, the connector can be made by

1. placing the fiber in a precision ferrule assembly [Shimizu et al. 1979; Tsuchiya et al. 1979; Suzuki N. et al. 1979; Nawata 1980; Nawata and Suzuki 1982; Minowa et al. 1982; Yoshizawa et al. 1982; Young W.C. and Curtis 1983; Saruwatari et al. 1983; Sankana et al. 1986a];
2. moulding a plastic ferrule concentrically to the *cladding* around the fiber [Cheung and Denkin 1980; Young W.C. et al. 1980; Cheung 1981].

For the more economical standard fibers with larger core eccentricity, it is necessary to center the ferrule relative to the core axis by

1. double eccentric tubes [Börner et al. 1972; Guttmann et al. 1975a; Tsuchiya et al. 1977; Shimizu and Tsuchiya 1978, 1979; Tsuchiya et al. 1979],
2. using a special aerostatic lathe [Khoe 1979a; Khoe et al. 1981b; Khoe 1982a,b].
3. grinding the ferrule coaxially to the fiber core [Kurochi et al. 1977; Morimoto et al. 1984].

These so-called core-centered ferrules do not require a precision fiber, precision parts in the ferrule, or an accurate matching between them. For cylindrical ferrules, a diameter of 2.5 mm is commonly used in Japan and in some other countries [Khoe et al. 1984a].

The ferrules or conical plugs at the ends of the two fibers to be connected are inserted into a cylindrical or biconical receptacle where they are aligned and retained. Most alignment techniques make use of an elastically deformable alignment sleeve.

Ideally, the alignment technique should allow frictionless ferrule insertion (reducing wear), and provide accurate alignment, protecting the mated ferrules against mechanical vibration and shock. These conflicting requirements have been met by using an alignment tube made from a material that exhibits the shape memory effect [Mallinson 1983]. For insertion of the ferrules, the inner diameter of the tube is simply increased by heating.

In a butt-joint connector, there is usually an air gap between the fiber end-faces and the light passes through two air-glass interfaces. A single interface has a power reflection factor of 3.5% corresponding to a transmission loss of 0.155 dB (5.131). The reflection factor of a connector is not simply two times 3.5% corresponding to an insertion loss of 0.31 dB. Because of interference effects, the reflection factor and the transmission factor of a connector depend on the separation between the fiber endfaces. With varying width of the gap between the fiber ends, the power reflection varies periodically between 0% and 13% (Fig. 5.21). Interference effects in fiber connectors have been investigated both theoretically and experimentally [Wagner and Sandahl 1982; Young W.C. et al. 1986b; Shah et al. 1987]. Since the endface separation varies with temperature, the transmitted and reflected power will fluctuate.

Reflections from connectors may affect the longitudinal mode spectrum and noise characteristic of laser transmitters and cause system degradation. Moreover, interference effects in connectors can introduce uncertainties into measurement procedures. It is thus important to minimize the reflection factor at connectors.

Methods for decreasing the Fresnel reflection from a single fiber endface have been discussed in Sect. 12.1.2.

Reflections from a connector can be reduced by

1. inserting an immersion fluid between the fiber ends. However, index matching liquids may cause handling problems, limited environmental stability, and aging of optical characteristics;
2. inserting a thin transparent film (index-matching button) [Tsuchiya et al. 1977] or silicon resin [Sankawa et al. 1986b] between the two fiber plugs or fastening a thin film to the plug endface [Suzuki N. et al. 1979; Nawata 1980];
3. physical contact of the fiber endfaces [Esposito 1979; Young W.C. et al. 1980; Young W.C. 1984, Suzuki N. et al. 1986]. Concave polishing of the fiber endfaces decreases multireflection and also contributes to a decrease in loss fluctuations by mating and remating, decreasing the wavelength dependence of the connection loss, and stabilizing the temperature characteristics [Minowa et al. 1982].
4. tilting the fiber endfaces to avoid coupling of the reflected beam back into the LP_{01} mode. In this scheme, the risk of end face damage is low as the fibers are not brought into physical contact. Insertion and return losses of 0.7 dB and 38 dB, respectively, have been reported [Rao and Cook

1986]. Introducing an angle between the fiber endfaces [Young W.C. et al. 1986b] does not reduce the Fresnel loss, but minimizes power fluctuations due to interference effects in the air gap [Shah et al. 1987].

In practice, there will always be a residual transverse offset of the core axis relative to the ferrule axis. This causes the insertion loss of a connector to vary periodically when one ferrule is rotated in the connector sleeve [Shimizu and Tsuchiya 1978].

The butt-jointed type is by far the most widely used connector and a wide variety of detachable single-mode connectors is now commercially available. The average insertion losses reported for butt-joint single-mode connectors are in the range of 0.13 dB [Sugita et al. 1986] to more than 1 dB.

Biconical single-mode connectors have been made [Young W.C. and Curtis 1983; Kaiser 1983] which have average losses of 0.23 dB if the fibers ends are in contact.

For obtaining very low losses, adjustable connectors have been made in which the fiber cores can be actively aligned. The fiber ends are fixed eccentrically in ferrules [Guttmann et al. 1975a; D'Auria et al. 1983]. The two ferrules are mounted in a V-groove with parallel but offsett axes. By rotating the ferrules around their axes, two angular positions can be found for which the core axes are aligned.

By making the ferrules from glass, the power lost at the connector which is radiated into the receiving ferrule can be detected and used to minimize the connector loss [Vucins 1977].

There is already one butt-joint connector available for use with more than one fiber [Satake et al. 1986]. Five fibers are embedded in a single plastic moulded ferrule with 6 mm diameter. The distance between the fiber axes is only 0.3 mm. The ferrules of a mating pair are aligned by two guiding pins. The average insertion loss is 0.5 dB.

As well as the butt-joint connectors considered so far, there are also expanded-beam connectors. These connectors use a first lens to expand and collimate the beam emitted by the first fiber (Sect. 12.2.1) and another lens to focus the beam onto the front face of the receiving fiber. The relative positions of each lens and its corresponding fiber are mechanically fixed. The two beam expanders are aligned and secured by a cylindrical receptacle. Since the beam diameter between the lenses is much larger than for butt-joints (Fig. 12.2), an expanded-beam connector is far less sensitive to lateral offsets (9.5), but is more sensitive to angular misalignments (9.8, 10).

Lens connectors have the following advantages:

1. The required accuracy in lateral adjustment is smaller than for butt-joint connectors.
2. The endface of the fiber is located within the connector and cannot be inadvertently damaged.
3. A dust particle on the surface of the lens or on a glass window absorbs only a very small fraction of the power transmitted by the wide beam.

The same particle located on the core endface would cause large power losses.

4. The cleaning of lens connectors is easy; it is even easier then cleaning electrical contacts.
5. The connector is very rugged and suited to use in the field, and its insertion loss is very reproducible.

On the negative side, however, lens connectors are bulkier than butt-joint connectors.

For the lenses, one can make use of separate ball lenses [Nicia 1978; Nicia and Tholen 1981; Masuda and Iwama 1982]. However, there are six glass-air interfaces which cause reflection losses. For obtaining small connector losses, the six glass surfaces must be antireflection-coated.

The number of reflecting glass-air interfaces can be reduced by attaching the fiber end directly to the plane endface of a graded-index or plano-convex homogeneous rod lens. Figure 12.2 shows the beam contours in an expanded beam connector using two graded- index lenses.

By applying a droplet of high-purity UV curing adhesive to the interstice between the fiber end and the rear face of the rod lens, the overall length of the lens can easily be optimized [Mallinson and Warnes 1985].

Alternatively, the lens can be connected permanently to the fiber in the form of a glass bead [Payne D.B. and Millar 1980; Millar and Payne 1980], or a glass asphere is cemented to the fiber [Brenner 1986].

The insertion loss of single-mode expanded-beam connectors using ball lenses, plano-convex rod lenses, and graded-index rod lenses has been analyzed by van der Veeken [1983].

Because of the large beam diameter, the beam diverges only very slowly, and the axial distance between the two lenses can be many millimeters before gap losses become significant. Thus, one can insert into the collimated wide beam devices such as half-silvered mirrors for beam splitting, full mirrors as switches, interference filters for wavelength selection, neutral density filters as optical attenuators or isolators for reducing optical feedback [Nicia and Rittich 1981; Nicia and Tholen 1981].

The insertion loss of lens connectors is comparable to that of butt-joint connectors. Average losses of 0.7 and 0.54 dB have been reported [Nicia and Tholen 1981; Masuda and Iwama 1982].

12.2.6 Fiber Tapers

Conical fiber tapers are fiber sections in which the fiber diameter changes continuously and monotonically in the longitudinal direction. Biconical tapers consist of a contracting tapered region of decreasing fiber diameter followed by an expanding taper of increasing diameter.

Tapered single-mode fibers are employed as gradual transitions between dissimilar fibers (Sect. 9.1), as beam to fiber couplers (Sect. 12.2.1), in laser to fiber couplers (Sect. 12.2.2), in fiber to integrated optic channel waveguide couplers (Sect. 12.2.3), or in fused biconical-tapered directional couplers (Sect. 12.2.11).

Wave propagation in tapered fibers has been analyzed theoretically by a number of authors [Marcuse 1970b; Snyder 1970a,b; Sporleder and Unger 1979; Tzoar and Pascone 1981; Baets and Lagasse 1982; Arnold J.M. and Felsen 1984; Casperson and Kirkwood 1985; Suchoski and Ramaswamy 1986; Marcuse 1987]. In the analysis of biconical tapers, one assumes either a parabolic [Burns et al. 1985, 1986] or a sinusoidal shape [Love and Henry 1986] for the core radius $a(z)$ as a function of longitudinal coordinate.

Since the spot size of the fundamental fiber mode is a minimum near the operating wavelength, the spot size can be increased either by increasing or by decreasing the core diameter along the fiber (Sect. 12.2.1). Usually, it is simpler to decrease the core diameter.

Fiber tapers can be made very simply: a fiber section is heated by a resistive filament, in a flame, or in an electrical arc to the softening point of the glass and pulled by applying axial tension. By this method, one obtains a biconical taper. Usually, the optical power transmitted through the fiber is monitored during the process of tapering. By breaking a biconical taper at the taper waist, one obtains two conical fiber tapers.

The local taper angle $\Omega(z)$ is the angle between the fiber axis and the intersection between the core-cladding boundary and a plane containing the fiber axis. It can be calculated as the derivative of the core radius $a(z)$ with respect to the longitudinal coordinate z:

$$\tan \Omega(z) = da(z)/dz \quad . \tag{12.19}$$

A taper is called adiabatic, if the taper angle is very small and changes very slowly along the fiber length.

For adiabatic tapers, a fundamental fiber mode incident on the taper input slowly changes its parameters such as phase constant and mode field diameter. The power remains in the fundamental mode; coupling to cladding modes or radiation modes is negligibly small [Marcuse 1987]. At a certain location z along the taper, the parameters of the LP_{01}-mode are approximately those of a fundamental mode propagating in a nontapered uniform fiber with a core diameter which equals the local value $a(z)$ of the core diameter in the taper.

For tapers in which the taper angle Ω or its derivative $d\Omega/dz$ are not very small, the incident fundamental mode loses power by radiation or by coupling to the HE_{12}-cladding mode [Snyder 1965, 1970a,b; Tzoar and Pascone 1981; Felsen 1986].

In a quickly contracting taper, the measured mode field diameter increases more slowly than expected from a theory in which the local value of the core radius is used to calculate the local value of the spot size [Keil et al. 1984a]. The reason for this is that the core becomes too thin to guide the wave beam (Sect. 2.2). In the taper, the beam cannot widen faster than it does due to diffraction when propagating in a homogeneous medium. Thus, in a strongly contracting taper, much guided power is converted into radiated power.

One is usually interested in achieving minimum insertion loss for a taper. Therefore, radiation and mode coupling must be avoided by choosing a slow taper. A simple criterion has been proposed for minimizing taper losses [Stewart and Love 1985; Love and Henry 1986]: The local value $\Omega(z)$ of the taper angle must be smaller than a maximum value $\Omega_{max}(z)$:

$$\Omega(z) < \Omega_{max}(z) = \frac{a(z)}{2\pi}[\beta_1(z) - \beta_2(z)] \quad , \tag{12.20}$$

which is determined by the difference between the local values of the phase constants $\beta_1(z)$ and $\beta_2(z)$ of the HE_{11} and HE_{12} modes. Equation (12.20) requires the taper angle to be smaller than the ratio of the core radius and the beat length (12.30) between the HE_{11} and HE_{12} modes. For small values of the normalized frequency $V(z)$, tapers have to be extremely slow [Nelson A.R. 1975; Baets and Lagasse 1982].

If one wishes to obtain a beam transformer of minimum length, a taper is not the optimum solution. Faster beam widening than in a taper can be obtained in a homogeneous or a defocusing medium (Sect. 9.1).

Since the local value of V is reduced in proportion to the fiber diameter, in very thin tapers, the fundamental mode field extends into the medium surrounding the cladding. Therefore, tapered single-mode fibers are devices that can be highly sensitive to external index variations [Lacroix et al. 1986a,c].

When a dispersion-flattened quadruple-clad fiber (Sect. 5.8) is tapered to a small diameter, the fundamental mode is no longer guided [Khoe et al. 1985], since it has a finite cutoff frequency (6.22). Therefore, a tapered pigtail between a laser diode and a dispersion-flattened fiber may not be made from the quadruple-clad fiber itself, but instead from a special matched cladding fiber, the field diameter of which is matched to the smaller field diameter (about $6\,\mu m$) of the dispersion-flattened fiber.

In a fast biconical taper, part of the power which is coupled into the HE_{12} mode in the contracting taper part is coupled back into the fundamental mode in the expanding part. Since the fundamental mode and the HE_{12} modes have different phase constants, mode interference causes the power transmitted through the biconical taper to depend on pull length [Burns et al. 1986; Lacroix et al. 1986c] and on wavelength [Cassidy et al. 1985, 1986; Lacroix et al. 1986a,d].

A slow biconic taper made of a depressed cladding fiber shows a transmission loss which oscillates rapidly with increasing taper length, whereas a similar double taper made of a matched cladding fiber shows only a slight monotonic decrease in the transmitted power.

This effect can also be explained by mode interference. In the depressed cladding fiber, waves can be guided both by the core and the outer cladding, which forms a dielectric tube waveguide [Boucouvalas and Georgiou 1985a,b;

Burns et al. 1986]. Normally, the core mode will not interact with the tube mode. But when the fiber diameter is reduced by tapering, the fields of the core and tube modes will begin to overlap. The modes will be coupled at those two points along the double taper where their phase velocities match (Sect. 7.4).

At the first point, power is transferred from the core mode to the tube mode. At the second point, power of the tube mode is reconverted into core power. Depending on the relative phase of the two modes at the second point, which depends on the taper length, the net power of the fundamental mode may be increased or decreased. Behind the biconic taper, the power guided by the tube waveguide will be absorbed in the coating. Thus, at the end of the fiber, one detects only the power guided in the core, which fluctuates, depending on the length of the tapered section.

Due to the absence of a secondary waveguide, these oscillations do not occur for a matched cladding fiber biconical taper when the index of the external medium is matched to the cladding index. However, the effect must be taken into account, when one tries to fabricate biconical-taper directional couplers (Sect. 12.2.11) from depressed cladding fibers.

Biconical fiber tapers have been applied for reducing the splice loss between fibers with different spot sizes and for obtaining splices between fibers with large core eccentricities which do not require core alignment [Ishikura et al. 1986b].

12.2.7 Optical Rotary Joints

For the purpose of optical communication between stationary and rotating parts, optical rotary joints are required. Rotary joints can be made by using the principle of the expanded beam connector (Sect. 12.2.5). The fiber axes coincide with the axis of rotation. One half-connector is stationary, while the other rotates. So far, rotary joints have been reported only for multimode fibers [Williams J.C. et al. 1984; Althouse et al. 1984; Shi et al. 1985; Ito and Numazaki 1985; Harstead et al. 1986]. However, utilizing the same principle, it should also be possible to make rotary joints for single-mode fibers.

12.2.8 Cladding Mode Strippers

In single-mode fiber measurements, in order to avoid errors caused by cladding modes (Sect. 6.5), these modes must be stripped from the fiber in front of the object to be investigated and in front of the detecting device.

A cladding mode stripper is a device that enhances the conversion of cladding modes to radiation modes. Cladding modes are stripped by avoiding reflections at the cladding-coating or coating-air interfaces.

If the index of the coating is equal or slightly larger than that of the cladding, the reflection at the outer coating surface can be reduced by immersing the coated fiber in a liquid with a higher index than the coating. The attenuation of cladding modes in a fiber immersed in a medium whose refractive index is higher than that of the core glass has been studied theoretically

[Arnold J.M. 1977a]. A list of 11 index matching liquids having refractive indices between 1.46 and 1.738 has been published by [Dakin et al. 1972]. Several hundred immersion liquids with refractive indices between 1.3 and 2.1 are commercially available. The liquids with $n > 1.8$ are both toxic and corrosive!

A method for measuring the refractive index of immersion liquids in the 500 to 1700 nm spectral range and experimental results for several liquids have been reported [Cooper 1982, 1983]. When an immersion oil is used in a cladding mode stripper, its loss data are unimportant, it is only required that its index is equal to or slightly higher than that of the fiber cladding. However, there are other applications of immersion liquids, in which the attenuation of the light in the liquid must be as low as possible, e.g. in fiber connectors to reduce Fresnel reflections or in the refracted near-field method for measuring the refractive-index profile (Sect. 13.10.2).

Absorption data for a variety of immersion liquids for long wavelength (1.2–1.6 μm) applications have been published [Melliar-Smith et al. 1980]. With respect to attenuation, Glycerin is a poor choice for long wavelengths. Halo-carbon polymers are essentially transparent between 800 and 2000 nm. Liquid paraffin is at least as good as silicone oil as a matching fluid, since its index is better matched to that of silica and because it is cheaper [Fox and Dennis 1980].

A simple cladding mode stripper can take the form of a flat plate on which a fiber loop is placed in a drop of immersion oil. If the index of the coating is smaller than that of the cladding, the primary fiber coating must be removed from the section inserted in the mode stripper in order to avoid total internal reflection at the cladding-coating interface.

Cladding modes can also be attenuated by painting the outside of the fiber black over a length of about 50 mm [Wang C.C. et al. 1983; Ross 1984].

Cladding modes usually become negligible after a few tens of meters due to their high loss. Furthermore, many modern fibers have lossy high-index coatings which act as effective mode strippers [Danielson B.L. 1985].

12.2.9 Attenuators

Optical attenuators are classified into fixed attenuators for adjusting the optical power level in repeater sections, and variable attenuators which are used e.g. for measuring the bit error rate as a function of the received optical power.

Fixed attenuators are commercially available. They consist of a flat plate made from a glass with high absorption losses. These so-called neutral density filters can also be used to attenuate the fundamental fiber mode by cutting the fiber, collimating the radiated beam with a first lens, inserting the plate, and focusing the beam onto the input end of the second fiber. These attenuation plates can also be inserted between the two plugs of an expanded beam connector.

Fixed attenuators can also be produced by utilizing the loss due to a gap between the fiber ends (Sect. 9.3). By selecting an appropriate distance between the plug ends of a butt-joint connector, one can obtain the desired amount of attenuation (in the range of about 3–30 dB) [Minowa et al. 1982]. Due to the wavelength dependence of the mode field diameter, the attenuation changes with wavelength.

For variable attenuators, high precision and a wide range of attenuation are required. The wide range can be obtained by combining fixed attenuation plates (in 10 dB steps, 0 . . . 50 dB) with a continuously variable rotatable attenuation plate (0 . . . 15 dB) [Minowa et al. 1982]. The attenuators are inserted into the free Gaussian beam between two lenses. Another construction of variable attenuator [Anslow et al. 1984] consists of two precisely aligned single-mode fibers, immersed in an index-matching liquid, with a longitudinal separation which can be varied by a servo-operated lead-screw to provide attenuations up to 40 optical decibels.

Instead of varying the gap between the fiber ends, one can also vary the tilt angle [Schulte 1983]. The ends of two single-mode fibers are constrained in a V-groove etched in the surface of a small block of silicon. Motion of the V-block changes the angle between the two fibers in a controlled manner. The optical loss from one fiber to another can be varied between 0.2 dB and several tens of decibels. In order to obtain a compact variable attenuator, one can simultaneously change the gap width and the tilt angle between the fiber axes by moving one fiber end in a curved V-groove [Masuda 1980].

12.2.10 Power Splitters and Combiners

The components described so far in Sects. 12.2.4–9 are optical two-port devices, that is, the signal enters via an input fiber and exits via an output fiber. There are functions within point-to-point systems and distribution systems which require devices with more than two ports to divide or combine waves.

Examples of multiport couplers are power splitters and combiners, directional couplers, star couplers, and fiber taps [Dalgleish 1980]. These optical multiport couplers will be described in Sects. 12.2.10–13. Wavelength selective optical multiports like wavelength division multiplexers and demultiplexers are described in Sect. 12.3.3.

A power splitter sends a single transmitted signal to two receivers. Power combiners can be used to superimpose two signals onto the same fiber.

We begin by considering two types of power splitter.

A micro-optic 3-dB power splitter can be made from two graded-index lenses bonded to a half-mirror in between. Figure 12.8 representing a micro-optic multi-demultiplexer can also be used to illustrate a beam splitter. One merely has to replace the interference filter between the two graded-index lenses by a half-mirror [Uchida T. and Sugimoto 1978].

A fiber Y-junction or T-coupler (Fig. 12.3) can be made by polishing two fibers to have semicircular cross-sections at one end [Weidel and Gruchmann 1979; Rittich 1981]. The semicircular fiber ends are cemented together so that the two semicircles form a circle, and butt-jointed to a third single-mode fiber.

Fig. 12.3. Single-mode fiber Y-junction

In order to avoid excessive power losses in the power splitter caused by radiation, the angle between the two legs of the Y must be very small ($< 1°$). A method for decreasing the radiation losses of wide Y-junctions consists in decreasing the refractive index near the branching region [Hanaizumi et al. 1985]. Wave propagation in single-mode Y-junctions has also been analyzed theoretically [Henry and Love 1985].

The micro-optic and the Y-junction power splitters can also be used as power combiners. When used as splitters, it is obvious that the minimum insertion loss is 3 dB, since 50% of the incident power is coupled to one output port and 50% to the other. It is also evident that there is a minimum insertion loss of 3 dB when the micro-optic power splitter (Fig. 12.8) is used as a power combiner.

However, it is not so obvious that there is also an inevitable minimum insertion loss of 3 dB between one leg of the Y and the common port [McMahon 1975; Izutsu et al. 1982; Rediker and Leonberger 1982; Someda 1984; Goodman 1985]. When a Y-junction is operated as a power combiner, at least fifty percent of the power incident in each leg is lost by radiation.

In the four-port optical directional couplers to be described in the next section, there are negligible radiation losses. Therefore, in comparison with Y-junctions, directional couplers are to be preferred, even if use is made only of three of the four ports.

12.2.11 Directional Couplers

An optical directional coupler is a four-port junction which is the optical analogue of a microwave directional coupler.

Single-mode fiber directional couplers are required for power splitting and combining. Directional couplers can also be exploited in optical communication systems for duplexing the signal and the local oscillator waves for coherent detection, for wavelength multiplexing/demultiplexing (Sect. 12.3.3), for laser redundancy, for monitoring backscatter (Sect. 13.4.4), and as a power tap.

Optical directional couplers can be made by utilizing either the evanescent field coupling between two parallel fibers, or the principle of the classical beam splitter. The former class of single-mode fiber directional couplers is based on the power exchange that occurs between two parallel dielectric fibers when the field carried by one fiber reaches the core of the other fiber. The fundamentals of the theory of optical directional couplers have been summarized in Sect. 7.4.

A directional coupler has four ports (Fig. 7.8). The power P_1 of the wave incident into port 1 is identical to the quantity $P_a(0)$ in the analysis in Sect. 7.4. The powers P_2 and P_4 leaving the throughput fiber 2 and the coupled fiber 4, correspond to $P_a(L)$ and $P_b(L)$, respectively, where L is the length of the coupled region. The power leaving port 3 is extremely small. Because power is coupled mostly in the forward direction, the coupler is called a directional coupler.

The division of the total power P_1 between the powers P_2 and P_4 leaving the output ports may be selected by varying the distance d between the fiber axes, i.e. the coupling coefficient κ (7.42), or by varying the coupling length L.

In practice, a four-port directional coupler is characterized by the following parameters [Tekippe and Willson 1985]: The coupling (splitting, branching) ratio in percent

$$r_c = 100 P_4/(P_2 + P_4) \ [\%] \tag{12.21}$$

which often is expressed in decibels

$$a_c = 10 \log[(P_2 + P_4)/P_4] \ ; \tag{12.22}$$

the directivity in decibels

$$a_d = 10 \log(P_1/P_3) \ ; \tag{12.23}$$

and the excess or insertion loss in decibels

$$a_e = 10 \log[P_1/(P_2 + P_4)] \ . \tag{12.24}$$

I have modified the definitions slightly in order that the coupling loss, the directivity, and the excess loss become positive decibel numbers. Note that the definitions used in the literature are not uniform.

The requirements for a directional coupler are a stable splitting ratio a_c, a large directivity a_d, and a low excess loss a_e. For some applications, it must also maintain a given state of polarization.

The fabrication techniques, performance, applications, and commercial availability of directional couplers for single-mode fibers have been reviewed [Tekippe and Willson 1985].

Single-mode fiber directional couplers can be fabricated by one of three different methods: by etching, by polishing, and by fusion-tapering. In the first method [Sheem and Giallorenzi 1979a; Sheem and Cole 1979; Liao and Boyd 1981], the plastic coating is stripped from two single-mode fibers, which are then twisted together. The claddings of the fibers are etched in a small container (bottle) to a diameter close to that of the core, giving an overlap of the evanescent fields in the cladding which is sufficient for power coupling. The coupling efficiency of these so called "bottle couplers" can be adjusted after the completion of etching by turning a threaded cap to control the number of twists between the two etched fibers as well as their separation. Besides 2×2 directional couplers, 3×3 bottle-couplers have also been made [Sheem 1980].

The first bottle couplers were filled with index matching oil. In more recent versions of the etching technique, the fibers are embedded in a suitable room-temperature vulcanized silicone [Tran et al. 1981] or in a gel glass [Tran and Koo 1981; Tran et al. 1981]. Etched couplers are delicate, non-ruggedized components, which are only suitable for use under laboratory conditions. The insertion loss of bottle couplers is of the order of 1 dB.

In the second method for producing fiber directional couplers, each fiber is cemented in a fused silica block with a curved groove cut in it and part of the cladding is removed by grinding, polishing, and lapping [Bergh et al. 1980a; Parriaux et al. 1981b; Bergh et al. 1982; Digonnet and Shaw 1982b; Nayar and Smith 1983; Pleibel et al. 1983; Jaccard et al. 1983, 1984; Georgiou and Boucouvalas 1985; Mao et al. 1986]. The directional coupler is then realized by placing one block on top of the other with a thin film of index-matching liquid between the two blocks (Fig. 12.4). The coupling can be varied by sliding one block laterally relative to the other.

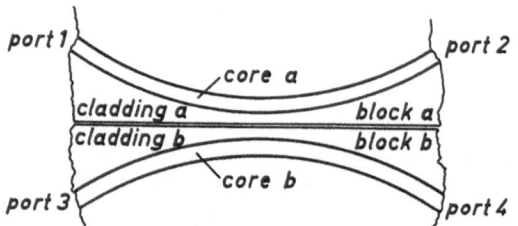

Fig. 12.4. Polished directional coupler

In a modification of the method, the fiber is embedded before polishing into a substrate glass having a low melting temperature [Beasley et al. 1983]. This method reduces the drift of coupling ratio with ageing.

Simple and accurate analytical expressions for the coupling coefficient, the effective interaction length, and the tunability of polished fiber directional couplers have been published by Tewari and Thyagarajan [1986]. A theoretical analysis of polished couplers with strong coupling where the fiber cores are touching each other, has been reported [Mao et al. 1986]. The losses of polished couplers caused by residual surface roughness and a lossy interstitial medium have been studied [Parriaux et al. 1984].

During production of a polished half-coupler, the distance between the polished surface and the core axis is an important parameter, since it determines the achievable coupling ratio. Because it is difficult to measure this distance using standard optical and mechanical instruments, it has been measured by observing the attenuation of the half-coupler when a droplet of liquid of appropriate refractive-index is placed on the polished fiber region [Digonnet et al. 1985; Nicholls 1985; Leminger and Zengerle 1985]. The influence of the refractive index of the overlay material on the attenuation and phase shift of the fundamental fiber mode transmitted through the half-coupler has been studied by using as external medium a variable-index glass block manufactured by lapping and polishing a MCVD preform [Brierley et al. 1986].

Polished-fiber directional couplers exhibit low loss, high directivity, and can be tuned to any required coupling ratio. However, elaborate fixturing is required to maintain a stable coupling ratio even for short periods of time.

Polished directional couplers made of fiber pairs with widely differing parameters couple light effectively between the two fibers only for that wavelength for which the phase velocities in the two waveguides is matched (7.54). A 3 dB bandwidth of 25 nm has been reported [Zengerle and Leminger 1986, 1987].

A polished directional coupler has also been made with a single-mode fiber in one half-coupler and a graded-index multimode fiber in the other [Whalen and Wood 1985]. Power can be coupled effectively from the single-mode fiber into the multimode fiber, but only a very small part of the power transmitted through the multimode fiber can be coupled into the single-mode fiber (Sect. 7.4).

In a polished half-coupler, the mode field is accessible and this fact can be exploited to fabricate in-line components like fiber Bragg reflectors (Sect. 12.3.1), fiber polarizers [Bergh et al. 1980b], or evanescent-field amplifiers [Sorin et al. 1983].

In the third and most important technique for making single-mode directional couplers, the fibers to be coupled are twisted, heated in an arc or a flame, and fused and tapered while the transmitted and coupled powers P_2 and P_4 are monitored. Tapering is stopped when the desired splitting ratio (12.21) is obtained [Villaruel and Moeller 1981; Kawasaki et al. 1981b; Slonecker 1982; Kawachi 1982b; Villaruel et al. 1983]. Reviews of fused single-mode coupler technology have been published by Kawasaki et al. [1983a] and Ragdale et al. [1984].

For taper-fused directional couplers, the tapering makes the core diameter and the local normalized frequency V so small that the electromagnetic field propagates in a waveguide consisting of the entire coupler cross section. The core regions then become insignificant and the coupling can be explained in terms of the coupling of two supermodes (Sect. 7.4) on a waveguide formed by the fiber cladding and the medium surrounding the fused fibers [Bures et al. 1983, 1984; de Fornel et al. 1984; Payne F.P. et al. 1985; Wright J.V. 1985; Snyder and Zheng 1985; Rodrigues et al. 1986; Chiang 1986]. The splitting ratio depends on the refractive index of any material that may be used to encapsulate the coupler for protection [Hill et al. 1984; Bures et al. 1984; Lamont et al. 1985b; Payne et al. 1986].

The technique for fabricating fused biconical-taper couplers has been described in a number of papers [Slonecker and Williams 1983; de Fornel et al. 1984; Bricheno and Fielding 1984; Slonecker et al. 1984; Georgiou and Boucouvalas 1985; Yataki et al. 1985; Bilodeau et al. 1986; Johnson D.C. and Hill 1986]. Fused biconical-taper single-mode directional couplers can be manufactured quickly and simply, with an arbitrary branching ratio. They have low excess loss (< 0.1 dB), and good thermal and mechanical stability. An automatic test facility for measuring the performance of fused directional couplers has been described by Kopera et al. [1984].

The coupling ratio of fused biconical-taper directional couplers can be tuned over a wide range by modest controlled bending of the tapered region with negligible effect upon the excess loss of the coupler [Schöner and Schiffner 1981; Kawasaki et al. 1983b]. Because of the finite cutoff frequency of the fundamental mode in depressed cladding fibers (6.22), it is difficult to fabricate fused couplers between fibers of this type [de Fornel et al. 1984]. For depressed cladding single-mode fibers, the outer cladding must be removed by controlled etching before fusing and tapering [Lamont et al. 1985a].

Low-loss biconically tapered fused fiber couplers have been applied as taps in a 15-terminal 13-coupler tapped tee data bus [Villaruel and Moeller 1982].

Since the transverse field extent a/W in the cladding changes with wavelength, the coupling ratio is wavelength dependent [Hill et al. 1985; Wright J.V. 1986]. For directional couplers made of equal fibers, the coupled power varies periodically with wavelength with a constant period, which is determined by the order of the coupler and the taper ratio. The order is the number of times the power has coupled from one fiber to the other during coupler fabrication [de Fornel et al. 1984a]. The wavelength dependence of the splitting ratio can be utilized for fabricating WDM multiplexers and demultiplexers (Sect. 12.3.3).

On the other hand, one often wants to have a beam splitter with a constant splitting ratio, e.g. a 3 dB splitter. By fusing two fibers of slightly different phase constants, tapered couplers can be made with a coupling ratio which is largely independent of wavelength [Mortimore 1985b].

Biconical-taper fused directional couplers made of dissimilar single-mode or few-mode fibers exhibit interesting and useful properties [Lamont et al. 1985a; Hill et al. 1985, 1987]. Only for that wavelength for which the phase constant of a mode in the first fiber equals that of a mode in the other fiber (tuned condition), does strong coupling occur with low insertion loss (7.54). This effect can be utilized for making WDM couplers (Sect. 12.3.3).

The performance data of fused directional couplers depend on the polarization of the input light.

Extremely long directional couplers can be made by using a fiber with two cores. In contrast to conventional fiber couplers, in which the coupling is achieved locally over a relatively short interaction length, two-core fibers couple the modes of the two waveguides over a long propagation length. Two-core fibers may be applied in directional couplers [Schiffner et al. 1980, Schöner and Schiffner 1981; Truesdale et al. 1986; Truesdale and Nolan 1986], or as wavelength selective filters in WDM systems or in nonlinear fiber optics [Kitayama K. et al. 1985; Kitayama K. and Ishida 1985]. A practical problem which remains to be solved is how to connect standard fibers easily and efficiently to a double-core fiber.

The directional couplers described so far are all made from two parallel dielectric waveguides located side by side at a small separation. Another class of optical directional couplers consists of a dielectric rod waveguide coupled to a coaxial dielectric tube waveguide. A coaxial coupler comprises a fiber in which a ring of raised index is present in the cladding. Such an index profile can be found in the depressed cladding fibers or in multiple-clad fibers, which are designed for optimized chromatic dispersion.

A coaxial fiber can guide at least two modes, the HE_{11} mode (LP_{01} mode) of the rod and the HE_{11} mode of the tube. When the evanescent fields of the rod and tube modes overlap, directional coupling occurs. The degree of overlap determines the coupling strength. When the waveguides are lossless and have equal phase constants (phase matching condition), complete periodic power exchange between the two modes occurs along the fiber.

For a given coaxial fiber and a given wavelength, phase matching can be obtained by biconically tapering the fiber [Boucouvalas and Georgiou 1985a,b,c]. Since the evanescent field of the tube mode may extend into the medium surrounding the cladding, the characteristics of tapered coaxial couplers depend on the refractive index of the external medium [Boucouvalas and Georgiou 1986a].

In contrast to conventional parallel couplers, coaxial couplers, because of their cylindrical geometry, can be analyzed and understood by exact solution of the wave equation [Cozens and Boucouvalas 1982; Boucouvalas 1985a,b; Böck 1986]. Coaxial fiber couplers have been investigated experimentally by Cozens et al. [1982] and Boucouvalas [1985b].

Possible uses suggested for coaxial couplers include: as directional couplers; as narrow bandpass filters; in external resonators for stabilizing the emission frequency of laser diodes; as fiber sensors; and as beam expanders [Boucouvalas and Georgiou 1986b].

Directional couplers for single-mode fiber systems can also be made from four graded-index lenses used as beam expanders and a classical beam splitter cube (Fig. 12.5). However, these microoptic directional couplers are much bulkier and lossier than fused tapered biconical fiber couplers.

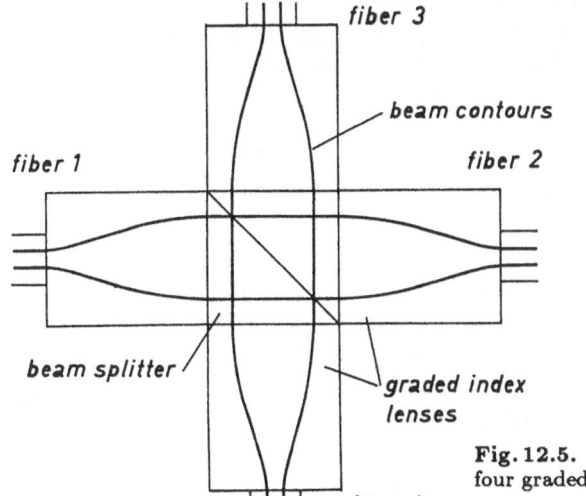

Fig. 12.5. Directional coupler made from four graded-index lenses and a beamsplitter

315

12.2.12 Star Couplers

A star coupler is an optical multiport with M input ports and N output ports. A transmissive star coupler is simply an optical power divider which – ideally – distributes the power incident on any input port equally between all output ports. For an ideal lossless star coupler the insertion or "furcation" loss is

$$a_f = 10 \log N \quad , \tag{12.25}$$

since the power is distributed to N output ports. Star couplers are applied in fiber-optic local area networks for interconnecting many terminals.

Single-mode star couplers have been fabricated by means of the encapsulated etching technique for making directional couplers (Sect. 12.2.11); both 3×3 couplers [Sheem 1980] and 10×10 couplers [Sheem and Giallorenzi 1979b] have been reported.

Because of the very low excess loss of fused biconical-tapered directional couplers, arbitrarily large transmissive star couplers can be made by connecting a number of 2×2 couplers [Marhic 1984]. Stable 8×8 (or 16 port) star couplers have been fabricated from eight lengths of single mode-fiber by forming twelve 3 dB fused coupling regions between appropriate fiber pairs [Mortimore 1985a]. By using broadband fused 2×2 couplers [Mortimore 1985b], wavelength-flattened 8×8 single-mode star couplers have been made [Mortimore 1986a,b]. The reported mean insertion loss (9.22 dB–9.47 dB) is only slightly larger than the furcation loss (12.25) ($10 \log 8 = 9.03$ dB) due to power splitting.

Fused 3×3 star couplers made by simultaneously fusing and tapering three fibers [Burns et al. 1982] and fused 9×9 star couplers using six 3×3 star couplers [Wang C.C. et al. 1985] have also been reported. By fusing four fibers in a "square" arrangement, an optical multiport can be made for determining the in-phase and quadrature components of a signal in a coherent homodyne detection system [Travis and Carroll 1985].

A new type of transmissive star coupler prevents an optical signal from returning to its own terminal [Healey and Faulkner 1985, 1986]. Ideally, the insertion loss is then only $10 \log(N - 1)$ instead of $10 \log N$, and two-way (full duplex) transmission becomes possible.

12.2.13 Fiber Taps

Fiber taps are optical three-ports by means of which a small part of the transmitted power can be extracted from the fiber without having to interrupt the fiber continuity. Taps can be used for identifying fibers carrying traffic [Nakazawa and Aoyama 1983; Finvers et al. 1986], for optimizing splices by local detection (Sect. 12.2.4), for providing a feedback signal for controlling the launched light power [Karr et al. 1978], and – of course – also for eavesdropping [Horgan 1985].

For characterizing a tap, the input, output, and tapped powers are denoted P_i, P_o, and P_t, respectively. The insertion loss of the tap is

$$a_\text{i} = 10\log(P_\text{i}/P_\text{o}) \quad , \tag{12.26}$$

and the coupling loss

$$a_\text{c} = 10\log(P_\text{i}/P_\text{t}) \quad . \tag{12.27}$$

The power tapped P_t is smaller than the power $P_\text{i} - P_\text{o}$ taken from the transmitted wave. A tapping (or collection) efficiency η is defined by

$$\eta = P_\text{t}/(P_\text{i} - P_\text{o}) \quad . \tag{12.28}$$

For small insertion loss a_i, the following relation holds between the coupling loss, the tapping efficiency in percent, and the insertion loss:

$$a_\text{i} = (434/\eta)\,10^{-a_\text{c}/10} \ [\%] \quad . \tag{12.29}$$

For a tap with 50% efficiency, a coupling loss of 19.4 dB corresponds to an insertion loss of only 0.1 dB.

A fiber tap can be made by bending the fiber and detecting the radiated power with a photodiode [Miller C.M. 1982a,b]. The insertion loss of this type of tap can easily be varied in the range 0 to 1 dB by changing the curvature of the fiber. Tapping efficiencies in the range 10–70% have been obtained with bend taps. For an optical identifier of this type [Nakazawa 1983], a coupling loss of about 25 dB has been measured. A hand-held fiber identifier utilizing the bent fiber principle has also been developed [Finvers et al. 1986].

The wave radiated from a bend can be coupled into an integrated optical channel waveguide by using a hologram [Arvidsson and Thylen 1980]. However, the power coupled into the channel waveguide was only 6×10^{-6} of the power transmitted through the fiber.

Fibers can also be tapped by transmitting an acoustic wave with a frequency of e.g. 3.5 GHz transversely through the fiber [Heffner et al. 1986]. The light diffracted by the acoustic wave is received by a photodetector. The coupling loss of this acousto-optic tap can be altered by varying the acoustical power.

At macrobends, power is radiated from a single-mode fiber guiding the fundamental fiber mode. Since the direction of wave propagation can be reversed, fiber taps can also be used for coupling light power into a continuous fiber, e.g. for local power injection prior to fusion splicing (Sect. 12.2.4).

When a macrobend is illuminated with a laser beam, part of the power transmitted by the beam can be launched into the fundamental fiber mode. However, efficient local injection is more difficult than local detection, since it is necessary to match the launching beam to the beam which the bend would radiate (Fig. 5.26).

As compared to a simple bend, an improvement of 10 dB in coupling efficiency has been reported both for launching and tapping [Hughes et al. 1985] by bending the fiber and simultaneously pressing it between two ridged glass blocks, so that the fiber forms a sharp bend which approximates a tilt.

With an injection laser as source, a power of about -40 dBm can be coupled into a macrobend by bending the fiber to a radius of 1–5 mm. Bending a fiber to such a small radius of curvature causes a serious problem, since the bending-induced stresses are 4 to 10 times the proof-test limit. Some fibers may even break during the splicing [Abe et al. 1987]. A better technique is to introduce periodic small-scale microbends into the fiber axis and to illuminate the fiber obliquely with a laser beam [Aberson and White 1986]. The powers injected are more than 20 dB higher than those reported for commonly used macrobend injectors. Proof-test measurements on fibers after application of the microbend injector showed no noticeable degradation in fiber strength from that of the virgin fiber [Aberson and White 1987].

12.2.14 Mode Filters

Mode filters are devices that ideally transmit one particular fiber mode without losses and in which all other modes are suppressed. Mode filters that transmit the fundamental LP_{01} mode but remove the LP_{11} mode are used e.g. in the bending method for measuring the effective cutoff wavelength (Sect. 13.8.1). They can also be inserted in short cable jumpers or cable repair sections in order to avoid modal noise caused by interference between the LP_{01} and LP_{11} modes (Sect. 11.6.1).

A simple LP_{01}/LP_{11}-mode filter consists of a fiber coil with e.g. two turns and a radius of curvature of 10 mm [Katsuyama 1979]. Since the bending loss of the LP_{11} mode near cutoff is much larger than that of the LP_{01} mode (Fig. 5.27), the LP_{11} mode can be effectivly removed by the fiber coil.

Another type of mode filter consists of a biconical fiber taper [Ozeki et al. 1975]. At the taper waist, the fiber must be single-moded. The biconical taper is immersed in index-matching liquid to eliminate cladding modes. The insertion loss of the mode filter for the LP_{01} mode is only 0.2 dB, whereas the rejection ratios are sufficiently large to suppress higher-order modes.

Higher-order modes can also be removed from the fiber by etching down part of the cladding and surrounding the fiber with an external higher index medium [Suematsu and Furuya 1975]. Due to their larger transverse field extent, higher-order modes are attenuated by power leakage (Sect. 6.4) whereas the attenuation of the fundamental mode remains negligibly small [Trommer 1985].

The LP_{11} mode propagating in a dual-mode fiber can be separated by a special type of polished directional coupler [Sorin et al. 1986b]. The dual-mode fiber is transversely coupled to a single-mode fiber. When the effective refractive index of the fundamental mode in the single-mode fiber is made to equal the effective index of the LP_{11} mode in the dual mode fiber, efficient power coupling between these two mode is possible, whereas, due to the mismatch of the phase velocities, little coupling occurs between the fundmental modes of the two fibers.

12.2.15 Mode Converters

In the two-mode wavelength region of a single-mode fiber, the LP_{01} and LP_{11} modes can be efficiently coupled by introducing periodic microbends into the fiber [Taylor 1984; Blake et al. 1986; Grebel and Herskowitz 1986]. In order to obtain maximum power transfer, the spacing Λ of the microbends should equal the beat length L_p between the modes to be coupled (condition for synchronous coupling or phase matching):

$$\Lambda = L_p = 2\pi/(\beta_{01} - \beta_{11}) \quad . \tag{12.30}$$

It has been suggested that these mode converters could be applied in switches, in polarization separators, and in displacement or force sensors.

12.3 Wavelength Selective Components

12.3.1 Wavelength Filters

Fixed or tunable optical bandpass filters are required for channel selection or wavelength suppression in multichannel wavelength division multiplexing (WDM) systems, for wavelength selection in several measuring methods, or as tuning elements in single-mode fiber lasers.

The classical methods for optical wavelength selection can be applied when the fundamental fiber mode is transformed into a collimated free wave beam by one of the methods described in Sect. 12.2.1. Classical filters use as wavelength selective elements multilayer interference filters [Lissberger and Roy 1985; McCartney 1986], holographic volume reflection filters [McCartney et al. 1985, 1986], Fabry-Perot interferometers [Mallinson 1985], or diffraction gratings.

The loss due to conventional grating monochromator inserted between two single-mode fibers is very large, about 30 dB. This loss can be reduced to about 10 dB by using an "imaging monochromator" which at its output forms a high resolution, well corrected 1:1 image of the output face of the first fiber, properly filtered in wavelength [Cisternino et al. 1985, 1986].

There have been a few suggestions for filters in which wavelength selection is accomplished in the single-mode fiber itself.

Wavelength selective Bragg reflectors can be made by introducing a periodic refractive index variation into the fiber in the longitudinal direction. To efficiently couple light into the backward direction, the spatial period Λ of the grating should be

$$\Lambda = \lambda/(2n_e) \quad , \tag{12.31}$$

where λ is the vacuum wavelength and $n_e = \beta/k$ (5.45) is the effective refractive index of the fundamental fiber mode.

A Bragg reflector can be produced optically in fibers with Ge-doped core, since this material is photosensitive, i.e. light causes a permanent index change [Kawasaki et al. 1978; Lam and Garside 1981; Parent et al. 1985].

Another method for producing a Bragg reflector in a single-mode fiber is to cement the fiber in a curved groove in a fused silica substrate and to polish until the evanescent field or the core field is accessible. This method is the same as is used in order to fabricate polished directional couplers (Sect. 12.2.11). A grating structure is then introduced into the field of the fundamental mode. The grating can consist of a commercial metal coated diffraction grating [Sorin and Shaw 1985a,b], it can be produced in a photoresist [Russell and Ulrich 1985], or it can be etched as a relief pattern directly into the core material [Bennion et al. 1986; Jauncey et al. 1986]. For the etched core version, a reflectivity of 92% and a bandwidth of 1.8 nm have been reported.

A tunable fiber-optic Bragg reflector has been made by using a fan-shaped diffraction grating in which the period varies continuously with position [Whalen et al. 1986]. The grating was pressed against a polished step-index fiber. By translating the fiber parallel to the major axis of the grating, the maximum of reflection could be tuned over the wavelength range from $1.46-1.54\,\mu$m. The power reflection coefficient was 11% and the 3 dB reflection bandwidth 0.8 nm. For a similar Bragg reflector [Sorin et al. 1987], the tuning range was 20 nm, the peak reflected power > 65%, and the bandwidth 0.6 nm.

All-fiber wavelength filters have also been fabricated by concatenating a number of directional couplers [Yataki et al. 1985]. A filter using two couplers had a throughput of 95% and a 3 dB bandwidth of 20 nm.

Instead of concatenating directional couplers, one can also cascade biconical fiber tapers (Sect. 12.2.6). Bandpass filters with peak transmission of 50% and half-power width of 6 nm have been made by cascading four biconical fiber tapers of different elongation [Lacroix et al. 1986b].

Optical fiber filters have also been made by doping the fiber core with rare earth ions, e.g. with holmium ions. At characteristic wavelengths, these fibers exhibit very high absorption peaks, e.g. 46 000 dB/km at 650 nm. Between the regions of large attenuation, there are regions of relatively small loss (e.g. 25 dB/km at 980 nm) [Farries et al. 1986].

By combining linear retarders and linear polarizers, two kinds of birefringent filters can be made: Lyot and Solc filters. In a birefringent filter, the variation of retardation with wavelength, in conjunction with input and output polarizers, is used to produce a desired spectral transmission function. Since both fiber retarders and fiber polarizers have been developed, all-fiber birefringent filters can now be made.

12.3.2 Dispersion Compensators

A dispersion compensator is an optical two-port with a strong wavelength dependence of the group delay and a low insertion loss. It can be used for compensating the chromatic dispersion and, thereby, for improving the transmission capacity of a single-mode fiber link. Dispersion compensators are also named pulse equalizers, pulse compressors, or dispersion transformers.

The condition for dispersion compensation is (11.102)

$$dt_g/d\lambda = -DL \quad , \tag{12.32}$$

where t_g is the group delay in the compensator, and D the average chromatic dispersion (11.98) of the fiber link of length L.

Group delay dispersion occurs when optical radiation is angularly dispersed by refraction or diffraction. Dispersion compensators can thus be made by using two prisms or two diffraction gratings [Martinez et al. 1984; Martinez 1987].

Another type of dispersion compensator consists of a concave holographic diffraction grating which spatially separates the spectral components [Bernard and Guillon 1986]. Each one is injected into a short multimode delay line fiber, the ends of which converge on a fast photodetector. The lengths of the multimode fibers are adjusted to exactly compensate the corresponding differential delay induced in the fiber link by its total chromatic dispersion.

Using this dispersion compensator, for a 36 km long single-mode fiber with a chromatic dispersion coefficient of 15.2 ps/(km nm) excited by a multilongitudinal mode 1.57 μm laser, the bandwidth has been enhanced by almost an order of magnitude. The insertion loss of the compensator was smaller than 3 dB.

Instead by an array of fiber delay lines, a single properly designed multimode fiber can also be used for compensating chromatic dispersion [Bhagavatula and Emig 1987]. The different wavelengths excite different modes in the compensating multimode fiber, and the intermodal delay times exactly compensate for the chromatic dispersion by the fiber-optic link. The bandwidth has been increased by a factor of 5–6.

12.3.3 Wavelength Multiplexers and Demultiplexers

A high-quality single-mode fiber has low attenuation in a broad wavelength range (Fig. 5.20). When only a single optical signal is transmitted through the fiber, its large transmission capacity is under-utilized.

Wavelength-division multiplexing (WDM) is a technique for transmitting several optical channels with different wavelengths over a single optical waveguide. At the end of the fiber link, the channels are separated by purely optical means.

When all channels are transmitted in the same direction, a WDM link is called unidirectional (one-way mode) and when signals are transmitted in both directions, bidirectional (two-way mode). The pros and cons of WDM-systems have been summarized by Winzer [1982]. Optical WDM systems are advantageous when very high bit-rate transmission is required, whereas time-division multiplexing (TDM) of many signals onto a single optical carrier wave is preferable, when lower bit-rate transmission is required [Nosu 1983].

In present WDM systems, the following factors limit the minimum channel separation to about 30 nm [Nicia 1981; Winzer 1982; Nosu 1983]:

1. tolerances in the emission wavelength of lasers,
2. temperature and current dependence of the laser peak wavelength,
3. laser aging, and
4. tolerances in the demultiplexer.

For one-way WDM, one needs optical multiports that combine waves of different wavelength in different fibers in one single-mode fiber at the beginning of the link (multiplexers, MUX), and other multiports that separate the different wavelengths into different fibers with only one wavelength per fiber at the end of the link (demultiplexers, DEMUX).

For bidirectional links, one needs a coupler, which acts as a multiplexer for some of the channels and as a demultiplexer for the others. Such a coupler is called a multi-demultiplexer (MUX/DEMUX) [Mahlein 1983] or muldexer for short [Winzer 1984]. For only one channel in each direction, the muldexer is called duplexer. The general term for multiplexers, demultiplexers, and muldexers is WDM couplers.

For the special case of only two channels with wavelengths λ_1 and λ_2, Fig. 12.6 shows schematically the operation of an optical three-port as a multiplexer, demultiplexer, and muldexer.

At the wavelength λ_1 of the first channel, an ideal WDM-coupler connects port I with the common link port L without losses, while disconnecting port II. For λ_2, an ideal coupler connects port II without losses with the link port L,

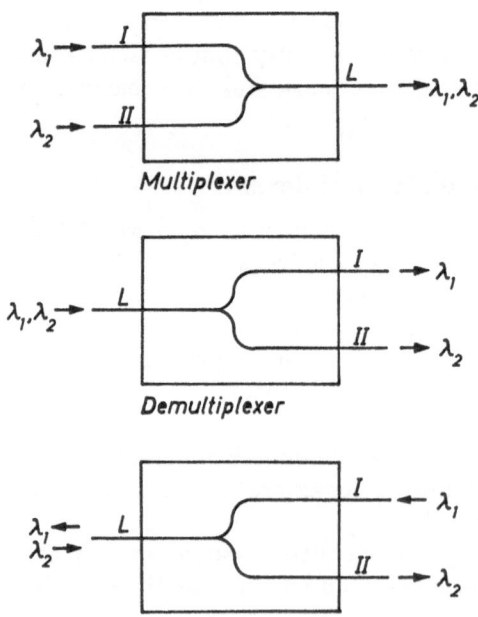

Multiplexer

Demultiplexer

Muldexer

Fig. 12.6. A WDM-coupler used as a multiplexer, demultiplexer, and muldexer

while disconnecting port I. In a practical WDM coupler, there are both losses as well as crosstalk between channels. The specific properties of a particular optical three-port determine whether it can be used as a multiplexer, demultiplexer, or muldexer.

WDM couplers are characterized by their loss and crosstalk properties. First, the ports have to be defined by certain cross-sections in fibers, mostly in single-mode fibers, but for detector ports, possibly also in multimode fibers. The transmission and reflection properties of a WDM coupler can, as for any multiport, be completely specified by a wavelength-dependent complex scattering matrix. For practical purposes it is, however, usually sufficient if a smaller number of parameters is known for all channel wavelengths.

For the special case of two channels, the definitions of these quantities [Mahlein 1983] can most easily be explained by considering the attenuations between the three ports I, II, and L. For simplicity, the three ports are assumed to be in single-mode fibers, since then the attenuations are independent of the direction of wave propagation.

For the wavelength λ_1 of channel 1 (Fig. 12.7a), the attenuation between ports L and I is the insertion loss a_{i1}. The insertion loss a_{i1} is the attenuation of channel 1 caused by inserting the WDM coupler. It must be as small as possible.

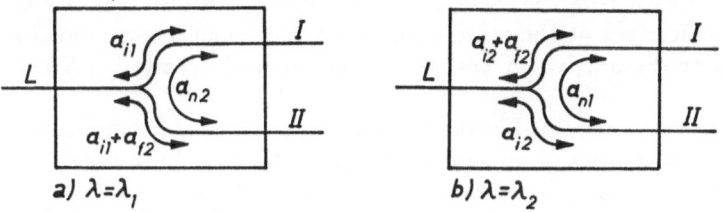

Fig. 12.7a, b. Definition of insertion losses and cross-talk attenuations for a three-port WDM-coupler

For $\lambda = \lambda_1$, the attenuation between L and II is larger than a_{i1} by an amount a_{f2}, which is called the far-end crosstalk attenuation in channel 2. The quantity a_{f2} determines the crosstalk between two channels in a unidirectional WDM system and, must therefore be as large as possible.

The attenuation a_{n2} between ports I and II at λ_1 is called the near-end crosstalk attenuation in channel 2. It determines the crosstalk between two channels in a bidirectional WDM system and, therefore, must also be as large as possible.

Thus, the far-end crosstalk attenuation characterizes the quality of a demultiplexer, and the near-end crosstalk attenuation the quality of a muldexer.

For the wavelength λ_2 of channel 2 (Fig. 12.7b), the attenuation between ports L and II is the insertion loss a_{i2}. The attenuation between L and I is larger than a_{i2} by the far-end crosstalk in channel 1, a_{f1}. The attenuation between ports II and I at λ_2 is the near-end crosstalk attenuation in channel 1, a_{n1}.

Notice, that the insertion loss and the crosstalk attenuations between two ports depend on the direction of transmission if one port is in a multimode fiber (Sect. 9.7). It should also be noted that the parameters of the WDM couplers are defined in optical decibels. When using direct photo-detection, the corresponding values of the electrical insertion loss or crosstalk attenuation are twice as large (5.118).

The above definitions can be generalized to WDM couplers with more than two channels.

The insertion losses a_i of the couplers should be as low as possible, since they must be compensated for by a corresponding reduction in link loss, i.e. in link length. For digital transmission, the signal to crosstalk ratio must not drop below 10–15 optical decibels. For one-way transmission assuming equal laser powers as well as equal route and coupler insertion losses, the signal to crosstalk ratio equals the far-end crosstalk attenuation a_f, which therefore must be larger than 10–15 dB.

For two-way transmission, the near-end crosstalk attenuation a_n must be larger than the sum of the minimum signal to crosstalk ratio plus the link attenuation. For example, for a link loss of 40 dB, the near-end crosstalk attenuation a_n must be 50–55 dB!

Some authors maintain that a multiplexer does not always demand a wavelength selective element. However, selective multiplexers can have smaller insertion losses and filter out at the system input side the spontaneous emission background of laser diodes, which can cause interchannel crosstalk [Winzer 1982].

Most WDM couplers reported in the literature are designed for multimode fiber systems. Several reviews of multimode WDM technology have been published [Watanabe 1983; Nosu 1983; Laude 1984]. Some reviews also give brief information about single-mode WDM devices [Tomlinson 1977; Nicia 1981; Winzer 1982; Mahlein 1983; Ishio et al. 1984].

Different physical effects can be exploited to produce WDM couplers:

1. the wavelength dependence of the angle of refraction for beam refraction at a prism,
2. the wavelength dependence of the angle of diffraction for beam diffraction at a grating,
3. the wavelength dependence of the transmission or reflection factor of a multilayer dielectric stack, and
4. the wavelength dependence of the coupling factor of a directional coupler.

According to whether these effects operate on freely propagating beams, on beams guided by integrated optical channel waveguides, or on beams guided by single-mode fibers, the WDM couplers are called micro-optical, integrated-optical, or all-fiber devices.

The first kind of WDM couplers are based on refracting prisms. Prism WDM couplers are made of a transparent material with large material dispersion $dn/d\lambda$. Prism WDM couplers were abandoned after the first experiments, because the high-dispersion materials required for prism fabrication exhibit high absorption at the same wavelengths. The insertion loss of WDM prism couplers is therefore too high for practical purposes [Mahlein 1983]. Moreover, the angular dispersion of prisms (2×10^{-4} radians/nm) is one order of magnitude smaller than that (10^{-3} radians/nm) of typical diffraction gratings [Nicia 1981].

The second kind of WDM couplers are based on diffraction gratings.

Grating WDM couplers utilize the wavelength dependence of the direction of the beam diffracted by a grating to combine or to separate the different WDM channels. A fiber to beam coupler (Sect. 12.2.1) produces a collimated wave beam, which is diffracted by the grating. Usually, blazed gratings are used. These preferentially diffract the light power into a specific diffraction order. Gratings produced by anisotropic etching of single-crystal silicon are superior to the conventional mechanically ruled gratings [Fujii et al. 1980; Watanabe 1983].

WDM couplers with diffraction gratings can multiplex or demultiplex many channels using only one angular dispersive element. Since the number of basic elements does not increase with the number of channels, the insertion loss is independent of the number of channels. Thus, grating couplers are attractive for WDM systems with a large number of channels.

For multiplexers, both the input and the output fibers must be single-mode fibers. A grating coupler has been reported [Hegarty et al. 1984] which concentrates the signals carried by nine single-mode fibers into one output single-mode fiber with insertion losses as low as 1.5 dB for a channel spacing of only 2 nm.

In demultiplexers, the short fibers guiding the demultiplexed channels to the photoreceivers may be multimode fibers without introducing noticeable pulse broadening. By using multimode output fibers, the insertion loss of the demultiplexer can be reduced. Grating demultiplexers with a single-mode input fiber and multimode output fibers have achieved insertion losses of 1 dB for 3 channels [Lipson and Harvey 1983] and 1–2 dB for 6 channels [Lipson et al. 1985]. A 20-channel micro-optic grating demultiplexer with a single-mode input and 20 multimode output fibers utilizing a graded-index rod lens and a silicon-etched reflection grating had an average channel spacing of 28 nm, insertion losses of 1.9–3.5 dB, and an isolation between adjacent channels of better than 20 dB [Seki et al. 1982]. For another single-mode WDM coupler, insertion losses as low as 0.5 dB have been reported [Laude et al. 1983].

Since the deflection angle is linearly related to wavelength, the fiber cores must be closely packed in order to maximize the ratio of channel width to channel separation. It is impractical to accomplish this with actual single-mode fibers because the outer diameters become unacceptably small. Therefore, an integrated optical waveguide concentrator has been used [Lipson et al. 1985]. The channel waveguides have an initial spacing appropriate for pigtailing which is gradually reduced to the desired value. However, the insertion of an integrated optic device introduced a high loss of about 6 dB.

Another method for broadening the channel bandwidth is to combine a diffraction grating with a special "dispersion-dividing" prism. For a nine channel single-mode multiplexer of this type, a channel bandwidth of 7 nm for a channel spacing of 10 nm has been obtained [Shirasaki et al. 1986].

The wavelength characteristics of grating demultiplexers can be controlled (tuning range 3.3 nm) by inserting an acousto-optic light deflector in the beam and changing the acoustic frequency [Kinoshita et al. 1986].

The third kind of WDM couplers are based on interference filters. An optical interference filter consists of a large number (20–30) of dielectric thin film layers which have alternate low (e.g. SiO_2, $n = 1.46$) and high (e.g. TiO_2, $n = 2.3$) refractive indices. Each layer has an optical thickness of approximately a quarter or a half wavelength in the material. Dielectric thin-film (DTF) filters can be tailored to have low-pass, band-pass or high-pass characteristics. The design and manufacture of dielectric multilayer thin-film filters has been described by Minowa and Fuji[1983].

Figure 12.8 illustrates one of many possible forms of WDM couplers using interference filters. The dielectric multilayer stack is located between the end-faces of two quarter-pitch graded index lenses (Sect. 12.2.1). The interference filter transmits wavelength λ_1 and reflects wavelength λ_2. This WDM coupler can be used as a multiplexer, a demultiplexer, or as a duplexer.

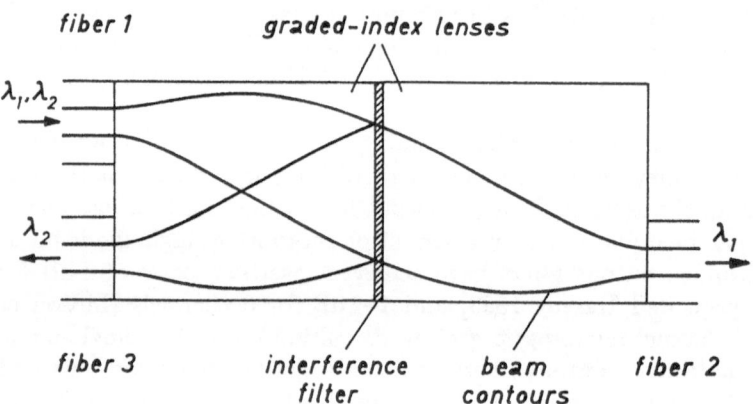

Fig. 12.8. Muldexer made from two graded-index lenses and an interference filter

Figure 12.8 explains its operation as a demultiplexer. Two optical channels with wavelengths λ_1 and λ_2 are received from fiber 1. In the first graded-index lens, the beams are widened and collimated. The beam with wavelength λ_1 is transmitted through the interference filter and focused by the second graded-index lens on the endface of output fiber 2. The beam with wavelength λ_2 is reflected by the dielectric multilayer and focused by the first lens on the endface of output fiber 3. For a multiplexer of this type with $\lambda_1 = 1275$ nm and $\lambda_2 = 1345$ nm, insertion losses of 0.5 dB for λ_1 and 1 dB for λ_2 have been measured [Lipson and Harvey 1983].

Another multi-demultiplexer consisting of three spherical lens collimators, two bandpass inference filters and a 100% mirror glued on a glass block had a pass band loss of 1 dB and a stop band attenuation of 45 dB for wavelengths of 1.2 μ and 1.3 μm [Tamura et al. 1985]. Since the three ports are single-mode fibers, the device can be used both as a multiplexer and as a demultiplexer.

A filter-type multi-demultiplexer for four channels (1.05 μm, 1.15 μm, 1.3 μm, 1.5 μm) had a total loss, which is the sum of the multiplexer and demultiplexer losses, of less than 2.3 dB [Watanabe et al. 1981].

Multiplexing more than two channels by means of interference filters is possible only by cascading the corresponding number of filters, resulting in a large insertion loss. Therefore, filter WDM couplers are attractive when the number of channels is small. On the other hand, filter devices cannot be manufactured for such narrow channel spacings as grating devices [Winzer 1982].

Single-mode WDM couplers have also been made by directly applying an interference filter to the obliquely polished fiber endface [Winzer et al. 1981; Reichelt et al. 1984].

A multi-demultiplexer combining a grating with interference filters for four channels (1.05 μm, 1.15 μm, 1.3 μm, 1.5 μm) had a total loss, which is the sum of the multiplexer and of the demultiplexer loss, of less than 3.2 dB [Watanabe et al. 1981].

The fourth kind of WDM couplers are based on optical directional couplers. Multiplexers and demultiplexers consisting of single-mode fiber directional couplers are especially promising due to their inherently low transmission losses. In contrast to micro-optic components, it is not necessary to transform the fundamental fiber mode into a collimated free beam wave and vice versa.

In symmetric fiber directional couplers (Sect. 12.2.11) which are built up of two identical fibers, the coupling efficiency as a function of wavelength oscillates between zero and maxima. The transfer function is similar to that of a comb filter. This effect can be used for multiplexing and demultiplexing by choosing e.g. λ_1 as a wavelength of maximum coupling and λ_2 as a wavelength of minimum coupling [Digonnet and Shaw 1982a, 1983; Köster 1984; Johnson D.C. and Hill 1986].

Symmetric directional couplers require a tight spacing of the fiber cores for a narrow channel separation, and thus the alignment tolerances are very stringent.

In asymmetric directional couplers, where two fibers with different core diameters and refractive indices, but with equal cladding refractive index are coupled, efficient power transfer is only possible around that wavelength for which the phase velocities of the coupled waves are equal (7.54) [Taylor 1973; Marcuse 1985]. Therefore, these asymmetrical couplers show true bandpass characteristics [Parriaux et al. 1981a, 1982].

In WDM filters using nonidentical fiber directional couplers, the crossover wavelength and the bandwidth can be determined from the wavelength dependence of the Petermann II spot size w_d (10.10) [Tewari et al. 1986]. Low-loss asymmetric single-mode fiber directional couplers applicable to WDM between transmission channels at $1.3\,\mu$m and $1.55\,\mu$m wavelength have been realized [Lawson et al. 1984; Zengerle and Leminger 1985; Whalen and Walker 1985; Georgiou and Boucouvalas 1986; Zengerle and Leminger 1986, 1987]. The insertion loss of the directional coupler can be made smaller than 0.05 dB.

The minimum channel separation of $\Delta\lambda = 30$ nm in present WDM systems corresponds to a frequency separation of

$$\Delta f = c\Delta\lambda/\lambda^2 = 5300\,\text{GHz} \tag{12.33}$$

for $\lambda = 1.3\,\mu$m. Since the bandwidth of the electrical signal is of the order of a few Gigahertz, at most, WDM systems still under-utilize the optical spectrum.

Therefore, in the future, we can expect the introduction of optical frequency-domain multiplexing (FDM) systems, in which the channel spacing is of the order of Gigahertz instead of Terahertz. FDM systems require coherent sources and heterodyne receivers to obtain electrical frequency selection through intermediate frequency bandpass filters. In principle, it is not necessary to differentiate between optical WDM and FDM systems [Mahr 1985]. However, it is customary to call systems in which the channels are demultiplexed by purely optical means WDM systems, and systems in which the channels are separated by electrical means FDM systems. With most optical WDM demultiplexers, it is not possible to separate FDM channels.

However, with a tunable fiber Mach-Zehnder interferometer, a FDM demultiplexer with a spectrum selectivity of 11 GHz and an insertion loss of 1.8 dB has been demonstrated [Inoue et al. 1985].

12.3.4 Fiber Resonators

Optical resonators can be made from single-mode fibers in two forms. In the first form, both ends of a section of fiber are made highly reflective (Sect. 12.1.3). In the second form, the ends of a section of fiber are spliced together to form a ring resonator.

Fiber resonators can be used for making devices such as optical spectrum analysers, tunable filters, wavelength multiplexers and demultiplexers, and sensors. A low-loss fiber with reflecting endfaces represents a transmission line resonator or, in optical terms, a fiber Fabry-Perot interferometer. As compared to the conventional bulky Fabry-Perot interferometers, fiber resonators are small, flexible, can be very long and do not require the two reflecting surfaces to be exactly parallel. The optical resonator can be tuned to resonance by changing the fiber length, e.g. by changing the temperature or by attaching the fiber to a piezoelectric ceramic rod and applying a variable voltage.

A fiber Fabry-Perot interferometer can be made by polishing the ends of a single-mode fiber and by bonding multilayer dielectric mirrors to each fiber end. Power can be coupled into and out of the resonator through the dielectric mirrors or by inserting a directional coupler with high coupling loss into the fiber.

The condition for resonance is that the fiber length L between the mirrors equals an integer number N of *half* wavelengths of the fundamental fiber mode:

$$L = N\lambda_{01}/2 \quad . \tag{12.34}$$

Since the fiber length is very large compared to the wavelength, the integer N is very large, and many resonances will be found in a narrow wavelength region. The transmission function is that of a comb filter. The difference between the frequencies Δf of two neighboring resonances is called the "free spectral range" Δf. From the resonance condition (12.34) the free spectral range can be calculated:

$$\Delta f = c/(2n_e L) \quad , \tag{12.35}$$

where c is the velocity of light in vacuum and n_e is the effective refractive index (5.45). For instance, a fiber length of 3.8 cm gives a free spectral range of 2.7 GHz.

The full-width-half-maximum (FWHM) width Δf_{FWHM} of a single resonance curve depends on the waveguide and mirror losses. The ratio of the free spectral range and the FWHM width is called the "finesse" F, of the interferometer

$$F = \Delta f/(\Delta f_{\mathrm{FWHM}}) \quad . \tag{12.36}$$

When the waveguide losses are very small as compared to the reflection losses, the finesse is [Yariv 1971]

$$F = \pi R^{1/2}/(1 - R) \quad , \tag{12.37}$$

where R is the power reflectivity of the mirrors. For $R = 0.99$, the calculated finesse is 313.

The maximum measured finesse of a fiber Fabry-Perot interferometer reported so far is $F = 500$, which, for a free spectral range of 250 MHz, corresponds to a minimum resolvable bandwidth $\Delta f_{\mathrm{FWHM}} = 0.5$ MHz [Stone J. 1985a,b; Stone and Marcuse 1986]. Within experimental error, the measured finesse corresponds to the calculated value, taking into account only reflection losses (12.37). Thus, a smaller bandwidth (and a narrower free spectral range) can be obtained by using a longer fiber.

Fiber Fabry-Perot interferometers have been used e.g. for producing temperature [Yoshino 1981] or strain [Ito 1986] sensors. The change of the temperature or length can be determined by counting the number of resonances.

In another application, a fiber resonator has been coupled as a so called external cavity to a semiconductor laser in order to minimize the emission linewidth [Favre and Le Guen 1983, 1985; Ihee et al. 1985].

A fiber-optic resonator can also be used as an optical spectrum analyzer applying temperature-tuning [Kist and Sohler 1983] or fiber length tuning [Hotate et al. 1983]. The frequency resolution of fiber resonators can be better than 10 MHz. In comparison, the frequency resolution of grating monochromators is worse than about 10 GHz and that of conventional Fabry-Perot interferometers poorer than about 1 GHz.

The second kind of fiber resonator is a ring or loop resonator. It can be made by splicing together the ends of a section of single-mode fiber. The waves multiply transmitted by the ring add in phase if the length L of the ring equals an integer number N of *full* wavelengths of the fundamental mode:

$$L = N\lambda_{01} \quad . \tag{12.38}$$

Since the length is very large compared to a wavelength, the order N of the resonance is very large, and there are a large number of resonances within a finite wavelength range. The frequencies of the resonances are approximately equidistant. For a fiber ring resonator, the difference of the frequencies of two neighboring resonances, the free spectral range Δf, is

$$\Delta f = c/(n_{e}L) \quad . \tag{12.39}$$

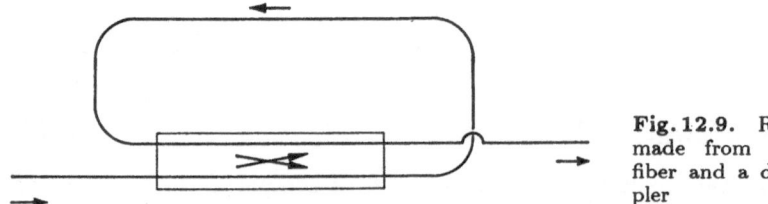

Fig. 12.9. Ring resonator made from a single-mode fiber and a directional coupler

For coupling light into and out of the ring resonator, one or two single-mode four-port couplers must be inserted in the ring waveguide (Fig. 12.9). For each resonance, an absorption dip can be observed at the through-port of a launching coupler and a transmission peak is observerd at the output port of a tapping coupler.

The characteristics of all-fiber ring resonators have been investigated both theoretically and experimentally for coherent light [Stokes et al. 1982, 1983], partially coherent light [Ohtsuka 1985], and light with arbitrary time variation of amplitude [Crosigniani et al. 1986]. Because of the very low attenuation of glass fibers and the low loss of fiber directional couplers, very sharp resonances are obtained. In one example [Bachus et al. 1983], the ring length was 36 cm, the wavelength 831 nm, the free spectral range 559 MHz, and the resonator bandwidth 62 MHz. The quality of this optical resonator defined as usual as the ratio of the resonant frequency to the bandwidth, is 5.8×10^{6}. A fiber-optic ring resonator has been applied for measuring the spectral linewidth of single-frequency semiconductor laser diodes with a resolution of better than 1 MHz [Tai et al. 1985].

Since the resonance dips or peaks are equidistant over the frequency axis, an optical Perot-Fabry or ring resonator is a very simple and inexpensive means for producing optical frequency markers, e.g. for optical sweep generators. A similar scheme with a dielectric waveguide Fabry-Perot interferometer has been used by the author in the millimeter wave region [Neumann 1964b].

In a fiber loop resonator, resonance is caused by coherent superposition of the partial optical waves which have passed the loop 1, 2, 3 ... times. Resonance can thus only be observed when the coherence length (3.36) of the source is very large compared to the loop length.

There is another single-mode fiber device which looks very similar to a ring resonator, and which likewise consists of a fiber loop and a directional coupler; this is the recirculating delay line. However, in this case, the coherence length of the light waves circulating in the loop is very short compared to the loop length. A typical example is short light pulses. Recirculating single-mode fiber delay lines have a number of applications. They can be used as transient buffer memories for analog or digital signals and as data rate transformers [Newton et al. 1982, 1983], as transverse notch filters [Bowers et al. 1982, Newton and Cross 1983], for producing high-speed pulse trains [Newton et al. 1983], for testing optical time-domain reflectometers (Sect. 13.4.4), or for optical signal processing operations [Jackson K.P. et al. 1985].

The frequency dependence of the transmission factor of a recirculating delay line is produced by interference of the modulation and not by interference of the optical carrier [Bowers et al. 1982]. The dynamic range of recirculating delay line filters is severely limited by intensity noise whose origin is laser phase fluctuations [Tur and Moslehi 1983]. The spectrum of this noise is structured by deep notches at zero frequency and at multiples of the inverse of the loop delay time [Tur et al. 1985].

13. Measuring Techniques

The measurement of the relevant fiber parameters is necessary during the manufacturing process, for system design, and for specification purposes. Since single-mode fiber technology is still advancing rapidly, it is difficult for researchers, manufacturers, and users to reach a consensus about the best way to measure a given parameter, even if the parameter is well defined. Since the recommendation of a few significant parameters, standard values, and simple and accurate measuring methods has technical and economical implications, standards committees like the CCITT are presently discussing how to define the quantities, how to measure them, and how to fix standard values for them.

Since the problems of the definition of the parameters and the identification of test methods for single-mode fibers are relatively new and since the experience gained so far is still very limited [Bonaventura 1984], at present, there exist only few recommendations by the standards committees. Therefore, different organizations often use different definitions and different measuring techniques, giving slightly different results. As long as there are no standard definitions and test methods, it is important always to state in reports and data sheets the definition and the measuring technique used.

The recommendations for single-mode fibers published by the CCITT [1984] are described and explained by Coppa et al. [1985b].

In this chapter on single-mode fiber measurements, we first describe the optical sources and detectors required for the different measurement procedures (Sect. 13.1). Methods for measuring optical power (Sect. 13.2) and fiber attenuation (Sect. 13.3) are then summarized. A special section (13.4) is devoted to the important backscattering method which allows one to measure fiber and splice loss as well as fiber length and fault location. Sections 13.5 and 13.6 describe methods for determining the transverse field distribution of the fundamental fiber mode and the radiation pattern of the fiber end, respectively. These functions are used in some of the various methods to measure the mode field diameter (Sect. 13.7). In Sects. 13.8–10, overviews are given of methods for measuring the single-mode limit, chromatic dispersion, and the refractive-index profile.

The CCITT classifies the measuring methods into reference test methods and alternative test methods. A reference test method (RTM) provides a measurement of the given characteristic strictly according to the definition which usually gives the highest degree of accuracy and reproducibility, whereas alternative test methods (ATM) provide results which are reproducible and relatable to the reference test method but more suitable for practical use [Coppa et al. 1984a].

The transmission properties of multimode fibers are characterized by the refractive-index profile (core radius, numerical aperture, form of the index profile), bandwidth, and attenuation. For the characterization of single-mode fibers, other parameters are essential: instead of the index profile, the mode field diameter (Chap. 10) and the effective cutoff wavelength of the second order mode (Sect. 6.3.3) are specified. Instead of the bandwidth, the chromatic dispersion parameter D (5.85) and the dispersion slope S (5.93) are measured. Only the attenuation (5.111, 112) is specified in the same manner for both multimode and single-mode fibers.

From multimode fibers, it is well known that transmission properties like attenuation and bandwidth are strongly dependent on the launching conditions, i.e. on the mode spectrum excited. For single-mode fibers, however, launching conditions lose their importance, provided that the fiber is operated in the single-mode region and provided that polarization effects are not important. Thus, from the experimental point of view, the measurement of single-mode fiber parameters is greatly simplified by the presence of a well defined electromagnetic field configuration. As a consequence, quantities like attenuation, dispersion, or joint loss are univocally defined, and the extrapolation of factory length characteristics to concatenated sequences of fibers is simply additive, thus helping the system designer very much. Unlike for multimode fibers, it is not necessary to guess a "concatenation factor" to calculate the bandwidth of a single-mode link from the bandwidths of the individual factory lengths (11.98).

On the other hand, measurements require very high accuracy and sensitivity, and a certain dependence of measured quantities on "extrinsic" test conditions (such as length of the sample, bending conditions, applied stress etc.) can be noticed, thus making interlaboratory comparisons more difficult [Coppa et al. 1984a].

In principle, most propagation parameters of single-mode fibers could be calculated from the refractive-index profile although direct measurement is often more appropriate.

Several reviews of methods for measuring the properties of single-mode fibers have been published [Inada 1981; White K.I. et al. 1981; Nosu 1982; Stern et al. 1982; Kapron and Lazay 1983; Eccleston and Dick 1983; Auge et al. 1983; Payne D.B. et al. 1984; di Vita et al. 1984; Srivastava and Franzen 1985; Lowe 1985; Anderson W.T. 1986; Uesugi et al. 1986; Costa and di Vita 1987].

13.1 Optical Sources and Detectors

13.1.1 Sources

The light sources used for the measurements should be stable in output power and center wavelength over a time period sufficiently long to complete the measurement procedure. For some techniques it is also necessary to have sources with adjustable output power, wavelength, and modulation.

For measurements, the following light sources are used: tungsten-halogen lamps, light emitting diodes, HeNe lasers, semiconductor lasers, fiber Raman lasers, and occasionally optical parametric oscillators.

Quartz-tungsten-halogen lamps usually consume 50 to 500 W of electrical power and produce high radiance incoherent light with a very wide spectrum. The width of the spectrum is usually reduced by a grating monochromator, which represents a tunable optical bandpass filter. Prism and grating monochromators have often been described in the literature, e.g. by Cancellieri and Ravaioli [1984]. The widths of the input and output gaps of the monochromator determine the linewidth of the output radiation. The widths of the slits must be chosen as a trade-off between two opposite requirements: a high wavelength resolution and a large transmitted optical power. As an alternative to a monochromator, a set of narrowband interference filters can also be used to tune the output wavelength.

The advantage of the lamp-monochromator combination is that the wavelength can easily be scanned in the interesting region of 0.5 to 1.8 μm; the main disadvantage is the very small amount of power that can be coupled into the fundamental mode. For a typical linewidth of 10 nm the power is only -50 dBm as compared to a power level of about -40 dBm coupled into a multimode fiber. Even if one chops the light and uses a lock-in amplifier for increasing the signal-to-noise ratio, the signal at the fiber input is only about 20 to 35 dB higher than the noise [Anderson W.T. 1986]. Therefore, the lamp-monochromator combination can be used to measure the wavelength dependence of fiber attenuation (Sect. 13.3.1), or of the spot size (Sect. 13.7.2), but not for measuring the near-field intensity distribution (Sect. 13.5) or the far-field radiation pattern (Sect. 13.6) where only a small fraction of total power is detected for obtaining the required spatial or angular resolution.

In order to avoid variations of the optical output power, the electric current through the tungsten lamp has to be stabilized.

Light emitting diodes couple higher powers into single-mode fibers. The theoretical fundamentals of coupling between an incoherent source and a single-mode fiber are described in Sect. 7.3. Practical values for the powers coupled from LED's into single-mode fibers are reported in Sect. 12.2.2. A surface emitting LED typically radiates about 0 dBm into air and couples about -10 dBm and -30 dBm into multimode and single-mode fibers, respectively. With edge-emitting and superradiant LED's, higher powers can be injected into the fiber. The light power emitted by an LED can easily be modulated by varying the injection current. Rise and fall times of some nanoseconds are typical. The radiation pattern of LED's is Lambertian (cosine, Sect. 7.3).

For wavelengths of 1.3 μm, the linewidth of the light emitted by an LED is of the order of 100 nm. It can be reduced by transmitting the light through a monochromator. With an LED as a source, the wavelength can be scanned in a small wavelength region, e.g. for determining the chromatic dispersion by measuring the wavelength dependence of the group delay (Sect. 13.9.2). If a larger variation of the wavelength is required, a set of LED's with different central wavelengths can be used.

The HeNe gas laser is a relatively cheap source, which can couple about 1 mW of coherent power into a single-mode fiber. Most HeNe lasers operate at the red wavelength of 0.633 µm, which is interesting for fiber sensors but not for optical communication with single-mode fibers because of the relatively large fiber loss at this wavelength. HeNe lasers operating at 633 nm are often used for aligning single-mode fiber measuring setups. Using mirrors that reflect at 1.15 µm, 1.32 µm, or 1.52 µm wavelength, HeNe lasers can also produce highly coherent light at these wavelengths. The output power of a HeNe laser cannot be modulated directly, for measuring purposes it is necessary to modulate the emitted light beam by a rotating wheel electromechanical chopper or a tuning fork chopper.

Semiconductor lasers radiate about 10 dBm into air and couple about 0 dBm of optical power into a single-mode fiber. The emitted light power can easily be modulated up to very high frequencies (several GHz) by changing the injection current. The laser wavelength can be tuned in a very small range either by changing the substrate temperature or by changing the injection current.

The total output of a semiconductor laser diode consists of a combination of unpolarized spontaneous emission and well-polarized coherent light. The coherent beam is linearly polarized parallel to the active zone of the laser, i.e. the electric field is parallel to the junction. For this polarization, the Fresnel reflection coefficient at the cleaved crystal faces is larger than for the orthogonal polarization, making the threshold current for self-oscillations smaller. As the coherent output power of the laser is increased, the degree of polarization of the output field increases, since the total spontaneous-emission rate is constant when the diode is operating above threshold. The maximum degree of polarization depends on the diode's lasing geometry. A quantum-well heterojunction laser exhibited the highest degree of polarization [Fuhr 1984].

High optical powers with wavelengths greater than 1.06 µm can be produced by using the nonlinear effect of stimulated Raman scattering in a single-mode fiber. A fiber Raman oscillator [Cohen L.G. and Lin 1978] represents a very useful source for measuring backscattering (Sect. 13.4), the near-field intensity distribution (Sect. 13.5), the far-field radiation pattern (Sect. 13.6), or chromatic dispersion (Sect. 13.9.1) at different wavelengths. But a Raman oscillator is expensive and delicate to operate.

A rather expensive pulse generator for the 1.18 µm to 1.6 µm wavelength region consists of a mode-locked Nd:YAG laser oscillator with output wavelength 1.064 µm followed by two Nd:YAG amplifiers, a KDP crystal for second harmonic generation (0.532 µm) and a LiNbO$_3$ parametric oscillator, which can be temperature tuned [Mochizuki et al. 1981b]. The pulse duration is 16 ps.

13.1.2 Detectors

A photodetector transforms the optical signal into an electrical signal. It must have a linear photocurrent versus optical power characteristic in the required measuring range. The spectral response should be compatible with the spectral characteristics of the source.

As soon as the optical signal is transformed into an electrical signal, one can use common electrical instruments to analyze the signal: voltmeters, oscilloscopes, spectrum analyzers, phase meters, lock-in amplifiers, boxcar integrators, transient recorders, signal processors, etc.

Photomultiplier tubes are high-gain, broadband detectors in the visible (0.4 to 0.7 μm) region, but their photoemissive efficiency falls sharply for wavelengths above 0.7 μm. Since the wavelengths of interest in optical communications are larger than 0.8 μm, photomultipliers are used only rarely for measuring purposes. By cooling the photomultiplier tube with dry ice the sensitivity can be improved [Wang C.C. et al. 1983].

The photodetectors commonly used in measuring techniques are based on the internal photoeffect and eventually use carrier multiplication by the avalanche effect. For simple photodiodes (PD's), the relation (5.113) between the incident optical power and the photocurrent is very linear over many decades. However, for avalanche photodiodes (APD's) at large powers, there are nonlinearities due to the drop in bias voltage caused by the photocurrent.

The product hf of Planck's constant h and the optical frequency f is the energy of one photon. Therefore, in optical communications, the optical frequency is sometimes also given in terms of the photon energy $W = hf$, where the unit for the energy is one electron volt. By definition, 1 eV is the amount of energy acquired or lost by an electron in moving through a potential difference of 1 V, i.e.

$$1\,eV = 1.602 \times 10^{-19}\,J \quad . \tag{13.1}$$

There is a simple rule for calculating the wavelength λ in micrometers from the photon energy W in electron volts:

$$\lambda[\mu m] W[eV] = 1.234 \quad . \tag{13.2}$$

Note that the number on the right hand side is easy to remember!

A photodiode can be used only at those optical wavelengths for which the photon energy is larger than the bandgap energy W_g of the material used in the photodiode:

$$W = hf > W_g \quad . \tag{13.3}$$

The material of the photodiode must therefore be matched to the wavelength to be detected. Silicon photodiodes can be used for wavelengths ranging from 0.4 to 1 μm, but because of the relatively large bandgap of silicon, they cannot be used in the wavelength region above 1.1 μm which is the most interesting for single-mode fibers. Instead, one has to use photodiodes made of germanium, ternary (InGaAs), or quaternary (InGaAsP) semiconductors. Germanium photodiodes are widely used between 1 and 1.8 μm, but they have a high dark current and are thus relatively insensitive. The dark current can be reduced by cooling the photodiode. InGaAs photodiodes exhibit lower dark leak currents than germanium types, but cost much more.

In order to improve the signal-to-noise ratio, the optical power should be intensity modulated, and synchronous detection with a lock-in amplifier (phase-sensitive amplifier) should be used. When the source cannot be modulated directly like an LED or LD, a mechanical light chopper (rotating disc with holes or vibrating slit) can be used. Moreover, by using modulation and synchronous detection, daylight causes no response, so that it is not necessary to work in a dark room.

For measuring two-dimensional intensity distributions, an IR vidicon is useful, since it avoids the mechanical movement of a scanning photodetector. In principle, an IR vidicon is a TV camera tube. Therefore, the invisible IR intensity distribution can be made visible on a conventional TV monitor. The image can also be digitized and processed further, e.g. to produce an intensity contour plot. In order to increase the signal-to-noise ratio, a large number of successive frames can be digitally averaged.

Since a vidicon is not specifically designed for precision measurements, it has several disadvantages: The sensitivity and magnification are not uniform across the face of the vidicon. The sensitivity and, most importantly, the linearity of infrared vidicons are marginal. Nonlinearities both in the deflection voltage vs location curve and in the optical intensity vs photocurrent curve must be corrected for. Because of a possible time jitter, the deflection voltages must be stabilized in frequency.

Another sensitive receiver for long optical wavelengths is a PbS photoconductor.

13.2 Measurement of Optical Power and Insertion Loss

For measuring optical power, commercial power meters are available. Most of these make use of a photodiode which produces a photocurrent proportional to the incident optical power P (5.113). They display either the optical power in mW, μW, or nW, or, alternatively, the power level a_r in decibels relative to one milliwatt (5.120)

$$a_r = 10 \log(P/1\,\text{mW}) \quad . \tag{13.4}$$

Powers of $1\,\text{mW}$, $1\,\mu\text{W}$, $1\,\text{nW}$, and $1\,\text{pW}$ correspond to relative power levels of $0\,\text{dBm}$, $-30\,\text{dBm}$, $-60\,\text{dBm}$, and $-90\,\text{dBm}$, respectively. Decibel units are useful for expressing data that have a wide dynamic range.

Besides power meters using quantum detectors, there are also meters using thermal detectors which convert the input optical radiation to heat, and sense the temperature increase through various physical mechanisms. Since quantum detectors have a wider dynamic range and faster response time, they are used most often in fiber power meters. However, optical power meters have limited accuracy. An interlaboratory test of different fiber power meters indicated [Gallawa and Yang 1986] that power meter readings taken in different laboratories exhibited a standard deviation of 0.71 dB. The accuracy and reproducibility of optical power measurements based on germanium detectors is discussed in a recent paper [Hentschel 1986]. Practical hints for achieving good accuracy are also given there.

Power meters measure optical power in absolute units, i.e. in microwatts, milliwatts, or dBm's. For measuring the insertion loss of an optical device by the so-called insertion or substitution technique, the optical receiver need not be calibrated absolutely, since only the ratio of two powers is to be determined.

For measuring the insertion loss of an optical two-port, the device under investigation is coupled to a source and a detector through connectors. The insertion loss is the difference between the received power levels without and with the device inserted. The insertion losses of the connectors used must be accurately reproducible. Interference effects in air gap connectors (Sect. 5.9.5) can be as large as 0.92 dB and thus have to be avoided, e.g. by tilting the fiber endfaces [Young W.C. et al. 1986a].

When the device under test is a connector with an air gap, the insertion loss varies sensitively with wavelength (Sect. 5.9.5). This causes a large variance in the measured insertion loss when lasers are used as sources. For obtaining reproducible values of the insertion loss, a source with a large linewidth like an LED should be used [Nosu 1982]. Techniques for precisely measuring the insertion loss of connectors have been described by Kaiser et al. [1982].

The insertion loss technique is the only viable method for measuring attenuation in field conditions. In order to make constant the coupling loss values of the single-mode fiber to the optical source, the use of a fusion method is suggested. When the loss of an installed cable section is to be measured by the insertion method, source and detector are separated by a large distance, and are required to have excellent long term stability.

13.3 Measurement of Fiber Attenuation

Fiber loss is a very important factor since, together with pulse broadening, it determines the maximum repeater spacing in a fiber-optic communication system.

In this section, methods are described for measuring total fiber attenuation (Sect. 13.3.1), the partial attenuations caused by absorption (Sect. 13.3.2), scattering (Sect. 13.3.3), macrobending (Sect. 13.3.4), and microbending (Sect. 13.3.5), as well as the attenuation of the unwanted LP_{11} mode (Sect. 13.3.6).

13.3.1 Cut-Back Method

The total fiber attenuation constant can be measured either by the cut-back method to be described in this section or by the backscattering method (Sect. 13.4).

The attenuation of a fiber section and the attenuation per unit length, i.e. the attenuation coefficient α in dB/km, have been defined in Sect. 5.9.1. The cut-back measuring technique [Tynes et al. 1971; Keck et al. 1972; Keck and Tynes 1972; Eccleston and Dick 1983] is a direct application of this definition (5.112). Basically, the setup is the same as that routinely used for measuring the attenuation of multimode fibers.

The fundamental mode is launched by the lamp-monochromator-chopper light source described in Sect. 13.1.1. In order to minimize the position sensitivity, it is necessary that the diameter of the incident light ray bundle is much larger than the mode field diameter (overfilled launch condition). First, the power $P(L)$ emerging from the far end of a fiber section of large length L (e.g. 1 km) is measured as a function of wavelength using a large area photodiode connected to a lock-in amplifier. Then, without changing the launching conditions, the fiber is cut-back to a short length of e.g. 2 m and the power $P(0)$ emerging from the new end is also measured as a function of wavelength. A cladding mode stripper (Sect. 12.2.8) must be inserted into the short fiber section in order to avoid errors caused by cladding modes. The attenuation a of the fiber in decibels is then given by (5.108)

$$a = 10 \log [P(0)/P(L)] \quad , \tag{13.5}$$

and the attenuation coefficient by (5.110)

$$\alpha = a/L \quad . \tag{13.6}$$

For multimode fibers, because of differential mode attenuation, the attenuation coefficient measured will depend strongly on the distribution of power over the several hundred guided modes, i.e. on the so-called mode spectrum, which can be influenced by the launching conditions. In order to be able to compare attenuation values measured in different laboratories, the standards committees therefore specify special launching conditions for measuring attenuation of multimode fibers.

Single-mode fibers guide only the two fundamental modes. With the exception of dichroic single-polarization single-mode fibers, there is no differential mode attenuation and the attenuation coefficient depends only on the waveguide properties and not on the launching conditions. In principle, one can use any launching conditions provided that the power coupled into the LP_{01} modes is large enough to produce a sufficient signal-to-noise ratio at the receiver output.

The dynamic range of the cut-back method is 20 to 35 dB when using a halogen lamp with monochromator [Anderson W.T. 1986], and about 70 dB when using laser diodes [Nosu 1982]. With the usual cut-back setup, loss increments of 0.05 dB can be detected [White K.I. 1985]. By replacing the tungsten lamp and monochromator by an LED or a diode laser, and packing both the source and the detector into a constant-temperature box, a test set stable to $+ 0.001$ dB over five hours has been developed [Namihira et al. 1982].

Measuring systems which use the cut-back technique for measuring the wavelength dependence of fiber attenuation are commercially available. At $\lambda =$

1.3 μm using a cooled ($-20°$C) Ge photodiode, for an integration time of 1 s, a loss measurement range of greater than 31 dB can be obtained for single-mode fibers.

Interlaboratory comparisons [Heitmann et al. 1980, 1981] of attenuation measurements performed by the cut-back technique showed that it is a very accurate method with loss variations of only $\pm 8\%$. In another round robin test [Gardner 1982], the cut-back technique yielded an interlocation standard deviation of 0.03 dB/km which is nearly an order of magnitude better than the σ-value achievable with typical multimode fibers.

Because of its simplicity and accuracy, the cut-back method has been appointed as the reference test method for fiber attenuation [CCITT 1984]. However, in many situations its destructive nature is a disadvantage.

In the two-mode region near cutoff of the LP_{11} mode, the powers transmitted by the LP_{01} and LP_{11} modes decrease exponentially along the fiber, but with different attenuation constants. Thus the total power does not decrease exponentially with fiber length, as is the basic assumption underlying the theory of the cut-back method. Hence, the simple cut-back technique does not permit evaluation of the attenuation constant of the fundamental mode for wavelengths below the cutoff wavelength of the LP_{11} mode.

If one nonetheless applies (13.5, 6) for calculating the fiber attenuation α, the attenuation versus wavelength curve of single-mode fibers may exhibit noticeable peaks at small wavelengths, due to the increased attenuation of higher order modes near their respective cutoff wavelengths. These peaks must not misinterpreted as absorption peaks caused by impurities. These peaks can be used for measuring the effective cutoff wavelengths of higher order modes (Sect. 13.8.1).

The total attenuation α is the sum of the partial attenuations caused by the various loss mechanisms (5.101, 110). For minimizing the fiber loss in a systematic manner, the fiber manufacturer wants to know the contributions of the different loss mechanisms such as UV and IR band-edge absorption, Rayleigh scattering, OH absorption, etc. When the various loss components are known, the wavelength dependence of the total attenuation can be represented by a closed form expression, which is useful for the designer of single-mode fiber communication systems.

Since the partial attenuations have characteristic wavelength dependences, it is possible to extract the partial attenuations from a spectral total attenuation profile. Three different methods for doing this have been published.

In the first and most simple method, only the absorption loss, the Rayleigh scattering loss and the wavelength independent loss can be obtained from spectral total loss data [Inada 1976]. In a wide wavelength region, the losses caused by UV and IR absorption are negligible, so that (5.124)

$$\alpha(\lambda) = \alpha_a(\lambda) + \alpha_s(\lambda) \quad , \tag{13.7}$$

where $\alpha_a(\lambda)$ is the loss caused by OH impurities and

$$\alpha_s(\lambda) = A\lambda^{-4} + \alpha_{si} \tag{13.8}$$

is the scattering loss caused by Rayleigh scattering plus α_{si} the wavelength independent loss caused by fiber imperfections. There are wavelength regions where the OH loss α_a can be expected to negligible. In these regions, a straight line is fitted to a plot of the measured total attenuation α as a function of λ^{-4}. The gradient of this line is equated to the Rayleigh scatter coefficient A (5.121) and the intercept with the loss axis with α_{si}. However, for GeO_2-SiO_2 single-mode fibers, it may be dangerous to fix the level of Rayleigh scatter from such an "Inada plot", unless a direct measurement has demonstrated negligible absorption [Garrett and Todd 1982].

A second more accurate method [Stone F.T. 1982] is a modification of the above method. Additionally, it takes into account the UV band-edge absorption by adding an exponential term $C\exp(D/\lambda)$ to (13.7).

In the most recent method [Walker 1986], the total spectral loss is very accurately described over the 1.2 to 1.7 μm band by an equation with 26 fitting parameters. The fitting parameters, which characterize the different loss components, are obtained from the total loss data by applying a nonlinear regression algorithm. For a more simple approximate description of the spectral loss curve, the same author proposes a function with only 5 fitting parameters.

Instead of determining the various loss components from the total spectral attenuation, one can also measure them directly. This will be described in the following sections.

13.3.2 Attenuation Caused by Absorption

The partial attenuation coefficient $\alpha_a = 8.69\alpha_a'$ caused by absorption can be determined selectively by measuring both the power dP_a absorbed in a fiber element of length dz with a calorimeter and the power P transmitted through this element, and by using the formula (5.102)

$$\alpha_a' = \frac{1}{2P}\frac{dP_a}{dz} \quad . \tag{13.9}$$

Absorption losses are measured by differential calorimetry using a thermocouple [Pinnow and Rich 1973, 1975; Zaganiaris and Bouvy 1974; White K.I. 1976], a platinum resistance [Cohen R.L. et al. 1974; Stone F.T. et al. 1978], or a pyroelectric thermometer [Kashyap and Pantelis 1982].

The sensitivity limit of bolometric techniques for measuring fiber losses has been studied by R.L. Cohen [1974]. With 1 W of optical power, absorption losses as low as 0.05 dB/km can be measured. This corresponds to a temperature rise of only 0.1 mK.

The calorimetric techniques can also be used to measure *total* attenuation by metallizing the fiber [Cohen R.L. 1974] or by threading the fiber through a blackened quartz tube [Cohen R.L. et al. 1974; Stone F.T. et al. 1978]. By metallizing or blackening, the scattered light is also absorbed, i.e. converted to heat.

13.3.3 Attenuation Caused by Scattering

The partial attenuation coefficient $\alpha_s = 8.69\alpha'_s$ caused by scattering can be selectively determined by measuring both the power dP_s scattered from a fiber element of length dz with a scattering detector and the power P transmitted through this element, and by using the formula (5.103)

$$\alpha'_s = \frac{1}{2P}\frac{dP_s}{dz} \quad . \tag{13.10}$$

One type of scattering detector consists of a cube formed of six solar cells [Tynes 1970; Tynes et al. 1971], or more simply, of two solar cells which sandwich the fiber to be investigated [Dakin 1974]. In order to measure the transmitted power P with the same detector, the fiber may be broken and the broken end introduced between the detectors.

The scattering loss can also be measured with an integrating sphere made from a solid piece of fused silica which is coated with high-reflectivity barium sulfate [Ostermayer and Benson 1974].

In contrast to the cut-back method which gives the average attenuation coefficient of a long fiber, the calorimetric and scattering techniques measure the losses in a short fiber section of 1 cm to 30 cm length. Hence, they can be used to investigate the uniformity of the loss along the fiber.

The attenuation α_s caused by scattering has also been determined [Ferman et al. 1986] by comparing with an OTDR (Sect. 13.4) the powers backscattered from the test fiber and from a reference fiber with known backscatter loss (13.38).

13.3.4 Attenuation Caused by Macrobending

A macrobend is a bend with a radius of the order of a centimeter or more and in which the curvature exists over a length of fiber at least equal to that radius [Garg and Eoll 1986b].

For measuring macrobending loss quantitatively, one can wind the fiber from a drum with large radius (> 500 mm) onto a drum with small radius R, and measure the transmitted power as a function of the fiber length wound up on the small drum. For the fiber section on the large stock drum, the macrobending loss is negligibly small. Since the absorption, scattering, and transition losses do not change during winding, one gets the pure bend attenuation coefficient α_b as the change in attenuation per change in fiber length wound up on the small drum.

When both the large and the small drums are made to rotate during the measurement, it is difficult to couple light into the fiber, since, at present, there are no rotating optical joints for single-mode fibers (Sect. 12.2.7), and since future rotating joints will have an insertion loss which depends on the angle of rotation.

An elegant solution of this problem is to use a stationary large drum, so that the launching end of the fiber is fixed [Bachmann et al. 1987]. While rotating around its own axis, the small drum orbits the large stock drum like a planet. The detector end of the fiber is coupled to a photodetector mounted on the axis of the planet drum.

This arrangement allows bending loss measurements with a large number ($n > 1000$) of fiber turns with various macrobending radii, depending on the radius of the interchangeable planet drum. Measurements of bending induced losses as low as $0.05\,\mathrm{dB/km}$, i.e. $0.00002\,\mathrm{dB/loop}$ ($R = 60\,\mathrm{mm}$) have been performed using this set-up. It is very important to be able to measure such small bending losses, since the linear relation between $\log \alpha_b$ and bending radius R observed at $\alpha_b > 100\,\mathrm{dB/km}$ cannot be extrapolated to bending losses below $100\,\mathrm{dB/km}$ [Bachmann et al. 1987]. Extrapolation predicts bending losses that are too small.

Qualitatively, the bending sensitivity of single-mode fibers can be determined by a simple experiment, the so called 3 dB loop test [Lazay and Pearson 1982]. The power transmitted through a short length of fiber (about 3 m) is monitored as the radius R of a single circular loop in the fiber is decreased. That radius $R_{3\,\mathrm{dB}}$ that corresponds to a 3 dB drop in transmitted power compared to the unbent fiber condition is a good measure of the mode confinement and thus of the curvature induced loss sensitivity of the fiber. Fibers with small curvature induced losses have $R_{3\,\mathrm{dB}} < 10\,\mathrm{mm}$.

In order to test the loss performance of single-mode fibers designed for 1300 nm operation at 1550 nm, the CCITT [1986] recommends a macrobending test: If 1300 nm-optimized single-mode fibers are intended for use in the 1550 nm wavelength region, the loss increase for 100 turns of fiber loosely wound with a 3.75 cm radius, and measured at 1550 nm, shall be less than 2.0 dB. 3.75 cm is the radius of the excess fiber deployed in a typical splice case.

13.3.5 Attenuation Caused by Microbending

A microbend is a bend where the axis of the fiber is displaced at most a few tens of micrometers over a length of at most a few millimeters [Garg and Eoll 1986b]. Several methods have been used for measuring the microbend sensitivity of single-mode fibers.

In one method, a kilometer of fiber is wound under tension on a 150 mm diameter reel such that, about once each meter, the fiber passes over itself. In that way, crossovers of successive layers are deliberately introduced [Kaiser et al. 1979; Tomita et al. 1982; Lazay and Pearson 1982; Kamikawa and Chang 1986]. The fiber is considered to be insensitive to microbends if no added loss at the wavelength of operation up to a tension by a weight of 90 grams is produced [Nagel 1984]. Although this so-called "basket weave" test is a rather extreme measure of microbending sensitivity, it produces a rapid and valuable design tool to evaluate fiber performance.

Microbending sensitivity has also been tested by pressing the fiber against one of three deforming surfaces: sandpaper [Nakahara et al. 1983], a wire mesh with e.g. 200 μm wire spacing, and a bed of parallel wires with a spacing of

7.8 mm, and by measuring the loss increase as a function of wavelength [Garg and Eoll 1986b]. The microbending loss increases more slowly with increasing wavelength than the macrobending loss [Garg and Eoll 1986b]; the extra microbending attenuation at 1300 nm, for example, can be comparable to that at 1550 nm [Garg and Eoll 1986a].

13.3.6 Attenuation of the LP$_{11}$ Mode

Since the single-mode limit is determined by the attenuation of the LP$_{11}$ mode (Sect. 6.3.3), there is significant interest in methods for measuring this attenuation. Several methods have been proposed for measuring selectively, i.e. in the presence of the fundamental mode, the attenuation of the second-order LP$_{11}$ mode. The first method is a modification of the cut-back method [van Leeuwen and Nijnuis 1984]. Light of variable wavelength is coupled into a multimode fiber which is spliced to the single-mode fiber to be investigated. It is important that cladding modes launched in the single-mode fiber are stripped efficiently. This is achieved by stripping the primary coating from the fiber and by drawing the fiber input end through a liquid cell filled with high-refractive-index oil (Sect. 12.2.8). First the power P_1 emerging from the end of the single-mode fiber with length L_1 (e.g. 1 m) is measured as a function of wavelength. Then the single-mode fiber is cut-back to a length L_2 of e.g. 5 cm and the power P_2 emerging from the new end is measured. Finally, the multimode fiber is broken and the power P_3 emitted from the multimode fiber end is measured as a function of wavelength. Since the fiber is cut-back twice, the method can be called the double-cut-back method.

Figure 13.1 schematically represents the ratios P_1/P_3 and P_2/P_3 of the powers in the single- and multimode fibers as functions of wavelength. For long wavelengths, the field in the multimode fiber only excites the fundamental mode of the single-mode fiber. The coupling efficiency between the two fibers is then of the order of only 1%, and, therefore, the coupling loss is about 20 dB (Sect. 9.7). For smaller wavelengths, the LP$_{11}$ mode becomes a guided wave and

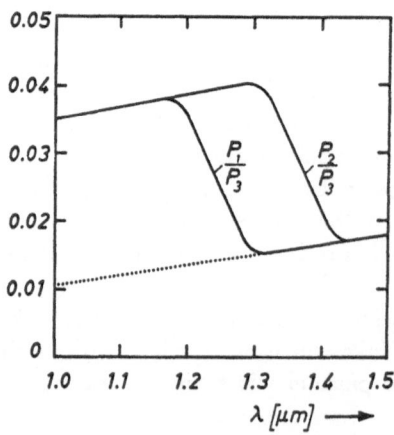

Fig. 13.1. Double cut back method for measuring the attenuation of the LP$_{11}$ mode: measured power ratios as functions of wavelength

will be excited by the fields in the multimode fiber. Therefore, the total power P_2 in the single-mode fiber will increase below the LP_{11} cutoff wavelength. Because of the relatively high attenuation of the LP_{11} mode, the total power P_1 for the larger length L_1 will be smaller than P_2.

The attenuation coefficient α_{11} of the LP_{11} mode can then be calculated from the three measured powers by

$$\alpha_{11} = \frac{4.34}{L_1 - L_2} \ln \left(\frac{P_1/P_3 - \eta_{01}}{P_2/P_3 - \eta_{01}} \right) . \tag{13.11}$$

The quantity η_{01} is the efficiency with which the multimode fiber excites the fundamental modes in the single-mode fiber. In the single-mode region, i.e. for large wavelengths, the total power in the single-mode fiber is the power of the LP_{01} mode only and the coupling efficiency is simply given by

$$\eta_{01} = P_1/P_3 = P_2/P_3 . \tag{13.12}$$

Because of the very small attenuation constant of the fundamental mode and the small difference $L_1 - L_2$ in fiber lengths, one finds experimentally $P_1 = P_2$. In the two-mode region, the coupling efficiency must be obtained by linear extrapolation as illustrated by the dashed straight line in Fig. 13.1.

Experimentally, one finds [van Leeuwen and Nijnuis 1984] that for small wavelengths the attenuation of the second order modes is very small. (From theory it follows that it is comparable to that of the fundamental mode). For wavelengths near cutoff, α_{11} increases very rapidly (Figs. 5.27 and 13.2), and for larger wavelengths the attenuation is practically infinitely large, i.e. the LP_{11} mode ceases to be guided by the fiber.

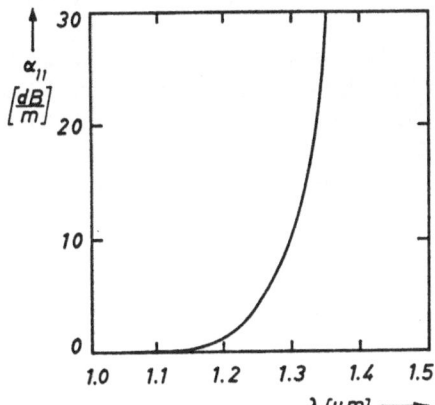

Fig. 13.2. Measured attenuation constant of the LP_{11} mode around cut off [van Leeuwen and Nijnuis 1984]

In a second method for measuring the LP_{11} mode attenuation [Ohashi et al. 1984], six optical powers have to be measured as functions of wavelength. First (Fig. 13.3.a), the LP_{11} mode power is completely attenuated near the input by a bend while the bending loss of the fundamental mode is negligible. P_1 is the output power after a length L and P_2 is the power after a cutback length of 2 m. In the second experiment (Fig. 13.3.b), there is no loop to

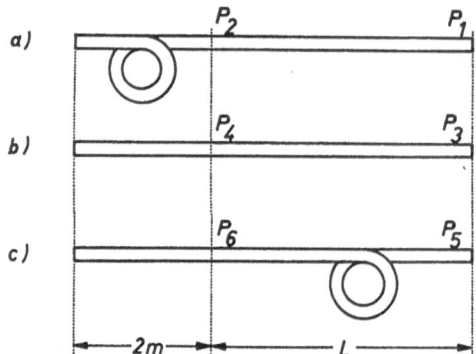

attenuate the second order mode. P_3 and P_4 are the powers measured after lengths of L and $2\,\mathrm{m}$, respectively. In the last experiment (Fig. 13.3.c), there is a bend at the fiber end used as a mode filter (Sect. 12.2.14) to eliminate the LP_{11} mode. The powers measured for a length L and a length of $2\,\mathrm{m}$ are P_5 and P_6, respectively.

The excitation at the fiber input end should be identical for experiments b) and c), i.e. $P_6 = P_4$. With this triple-cut-back method, one can separately determine the LP_{01} mode attenuation

$$\alpha_{01} = \frac{10}{L} \log\left(\frac{P_2}{P_1}\right) \quad , \tag{13.13}$$

and the LP_{11} mode attenuation

$$\alpha_{11} = \frac{10}{L} \log\left(\frac{P_1 P_4 - P_2 P_5}{P_1 P_3 - P_1 P_5}\right) \quad . \tag{13.14}$$

With a third method [Nijnuis and van Leeuwen 1984], one can measure the bending loss of the LP_{11} mode. First (Fig. 13.4), the output power P_1 is measured for the desired bend radius R. Then, a few small loops (1 to 2 cm radius) are inserted in the fiber as a mode filter (Sect. 12.2.14) to remove all light in the LP_{11} mode (output power P_2). Finally, one measures the output power P_3 of the straight fiber. When the launching conditions are maintained and when the fundamental mode loss is the same for the three experiments, the loss of the LP_{11} mode caused by bending the fiber to a radius of R can be calculated from the three measured powers:

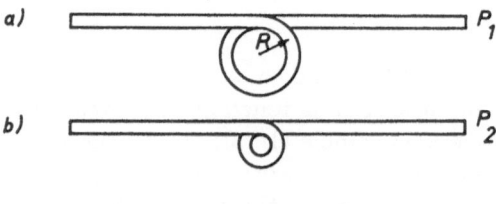

Fig. 13.4. Method for measuring the bending loss of the LP_{11} mode

$$\alpha_{11} = \frac{10}{L} \log \left(\frac{P_3/P_2 - 1}{P_1/P_2 - 1} \right) \quad . \tag{13.15}$$

In a fourth, heterodyne interferometric method [Shibata et al. 1985; Tsub-okawa et al. 1986], a variant of the interferometric technique for measuring chromatic dispersion (Sect. 13.9.3) is used to measure the bend loss of the LP_{11} mode. By an acousto-optic modulator driven at a frequency f_m, laser light is split into a reference beam with optical frequency f and a signal beam with frequency $f + f_m$. The reference beam is transmitted through a strictly single-mode fiber and a variable optical delay line, and the signal beam through the dual-mode fiber under test. The two beams are then combined and detected by a Ge APD. For a certain length of the delay line, the delays of the LP_{01} mode in the reference arm and the LP_{11} mode in the test arm of the interferometer are equal, and the coherent beams produce a beat signal at the modulation frequency f_m. The intermediate frequency voltage is proportional to the amplitude of the transmitted LP_{11} mode. By winding different lengths of the test fiber onto drums with various bend radii and observing the change in the amplitude of the transmitted LP_{11} mode, its bend loss can be determined.

For another length of the delay line, the fundamental mode transmitted through the reference fiber can be made to beat with that of the test fiber. From the change in length of the delay line, the difference in the group delays of the LP_{01} and LP_{11} modes can be evaluated.

In a fifth method [Biet and Pocholle 1983], the attenuation of the LP_{11} mode is measured with an optical time domain reflectometer (Sect. 13.4).

A sixth direct measurement of the LP_{11} mode attenuation can be achieved provided one has means to consistently generate and selectively detect the LP_{11} mode. The LP_{11} mode can be generated at a splice with transverse offset. A suitable detector for the LP_{11} mode is a mandrel wrap local detector which selectively taps and detects a constant fraction of the LP_{11} power while not disturbing the fundamental mode power. The attenuation coefficient can be calculated from the tapped powers $P_{11}(L_1)$ and $P_{11}(L_2)$ measured at two different distances L_1 and L_2 from the splice [Sears et al. 1986a].

There is a close relation between the attenuation of the LP_{11} mode and its effective cutoff wavelength (Sect. 6.3.3). At the effective cutoff wavelength λ_{ce} as defined by the CCITT [1984] and measured by the bending method (Sect. 13.8.1), the attenuation a_{11} of the LP_{11} mode in the 2 m test fiber is 19.34 dB (6.48) and the attenuation related to the length of the loop with 14 cm radius (6.50) is

$$\alpha_{11} = 22\,000\,\text{dB/km} \quad . \tag{13.16}$$

13.4 Backscattering Method

The backscattering method is one of the most important measuring techniques in optical communications, since it allows one to determine nondestructively local fiber attenuation, splice loss, fiber length, and the location of connectors, splices, and faults.

The backscattering method is similar to the time-domain reflectometry well known from radio-frequency techniques: a short light pulse is injected into the fiber input end, and the light coming back out of the fiber is detected by a receiver also located at the fiber input end. But there is one important difference: the reflection factor of the open end of a coaxial cable or a short circuit is 100%, whereas the power reflection from an ideal fiber end is only 3.5%. (5.131). For an oblique or ragged endface, the power reflection factor can be much smaller than 3.5% [Marcuse 1975c]. At a splice with a transverse core offset, there will be a loss (Sect. 9.2), but no reflection, because there is no interface between two media with appreciably different refractive indices. Thus, the location of a splice cannot be found by observing reflections. However, there is a remedy, as has been suggested by Kapron et al. [1972] and first implemented by Barnoski and coworkers [Barnoski and Jensen 1976, Barnoski et al. 1976, 1977], and Personick [1977]: the backscattering method.

In the fiber, the guided mode continuously loses a small part of its power by Rayleigh scattering (Sect. 5.9.2). This scattered power is radiated roughly isotropically. A very small part of the scattered power is scattered into the backward direction and launches a fundamental mode guided backwards to the receiver at the input end. This backscattered wave can be detected by a very sensitive photoreceiver and the quantities mentioned at the beginning of this section can be deduced.

Several reviews of the backscattering method can be found in the literature [Danielson B.L. 1981, 1982; Wright S. et al. 1983a; Hartog 1984; Healey 1985b, 1986; DuPuy 1986].

In the following, we describe the measuring setup (Sect. 13.4.1), the fundamentals of the theory (Sect. 13.4.2), methods for evaluating measured backscattering signatures (Sect. 13.4.3), and the design and performance of practical backscattering apparatus (Sect. 13.4.4).

13.4.1 Backscattering Setup

The main components of an optical backscattering measurement setup are (Fig. 13.5): an optical pulse generator, an optical directional coupler which couples the pulse into the fiber sample and couples the backscattered light onto a photoreceiver, and electronic circuits which amplify and average the backscattered signal, take the logarithm of it, and display it as a function of the location along the fiber.

Although one detects not only reflections, but also the much weaker backscattered signal, the common name for the backscattering measurement setup is Optical Time Domain Reflectometer, or OTDR for short.

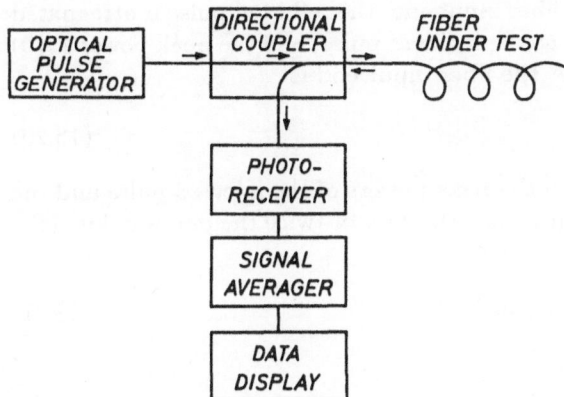

Fig. 13.5. Basic building blocks of an optical OTDR

13.4.2 Theory of the OTDR

To understand the backscattering method, it is best to begin with an analysis of the mode of operation in which a reflection is observed [Guttmann and Krumpholz 1975; Ueno and Shimizu 1976]. A rectangular pulse with peak power $P(0)$ and duration τ is injected into the fiber input located at $z = 0$. A discontinuity with power reflection factor R is assumed to be located at a distance z_d from the fiber input end, e.g. a connector with an air gap or the fiber end. During propagation from the fiber input end to the discontinuity, the pulse experiences an attenuation in decibels $a(z_d)$, which is the sum of the attenuations of the individual fiber sections and any splices and connectors. The attenuation $a(z)$ from the fiber input to an arbitrary point at distance z from the input is often named the "one-way loss".

Neglecting pulse broadening by chromatic dispersion, the peak power of the pulse incident on the discontinuity is (5.112)

$$P(z_d) = P(0)10^{-a(z_d)/10} \quad , \tag{13.17}$$

and the power of the pulse reflected from the discontinuity

$$P_r(z_d) = RP(z_d) \quad . \tag{13.18}$$

It is useful to define a "return loss", which in this mode of operation is a "reflection loss",

$$a_r = -10\log R \quad . \tag{13.19}$$

For an ideal fiber endface in air, the power reflection coefficient is $R = 0.035$ (5.131) and the reflection loss $a_r = 14.6\,\mathrm{dB}$.

In single-mode fibers, the insertion loss of fiber sections, connectors, or splices is independent of the direction of wave propagation (this is in contrast to multimode OTDR's, where the loss depends on the mode spectrum, which in general will be different for the two directions of wave propagation). Therefore, during propagation back to the fiber input end, the reflected pulse is attenuated by the same one-way loss $a(z_d)$ as the incident pulse, and the peak power $P_r(0)$ of the reflected pulse arriving at the fiber input end is

$$P_r(0) = P_r(z_d)10^{-a(z_d)/10} .\qquad(13.20)$$

The level difference between the peak powers of the injected pulse and the reflected pulse at the fiber input equals the sum of twice the one-way loss (the "two-way loss") and the return loss:

$$10\log[P(0)/P_r(0)] = 2a(z_d) + a_r \quad .\qquad(13.21)$$

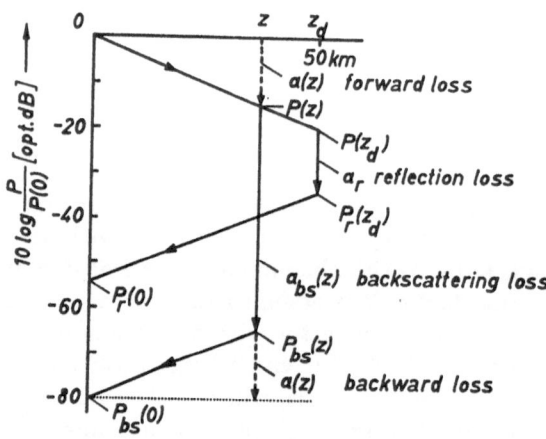

Fig. 13.6. Power levels $10\log[P(z)/P(0)]$, $10\log[P_r(z)/P(0)]$, and $10\log[P_{bs}(z)/P(0)]$ of the incident, the reflected, and backscattered waves, respectively, along the fiber. The reflected wave is due to the fiber end at z_d, and the backscattered wave due to scattering in a length element at z

Figure 13.6 graphically represents the power levels $10\log[P(z)/P(0)]$ and $10\log[P_r(z)/P(0)]$ of the incident and the reflected wave, respectively, as functions of the longitudinal coordinate. A fiber with an attenuation of 0.4 dB/km, and a reflecting discontinuity with a reflection loss of 14 dB (an ideal fiber end) at a distance $z_d = 50$ km from the fiber input end have been assumed. The one-way loss is

$$a(z_d) = 0.4\,\text{dB/km} \cdot 50\,\text{km} = 20\,\text{dB} \quad .\qquad(13.22)$$

At the fiber input, the reflected signal (13.21) is $2 \times 20\,\text{dB} + 14\,\text{dB} = 54$ optical decibels smaller than the injected signal.

Relative to the injected pulse, the reflected pulse observed at the fiber input end is delayed by the time

$$t = 2v_g z_d \qquad(13.23)$$

needed for the light to propagate to the reflecting discontinuity and back to the fiber input. Therefore, for known group velocity (5.68)

$$v_g = c/n_{ge} \tag{13.24}$$

of the LP_{01} mode, the distance z_d of the discontinuity from the fiber input end can be determined from the time delay.

The fiber group index n_{ge} is best determined by measurement, rather than from a calculated or measured "core refractive index". Since the fiber is stranded within a cable structure, it is typically 0.1% longer than the cable, thereby increasing the delay per unit cable length [Kapron et al. 1986]. For the cable situation, the excess length of the fiber over that of the cable is implicitly contained in the effective group index n_{ge}.

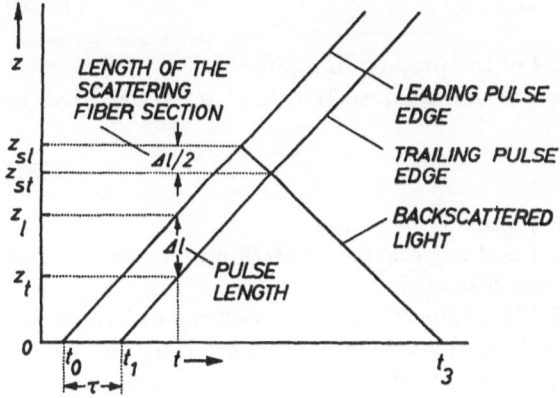

PULSE DURATION   Fig. 13.7. Timetable for an OTDR

Now, for understanding the backscattering method, instead of considering the pulse reflected from a discontinuity, we consider the light backscattered from a short fiber section. Figure 13.7 is a graph of location versus time for the light propagating in the fiber. The transmitted pulse begins at time t_0 and ends at $t_1 = t_0 + \tau$ where τ is the pulse duration. The fiber input end is assumed to be located at $z = 0$. Since the pulse propagates with the group velocity v_g, at time t the leading and trailing pulse edges are at the locations

$$z_l = v_g(t - t_0) \quad , \quad \text{and} \tag{13.25}$$

$$z_t = v_g(t - t_1) \quad , \tag{13.26}$$

respectively. Thus, the spatial length of the pulse propagating through the fiber is

$$\Delta l = z_l - z_t = v_g(t_1 - t_0) = v_g\tau \quad . \tag{13.27}$$

The backscattered signal is observed at the time t_3. During the time interval $t_3 - t_0$, the leading pulse edge propagates a distance $v_g(t_3 - t_0)$. Since the light must propagate to the scattering point and back to the fiber input end, the light from the leading pulse edge is being scattered at a point with distance

$$z_{sl} = v_g(t_3 - t_0)/2 \qquad (13.28)$$

from the fiber input end. Correspondingly, light from the trailing pulse edge is scattered at

$$z_{st} = v_g(t_3 - t_1)/2. \qquad (13.29)$$

The length of the backscattering fiber section is

$$z_{sl} - z_{st} = v_g(t_1 - t_0)/2 = \Delta l/2 \quad , \qquad (13.30)$$

that is only half of the length Δl of the propagating pulse!

The location of the backscattering fiber section is characterized by the coordinate

$$z = (z_{st} + z_{sl})/2 \qquad (13.31)$$

of the point in the center of the section.

Between the fiber input end and the scattering section at z, the incident pulse is attenuated by the one-way loss $a(z)$.

Because of the definition (5.103) of the attenuation coefficient α'_s caused by scattering, the power $P_s(z)$ Rayleigh scattered out of the scattering fiber section with length $\Delta l/2$ is two times the product of the power $P(z)$ transmitted, the attenuation coefficient α'_s caused by scattering in Nepers/km, and the length $\Delta l/2$ of the backscattering section:

$$P_s(z) = P(z)2\alpha'_s(z)\Delta l/2 = P(z)\alpha'_s(z)\Delta l \quad . \qquad (13.32)$$

In general, the attenuation coefficient caused by scattering will depend on the location z.

A scattering loss $a_s(z)$ is defined by the level difference between the pulse power $P(z)$ transmitted through the backscattering fiber element and the power $P_s(z)$ scattered out of this element:

$$a_s(z) = 10\log[P(z)/P_s(z)] \quad . \qquad (13.33)$$

Using (13.32), the scattering loss can be expressed in terms of the scatter attenuation coefficient α'_s and the length Δl of the propagating pulse

$$a_s(z) = -10\log[\alpha'_s(z)\Delta l] \quad . \qquad (13.34)$$

As an example, we take $\lambda = 1.3\,\mu\text{m}$. From (5.121) for $A = 1\,\text{dB}\mu\text{m}^4/\text{km}$, the attenuation constant caused by scattering can be estimated to be $a_s =$

0.35 dB/km or $\alpha_{s}' = 0.04$ Nepers/km. For a pulse duration of $\tau = 1\,\mu$s, which seems to have become a de facto standard in single-mode reflectometry [Healey 1985b] and a group velocity v_g of about 200 000 km/s, the pulse length is $\Delta l = 200$ m and the length of a backscattering fiber element 100 m. From (13.34), the scattering loss is found to be $a_s(z) = 21$ dB.

Only a small part $P_{bs}(z)$ of the scattered power $P_s(z)$ is scattered into the backward direction and coupled into the backward propagating fundamental mode. A capture fraction $S(z)$ is defined [Neumann 1980] by

$$S(z) = P_{bs}(z)/P_s(z) \quad , \tag{13.35}$$

and a capture loss by

$$a_c(z) = -10 \log S(z). \tag{13.36}$$

In general, the capture fraction and the capture loss will depend on the location z along the fiber axis.

The capture fraction S (5.123) depends on the ratio of the spot size w_G and the vacuum wavelength λ

$$S = 0.038[\lambda/(n_2 w_G)]^2 \quad . \tag{13.37}$$

For a wavelength of $1.3\,\mu$m and a spot size of $5\,\mu$m, the capture fraction is $S = 0.0011$ and the capture loss 29 dB.

The level difference $a_{bs}(z)$ between the peak power $P(z)$ of the incident pulse and the backscattered power $P_{bs}(z)$ is then the sum of the scattering loss (13.34) and the capture loss (13.36):

$$a_{bs}(z) = 10 \log[P(z)/P_{bs}(z)] = a_s(z) + a_c(z) \quad . \tag{13.38}$$

Formally, this "backscattering loss" or "Rayleigh scatter capture level" [Healey and Smith 1982] corresponds to the reflection loss introduced by (13.19). Reflection loss and backscattering loss are two kinds of return loss. The backscattering loss is very large, in the above numerical example 50 dB.

During propagation back to the fiber input end, the backscattered light is attenuated by the one-way loss $a(z)$.

The level difference between the peak power $P(0)$ of the injected pulse and the backscattered power $P_{bs}(0)$ at the fiber input is twice the one-way loss plus the backscattering loss:

$$10 \log[P(0)/P_{bs}(0)] = 2a(z) + a_{bs}(z) \quad . \tag{13.39}$$

Formally, this equation corresponds to the total loss in the mode of operation in which one observes a reflecting discontinuity (13.21).

Figure 13.6 also represents the power level $10 \log[P_{bs}(z)/P(0)]$ of the wave backscattered from a length element at a distance $z = 37.5$ km from the fiber input end. A fiber attenuation of 0.4 dB/km and a backscattering loss of 50 dB have been assumed. At the fiber input, the backscattered signal is (13.39)

$$2 \times 0.4\,\text{dB/km} \ \times 37.5\,\text{km} \ + 50\,\text{dB} \ = 80\,\text{dB} \tag{13.40}$$

smaller than the injected signal!

Comparing the loss formulae for the reflected signal (13.21) and the backscattered signal (13.39), the backscattered signal is found to be $a_{\text{bs}}(z) - a_{\text{r}}$ decibels smaller than the reflected pulse [Neumann 1978]. Comparing the backscattering loss of our example (50 dB) and the reflection loss of an ideal fiber endface (14 dB), the backscattered signal is 36 optical dB smaller than the Fresnel reflection. Since this number corresponds to 72 electrical decibels (5.118), one sees that amplifiers with very high dynamic range must be used when one wants to evaluate a Fresnel reflection and the backscattered signal simultaneously. The strong Fresnel reflections caused by the fiber input face might in some cases saturate the receiver.

The light backscattered from a fiber element at a distance z (13.31) arrives back at the fiber input end at the time

$$t_3 - t_0 = 2z_{\text{sl}}/v_{\text{g}} \tag{13.41}$$

after the leading edge of the input pulse. By adding (13.30) and (13.31), z_{sl} can be expressed in terms of the location z of the backscattering element and the length Δl of the propagating pulse

$$t_3 - t_0 = 2(z + \Delta l/4)/v_{\text{g}} \quad . \tag{13.42}$$

This relation can be used to determine the location z of the scattering element from the time delay $t_3 - t_0$. Note however, that in our analysis, a rectangular light pulse has been assumed. For pulses with a finite rise time, the deduced fiber length depends on whether the fiber end is reflecting or not. Therefore, for accurately locating a nonreflecting fault, one has to subtract an empirical correction factor from the distance detected by the OTDR [So et al. 1986].

An OTDR displays a so-called backscatter signature $B(z)$, which is five times the logarithm of the backscattered power $P_{\text{bs}}(0)$ as a function of the distance z of the backscattering fiber section from the fiber input end. Using (13.42), the time axis is calibrated in terms of z. The actual value of the group velocity v_{g} or the equivalent quantities specific group delay (5.69)

$$\tau_{\text{g}} = 1/v_{\text{g}} \tag{13.43}$$

or effective group index (5.68)

$$n_{\text{ge}} = c/v_{\text{g}} \tag{13.44}$$

must be supplied (inserted into the OTDR). From (13.39) one obtains for the backscattering signature

$$B(z) = 5 \log P_{bs}(0) = C - a(z) - 0.5 a_{bs}(z) \quad , \qquad\qquad (13.45)$$

where C is a constant. Equation (13.45) is the fundamental formula for the backscattering method.

In the preceding section, the formula for the backscattering signature has been derived in simple terms. Several authors made a more complete analysis of the backscattering method [Barnoski 1976; Personick 1977; Neumann 1980; Brinkmeyer 1980b; Rogers 1981; Hartog and Gold 1984]. In one analysis of single-mode OTDR's [Philen et al. 1982], a factor of $\frac{1}{2}$ is missing in the formula for the backscattered power [Brinkmeyer and Neumann 1984; Philen et al. 1984].

The backscattered power calculated previously is a time averaged value. When the injected light is monochromatic, the backscattered power fluctuates slowly due to slow variations of temperature, pressure, strain, or optical frequency [Eickhoff and Ulrich 1981a].

13.4.3 Evaluation of Backscattering Signatures

We now address the question of how to determine the fiber attenuation or splice loss from a measured backscattering signature $B(z)$ (13.45). Most often, the backscattering signature is evaluated by assuming the backscattering loss $a_{bs}(z)$ (13.38) to be constant along the fiber waveguide, i.e. independent of the coordinate z of the scattering section. Later we will explain methods for determining the fiber attenuation or the splice loss if this assumption cannot be made.

For constant backscattering loss a_{bs}, neglecting the peaks caused by reflections from the fiber input (S) and output (F) ends and other reflecting discontinuities (R), it can be seen from (13.45) that the backscattering signature simply represents the relative signal level $-a(z)$ of the incident wave along the fiber length (Fig. 13.8).

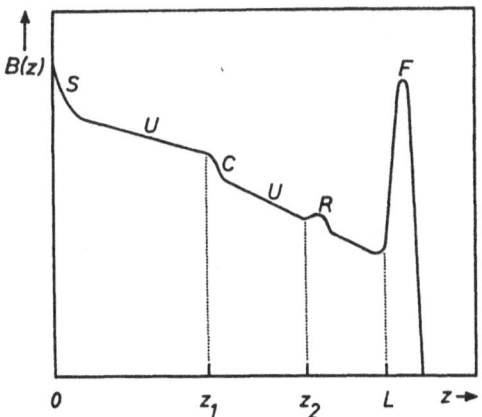

Fig. 13.8. Typical backscatter signature:
U: uniform fiber section
C: connector or splice without reflection
R: reflecting fiber discontinuity
F: Fresnel reflection from the fiber end

The one-way loss up to a point z_1 is $a(z_1)$, and the one-way loss up to another point z_2 with $z_2 > z_1$ is $a(z_2)$. The attenuation a of the incident wave between points z_1 and z_2 is the difference in the one-way losses, i.e.

$$a = a(z_2) - a(z_1) \quad . \tag{13.46}$$

Using the formula (13.45) for the backscattering signature, the attenuation can be expressed in terms of the backscattering levels $B(z_1)$ and $B(z_2)$ which can be read from the OTDR:

$$a = B(z_1) - B(z_2) \quad . \tag{13.47}$$

Thus, the fiber attenuation between two cross sections z_1 and z_2 along the fiber is then equal to the difference in the backscattering levels at the two points.

In uniform fiber sections (U in Fig. 13.8), the attenuation a is proportional to the distance $z - z_1$ between the two points, where z_1 is fixed and $z_2 = z$ is an independent variable:

$$a(z) = B(z_1) - B(z) \quad . \tag{13.48}$$

Uniform fiber sections give straight backscattering signatures. Taking the derivative of (13.48) with respect to the variable z gives

$$\alpha(z) = da(z)/dz = -dB(z)/dz \quad , \tag{13.49}$$

showing that the negative slope of the backscattering signature gives the local attenuation coefficient $\alpha(z)$.

Good agreement between the attenuation values measured with the backscattering and cut-back methods has been reported [Cisternino et al. 1985, 1986]. In a commercial single-mode fiber transmission system, the difference between attenuation values measured by the backscattering and the cut-back methods was only 0.02 dB/km with a standard deviation of 0.09 dB/km [Ishihara et al. 1983].

The presence of a connector or a splice (C in Fig. 13.8) causes a sudden decrease in the backscattering signature, since both the incident wave and the backscattered wave are attenuated by the joint. For very short pulses, a lossy joint causes a step in the backscattering signature. The location of the discontinuity can be read from the position z_1 of the beginning of the steep decrease, and the insertion loss of the discontinuity can be obtained by extrapolating the straight sections of the backscattering curve and reading the level difference at the splice location. Several commercial OTDR's automatically calculate the joint loss. The operator only has to move the cursor to 3 or 4 characteristic points on the backscattering curve.

Unlike more conventional TDR measurements, a reflection from a discontinuity is not required for determining the position and the loss of the discontinuity.

If the fiber ends in an index matching fluid, the Fresnel reflection F from the fiber end shown in Fig. 13.8 will be absent. Nevertheless, the fiber end can be recognized from the cessation of the backscattered signal, and the fiber length L can be read from the beginning of the steep decrease in backscattered signal.

The spatial resolution of an OTDR, i.e. its ability to distinguish between two defects which occur in close proximity, is determined not only by the emitted pulse length τ, but also by the postdetection bandwidth and the digital sampling interval. B.L. Danielson [1985] proposed a working definition of the spatial resolution limit R as a measure of the smallest spacing of *two* reflecting elements which can just be resolved:

$$R = \tau_\mathrm{m} v_\mathrm{g}/2 \quad , \tag{13.50}$$

where τ_m is the measured full width at half-maximum (FWHM) power of a discrete reflection as observed on the OTDR display. On the usual logarithmic scale, this corresponds to the pulse width at the $5\log(1/2) = -1.5\,\mathrm{dB}$ backscattering signature level.

Since the duration τ_m of the displayed pulse is somewhat longer than that (τ) of the transmitted pulse, the spatial resolution limit R is somewhat larger than the length $\Delta l/2$ (13.30) of a backscattering fiber section. For a displayed pulse width of $\tau_\mathrm{m} = 1\,\mu\mathrm{s}$, the spatial resolution is 100 m. Using pulses of 100 ps duration, a two-fault resolution of the order of 1 cm has been demonstrated [Ripamonti and Cova 1986].

The location of a *single* fault can, in principle, be measured with resolution better than $\Delta l/2$. Assume the laser pulses to have a short rise time, the receiver bandwidth and the sampling rate to be very large, and the noise to be very small. The beginning of the steep transition in the backscattering signature caused by the fault can then be determined with an uncertainty which is much smaller than the length $\Delta l/2$ of the transition in the backscattering signature. Thus one has to distinguish between two kinds of resolution: that for recognizing two adjacent faults, which is $\tau_\mathrm{m} v_\mathrm{g}/2$ (13.50), and that for locating a single fault, which can be smaller [Kimura et al. 1985a]. Note, however, that the accuracy with which a single fault can be located also depends on informing the OTDR about the exact value of the group velocity. Due to the uncertainties in the excess length of the fiber in a cable, in the exact cable route, and in the spare fiber lengths in splicing cabinets, it is difficult to determine the group velocity to better than several tenths of one per cent.

When the backscattering loss $a_\mathrm{bs}(z)$ (13.38) depends on the location z, the simple correspondence (13.48) between the backscattering signature $B(z)$ and the signal level $-a(z)$ no longer holds. For instance, when jointing different fibers, backscattering curves are sometimes found in which the splice or connector causes the backscattered signal to increase, instead of decreasing. Of course, the discontinuity cannot amplify the signal. The explanation for this observation is simply that the backscattering loss of the fiber behind the splice is smaller than that before the splice. Stated in other terms, the product of the attenuation coefficient caused by scattering α_s and the capture fraction S is larger in the second fiber.

When the spot size changes monotonically along the fiber, the capture fraction S (13.37) is also z-dependent. Since the backscattering loss $a_{bs}(z)$ is not constant, the slope of the backscattering curve (13.49) does not give the true fiber attenuation α!

Several methods have been reported for evaluating backscattering signatures when the backscattering loss $a_{bs}(z)$ is not constant. In the first method, two backscattering curves must be measured, one from each fiber end [Di Vita and Rossi 1979, 1980]. Since the backscattering loss for single-mode fibers is independent of the direction of wave propagation, one can write for the backscatter signatures (13.45) $B_a(z)$ and $B_b(z)$ measured from end A and B, respectively:

$$B_a(z) = C_a - a(z) - 0.5a_{bs}(z) \quad , \tag{13.51}$$

$$B_b(z) = C_b - [a(L) - a(z)] - 0.5a_{bs}(z) \quad . \tag{13.52}$$

Note that the backscattered signature observed at end B has a one-way loss $a(L) - a(z)$.

Simply by adding and subtracting the two backscatter signatures, the information about the length dependence of the backscattering loss $a_{bs}(z)$ and about the signal level $-a(z)$ can be obtained separately:

$$a_{bs}(z) = -B_a(z) - B_b(z) - a(L) + C_a + C_b \quad , \tag{13.53}$$

$$a(z) = 0.5[B_b(z) - B_a(z) + a(L) + C_a - C_b] \quad . \tag{13.54}$$

To graphically relate the signals backscattered from the *same* fiber element, it is necessary to plot the signature $B_b(z)$ from the right-hand side to the left-hand side of the diagram. In equations (13.53, 54), the last three terms are unimportant constants.

The attenuation constant of a uniform fiber section can be obtained from the slope of the signal level $-a(z)$ (13.54) as has been explained before for constant backscattering loss, but without having to assume that the backscattering loss is a constant.

When two fibers with different mode field diameters are spliced together, backscattering is stronger from the fiber with the smaller spot size, since the capture fraction S is inversely proportional to the square of the spot size (13.37). The mismatch of the mode field diameter also causes a splice loss, which, however, is small compared to the step in the backscattering signature caused by the change in capture fraction [Kapron et al. 1986]. The true splice loss caused by offset, tilt, core deformation, or other effects is simply the arithmetic average of the apparent splice loss values measured at ends A and B [Matthijsse and de Blok 1979a]. This rule can be derived from (13.51, 52). The splice loss measured by this method agrees to within $0.03\,\mathrm{dB}$ with the splice loss measured by the cut-back method [Anderson W.T. et al. 1986].

From the curve $a_{bs}(z)$ (13.53) of the backscattering loss versus length, the fiber manufacturer can recognize whether the fiber is uniform, i.e. has constant product $\alpha_s S$, or whether there are changes in the structural parameters, e.g. relative refractive index difference, along the length of the fiber [Gold and Hartog 1982a,b].

If the length L of an installed fiber cable is not too large and if a long-range OTDR is available, for measuring the backscattering curves from both ends, it is not necessary to carry the OTDR from one fiber end to the other, which may be many kilometers away. One simply has to connect fiber end A to the backscattering apparatus and to apply a reflector at the far end B [Gold et al. 1984]. Then, during the time interval from 0 to $2L/v_g$ the backscatter signature $B_a(z)$ is observed, and during the time interval $2L/v_g$ to $4L/v_g$ the signature $B_b(z)$ will be displayed. This "folded-path OTDR" technique also provides the information necessary for the alignment of the fiber ends before splicing.

An alternative technique for measuring the real splice loss from only one fiber end consists in measuring the apparent splice loss at two different wavelengths, i.e. $1.3\,\mu\mathrm{m}$ and $1.55\,\mu\mathrm{m}$. The splice loss at $1.3\,\mu\mathrm{m}$ is obtained by multiplying the difference of the apparent splice losses at 1.3 and $1.55\,\mu\mathrm{m}$ by an empirical factor of 4 [Furukawa and Koyamada 1986].

Finally, the true splice loss can be measured from only one end of the fiber link if the ratio of the backscattering losses a_{bs} of the two fibers is measured separately beforehand [Matthijsse and de Blok 1979b].

An OTDR can also be used to measure the attenuation of the unwanted second order LP_{11} mode [Biet and Pocholle 1983]. In the dual mode region $(\lambda < \lambda_c)$, the total backscattered power is the sum of the powers backscattered from the LP_0 and LP_{11} modes. By inserting a mode filter (Sect. 12.2.14), e.g. a fiber loop with small diameter, the LP_{11} mode can be suppressed and the backscattering signature of the fundamental mode measured separately. Then, by subtracting the LP_{01} mode power from the total backscattered power, the backscattering signature of the LP_{11} mode can be determined and evaluated with respect to attenuation.

13.4.4 Design and Performance of Practical OTDR's

In this section, we review the design and performance of practical single-mode OTDR's. First, we address the problems of polarization sensitivity and of the extremely low level of backscattered light. The range of an OTDR can be increased by signal averaging, by using higher pulse powers, by correlation techniques, or by coherent detection. Saturation of the optical receiver due to Fresnel reflections can be avoided by optical gating with an acousto-optic deflector. Finally, some performance data of OTDR's will be given.

The design of single-mode OTDR's has been described both in review papers [Wright S. et al. 1983a; Healey 1985b, 1986] and in a large number of original papers [Healey 1981b,c; Heckmann et al. 1981; Okada 1982; Healey and Smith 1982; Gold and Hartog 1983a,b, 1984a,b; Bernard et al. 1984b; Tokuda et al. 1984; Horiguchi T. and Tokuda 1984; Stone J. et al. 1985; Cottrell and Brain 1986].

In general, the state of polarization (Sect. 5.3.5) of the detected backscattered light is different from that of the laser pulse coupled into the fiber input end and depends on the distance z of the backscattering fiber element from the fiber input end. On the one hand, this effect can be utilized to study the evolution of the polarization in the fiber with the so-called Polarization Optical Time Domain Reflectometry (POTDR). On the other hand, the fluctuation of the polarization of the backscattered light can cause errors in single-mode OTDR's. Polarization effects in single-mode optical time domain reflectometry have been analyzed by several authors [Brinkmeyer 1981; Rogers 1981; Ross 1982; Nakazawa et al. 1983; Nakazawa 1983a].

With conventional OTDR's, one is usually not interested in the birefringence of the fiber, but only in the decay $-a(z)$ of the total transmitted power along the fiber, which can be deduced from the total backscattered power $P_{bs}(0)$ received (13.45). Since the polarization of the received backscattered light depends on the distance z of the scattering element from the fiber input, it is very important to avoid devices between the fiber and the detector whose insertion loss depends on the state of polarization [Hartog et al. 1981]. The backscattering signature of a uniform fiber will not be a straight line if the detection system is polarization sensitive. Thus, the polarized nature of the backscattered signal can cause errors in conventional backscattering measurements.

At oblique incidence, the reflection and transmission factors of single (fiber endface) or multiple (interference filters or mirrors) dielectric interfaces depend on the polarization. At normal incidence, the reflection and transmission factors of the surface of an anisotropic material (a crystal) are also polarization dependent. In acousto-optic deflectors, which are often used in OTDR's for reducing the losses below that of a simple beam-splitter and for masking the Fresnel reflection from the fiber input face, both the diffraction efficiency and the surface reflection are polarization sensitive [Uchida N. and Niizeki 1973]. Coherent detectors are also polarization sensitive, since the local oscillator wave must be matched in polarization to the received wave in order to obtain maximum signal.

In order to reduce errors caused by unwanted fluctuations superimposed on the backscattered signal detected by a conventional OTDR, one must therefore avoid dichroic elements like polarizers or polarizing beam-splitters. A nonpolarizing beam-splitter should be used at near-normal incidence to ensure that its reflectivity is almost polarization independent [Hartog et al. 1981]. For acousto-optic deflectors, a low birefringence crystal has to be chosen with a figure of merit that changes only slightly with polarization. The residual unwanted superimposed fluctuations can be reduced by a polarization scrambler [Aoyama K. et al. 1981]. The polarization is scrambled by continuously twisting a part of the fiber near the input end during the measurement.

Another technique for reducing the polarization noise in single-mode OTDR's is based on the depolarization of light in a highly birefringent fiber which is inserted between the OTDR and the test fiber [Horiguchi T. et al. 1985a,b;

Brinkmeyer and Streckert 1986]. For proper operation of the depolarizer, the group delay difference of the polarization eigenmodes in the polarization-maintaining fiber must be much larger than the coherence time of the laser, and the laser light must be linearly polarized at an angle of 27.4° or 62.6° relative to the principal axes of the birefringent fiber.

Besides the problem of polarization sensitivity, the main problem with the backscattering technique is that the backscattered power is extremely small. To show this, we continue the numerical example of Sect. 13.4.2, in which the backscattering loss (11.38) was found to be $a_{bs} = 50\,dB$. Assume that the signal backscattered from a fiber element with one-way loss $a(z)$ of 25 dB is to be detected. Then, from (13.39), the power of the received backscattered wave is calculated to be 100 optical decibels below the peak power of the injected pulse. For a semiconductor laser, the injected power level is of the order of 0 dBm and the received backscattered power $-100\,dBm$ or $10^{-13}\,W$. The energy received from a scattering section of length $\Delta l/2$ (13.30) equals the power multiplied by the pulse duration τ. For $\tau = 1\,\mu s$, it is $10^{-19}\,Ws$, which is smaller than the energy transmitted by a single 1.3 μm photon ($1.5 \times 10^{-19}\,Ws$). Thus, on the average, less than one photon per pulse is received from the scattering fiber section of length $\Delta l/2 = 100\,m$ [Wright S. et al. 1983a].

It is much more difficult to build an OTDR for 1.3 μm single-mode fibers than for 0.85 μm multimode fibers, since (i) less power can be coupled into the fiber; (ii) the Rayleigh scattering is smaller by a factor of $(0.85/1.3)^4 = 0.18$ (5.121), and (iii) the refractive index difference and therefore the capture fraction S (5.123) is smaller. In practice, the backscattered power level is 10–15 dB below that of multimode fibers for the same light-source power [Nakazawa and Aoyama 1983; Nakazawa et al. 1984a]. Thus it is no surprise that the first experimental OTDR's for single-mode fibers were reported as recently as 1980 [Philen 1980; Healey and Hensel 1980].

In real time, the backscattering signature can only be observed from near the fiber input end, where the one-way attenuation $a(z)$ is still small. At a certain maximum distance z_m from the input end, the observed backscattered signal will equal the noise and for larger z-values it is not possible to detect the backscattering signature.

The "dynamic range" D of a backscattering apparatus is usually expressed by the maximum possible measurable one-way loss at which the signal to noise ratio is unity [Healey and Smith 1982; Healey 1984b; Danielson 1985]:

$$D = 5\log[P_{bs}(0)/P_n] \quad , \tag{13.55}$$

where $P_{bs}(0)$ is the backscattering signal from points near the input end of a reference test fiber (for which the one-way loss $a(z)$ is negligibly small) and where P_n is the RMS noise power after signal averaging. The location of the RMS noise level on a logarithmic display may be determined by noting that 16% of the sampled noise amplitudes will lie above the RMS value [Danielson 1985]. The dynamic range of an OTDR can be measured either with a long fiber, a fiber and an attenuator, or a fiber recirculating delay line (Sect. 12.3.4) [Danielson 1985; Nazarathy and Newton 1986; Newton et al. 1986].

If the backscattered power is to be determined with an accuracy of 0.1 dB, the electrical signal-to-noise ratio must be 16 dB. Because of (5.118) and (13.39), the change in dynamic range is one quarter of the change in electrical SNR. Therefore, the dynamic range specified for 0.1 dB error is 4 dB shorter than the dynamic range D (13.55) defined for unit signal-to-noise ratio.

A more general study of the relation between the signal-to-noise ratio and the measurement accuracy for local attenuation, link loss, joint loss, and fault location has been made by R.M. Howard et al. [1986].

In our numerical example, the backscattering loss is about 50 dB, and thus much larger than the reflection loss of an ideal fiber endface, which is only 14 dB. Therefore, in the backscattering signature the Fresnel peak from the fiber output end is much higher than the signal backscattered from fiber elements near the fiber end. Thus the dynamic range of an OTDR for locating reflections is much larger than the dynamic range for observing backscattering signals. The dynamic range for locating reflections can be estimated to be about $(50 - 14)/2\,\mathrm{dB} = 18\,\mathrm{dB}$ larger than the range in the backscattering mode. However, in the absence of a Fresnel reflection, for accurately locating the fiber end through the cessation of the backscattered signal, the actual fiber loss must be about 3 dB lower than the one-way-loss range D defined (13.55) in terms of unity SNR [Brain and Cottrell 1986].

In order to increase the dynamic range of the OTDR it is necessary to repeat the pulse periodically and to average successive backscattering curves. If the averaging is done by adding N curves, the electrical signal-to-noise voltage ratio is increased by a factor of $N^{1/2}$, since the signal power increases in proportion to N^2, whereas the noise power increases in proportion to N. Because of the relation between electrical and optical decibels (5.118) for direct photo-detection (a factor of 2), and since the observed radiation must travel in the forward direction to the scattering point and then return in the backward direction to the receiver (another factor of two), averaging increases the one-way dynamic range D of the OTDR by [Healey 1984b]

$$(1/4)20\log N^{1/2} = 2.5\log N \quad . \tag{13.56}$$

Thus, it is important to choose the number N of additions as large as possible. This can be done by increasing the pulse repetition frequency and by increasing the observation time. But there are limits for both quantities: Since the next pulse may not be emitted before the cessation of the backscattering curve (to prevent overlap), the maximum possible pulse frequency is $v_{\mathrm{ge}}/(2L_{\mathrm{max}})$, where L_{max} is the maximum fiber length to be investigated. For a maximum unambiguous range $L_{\mathrm{max}} = 100\,\mathrm{km}$, the maximum possible pulse repetition frequency is only 1 kHz. For practical reasons, the time of observation cannot be too long. If, e.g., the pulse repetition frequency is 1 kHz and the curves are averaged over a time of 10 minutes, the number of additions is $N = 600\,000$ and the gain in range (13.56) is 14.4 dB.

For the figures quoted, it was tacitly assumed that full backscattering curves had been added. This can be done using a fast analog-to-digital (AD) converter and adding the digital values representing the backscattering power at discrete points in time [Jeffery and Hullett 1982; Wright S. et al. 1983b; Nakazawa and Aoyama 1983], i.e. by using multichannel averaging.

In this connection it is interesting to note that the quantization error of the AD converter can be decreased by adding a noise voltage to the signal and by averaging the sum of both signal and noise [Oliver and Cage 1971; Diebold 1977; Nakazawa and Aoyama 1983]. The optimum RMS voltage of the added noise is of the order of the voltage corresponding to the least significant bit. For small backscattering signals, the necessary noise is automatically supplied by the photoreceiver. The method allows one to use, e.g., an 8 bit AD converter, although without noise and averaging its dynamic range is only $48\,\mathrm{dB_{el}}$. If the signal is very small in comparison to the noise, even a 1 bit AD converter can be used [Healey and Smith 1982].

In digital OTDR's, the spatial resolution is limited by the sampling rate which is determined by the conversion speed of the AD converter. The effective sampling rate can be increased by time division multiplexing k backscattering curves, each being time shifted by a fraction $1/k$ of the original sampling period [Kimura et al. 1985b].

The peak output pulse power of the laser fluctuates from pulse to pulse. It is one of the advantages of multipoint sampling and averaging that this laser noise does not produce an error in the backscatter signature [Hullett and Jeffery 1981b; Nakazawa and Aoyama 1983]. The same advantage also applies to a more simple forerunner of the multipoint sampling technique, the so-called two-channel or two-point processing technique [Conduit et al. 1980a,b,c; Hullett and Jeffery 1981a, 1982], in which each individual backscattering curve is sampled at two points.

More simple OTDR's use an analogue boxcar integrator for averaging instead of a multichannel digital averager [Barnoski and Jensen 1976; Barnoski et al. 1977]. Since this device samples only one point per backscattering curve, a point which is slowly shifted through the backscattering signature, much information is wasted and the number of additions and the improvement in signal-to-noise ratio is much smaller. For the same dynamic range, the measurement time with analogue averaging is much longer than with multipoint digital averaging.

The backscattered power is proportional to the peak power $P(0)$ coupled into the fiber (13.17) and to the pulse duration τ (13.32), i.e. proportional to energy of the pulse injected into the fiber. The dynamic range of an OTDR can thus be increased by increasing the pulse energy. Increasing the pulse width τ decreases the backscattering loss a_{bs} (1.34, 38) and increases the range. But this method will result in an increase of the length $\Delta l/2$ of the scattering fiber section, i.e. a loss in spatial resolution. Thus, there is a trade-off in resolution and range. Long pulses give large backscattered power but degrade the spatial resolution.

Another method for increasing the pulse energy is to use a laser with higher output peak power, e.g. a neodymium-YAG laser operating at a wavelength of $1.32\,\mu\mathrm{m}$ [Philen 1980; Nakazawa and Tokuda 1981; Nakazawa and Aoyama

1983; Gold and Hartog 1984a] or a Q-switched erbium glass (Er^{3+}:glass) laser operating at 1.55 μm [Nakazawa et al. 1984a,b]. But these lasers are bulky and the OTDR not suitable for field use. Moreover, at powers higher than about 1 to 10 W [Smith R.G. 1972; Nakazawa and Tokuda 1981], nonlinear effects like stimulated Brillouin or stimulated Raman scattering will occur (Sects. 5.4, 5.9.6). The onset of nonlinear effects like stimulated Brillouin or stimulated Raman scattering can be observed by a splitting of the return pulse as the power level is increased [Philen 1980]. The splitting is due to the shift in wavelength of the scattered light and chromatic dispersion of the fiber. A power of 10 W is the upper limit for achieving a precision attenuation loss measurement.

On the other hand, if one is solely interested in locating faults, but not in measuring losses, one can make positive use of the light which, for powers larger than 10 W, is Raman shifted in the fiber sample itself [Stensland and Borak 1981; Noguchi et al. 1982; Murakami et al. 1982; Noguchi 1984]. The pulses from a Nd:YAG laser operating at a wavelength of 1.064 μm are coupled into the fiber via a mirror that reflects the wavelength 1.064 μm, but transmits longer wavelengths. The Stokes light pulses generated by stimulated Raman scattering are Rayleigh backscattered in the fiber and transmitted through the mirror onto an avalanche photodiode or cooled Ge-PIN photodiode. By observing the backscattering signature of the broadband Stokes light, reflecting or attenuating discontinuities can be located. One advantage of this Raman Optical Time Domain Reflectometry (ROTDR) is that the mirror suppresses the Fresnel reflection from the fiber input since its wavelength is equal to that of the incident light pulse.

Healey [1985b] compared ROTDR's with conventional OTDR's and concluded that the ROTDR technique offers no performance advantage over linear OTDR systems using more suitable lasers operating at the correct wavelength.

Most OTDR's analyze the fiber properties only at one or a few fixed wavelengths. Wavelength tunable OTDR's [Kawasaki et al. 1981a] can be made by using a fiber Raman laser. In order to prevent the onset of nonlinear phenomena in the fiber under test, the wavelength selection by a monochromator should be made at the input end. But the loss of a conventional monochromator inserted between two single-mode fibers is very large, about 30 dB (Sect. 12.3.1). This loss can be reduced to about 10 dB by using an "imaging" or "folded-path" monochromator which at its output forms a high resolution, well corrected 1:1 image of the output face of the Raman fiber, properly filtered in wavelength [Cisternino et al. 1985, 1986]. In another tunable OTDR [Nakazawa and Tokuda 1983], the Raman fiber was pumped by a Nd^{3+}:YAG laser simultaneously oscillating at wavelengths of 1.32 and 1.34 μm. Tunable OTDR's can be used to measure the wavelength dependence of attenuation, or to study the influence of the LP_{11} mode on the backscattering curve.

Another method to increase the energy injected into the fiber uses a pseudo-random pulse sequence instead of an isolated pulse and correlates the received backscattered signal to a delayed version of the injected pulse sequence, in order to get the backscattering signature [Okada et al. 1980; Sudbø1983]. Correlation OTDR's have not led to any improvement in dynamic range [Healey

1985b]. However, they have been of use in the location of reflecting breaks at distant points [Healey 1981a], and in short-range high resolution reflectometry [Bernard et al. 1984a,b].

For increasing the dynamic range of an OTDR one must also improve the receiver sensitivity to the ultimate quantum limit. A relatively simple method for decreasing the photodiode dark current and thus increasing the receiver sensitivity is to cool it either with a thermocouple to $-18°C$ [Gold and Hartog 1983b], or with liquid nitrogen to $-196°C$ [Healey 1981b; Fujise and Kuwazuru 1984].

Since coherent detection is, in principle, more sensitive than the usual direct detection, homodyne [Healey et al. 1984] and heterodyne [Healey and Malyon 1982; Wright S. et al. 1983a,b, 1985] detection have been tried to increase the range. For coherent detection, in contrast to direct detection (5.118), the relative electrical power level equals the relative optical power level. Therefore, coherent detection radically reduces (by a factor of 2) the dynamic range of the electrical signal in the receiver compared with direct detection.

In order to avoid an excessive intermediate frequency bandwidth, a narrow-linewidth single-mode laser must be used for coherent OTDR's. The linewidth of the laser can be narrowed down to 50 kHz by coupling the laser to a length (e.g. 2 km) of optical fiber [Epworth et al. 1984, Heckmann et al. 1985; Mark et al. 1985a,b]. The backscattered signal acts to narow the linewidth, but the stabilization mechanism is not yet completely understood. The center frequency of the laser is constantly jumping by some hundred MHz at a rate of a few times per second. But in an OTDR this effect is not detrimental. Linewidth narrowing and frequency hopping caused by Rayleigh backscattering has recently been studied both experimentally and theoretically [Chraplyvy et al. 1986].

With narrow-linewidth single-mode lasers, a fading phenomenon is observed [Healey 1984a]. It is due to a combination of an optical mixing penalty caused by the "random" state of polarization of the backscattered wave and a speckle-like phenomenon due to the addition of a vast number of scattered waves. The backscattering signature displays a jagged appearance, which changes constantly from pulse to pulse as a result of mechanical and thermal changes in the scattering volume and due to variations in the laser frequency. The statistics of the backscattered signal, i.e. its probability density function and its variance, have been analyzed [Eickhoff and Ulrich 1981a; Healey 1985a] and it has been found that with decreasing source coherence time the backscattering signal becomes less random in appearance. The speckle-like appearance can be removed by averaging many independent backscattering curves, if the fading rate is increased by wavelength modulation [Healey 1984c]. Because of these features, coherent detection has not yet led to any increase in dynamic range [Gold and Hartog 1984b].

In a common heterodyne OTDR, two power splitters and one acousto-optic modulator are used and these introduce a large insertion loss corresponding to a loss of 5 dB in one-way dynamic range. A new scheme, which eliminates the power splitters by using two acousto-optic modulators, reduces this loss to about 1 dB [Rybach and Heckmann 1986].

The sensitivity of the photoreceiver can also be increased using the so-called photon counting technique [Healey and Hensel 1980; Healey 1980, 1981b, c; Levine et al. 1985c]. In the photon counting technique, an avalanche photodiode (APD) is biased to just below breakdown. In principle, a single photon incident on the detector could be sufficient to initiate a local plasma breakdown, and produce an output current pulse. The number of output pulses per second is proportional to the number of photons incident per second, i.e. to the optical power. Practical germanium APD's achieve sensitivities of about 1000 photons per current pulse [Wright S. et al. 1983a]. In the first OTDR's using photon counting, the photodetector had to be cooled by liquid nitrogen (77 K). Recently, however, photon counting OTDR's operating at room temperature have been demonstrated [Levine et al. 1985a,b; Bethea et al. 1986].

Another approach for increasing the receiver sensitivity is to use a semiconductor laser as an optical preamplifier for the backscattered signal [Suzuki K. et al. 1984]. By switching the injection current above and below lasing threshold, the same laser can be used both as an optical pulse generator and as an optical amplifier. The improvement in one-way dynamic range was found to be 4.5 dB, 3 dB of which were attributed to removal of the directional coupler. But note that the gain of a laser amplifier depends on polarization [Mukai and Yamamoto 1981]: it is larger for the electric field vector parallel to the active zone. This, incidentally, is one of the reasons why the semiconductor output wave is linearly polarized parallel to the active zone (Sect. 13.1.1).

A simplified version of an OTDR can be implemented in an installed fiber communication link [Daino and Spano 1984a,4b]. During a test phase, the semiconductor laser of the transmitter is operated first as a pulse source and thereafter as an optical amplifier for the signal reflected or backscattered from a fault. The monitoring photodiode, placed at the rear face of the laser, is used to detect the backscattered signal. In a similar scheme [Nakazawa et al. 1985], a single semiconductor laser diode was used both as a pulse emitter and a photodetector.

In addition to the extremely small backscattered power, there is the problem of the large level difference between the injected pulse and the backscattered light. At the fiber input end, this was previously estimated to be of the order of 50 optical decibels. If part of the transmitted pulse is detected by the photodetector, the subsequent amplifier will be saturated for a short time, during which the backscatter signature cannot be observed. Thus, direct source-detector coupling must be avoided by using an optical directional coupler with a very high directivity (12.23) and by avoiding reflections from the fiber input end. For the optical directional coupler, a conventional beam splitter or a fiber directional coupler (Sect. 12.2.11) can be used. Note, however, that the detector and the amplifier can also be saturated by the Fresnel reflection from a connector or from the fiber end [Nakazawa and Aoyama 1983].

Instead of a directional coupler, one can also use an acousto-optic light deflector [Gordon E.I. 1966; Uchida and Niizeki 1973], which relies on Bragg reflection from a sound wave [Nakazawa et al. 1981; Nakazawa and Aoyama 1983; Horiguchi T. et al. 1984; Gold and Hartog 1984b]. The deflector has a smaller insertion loss (about 2 dB) than a beam splitter (>3 dB) and allows one to gate out the Fresnel reflection from the front end of the fiber and other discontinuities and to shut off the backscatter signal from the near part of the fiber while the far end is being measured, thus avoiding saturation of the amplifier.

It is also possible to gate the photoreceiver during the time of a Fresnel reflection [Personick 1977] or to cancel the Fresnel return using two APD's, one of which receives the backscattered light whereas the other receives the input pulse [Aoyama K. et al. 1981].

The dynamic range obtainable with OTDR's depends on the expenditure made. In a review paper, Healey [1985b] compared the different techniques for maximizing the dynamic range of OTDR's for single-mode fibers and listed the performance parameters of the various experiments reported so far in the literature. In order to obtain maximum dynamic range, it is necessary to simultaneously optimize the launch power, receiver sensitivity, and noise reduction by averaging. In the following, only some of the best performance results will be reported.

Using a semiconductor laser at $1.3\,\mu$m coupling $1\,\mu$s pulses with a power level of $+5$ dBm into the fiber, an acousto-optic deflector in place of the conventional beam splitter, a HgCdTe photodiode, and multichannel digital averaging ($N = 10^6$), a receiver sensitivity of -105 dBm and a one-way dynamic range of 30 dB have been reported [Gold and Hartog 1984a,b]. With the same apparatus, the dynamic range has been further increased to 41 dB by replacing the semiconductor laser by a Nd:YAG laser operating at a wavelength of $1.32\,\mu$m; this coupled pulses of $1\,\mu$s duration and 1 W peak power into the fiber. A one-way loss of 41 dB corresponds to a fiber length in excess of 100 km for state-of-the-art single-mode fibers at $1.3\,\mu$m. This performance is greater by nearly 10 dB one-way than any reported previously.

Using a $1.55\,\mu$m Q-switched Er^{3+}:glass laser, with a pulse width of $1\,\mu$s and a peak power of 4 W in the fiber, a TeO_2 acousto-optic light deflector, a Ge APD and multisampling averaging ($N = 2^{12} = 4096$), a one-way dynamic range for finding nonreflecting faults of about 30 dB has been obtained [Nakazawa et al. 1984a]. Using a fiber Raman laser consisting of a Q-switched Nd:YAG laser operating at $1.32\,\mu$m and a highly P_2O_5-doped silica single-mode fiber producing $0.8\,\mu$s pulses with a peak power of 2.4 W at $1.59\,\mu$m, a one-way dynamic range of 35 dB has been achieved [Suzuki K. et al. 1986].

Besides the conventional OTDR's which use pulsed optical sources in their operation, there have been some proposals for reflectometers based on modulating either the optical frequency or the frequency of the sinusoidal intensity modulation.

The first scheme operates in analogy to the frequency-modulated continuous-wave (FM-CW) radar. The laser power is constant, and the instantaneous *optical* frequency varies like a sawtooth function [Eickhoff and Ulrich 1981c; Uttam and Culshaw 1985]. The sweeping of the optical frequency f can also be replaced by a sweep of the phase constant β caused by a slow variation of the fiber temperature, strain, or ambient pressure [Eickhoff and Ulrich 1981b].

These optical frequency domain reflectometers (OFDR) cannot analyze long fibers but have the advantage of high spatial resolution.

The second scheme uses a source with constant optical frequency which is sinusoidally modulated in intensity. The frequency of the intensity modulation is swept through a range and the backscattered data are inverse Fourier transformed to produce the usual temporal backscattering signature [Kapron et al. 1981; MacDonald 1981; Ghafoori-Shiraz and Okoshi 1985, 1986a,b].

As yet, however, the performances of reflectometers using either optical or modulation frequency modulation are inferior to the performance of the conventional pulsed OTDR's.

13.5 Measurement of the Near-Field Intensity Distribution

From the radial field distribution $E(\varrho)$ of the fundamental mode, one can calculate the various spot sizes (Sect. 10.1), the refractive index profile (Sect. 13.10.2), the effective refractive index $n_e = \beta/k$ (5.47), and the effective group index n_{ge} (5.75). Aside from a multiplicative factor caused by reflection, the function $E(\varrho)$ is equal to the field distribution in the fiber endface, the so-called near-field amplitude distribution. Thus, it is very important to have methods for measuring the near-field intensity distribution from which the amplitude distribution can be obtained by taking the square root (Sect. 8.2).

The transverse extent of the fundamental mode near field is very small: At a distance w_G from the fiber axis, the intensity is only about 13% of its value on axis, i.e. the level is down by 8.7 dB. Within a cylinder of radius w_G, approximately 86% of the total power is transmitted (5.39). For standard fibers with the dispersion minimum near $1.3\,\mu m$, the mode field diameter is only $10\,\mu m$.

Since we have no optical probe antennas whose effective area has a diameter which is very small compared to the diameter of the near-field distribution, it is not possible to measure the near-field intensity distribution directly. One has to magnify the near-field pattern using a microscope objective.

The technique for measuring the near-field intensity distribution has been described in the literature [Murakami et al. 1979; Anderson W.T. and Philen 1983a; Kim and Franzen 1981, 1983; Srivastava and Franzen 1985; Fleming S.C. et al. 1985; Ohashi et al. 1986]. The apparatus (Fig. 13.9) essentially consists of a calibrated flat-field microscope objective imaging the near-field intensity distribution onto an image plane. The magnification of the optical system must be selected to be compatible with the desired resolution. The magnification can be measured by previously scanning the length of a specimen whose dimensions are already known with suitable accuracy. The focusing must be performed with maximum accuracy, in order to reduce the error in magnification due to the scanning of a misfocused image.

The magnified near-field intensity distribution can be scanned by a vidicon (Fig. 13.9), charge coupled devices, or other pattern recognition devices. If

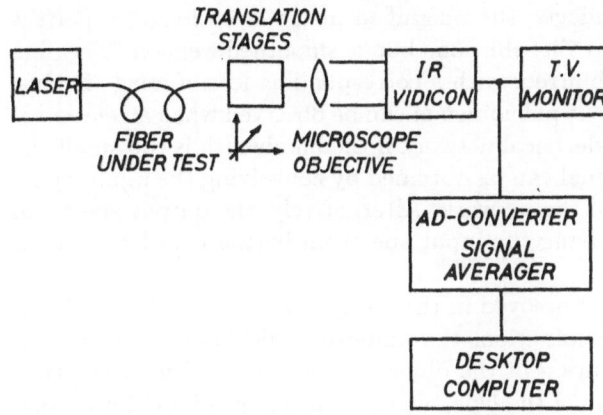

the near-field distribution is known to be circular, it is sufficient to scan the intensity along a diameter, e.g. using a linear photodiode array, a scanning photodetector with pinhole aperture, or a scanning fiber end with photodetector at the other end. The detector characteristics must be linear over the range of intensities encountered. The sensitive area of the detector should be small with respect to the enlarged image of the near-field distribution.

Note that all photodetectors listed measure the near-field *intensity* distribution, from which the near-field *amplitude* distribution is calculated by taking the square root. It is common practice to transfer the near-field data to a desktop computer, which then calculates the quantities of interest.

A simple method [Millar 1984a] for obtaining a real-time plot of the radial near-field intensity distribution $E^2(\varrho)$ on an oscilloscope consists in imaging the near field by a microscope objective onto a slit, the length of which is changed at a constant rate with time by a rotating chopper blade. The time derivative of the light power transmitted through the slit is proportional to the radial intensity distribution $E^2(\varrho)$.

The near-field technique is mainly used to measure the radial field distribution of the fundamental fiber mode. Additionally, it can be used to determine geometrical parameters such as cladding diameter, cladding ellipticity, and mode field concentricity error. To determine these parameters, one must also launch cladding modes in the fiber and measure the two-dimensional near-field intensity distribution.

Having described the near-field technique to determine the mode pattern, some disadvantages of this method should now be mentioned. In order to obtain the spatial resolution required, the linear magnification must be of the order of 100. Since the power is then distributed over an area which is 10 000 times the original area in the fiber, the intensity in the maximum of the magnified near-field pattern lies 40 dB below that on the fiber axis. Since the tails of the near-field distribution must be measured down to −40 dB in order to determine e.g. the spot size with an error of 1% [Streckert 1985], high power sources, i.e. lasers, and very sensitive photoreceivers must be used. In practice, it is impossible to measure the near-field intensity distribution using incoherent sources.

Because of diffraction effects, the magnified image is no longer a perfect replica of the true near-field distribution, but a smoothed version. The fine details of the near-field are blurred, with a corresponding loss of image fidelity resulting. An exactly analogous phenomenon can be observed when an electrical signal is applied to a linear electrical network if the bandwidth is too small. As is well known, the output signal can be obtained by convolving the input signal with the impulse response of the network. Alternatively, the output spectrum can be calculated by multiplying the input spectrum by the transfer function of the network.

Similar concepts can be employed in the study of imaging systems [Goodman 1968]: For coherent illumination, the amplitude distribution in the image is the amplitude distribution in the object plane convolved with the two-dimensional impulse response of the lens, which is the image of a point object. Alternatively the spectrum of the image, which is the two-dimensional Fourier transform of the amplitude distribution (8.7), is the spectrum of the object multiplied by the (coherent) transfer function of the optical system.

Using the Rayleigh criterion of resolution (two point sources can be resolved if the center of the image of one point falls on or beyond the first zero of the diffraction pattern of the other point), the minimum resolvable separation of two points in the object plane, i.e. the resolution, is $0.61\lambda/N_A$, where N_A is the numerical aperture of the imaging system used. For example, for $\lambda = 1.3\,\mu m$ and $N_A = 0.3$ the resolution is $2\lambda = 2.6\,\mu m$, and for an object with a width of e.g. $20\,\mu m$, only 8 points can be resolved. This example shows that it is very important to use optical systems with a high numerical aperture.

The magnification of a microscope objective is not strictly uniform across the field of view. The lens therefore needs to be highly corrected.

The focal length of the lens depends slightly on the wavelength. Thus it is necessary to readjust the system when a new wavelength is chosen. This procedure requires great care, both in focusing and in calibration of the magnification.

Because of these disadvantages, the near-field technique is not very popular. There is an alternative method for determining the near-field intensity distribution, which avoids some of the disadvantages: the far-field radiation pattern is measured as described in the Sect. 13.6 and the near-field distribution is calculated from it by a Hankel transform (8.18, 20). Instead of performing the Hankel transform numerically, one can also determine the near field by expanding the far-field data as a series of Gauss-Laguerre functions using a linear fitting procedure [Freude and Sharma 1984, 1985, 1986]. These functions maintain their structure after Fourier transformation, and, therefore, the near field is known in terms of Gauss-Laguerre functions. Stated in physical terms, the radiation field is considered as a superposition of Gaussian beams with different orders (Sect. 4.8). The main advantage of this method lies in its inherent insensitivity to random measurement errors.

13.6 Measurement of the Far-Field Radiation Pattern

At the fiber end, the LP_{01} mode is transformed (Sects.2.4, 8.1) into a free beam which is very similar to the fundamental Gaussian beam discussed in Sects.4.1–4.4. By measuring the intensity distribution in the radiated beam, one gains useful information which can be used to calculate e.g. the near-field amplitude distribution or the mode field diameter.

For measuring the far-field radiation pattern, the fiber is allowed to radiate freely from its end [Hotate and Okoshi 1979b]. At a distance r from the center of the radiating fiber endface, the intensity (optical power per unit area) is measured on a spherical surface as a function of the polar angle θ. Figure 8.1 shows the coordinates used in this measuring method.

At this point, it is necessary to review and compare the different functions used to characterize the far-field distribution. The far-field amplitude distribution $E_f(r, \theta)$ of the LP_{01} mode has been defined (8.13) as the electric field strength of the beam radiated from the end of a single-mode fiber at a point with polar angle θ at a distance r from the fiber end. The distance r is assumed to be large compared to the Rayleigh distance $z_R = \pi w_G^2 / \lambda$ (4.6) of the beam. For arbitrary single-mode fibers, the far-field amplitude is inversely proportional to r and the phase decreases linearly with r. Therefore, when discussing the far-field amplitude distribution, one usually omits a factor

$$j(\lambda r)^{-1} \exp(-jkr) \qquad (13.57)$$

and considers a function $E_H(q)$ (8.18):

$$E_H(q) = -j\lambda r \exp(jkr) E_f(r, \theta) \quad , \qquad (13.58)$$

where the spatial frequency (8.14)

$$q = (1/\lambda) \sin \theta \qquad (13.59)$$

is introduced as a new kind of angular coordinate. The normalized far-field amplitude distribution $E_H(q)$ and the near-field amplitude distribution $E(\varrho)$ form a pair of Hankel transforms (8.16, 20). Note however, that the simple relations (8.2, 6, 12, 13) between the near-field amplitude distribution and the far-field radiation pattern only apply when there is no radiation for polar angles much larger than about $10°$. For single-mode fibers, this condition is satisfied only approximately. Therefore, the simple relations are not very accurate.

With a direct photodetector, one measures not the amplitude distribution, but the intensity. A far-field radiation pattern $P_p(\theta)$ is defined as the angular distribution of the power detected by a photodetector with a pinhole aperture moved on a circle around the fiber end. The far-field radiation pattern $P_p(\theta)$ is proportional to the square of the normalized amplitude distribution

$$P_p(\theta) = c|E_H(q)|^2 \quad , \qquad (13.60)$$

with c being an unimportant constant and with the variables θ and q related through (13.59).

Instead of measuring the intensity on a circle ($r =$ const., θ variable), it can also be measured on a *transverse straight line* ($z =$ const., ϱ variable) (Fig. 8.1). The power $P'_\mathrm{p}(\varrho)$ received from a photodetector scanned along a straight line through an axis point with coordinate z is related to the power $P_\mathrm{p}(\theta)$ measured by a photodetector scanned along a circle through the same point by

$$P'_\mathrm{p}(\varrho) = P_\mathrm{p}(\theta) \cos^3 \theta \quad , \tag{13.61}$$

where the variables θ and ϱ are related by

$$\varrho = z \tan \theta \quad . \tag{13.62}$$

Relation (13.61) is obtained by taking into account that the power received is the product of the intensity $|E_\mathrm{f}(r, \theta)|^2$ and the projected sensitive area of the photodetector. A factor of $\cos^2 \theta$ in (13.61) is caused by the fact that, for the linear scan, the distance r increases in inverse proportion to $\cos \theta$, and another factor of $\cos \theta$ by the fact that the projected area decreases as $\cos \theta$. A possible dependence of the reflection loss on the angle θ has been neglected in (13.61).

In order to measure the two-dimensional intensity distribution in the far field, a vidicon, charge coupled devices, or other pattern recognition devices can be used. Since the far field of the fundamental mode is approximately circularly symmetric, i.e. independent of the azimuth ϕ (Fig. 8.1), it is often sufficient to measure the intensity as a function of the polar angle θ only. There are several alternatives for measuring the function $P_\mathrm{p}(\theta)$ or $P'_\mathrm{p}(\varrho)$:

1. fixed fiber end with a small-area detector which is displaced on a circle (Fig. 13.10);
2. fiber end rotatable with fixed small-area detector;
3. fixed fiber end with a small-area detector which is displaced linearly in a transverse direction;
4. fiber end transversely displaced with fixed small-area detector;

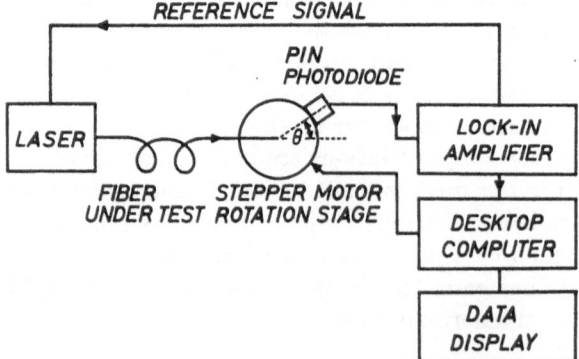

Fig. 13.10. Experimental apparatus for measuring the far-field radiation pattern

The apparatus for measuring the near-field distribution (Sect. 13.5) can also be used to measure the far-field distribution simply by removing the imaging optics.

Far-field measurement setups have been described in the literature [Kim and Franzen 1981; Anderson W.T. and Philen 1983a; Nishimura and Suzuki 1984; Richter 1985; Srivastava and Franzen 1985; Ohashi et al. 1986].

Note that all photodetectors measure the far-field *intensity* distribution, from which the far-field *amplitude* distribution is to be calculated by taking the square root and taking into account the sign reversal at the zeros of the far-field pattern (Sect. 8.1). Experimentally, one finds minima of intensity instead of zeros. However, these minima can be interpreted as zeros of field amplitude [Richter 1985].

A desktop computer can be used to control the movement of the scanning photodetector and to further process the far-field data to obtain the various far-field widths (Sect. 10.2), or the near-field distribution from a Hankel transform (8.20).

The major advantage of the far-field method is simplicity. The far field can be measured very accurately, since no calibration within fractions of a micrometer is required. Since the far-field pattern is much larger than the near-field pattern, extreme positioning stability is not required. This simplifies the fiber alignment apparatus and relaxes the requirements on instrument recalibration.

For the distance r between the radiating fiber end and the scanning detector, a compromise must be found: for a short distance, the received signal is large, but the angular resolution is small. For a large distance, the resolution is better but the dynamic range is small. The CCITT recommends a distance of at least 2 cm between the fiber end and the detector. The detector's active area should not subtend an angle greater than $0.5°$.

If for the angular resolution a value of e.g. $0.57°$ corresponding to 0.01 radians is chosen, the diameter of the photodetector is $0.01\,r$. By using the theory of the fundamental Gaussian beam (Sect. 4.1), one can calculate the intensity on the beam axis, which, multiplied by the area of the photodetector, gives the power received. For $\lambda = 1.3\,\mu$m and $w_G = 5\,\mu$m, the received power turns out to be about 27 dB smaller than the power guided by the fundamental fiber mode and radiated from the fiber end. However, it is not sufficient to measure the intensity on the beam axis. For determining e.g. the far-field width to within an error of about 1%, the radiation pattern must be measured at least down to $-30\,$dB [Nishimura and Suzuki 1984; Streckert 1985].

Computer simulations [Klemas et al. 1985a,b] have shown that the far field must be measured to angles up to $25°$ ($-50\,$dB) with respect to the fiber axis to obtain the spot size w_G with an error of smaller than 1% by using the inverse Hankel transform.

There are also proposals to evaluate the angle of the first zero of the far-field radiation pattern (Sect. 13.8.10). Since the first side lobe is between 28 dB and 68 dB lower than the central maximum (Sect. 8.1), an extremely large dynamic range of the measuring apparatus is required.

For the above reasons, laser sources are necessary for measuring the far-field pattern. Using injection lasers, dynamic ranges in the far field of 40 dB [Nishimura and Suzuki 1984], 50 dB [Anderson W.T. and Philen 1983a], and 70 dB [Richter 1985] have been obtained. If the far-field pattern is to be measured as a function of wavelength, an expensive set of lasers is needed.

13.7 Measurement of the Spot Size

The spot size is one of the most important parameters characterizing a single-mode fiber since it determines launching efficiency (Sect. 7.2.2), splice loss (Sects. 9.1–5), transition loss (Sect. 5.9.8), microbending loss (Sect. 5.9.9), waveguide dispersion (Sect. 5.8), and the backscattering capture fraction (Sect. 5.9.2). The mode field diameter is two times the spot size.

At present, there is no undisputed recommendation for the definition of the spot size. Therefore, in Sect. 10.1 five definitions for the spot size have been reviewed: w_s is the simple $1/e$ radius (Sect. 10.1.1), w_G is the spot size of the Gaussian beam which maximizes the launching efficiency (Sect. 10.1.2), w_e is the effective spot size related to microbending loss (Sect. 10.1.3), w_d is the "strange" spot size related to waveguide dispersion (Sect. 10.1.4), and w_a is the autocorrelation spot size related to the transverse offset loss (Sect. 10.1.5).

At present, there are also no final recommendations of methods to measure the spot size. Thus in this section, several published measuring methods will be reviewed.

The status of methods for measuring the mode field diameter has been described several times [Anderson W.T. and Philen 1983a; Coppa et al. 1985a; Anderson W.T. 1986; Dick and Shaar 1986].

13.7.1 Spot Size Determined from the Field Distribution

This section describes methods for determining the spot size either from the measured near-field distribution or from the measured far-field radiation pattern.

If the near-field amplitude distribution $E(\varrho)$ is known, the five spot sizes w_s, w_G, w_e, w_d, and w_a can be calculated from the definitions [Sect. 10.1.1; Eqs. (10.3, 8, 10); Sect. 10.1.5] [Anderson W.T. and Philen 1983a]. The method for measuring the near-field intensity distribution and its inherent problems have been described in Sect. 13.5. In order to avoid errors caused by the truncation of the tails of the near-field distribution, a considerable dynamic range is required. To obtain the spot size to within a systematic error of 1%, the near-field distribution must be measured down to levels approximately 40 dB below the maximum level [Streckert 1985].

If the Petermann II spot size w_d is to be calculated from a measured near-field distribution (10.10), a function measured at discrete arguments has to be differentiated. This will result in relatively large errors. A better method seems to be to approximate the measured near-field pattern by a Gaussian function for small distances from the axis and by a modified Hankel function of zero order for larger distances [Sharma E.K. and Tewari 1984]. This two-parameter Gaussian-Hankel approximation (Sect. 5.3.2) is more accurate than the simple one-parameter Gaussian approximation. The two fit parameters are obtained by a simple least-squares fitting procedure. The Petermann II spot size w_d is calculated by inserting the fitting parameters into a simple closed form expression. In order to test this method in a numerical simulation, the authors calculated the strange spot size for two special cases of power-law index profiles (5.6, 7) both directly corresponding to the definition (10.10) and using the Gaussian-Hankel approximation. They found that the spot size w_d can be obtained with high accuracy. It remains to investigate whether this method is sufficiently accurate for other index profiles too.

As has been explained in Sect. 8.1, the near-field distribution can also be obtained by measuring the far-field distribution (Sect. 13.6) and Hankel transforming it (8.20) [Anderson W.T. and Philen 1983a; Samson 1984]. But then a large dynamic range is required for the far-field apparatus, too [Klemas et al. 1985a,b]. In the far-field method for determining the spot size, errors are caused by a finite dynamic range of the optical receiver, a finite angular increment, fiber end position accuracy, stability of light source power, etc. The required conditions have been clarified by Myogadani and Tanaka [1985].

The effective spot sizes w_e of three single-mode fibers with nearly step-index profiles have been determined both directly from the near field and from the Hankel-transformed far field [Ohashi et al. 1986]. By correcting for the nonlinearity of the vidicon, by using a high-dynamic-range far-field setup and a small sampling step angle, and by extending the range of numerical integration in (10.8) to $\varrho_{max} = 15\,\mu m$, the spot sizes determined by the two methods could be made to deviate by less than $0.2\,\mu m$ from the theoretical values calculated from the refractive-index profile.

Because of the large dynamic range required for calculating the near field from the far field, a better method for determining the mode field diameter is to measure one of the various far-field widths, for which a smaller dynamic range is sufficient, and to make use of the relations (10.29–31) between the far-field widths (W_d, W_G, W_e) and the fundamental mode spot sizes (w_e, w_G, w_d). For instance, the effective spot size w_e (10.8) has been determined by measuring the "strange" far-field width W_d (10.27) and by using the relation $w_e = 1/(\pi W_d)$ (10.29) [Nishimura and Suzuki 1984]. In this far-field method, a dynamic range of 32 dB is sufficient to obtain the spot size w_e with an accuracy of better than 0.5%.

Similarly, the spot size w_G related to launching (Sect. 10.1.2) has been determined by measuring the far-field width W_G related to launching (Sect. 10.2.2) and by applying the relation $w_G = 1/(\pi W_G)$ (10.30) [Anderson W.T. 1984]. If the fiber endface is not normal to the fiber axis, the far-field radiation pattern will be distorted. But the far-field width W_G is not changed significantly, provided the angle between the normal to the endface and the fiber axis is smaller than 5° [Nicholson 1984]. In practice, this condition can easily be satisfied.

The Petermann II spot size w_d (10.10) can also be obtained by measuring the effective far-field width W_e (10.28) and by using the relation $w_\mathrm{d} = 1/(\pi W_\mathrm{e})$ (10.31) [Pask 1984; Anderson W.T. et al. 1987].

The main disadvantage of all methods to determine the spot size from the measured near- or far-field intensity distribution is that the insertion loss of the measuring apparatus is very large and that, therefore, lasers have to be used, making it very expensive to make measurements at varying wavelengths.

This disadvantage is avoided in spot size determinations that essentially measure the total power transmitted and power levels which are only a few decibels smaller. These so-called "integral" methods include the transverse offset method (Sect. 13.7.2), various aperture methods (Sect. 13.7.3), and the so-called mask methods (Sect. 13.7.4).

13.7.2 Transverse Offset Method

Because of the drawbacks associated with methods for measuring the near- and far-field distributions, the transverse offset method first proposed by Streckert [1980] has become very popular [di Vita et al. 1984]. The implementation of the method has been described in the literature [Anderson W.T. and Philen 1983a; Srivastava and Franzen 1985].

The hardware for this method, which is also called the traversing joint method, is shown schematically in Fig. 13.11. The light emitted by a quartz halogen lamp (QH) is filtered by a monochromator (M), chopped (CH) and coupled into the fiber. The fiber is cut and the coatings are stripped for approximately 10 cm at each end. Two endfaces are prepared, which are flat and normal to the fiber axis to within 1° (Sect. 12.1.1). Each fiber end is drawn through a cladding-mode stripper (CMS) (Sect. 12.2.8) to ensure rapid attenuation of cladding modes excited at the joint or by mode coupling due to bends.

One fiber end is fastened to the clamp of a precision three-axis translation stage (xyz-micromanipulator) with a positioning accuracy of about 0.1 μm. One transverse axis is driven by a precision translator and the location of the stage is determined by an accurate electronic gauge. A microscope is provided for visual observation of the joint. The transmitted power is detected by a photodiode (PD), rectified by a lock-in amplifier (LI), and plotted on an xy-recorder (RC).

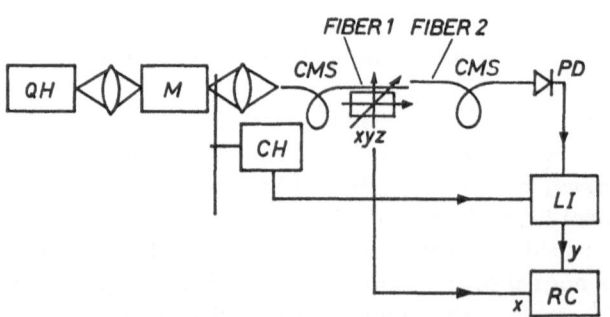

Fig. 13.11. Experimental apparatus for measuring the transverse offset loss:
QH: quartz-halogen lamp
M: monochromator
CH: chopper
CMS: cladding mode stripper
xyz: micromanipulator
PD: photodiode
LI: lock-in amplifier
RC: xy-recorder

The fiber ends are initially positioned for maximum transmitted power $P_t(0)$. The longitudinal separation of the fiber endfaces should be less than $5\,\mu m$. This is necessary to prevent the mode field diameter of the beam radiated from the end of the first fiber from increasing due to diffraction beyond a certain limit [about 0.4% for typical fibers (4.5)]. The second fiber end is progressively displaced in a direction transverse to the first fiber end, and the power transmission $P_t(s)$ is measured as a function of the transverse offset s. The coupling efficiency $\eta(s)$ is calculated as the ratio

$$\eta(s) = P_t(s)/P_t(0) \quad . \tag{13.63}$$

The function $\eta(s)$ is very nearly Gaussian. For a convenient evaluation of the results, the optical power versus offset curve can also be digitized and transferred to a desktop computer.

Corresponding to its definition (Sect. 10.1.5), the autocorrelation spot size w_a is the offset for which the power transmission factor is $1/e$ of its maximum value:

$$\eta(w_a) = 1/e \quad . \tag{13.64}$$

Since the slope of the transmission curve is zero at the maximum, it is difficult to determine the position of the maximum precisely, and instead of measuring the spot size w_a it is better to measure the mode field diameter $2w_a$, which is the distance between the two points for which the coupling efficiency $\eta(s)$ is $1/e = 0.37$ of its maximum value. Note that no curve fitting procedure need be used in determining the spot size w_a from the coupling efficiency function $\eta(s)$. Fitting a Gaussian function to the power transmission curve $\eta(s)$ and taking the $1/e$ radius (w_{aG}) of this Gaussian fitting function [Anderson W.T. and Philen 1983b; Srivastava and Franzen 1985] corresponds neither to the definition of the autocorrelation spot size w_a nor to any other spot size defined in this book. If a Gaussian function is least-squares fitted to the far-field *intensity* distribution, the $1/e$ width W_{IG} of this Gaussian function is related to w_{aG} by [Samson 1986]:

$$w_{aG} = 1/(\pi W_{IG}) \quad . \tag{13.65}$$

We have described how the transverse offset method is used to determine the autocorrelation spot size w_a. However, it can also be used to determine the other versions of spot size from the measured function $\eta(s)$.

The Petermann II spot size w_d (10.10) is related to the second derivative of the coupling efficiency function $\eta(s)$ with respect to the offset s taken at zero offset (10.11). Using this relation, the strange spot size w_d has been determined experimentally from the normalized curvature of power transmission coefficient as a function of offset at the point of zero offset [Sha et al. 1986]. However, taking the second derivative of a measured function will result in large errors.

The most convenient method to determine w_d consists in fitting a suitable function $f(s)$ to the measured $\eta(s)$ values for $s \ll w_d$ and taking the second derivative of the fitting function. Different forms of the fitting function have been proposed, a simple parabola, a Gaussian function, and a product of a parabola and a Gaussian.

The Gaussian function yields a better fit than the simple parabola [Anderson W.T. et al. 1987]. Thus, for small offsets s, the power transmission function can be well approximated by a simple Gaussian function [Pask 1984; Anderson W.T. 1984] and the joint loss by (9.5)

$$a = 4.34(s/w_d)^2 \quad . \tag{13.66}$$

Plotting the measured insertion loss as a function of s^2 initially (up to $s/w_d = 0.4$ corresponding to an insertion loss of $0.7\,\mathrm{dB}$) gives a straight line, the slope of which is $4.34/w_d^2$. Computer simulations showed that this evaluation of transverse offset data supplies very accurate values for the spot size w_d related to dispersion [Streckert 1986] . An even better fit is claimed [Calzavara et al. 1986a] to be achieved by a product of a parabola and a Gaussian function.

It is also possible to use the transverse offset method to measure the transverse widths of noncircular fundamental mode fields (Sect. 13.7.5). The field distribution can be approximated by the product of two Gaussian functions [Snyder 1981]:

$$E_x(x,y) = E_0 \exp(-x^2/w_{a\,\mathrm{max}}^2) \exp(-y^2/w_{a\,\mathrm{min}}^2) \quad . \tag{13.67}$$

The maximum and minimum autocorrelation spot sizes $w_{a\,\mathrm{max}}$ and $w_{a\,\mathrm{min}}$ can be measured by cutting the fiber, preparing endfaces and making a joint without rotating the fiber ends about their axes. Scanning in two orthogonal directions gives $w_{a\,\mathrm{max}}$ and $w_{a\,\mathrm{min}}$.

In principle, one can also determine the other spot sizes w_s, w_G, and w_e by using the transverse offset technique. Use must be made of the fact that the near-field autocorrelation function $E*E$ and the far-field intensity pattern $|E_H|^2$ form a pair of Hankel transforms (Fig. 8.5). One has to measure the normalized transverse offset coupling function $\eta(s)$, to take the square root of $\eta(s)$, which gives the near-field autocorrelation function $E*E$, and to Hankel transform $E*E$ in order to get the far-field radiation pattern $|E_H|^2$ (8.51). From the far-field radiation pattern, the other spot sizes can be determined as described in Sect. 13.7.1. However, to the author's knowledge, this method for determining w_s, w_G, and w_e from transverse offset data has not yet been tested.

The transverse offset technique has several advantages:

1. The most important advantage is that, without offset, the insertion loss of the measuring apparatus is negligibly small and the received photocurrent is relatively high. Moreover, in order to determine w_a, it is sufficient to measure the coupling efficiency $\eta(s)$ down to, say, 10%. Therefore, it is possible to use the simple tungsten-lamp/monochromator combination (Sect. 13.1.1) as a tunable optical source, and it is easy to measure the spot size as a function of wavelength.

2. The transverse offset method corresponds strictly to the definition of the autocorrelation spot size w_a.
3. The technique can readily be applied using existing attenuation measuring equipment by adding a micromanipulator. Therefore, the minimum necessary equipment is relatively simple and inexpensive.
4. The measurement is relatively rapid [Anderson W.T. and Philen 1983a] and it gives valuable data directly related to splice loss.
5. The method is quite accurate; the error in the spot size was found to be less than 2% [Samson 1985a; Streckert 1985].
6. It is not necessary to know the absolute value of the power transmission factor of the gap for zero transverse offset, only the change in splice loss versus offset needs to be measured. This is much easier than measuring absolute splice loss (Sect. 13.2).
7. With the transverse offset method, no complex mathematical evaluation is necessary to determine the spot sizes w_a and w_d: no overlap integral, no RMS value, no derivative, and no Hankel transform have to be calculated.
8. Unlike near-field scan data, transverse offset data are very nearly Gaussian [Anderson and Philen 1983a]. Thus fewer data points are needed if one wishes to approximate the transmission factor versus offset curve by a Gaussian function.

Of course, there are problems with the transverse offset method, too. Some of these problems result from the underlying theory, which assumes that there is no angular misalignment, no end separation, and no fixed transverse offset orthogonal to the scan direction.

Angular misalignment, due to misaligned vacuum chucks or translation stages, can be reduced to less than half a degree very easily [Anderson W.T. and Philen 1983a; Samson 1985a] and, hence, presents no problem. End separation, on the other hand, must be allowed for, because the fiber ends are not perfectly flat and perpendicular to the fiber axis (Sect. 12.1.1) and because one fiber end must be scanned past the other without touching it. But from the theory of Gaussian beams (Chap. 4), it can be shown [Streckert 1985] that the error in the spot size w_a remains smaller than 0.5% if the gap width is smaller than 10% of the Rayleigh distance, i.e. $6\,\mu$m for $\lambda = 1.3\,\mu$m and a spot size of $5\,\mu$m. A gap of this width is easily achieved in practice, if care is taken to break good ends and if the ends are brought close together while observing the gap with a microscope [Anderson W.T. and Philen 1983a].

Most authors [Millar 1981b, Samson 1985a] recommend filling the gap with a low-viscosity index-matching liquid to avoid interference effects caused by multiple reflections (Sect. 5.9.5). The surface tension of this droplet has a tendency to bend the fibers toward each other as one fiber is scanned across the other. To reduce this effect, the length of fiber protruding from the fiber holders is minimized and a small amount of silicon resin is placed behind the fiber end to support the fiber [Samson 1985a].

A residual transverse offset normal to the scan direction would not produce any error if the field distribution were truly Gaussian [Samson 1985a]. For actual field distributions, the error is small, e.g. for a step index fiber at $V = 2.4$

it is smaller than 1% provided the loss caused by the fixed vertical offset is smaller than 0.1 dB [Streckert 1986].

A criticism that has been levelled against the transverse offset method concerns the need to prepare two endfaces at a time. This difficulty can be overcome by always using the same reference fiber with a mirror polished endface [Streckert 1980, 1985]. Again, one has to measure the power transmission factor as a function of the offset. From (9.12), a formula for the spot size w_a of the fiber sample can be derived

$$w_a = \sqrt{2s^2 - w_{ar}^2} \quad , \tag{13.68}$$

in which s is the offset for $1/e$-coupling efficiency and w_{ar} the known spot size of the reference fiber. For strictly Gaussian near fields of both fibers, this kind of measurement introduces no error at all. For real field distributions, there is a slight error in determining w_a, which for similar fibers and for the most interesting cases is only small fractions of one percent [Streckert 1985]. The relation (13.68) can also be used for determining the field width of a fundamental Gaussian laser beam in air or the lateral and transverse field widths in the exit plane of a single-mode semiconductor laser. The unknown field pattern need only be scanned with a single-mode fiber of known spot size.

Since 1985, commercial apparatus is available for measuring the spot size using the transverse offset technique.

There are several variations of the original transverse offset method. Instead of increasing the offset stepwise, it is also possible to mechanically vibrate the end of the input fiber with a frequency of e.g. 50 Hz and with an amplitude of about 20 μm. By displaying on an oscilloscope screen the transmitted power as a function of the instantaneous offset, one can conveniently adjust the secondary fiber for maximum transmitted power and read the spot size w_a from the screen [Streckert 1984].

In the transverse offset method described, the offset is varied continuously or in very small steps at some different discrete wavelengths. There is a variant of the method [Millar 1981b, 1982a,b], in which the wavelength is scanned continuously for the case of no offset and then for one fixed value of the offset. For obtaining the spot size by this method, one has to assume that the coupling efficiency versus offset curve is described by a true Gaussian function. Implementations of this technique, which requires an accurate knowledge of the offset and of the position of the peak transmission, are not trivial [Srivastava and Franzen 1985].

In another variant of the transverse offset method [Calzavara et al. 1985, 1986a–c], the fiber end is placed in the center of a spherical mirror. The image of the near-field distribution launches a reflected wave in the fiber. If the fiber end is displaced transversely by a vector $\boldsymbol{x}$, the image is displaced by $-\boldsymbol{x}$ and the transverse offset between the near field and its image is $2|\boldsymbol{x}|$. The reflected power is measured as a function of $s = 2|\boldsymbol{x}|$. The full $1/e$-width of this curve equals the mode field diameter $2w_a$.

As compared to the traditional transverse offset method, this "self-imaging offset technique" offers several advantages:

— only one endface has to be prepared,
— there is no gap between two endfaces,
— there is no need to align the axes of two fiber ends,
— for fibers with elliptical intensity distribution, no azimuthal adjustment is to be made.

Difficulties with this method are that the fiber end has to be preliminarily centered with respect to the mirror by using a HeNe laser and a microscope, and that the detected power includes a contribution from Fresnel reflections at the fiber end, which must be measured and subtracted.

To close this section, we note that attempts have also been made to determine the spot size by measuring the insertion loss of a fiber junction with only a variable longitudinal gap [Streckert 1980; Masuda et al. 1981]. The disadvantage of this method is that the transmitted power depends only weakly on the distance between the fiber ends but strongly on the transverse offset. Therefore, a residual transverse offset causes substantial measurement errors.

13.7.3 Aperture Methods

Several methods have been proposed for determining the spot size from the optical power transmitted through an aperture when either the dimension or the location of the aperture is changing. Figure 13.12 shows some apertures which can be used: a circular hole, a slit, a half-plane, and an array of parallel slits and opaque strips known as a Ronchi ruling.

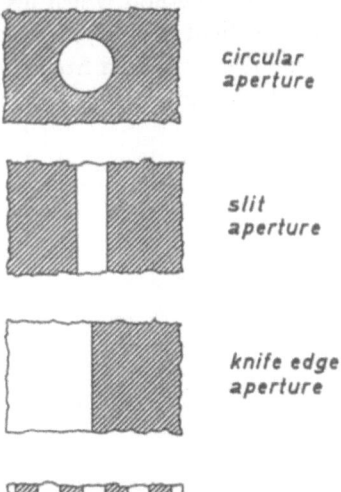

circular aperture

slit aperture

knife edge aperture

Ronchi ruling

Fig. 13.12. Apertures which can be used to determine the far-field radiation pattern and the spot size

In the so-called variable aperture launch (VAL) method [Alard et al. 1981, 1982c; Alard and Sansonetti 1983; Srivastava and Franzen 1985], a circular aperture varies the aperture of the lens which focuses the output slit of the monochromator onto the fiber input face. One measures the power coupled into the fiber. Assuming the source to be Lambertian, i.e. the radiance (Sect. 7.3) to be independent of the angle θ relative to the normal, and assuming the near-field distribution of the fundamental fiber mode to be truly Gaussian, the launched power $P_v(\theta)$ depends on the numerical aperture $\sin\theta$ of the input beam as follows [Alard et al. 1981]:

$$P_v(\theta) = P_{max}[1 - \exp(-2\pi^2 w_G^2 \sin^2\theta/\lambda^2)] \quad , \tag{13.69}$$

where P_{max} is the value of $P_v(\theta)$ when $\sin\theta$ approaches unity.

Measuring the power $P_v(\theta)$ launched into the fundamental fiber mode and plotting $-\ln[1 - P_v(\theta)/P_{max}]$ as a function of $\sin^2\theta$ should give a straight line, the slope of which is $2\pi^2 w_G^2/\lambda^2$. Thus, for a truly Gaussian field distribution, the spot size could be determined simply from the measured power vs aperture diameter curve.

The VAL method has been criticized, since the combination of light source, monochromator, and camera lens is not Lambertian [Anderson W.T. and Philen 1983a; Samson 1985a] and since the simple relation (13.69) between the launch numerical aperture $\sin\theta$ and the power launched does not hold for non-Gaussian fundamental mode fields [Nicolaisen and Danielsen 1983; Dick et al. 1984; Srivastava and Franzen 1985].

The effective spot size w_e can be calculated from the VAL data $P_v(\theta)$ [Alard and Sansonetti 1983]. However, this technique has not been successfully applied, mainly because the equation for w_e involves the second derivative of measured discrete values [Nishimura and Suzuki 1984].

There is another integral method similar to the VAL method which also uses a variable circular aperture, the variable aperture far-field (VAFF) method [Nicolaisen and Danielsen 1983; Dick et al. 1984]. The beam radiated from the end of the fiber is transmitted through a lens with variable aperture (Fig. 13.13).

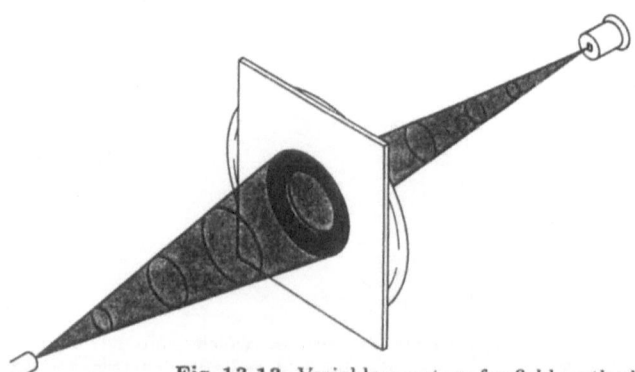

Fig. 13.13. Variable aperture far-field method for measuring the spot size

The lens focuses the transmitted light onto a photodetector. The power P_v transmitted through the aperture is measured as a function of the aperture diameter $2\varrho_a$. The variable aperture far-field method is mathematically equivalent to the variable aperture launch method, but does not require the launching optics to be Lambertian. Usually, the diameter of the aperture is not changed continuously but stepwise. At least twelve apertures spanning the half-angle range of numerical apertures from 0.02 to 0.25 should be used.

Instead of an array of apertures with different diameters, one can also use a single circular aperture which is translated along the fiber-detector optical axis in order to change the numerical aperture $\sin \theta$ [Saravanos and Lowe 1986a–c]. With this so-called far-field axial scanning technique (FFAST), a larger number of acceptance angles can be selected.

In another variant of the variable aperture far-field technique [Meunier J.-M. et al. 1985], instead of the power transmitted through the aperture, one measures the light power reflected by the screen which contains the circular aperture. The circular aperture is cut in the surface of an integrating sphere. The numerical aperture is varied by moving the fiber end axially in front of the aperture along a sphere diameter.

At present, standards bodies agree that for conventional single-mode fibers operating close to the LP_{11} cutoff, the Gaussian assumption is satisfactory for evaluating the mode field diameter from the multiple aperture data [Lowe 1985]. Note, however, that for fibers with non-Gaussian radial field distribution, it is not sufficient to determine the spot size directly from the power transmission function of the variable aperture by using (13.69). Since the assumption of a truly Gaussian far-field distribution is invalid, the obtained spot size depends on how this parameter is extracted from the measured data. This has been found using computer simulations [Nicolaisen and Danielsen 1983; Samson 1985a]. Even for a step index fiber with nearly Gaussian field distribution, the spot size w_G calculated via the far-field variable aperture method differs systematically from the exact Gaussian spot size w_G by 5% at $V = 2.4$ and by 14% at $V = 1.5$.

Therefore, for standard fibers operating in the third window ($\lambda = 1550\,\text{nm}$), and for dispersion-shifted and dispersion-flattened fibers, an algorithm must be used which does not rely on a Gaussian approximation for the radial field distribution [Saravanos and Lowe 1985].

From the transmitted power P_v as a function of numerical aperture $\sin \theta$, one can determine the far-field radiation pattern. For an aperture with radius ϱ_a at a distance z from the fiber end, the angle subtended by the aperture is

$$\theta = \tan^{-1}(\varrho_a/z) \tag{13.70}$$

corresponding to a spatial frequency (8.14)

$$q = (1/\lambda) \sin \theta \quad . \tag{13.71}$$

If $P_v(q)$ denotes the power transmitted through the aperture, the far-field intensity pattern $|E_H(q)|^2$ can be obtained by differentiation [Alard and Sansonetti 1983; Saravanos and Lowe 1985]

$$|E_{\mathrm{H}}(q)|^2 = cq^{-1}dP_{\mathrm{v}}(q)/dq \quad , \tag{13.72}$$

where c is an unimportant constant. The one-dimensional far-field pattern is thus obtained without resorting to the usual one-dimensional techniques (Sect. 13.6) where small signals are measured and high-power sources are required. From the far-field radiation pattern, one can then calculate the different far-field widths, from which the spot sizes can be obtained using (10.29, 30 or 31). Alternatively, one can determine the radial near-field distribution by a Hankel transform [Alard and Sansonetti 1983] and calculate the spot sizes from the near field.

The effective far-field width W_{e} can be directly expressed in terms of the function $P_{\mathrm{v}}(q)$ measured with the variable aperture far-field method. By replacing the far-field intensity $E_{\mathrm{H}}^2(q)$ by $(1/q)dP_{\mathrm{v}}(q)/dq$ (13.72) in the definition of the effective far-field width W_{e} (10.28), one gets [Saravanos and Lowe 1985]

$$W_{\mathrm{e}}^2 = 2 \int\limits_0^{1/\lambda} \frac{d}{dq}[P_{\mathrm{v}}(q)/P_{\mathrm{v\,max}}]q^2 \, dq \quad , \tag{13.73}$$

where $P_{\mathrm{v\,max}}$ is the maximum power transmitted through the largest aperture.

To use (13.73) for calculating the effective far-field width, one has first to take the derivative $dP_{\mathrm{v}}(q)/dq$ of the variable aperture data, and then to evaluate the integral. The derivative in the calculation can be eliminated by defining a complementary aperture power transmission function

$$a(q) = 1 - P_{\mathrm{v}}(q)/P_{\mathrm{v\,max}} \quad , \tag{13.74}$$

which is the relative power *not* transmitted through the circular aperture. Replacing $P_{\mathrm{v}}(q)/P_{\mathrm{v\,max}}$ in the integrand of (13.73) by $1 - a(q)$, and integrating by parts, gives

$$W_{\mathrm{e}}^2 = 4 \int\limits_0^{1/\lambda} a(q)q \, dq \quad . \tag{13.75}$$

Thus the effective far-field width W_{e} and the Petermann II spot size $w_{\mathrm{d}} = 1/(\pi W_{\mathrm{e}})$ (10.31) can be obtained directly from the variable-aperture far-field data by a simple integration [Fox 1986].

Measuring systems which use the variable aperture in the far-field technique (VAFF) for determining the mode field diameter are commercially available. In the wavelength range 0.8–$1.6\,\mu\mathrm{m}$, the mode field diameter can be measured with a resolution of e.g. $0.01\,\mu\mathrm{m}$ and a repeatability (2σ) of better than 1%.

The second form of aperture is a narrow transparent slit in an opaque diaphragm (Fig. 13.12). One can either shift the slit transversely through the beam or change its width b.

In the first technique [Samson 1985c, 1986], a thin vertical slit in a diaphragm is scanned horizontally across the far-field of the fiber (Fig. 13.14).

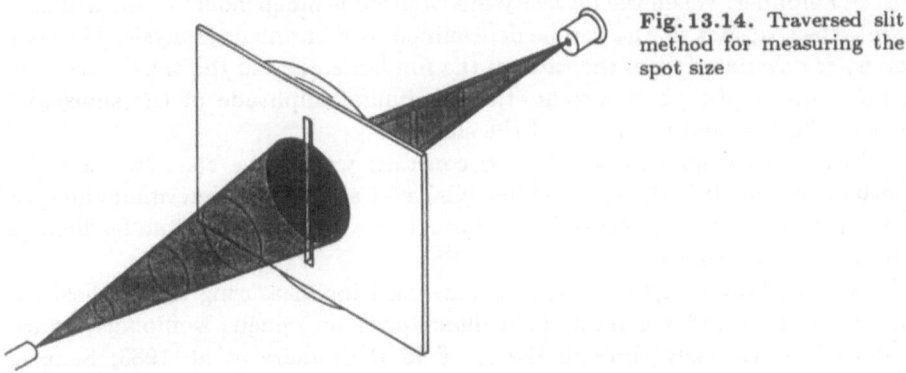

Fig. 13.14. Traversed slit
method for measuring the
spot size

The total light power P_s transmitted through the slit is focused onto a stationary photodetector by a spherical lens. In this "strip integrated far-field method", the light power $P_s(p)$ is measured as a function of the variable

$$p = (1/\lambda) \sin \theta_x \quad , \tag{13.76}$$

where θ_x is the angular position of the slit in the horizontal plane. The function $P_s(p)$ is essentially the forward Abel transform [Bracewell 1956] of the far-field intensity distribution. Thus, by measuring the power $P_s(p)$ transmitted through the slit, and taking the inverse Abel transform of it, one can determine the far-field radiation pattern and, thence, the different spot sizes.

The spot size $w_d = 1/(\pi W_e)$ (10.31) related to dispersion can also be calculated directly from the function $P_s(p)$ since [Samson 1985c, 1986]

$$W_e^2 = 4 \frac{\displaystyle\int_0^{1/\lambda} P_s(p) p^2 \, dp}{\displaystyle\int_0^{1/\lambda} P_s(p) \, dp} \quad . \tag{13.77}$$

Using this formula, it is not necessary to calculate the inverse Abel transform which is a computationally involved process and which has problems owing to the amplification of noise in the measurement.

There is also a relation between the power transmission function of the scanned slit and the amplitude transmission function of a joint with transverse offset: The amplitude transmission function $E*E = \eta^{1/2}$ of the offset joint (Sect. 13.7.2) is equal to the Fourier cosine transform of the function $P_s(p)$ [Samson 1986].

In a technique similar to the scanned-slit method, a magnified image of the near field is scanned over a narrow slit in a diaphragm by sinusoidally oscillating the fiber end [Coppa et al. 1985c, 1986c]. The light source, an LED, is switched on for one direction of scan only. The light power transmitted through the slit is detected by a photodiode. The spectrum of the photocurrent is analyzed by a lock-in amplifier. When the intensity distribution is independent of the azimuth ϕ, the effective spot size w_e can be determined by a harmonic analysis. The spot size w_e is calculated from the ratio of the fundamental and the third-harmonic components of the photocurrent, the maximum amplitude of the sinusoidal fiber oscillation, and the width of the slit.

Instead of displacing a slit with constant width, one can also vary the width of the slit. It is thereby possible [Calvo et al. 1986] to determine the spot size w_d from the power transmitted through the slit located in the far-field as a function of its width.

A third kind of aperture (Fig. 13.12) used for measuring the far-field radiation pattern and the mode field diameter is an opaque semiplane, which is moved transversely through the far field [Kuwahara et al. 1983; Samson 1985b, 1986; Otten and Kapron 1986; Anderson W.T. et al. 1987]. In this so-called knife-edge method, one measures the power $P_k(p)$ transmitted past the semiplane (Fig. 13.15) as a function of the variable

$$p = (1/\lambda) \sin \theta_x \quad , \tag{13.78}$$

where θ_x denotes the angular position of the edge. The power P_k transmitted in the knife-edge experiment is essentially the integral of the power P_s transmitted through the slit with width b

$$P_k(x) = \frac{1}{b} \int\limits_{-\infty}^{x} P_s(\xi)\, d\xi \quad . \tag{13.79}$$

Conversely, P_s can be obtained from P_k by differentiation

$$P_s(x) = b\, dP_k(x)/dx \quad . \tag{13.80}$$

The variable x is the distance of the slit or knife edge from the beam axis. This new variable x is related to the old variable p (13.78) via

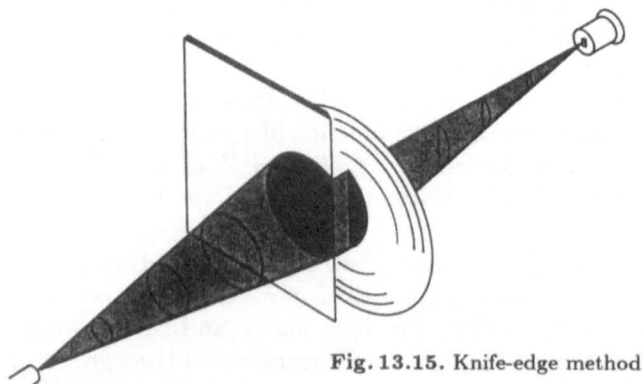

Fig. 13.15. Knife-edge method for measuring the spot size

$$\tan \theta_x = x/z \quad , \tag{13.81}$$

where z is the distance of the plane of the aperture from the fiber end.

With this "area integrated far-field method", the Petermann II spot size $w_d = 1/(\pi W_e)$ (10.31) may also be calculated explicitly from the measured data, either by taking the derivative of the knife-edge function P_k in order to obtain the slit function P_s and then inserting P_s into (13.77), or by inserting the knife-edge data directly into the expression

$$W_e^2 = 8 \int\limits_0^{1/\lambda} [P_k(p)/P_k(0)]p\,dp \quad . \tag{13.82}$$

Equation (13.82) can be derived from (13.77) via integration by parts [Samson 1986]. The power $P_k(0)$ is half of the maximum transmitted power.

It has also been proposed [Dick and Shaar 1986] that the spot size be determined from the $P_k(p)$ function either by directly fitting an error function to the measured data or by differentiating $P_k(p)$ and fitting a Gaussian function to the derivative. However, it has not been clarified whether and how the values determined by these evaluation methods are related to one of the different spot sizes defined in Chap. 10.

A fourth kind of aperture is the Ronchi ruling (Fig. 13.12) which consists of alternate transparent and opaque strips. Ronchi rulings are widely used for measuring the diameter of freely propagating Gaussian beams [Dickson 1979]. There are two methods for determining the fiber spot size using Ronchi rulings.

In the first method, the Ronchi ruling is moved transversely through the beam in the far-field and the power of the light transmitted through the ruling is measured. Assuming a Gaussian amplitude distribution, the spot size w_G can be calculated from the ratio of the maximum and the minimum transmitted power [Pocholle and Auge 1983; Auge et al. 1983].

In the second method, a Ronchi ruling with variable spacing is used. By measuring the transmitted power as a function of the spacing of the bars and spaces, the spot size w_G can be determined [Stewart et al. 1983].

Since the radial field distribution of the fundamental fiber mode is not strictly Gaussian, it is questionable whether the Ronchi ruling methods, which assume a Gaussian distribution for evaluation, may be applied for determining the mode field diameter.

13.7.4 Mask Methods

The mask methods to be described next can be used to measure either the effective spot size w_e or the Petermann II spot size w_d [di Vita et al. 1984; Stewart et al. 1984; Caponi et al. 1984].

The methods are best explained by considering the definition of the effective spot size w_e (10.8):

$$w_e^2 = 2 \frac{\displaystyle\int_0^\infty \varrho^2 E^2(\varrho)\varrho \, d\varrho}{\displaystyle\int_0^\infty E^2(\varrho)\varrho \, d\varrho} \quad . \tag{13.83}$$

Since $E^2(\varrho)$ is proportional to the intensity at a distance ϱ from the fiber axis, the product $E^2(\varrho)\varrho \, d\varrho$ is proportional to the power transmitted through a circular ring with radius ϱ and width $d\varrho$ and the integral in the denominator is proportional to the total power transmitted by the fundamental mode.

Similarly, the integral in the numerator is proportional to the power transmitted through a mask, which has a power transmittivity $T(\varrho)$ proportional to the square of the distance ϱ from the axis. This can be seen as follows: $E^2(\varrho)\varrho \, d\varrho$ is the power incident on the mask through the ring described above, and $\varrho^2 E^2(\varrho)\varrho \, d\varrho$ is proportional to the power in the ring leaving the mask.

There are several masks with a transmittance proportional to ϱ^2. Besides a grey distribution, the circular symmetry of the LP_{01} field means that one can also use the mask shown in Fig. 13.16, which consists of two totally opaque fields, separated from a totally transparent field by a Fermat spiral $\varrho(\phi)$ described by the function

$$\phi = (\pi/2)(\varrho/\varrho_0)^2 \quad . \tag{13.84}$$

Because of the tiny dimensions of the field distribution in the fiber endface, the mask must be located in the magnified image of the near-field distribution

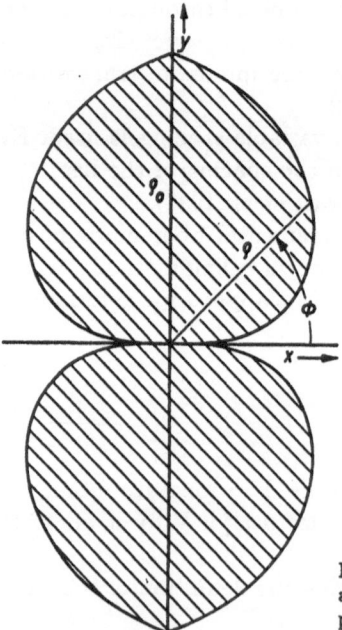

Fig. 13.16. Mask method for measuring the spot sizes w_e and w_d. Mask whose azimuth-averaged transmittivity is proportional to ϱ^2

(Sect. 13.5). The constant ϱ_0 is the largest diameter of one of the opaque regions; it must be equal to three spot sizes of the magnified near field [Caponi et al. 1985]. A lens focuses the power transmitted through the mask onto the detector of an optical power meter.

The method to determine w_e simply consists in measuring the powers P_m and P transmitted with and without mask placed in the magnified *near field*, respectively, which are proportional to the integrals in the numerator and denominator in the defining equation (13.83). The spot size w_e can then be found from:

$$w_e = \frac{\varrho_0}{M} \left(\frac{2P_m}{P} \right)^{1/2} , \qquad (13.85)$$

where M is the magnification.

Centering the mask relative to the beam axis is a simple matter, since the received power has a minimum when shifting the mask in the x-direction and a maximum when shifting it in the y-direction (Fig. 13.16).

The Petermann II spot size w_d (10.10) can also be measured by the mask method since $w_d = 1/(\pi W_e)$ (10.31) is inversely proportional to the effective far-field width W_e, the definition of which (10.28) formally corresponds to the definition of the effective near-field width w_e (10.8): The near-field distribution $E(\varrho)$ is replaced by the far-field distribution $E_H(q)$ and the radial coordinate ϱ by the spatial frequency q. Therefore, the integrals in the denominator and numerator of the equation (10.28) defining W_e are proportional to the total power P transmitted by the wave beam in the far field and to the power P_m transmitted through a mask with a transmittance proportional to $q^2 = (1/\lambda)^2 \sin^2 \theta$ located in the *far field* at a distance z from the radiating fiber end. The ratio of the powers gives the effective far-field width:

$$W_e = \frac{\varrho_0}{z\lambda} \left(\frac{2P_m}{P} \right)^{1/2} . \qquad (13.86)$$

If the "opaque" areas of the mask have a small residual transmittance, the distance z between the mask and the fiber end must be varied. The spot size can then be calculated from the slope of the transmitted power versus z^2 curve [Stewart et al. 1984].

The advantage of the mask methods is that the spot sizes can be determined directly by measuring the ratio of the photocurrents detected with and without the mask. Thus a tungsten lamp with monochromator is sufficient as optical source, and the wavelength dependence of the spot size can be determined easily.

Since the mask methods have been proposed only recently, it remains to be seen whether they will be accepted in practice. First preliminary results [Caponi et al. 1985; Samson 1986] seem to indicate systematic errors caused by the nonideal form of the cusps in the center of the mask and by the finite radius of the mask.

13.7.5 Noncircular Mode Fields

A large number of definitions have been proposed for the spot size of circular
fundamental mode fields and an even larger number of methods exist for measuring these quantities. Dielectric waveguides with noncircular mode fields are
widely used in integrated optics, and some varieties of noncircular polarization-
maintaining fibers (bow-tie, PANDA, elliptical core types) are now in use in
the sensor field and for coherent communication. Only a few definitions and
measurement methods have been proposed for optical waveguides with noncircular fundamental mode fields [Hayata et al. 1986; Goully et al. 1986]. One
method for measuring the autocorrelation spot sizes $w_{a\,max}$ and $w_{a\,min}$ with
the transverse offset method has been mentioned in Sect. 13.7.2.

13.7.6 Comparison of Methods for Measuring the Spot Size

There are various functions which can be measured in order to determine the
mode field diameters corresponding to the different definitions. In principle,
each of these functions is obtained by measuring an optical power as a function
of a geometrical parameter. The geometrical variable can be the radial coordinate of a pinhole in the magnified near field or in the far field, the transverse
offset of a fiber joint, the radius of a circular hole in a screen, the coordinate
of a slit, or the coordinate of a knife edge.

The six functions which can be measured are:

1. the near-field intensity distribution $|E(\varrho)|^2$,
2. the far-field radiation pattern $|E_H(q)|^2$,
3. the transmission factor $\eta(s) = |E*E|^2$ of a joint with transverse offset,
4. the power $P_v(q)$ transmitted through a variable circular aperture,
5. the power $P_s(p)$ transmitted through a slit aperture, and
6. the power $P_k(p)$ transmitted through a knife-edge aperture.

These functions are related to one another by a number of functional transforms such as derivatives, integrals, autocorrelation functions, Hankel transforms, Abel transforms, Fourier cosine transforms .... . The relations between
the functions 1, 2, and 3 have already been illustrated in Figs. 8.5. The relations
between functions 2 to 6 have been reviewed in Sect. 13.7.3 and are represented
schematically in Fig. 13.17 [Anderson W.T. et al. 1987]. Because of the relationships between the functions 1 to 6, theoretically or experimentally derived
functions for any domain may be transformed into the other domains, and the
mode field diameters can be calculated from the definitions [Anderson W.T.
and Kilmer 1986].

There have been some investigations to compare the different methods for
measuring the spot size of the fundamental fiber mode [Anderson W.T. and
Philen 1983a; Franzen 1985a; Franzen and Srivastava 1985; Samson 1986; Dick
and Shaar 1986; Anderson W.T. et al. 1987]. The first paper [Anderson W.T.
and Philen 1983a] compares four methods for determining the spot size w_G
related to launching (Sect. 10.1.2):

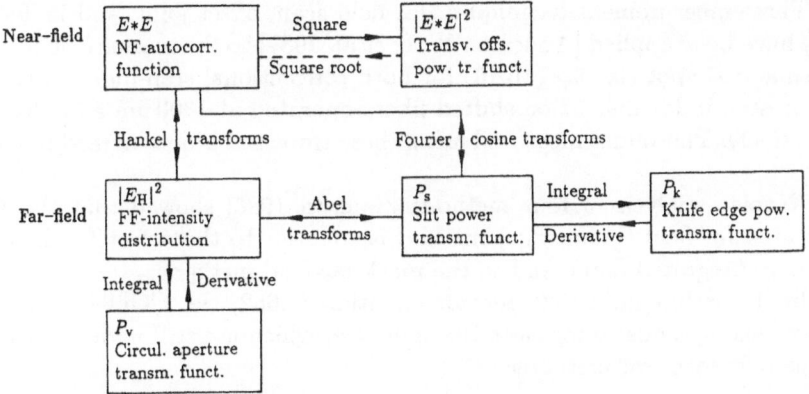

Fig. 13.17. Relations between transverse offset transmission function $|E*E|^2$, far-field intensity distribution $|E_H|^2$, slit power transmission function P_s, knife-edge power transmission function P_k, and circular aperture transmission function P_v

1. measuring the near-field intensity distribution (Sect. 13.5) and maximizing the launch efficiency integral (10.1),
2. measuring the autocorrelation spot size w_a (Sect. 10.1.5) by the transverse offset method (Sect. 13.7.2) and approximating w_G by w_a,
3. measuring the far-field distribution (Sect. 13.6), performing a Hankel transform (8.20) to get the near-field distribution and maximizing the near-field overlap integral (10.1), and
4. measuring the far-field distribution and maximizing a far-field overlap integral which is similar to (10.25) but uses for the integration variable the polar angle θ instead of the spatial frequency q (10.23). (In a later publication, W.T. Anderson [1984] showed that using q instead of θ makes the approximate relation between the far-field width W_G and the spot size w_G an exact one (10.30).)

For the three different standard fibers investigated, the values of the spot size found using the four measuring methods agreed quite well (to within 2%). The authors found the far-field overlap technique to be the easiest and fastest of the four techniques. But one has to remember that a laser must be used to measure the far-field distribution. If the wavelength dependence of the spot size is required, the transverse-offset technique is only slightly more difficult than the far-field overlap technique.

In an interlaboratory experiment, four techniques were used to determine the spot size [Franzen 1985a; Franzen and Srivastava 1985]: the near-field and far-field methods (Sect. 13.7.1), the transverse splice offset method (Sect. 13.7.2), and the variable aperture far-field method (Sect. 13.7.3). Five fibers, intended for 1300 nm operation, were investigated at different laboratories. At $\lambda = 1.3\,\mu$m, the standard deviation of all values of the field diameter (of 8–10 μm) measured with the different techniques was 0.15 μm. At $\lambda = 1.55\,\mu$m, the mode shapes are more non-Gaussian and systematic offsets between the techniques are more likely. This explains the larger spread of 0.28 μm found at the longer wavelength where the mode field diameter is in the range 9–12 μm.

Three measurement techniques (far-field scan, offset joint, and knife-edge scan) have been applied [Anderson W.T. et al. 1987] to the measurement of the Petermann II spot size w_d (10.10) for both conventional step-index fibers and for non-step-index dispersion-shifted fibers operated at 1300 nm and 1540 nm, respectively. The values measured using these three techniques agreed to within 1.6%.

A noise analysis of four methods [Samson 1986] showed that the area-integrated far-field (knife-edge) method is superior to the direct (pinhole), to the strip integrated (slit), and to the mask far-field methods.

In its preliminary 1984 recommendation G.652, the CCITT [1984] describes two methods to measure the spot size, which are still under consideration as reference test methods:

1. determining w_G by maximizing the overlap integral with a Gaussian function either in the near field or the far field (Sect. 13.7.1) and
2. determining w_a by the transverse offset method (Sect. 13.7.2).

Both these methods have advantages and disadvantages, and both of them are left for future consideration in the CCITT [Bonaventura 1984; Coppa et al. 1985b].

Currently (1987), five techniques are being considered by the CCITT for standardization [Anderson W.T. 1986]:

1. magnified near-field scanning,
2. far-field scanning,
3. measuring power transmission through an offset joint,
4. measuring power transmission through a circular aperture in the far field, and
5. knife-edge scanning.

The choice of mode field diameter is, in principle, only a matter of convention. But any one definition of spot size is not sufficient to uniquely describe all properties of fibers with non-Gaussian field distribution. The intrinsic reason for this is that it is generally impossible to characterize a function (the radial field distribution $E(\varrho)$) by a single parameter (the spot size w). As long as there is no recommended definition for the mode field diameter, the best method would be to record a complete distribution curve, from which any measure of spot size can be calculated. The best commercial test apparatus would measure one of the six optical power versus geometrical parameter functions, calculate, display and store both the near-field distribution and the far-field radiation pattern as well as the values of the spot sizes corresponding to the different definitions.

13.8 Measurement of Cutoff Wavelength

The cutoff wavelength of a LP_{lm} mode is the maximum wavelength for which the mode is guided by the fiber (Sect. 6.2). Usually, it is the LP_{11} mode which has the highest cutoff wavelength and which makes a single-mode fiber double-moded for short wavelengths. Thus the cutoff wavelength of the LP_{11} mode limits the region of single-mode operation at short wavelengths and is a very important quantity to know.

One has to distinguish between the theoretical cutoff wavelength λ_c (Sect. 6.3.2) and the effective cutoff wavelength λ_{ce} (Sect. 6.3.3) of the LP_{11} mode. The theoretical cutoff wavelength λ_c is a well-defined quantity. At

$$\lambda = \lambda_c \quad , \tag{13.87}$$

the cladding parameter is zero (6.14):

$$W = 0 \quad , \tag{13.88}$$

and because of (6.5), the phase constant equals the wavenumber in the cladding:

$$\beta_{11} = n_2 k \quad . \tag{13.89}$$

The theoretical cutoff wavelength can be calculated from the refractive-index profile, but is difficult to measure. The theoretical cutoff wavelength λ_c can be obtained by measuring the effective cutoff wavelength λ_{ce} for various bending radii R of a short test fiber, plotting λ_{ce} versus the curvature $1/R$, and reading the λ_{ce} value for very small values of $1/R$ [Lazay 1980].

In contrast to the theoretical cutoff wavelength, the effective cutoff wavelength λ_{ce} is not so well defined, but is easy to measure. The effective cutoff wavelength depends on the measurement method used to define it, on the fiber length, on the radius of curvature of the fiber axis, and on the launching conditions at the fiber input.

Many methods have been proposed in the literature for measuring (and defining) the effective cutoff wavelength. Because different measurement techniques can give significantly different results, the measurement has become somewhat controversial [Payne D.B. et al. 1984].

Provisionally, the CCITT [1986] defines the (effective) cutoff wavelength as follows (Sect. 6.3.3):

"The cutoff wavelength is the wavelength greater than which the ratio between the total power, including launched higher order modes, and the fundamental mode power has decreased to less than a specified value.

Note: By definition, the specified value is chosen as 0.1 dB for a substantially straight 2 m length of fiber including one single loop of radius 140 mm".

For measuring the effective cutoff wavelength defined in this manner, the CCITT [1986] recommends two transmitted power techniques as reference test methods: the bending method (Sect. 13.8.1), and and the power step method (Sect. 13.8.2). As an alternative test method, the CCITT [1986] recommends that λ_{ce} be determined from the wavelength dependence of the spot size (Sect. 13.8.3).

However, the definition of the cutoff-wavelength and its related values still needs further study [Bonaventura 1985]. Since the present definition of the cutoff wavelength (on 2 m of fiber) supplies only limited information on the cable characteristics and since in long cables, the operating wavelength can be smaller than the CCITT cutoff wavelength without causing unwanted modal noise or bimodal dispersion effects (Sect. 11.6.1).

Besides the three methods recommended by the CCITT for measuring the effective cutoff wavelength, several other methods have been proposed in the literature, which utilize different effects to detect the presence of the disturbing higher order modes. These alternate methods will generally result in values of the effective cutoff wavelength which are different from those defined by the CCITT.

Nevertheless, it is interesting in itself to know the various effects caused in single-mode fibers due to the presence of higher order modes. Thus in Sects. 13.8.4–13, these other methods for measuring cutoff will be briefly reviewed.

13.8.1 Bending Method

The cutoff wavelength can be determined from the increase in attenuation caused by bending the fiber [Katsuyama et al. 1976; Srivastava and Franzen 1985; Franzen 1985b]. The bending method is recommended by the CCITT [1986] as one of two reference test methods for the effective cutoff wavelength which has been defined in Sect. 6.3.3. The theoretical basis of the bending method for measuring the single-mode limit has been explained in Sect. 6.3.3.

For wavelengths a little shorter than cutoff, the field of the LP_{11} mode extends far into the cladding, which results in large bending losses as compared to those of the fundamental mode. This effect is used to measure the cutoff wavelength.

The test apparatus used is the same as that for measuring fiber attenuation by the cutback method (Sect. 13.3.1). However, the launch conditions used must be sufficient to excite both the fundamental and the LP_{11} modes (Sects. 7.2, 3). First, a 2 m length of test fiber is inserted into the apparatus and bent to form one loosely constrained single loop with a constant radius of 140 mm. Care should be taken to avoid any bends of radius smaller than 140 mm. Cladding modes must be stripped from the fiber by inserting a cladding mode stripper (Sect. 12.2.8). The power $P_1(\lambda)$ transmitted through the fiber sample is then measured as a function of wavelength λ. The quantity $P_1(\lambda)$ corresponds to the "total power, including launched higher order modes" used in the definition of the cutoff wavelength.

Secondly, keeping the launch conditions fixed, at least one additional loop of sufficiently small radius is introduced into the test sample as a mode filter (Sect. 12.2.14) to suppress the LP_{11} mode without attenuating the fundamental mode at λ_{ce}. A typical value for the radius of this loop is 30 mm, but the cutoff determined by this method is relatively insensitive to loop diameter [Franzen 1985b]. In this case, a smaller transmitted power $P_2(\lambda)$ is measured, which is the "fundamental mode power" mentioned in the definition of λ_{ce}.

From the powers $P_1(\lambda)$ and $P_2(\lambda)$ transmitted without and with the loop, respectively, the level difference Δa (6.45) between the total power and the fundamental power is calculated as

$$\Delta a(\lambda) = 10 \log[P_1(\lambda)/P_2(\lambda)] \quad . \tag{13.90}$$

The level difference will exhibit a peak in the wavelength region where the radiation losses due to the small loop are much higher for the LP_{11} mode than for the fundamental LP_{01} mode (Fig. 13.18). At wavelengths which are much shorter than the LP_{11} mode cutoff wavelength, the LP_{11} mode power is well confined within the fiber core and negligible loss is induced by the 3-cm loop resulting in $\Delta a(\lambda) = 0$. For long wavelengths, the LP_{11} mode is not guided by the fiber and if the loop radius is large enough not to add any curvature loss to the fundamental mode, there also is no loss increase.

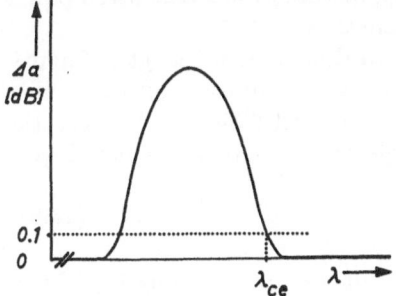

Fig. 13.18. Bending method for measuring the cutoff wavelength of the second order mode: level difference (13.90) versus wavelength

The shape of the loss versus wavelength curve has been calculated taking into account both micro- and macrobending losses, and has been found to be in excellent agreement with experimental results [Gambling et al. 1977d; Srivastava and Franzen 1985].

Corresponding to the CCITT definition of the effective cutoff wavelength given in Sect. 6.3.3, λ_{ce} is determined as the longest wavelength at which the level difference is $\Delta a = 0.1 \, \text{dB}$ (Fig. 13.18). Assuming that at the fiber input end, equal powers are launched into each of the two LP_{01} and each of the four LP_{11} modes, at λ_{ce}, the loss of the LP_{11} mode in the 2 m of fiber is 19.34 dB (6.48). The loss is predominantly due to the fiber loop with 140 mm radius. Thus, the bend attenuation coefficient for $\lambda = \lambda_{ce}$ is 22 000 dB/km (6.50).

For the reasons discussed in Sect. 6.3.3, the measured cutoff wavelength λ_{ce} is smaller than the theoretical cutoff wavelength λ_c. When the bending method is applied to fibers with values of length and curvature that are different from those ($L = 2 \, \text{m}$, $R = 140 \, \text{mm}$) recommended by the CCITT, λ_{ce} is found to depend both on the fiber length L and on the curvature $1/R$ (Sect. 6.3.3)

The bending method is experimentally very simple and can be combined with the measurement of spectral attenuation by the cut-back method. Commercial setups which measure the effective cutoff wavelength using the bending method are available.

13.8.2 Power Step Method

The second reference test method recommended by the CCITT [1986] for the effective cutoff wavelength is the so-called power step method [Srivastava and Franzen 1985].

The first part of the method is identical to the bending method (Sect. 13.8.1): The test apparatus is again the same as that to measure fiber attenuation by the cutback method (Sect. 13.3.1) and the launch conditions used must be sufficient to excite both the fundamental and the LP_{11} modes (Sects. 7.2, 3). A 2 m length of single-mode fiber under test is inserted into the test apparatus and bent to form one loosely constrained single loop with a constant radius of 140 mm. Care must be taken to avoid any bends of radius smaller than 140 mm and cladding modes must be stripped from the fiber by inserting a cladding mode stripper (Sect. 12.2.8). As in the bending method, the transmitted power $P_1(\lambda)$ is measured as a function of the wavelength λ.

The single-mode fiber is then replaced by a short (1–2 m) length of multimode fiber and the power $P_3(\lambda)$ emerging from the end of the multimode fiber is measured over the same wavelength range. The level difference between the powers launched into the multimode and single-mode fibers is calculated as

$$\Delta a(\lambda) = 10 \log[P_3(\lambda)/P_1(\lambda)] \quad . \tag{13.91}$$

A typical plot of the level difference as a function of wavelength is shown in Fig. 13.19. The power launched from the monochromator into the multimode fiber is about 15 dB larger than that launched into the single-mode fiber (Sect. 9.7). For long wavelengths, the source launches only two fundamental modes in the single-mode fiber, and the level difference Δa is large. For wave-

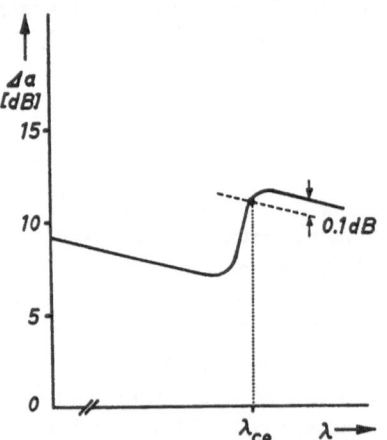

Fig. 13.19. Power step technique for measuring the cutoff wavelength of the second order mode: level difference (13.91) versus wavelength

lengths smaller than the cutoff wavelength of the LP_{11} mode, power is additionally launched into the four LP_{11} modes, giving a smaller level difference between the powers in the multimode and single-mode fibers.

Assuming that each of the two LP_{01} and four LP_{11} modes carries the same power, the output power at the end of the single-mode fiber increases by a factor of $(2 + 4)/2 = 3$ in going through cutoff [Srivastava and Franzen 1985]. This consideration explains the reduction in level difference of about 4.8 dB around cutoff (Fig. 13.19). Because of the "step" in the transmitted power versus wavelength curve, the method is called the "power step" method [Franzen 1985b]. Another name is "multimode reference" method.

Normalizing the power coupled into the single-mode fiber to the power coupled into the multimode fiber eliminates possible large spectral variations in system response, caused e.g. by the source, a grating, order-sorting filters, the detector, etc.

The effective cutoff wavelength λ_{ce} is determined by the intersection of the $\Delta a(\lambda)$ curve with a straight line displaced 0.1 dB and parallel to the straight line fitted to the long wavelength portion of $\Delta a(\lambda)$ [CCITT 1986] (Fig. 13.19).

It can be shown that at λ_{ce}, the attenuation of the LP_{11} mode is 19.34 dB as in the case of the bending method (6.48). Therefore, both methods approximately yield the same values for λ_{ce} as has been found in a round-robin test [Franzen 1985b].

Originally, the $\Delta a(\lambda)$ curve was determined as the loss of a splice between a multimode fiber and the single-mode fiber under test [Yamauchi et al. 1982].

Recently, it has been observed [Shaar et al. 1987] that the transmission factor of the reference multimode fiber does not change smoothly with wavelength, but exhibits ripples. This interference-like phenomenon can cause significant problems in measurement of the cutoff wavelength by the power step method. These problems can be solved by underfilling the multimode fiber, i.e. by using a small launching numerical aperture and a small launching spot diameter.

A variant of the power step technique is the "multiple bend method" [Lazay 1980; Franzen 1985b]: curvature is introduced into the fiber using a single turn around reels with different radii R. The cutoff wavelength for each curvature is determined from the transmitted power ratio $P_b(\lambda)/P_m(\lambda)$, where $P_b(\lambda)$ is the power transmitted through the curved single-mode fiber and, for purpose of normalization, $P_m(\lambda)$ is the optical power transmitted through a multimode fiber. The cutoff $\lambda_{ce}(R)$ for a drum radius R is the wavelength where the tangent to the power steps intersects the base line. $\lambda_{ce}(R)$ depends on the drum radius (1 cm ... 10 cm) approximately according to the following power series expansion (6.53)

$$\lambda_{ce} = \lambda_{ce0} - A/R + B/R^2 \quad , \tag{13.92}$$

where λ_{ce0}, A, and B are constants.

The multiple bend method requires several measurements and a more detailed data analysis. Moreover, the spread of the measured data is larger than for the single bend method [Franzen 1985b].

13.8.3 Spot Size Method

A third method for determining the effective cutoff wavelength is to measure the change of spot size with wavelength [Millar 1981b]. The CCITT [1986] recommends this method as an alternative test method for the measurement of the effective cutoff wavelength. However, it has not yet acquired any popularity in the United States [Srivastava and Franzen 1985].

The spot size $w_a(\lambda)$ is measured by the transverse offset method (Sect. 13.7.2) as a function of wavelength. A two-meter length of fiber is used on each side of the joint, and a single loop of radius 140 mm is then formed in each of these two-meter lengths. Cladding modes should not propagate.

In the single-mode region, the spot size is found to decrease almost linearly with decreasing wavelength (Fig. 13.20). As cutoff is approached, the contribution from the second-order mode causes the spot size to deviate significantly from the expected values for a single mode. Strictly speaking, the parameter w_a as determined by the transverse offset method represents the spot size in the single-mode region only, but has no physical interpretation in the multimode regime [Alard et al. 1982a].

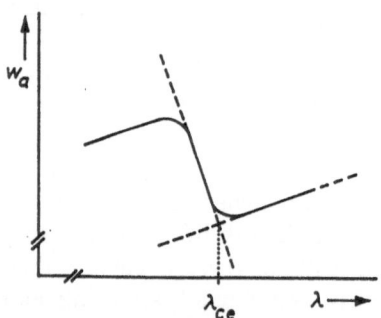

Fig. 13.20. Spot size technique for measuring the cutoff wavelength: wavelength dependence of the spot size w_a measured with the transverse offset method

Two straight lines with a positive and a negative slope are fitted through the measured points. The point of intersection of these two lines determines the effective cutoff wavelength.

It is advantageous that the same apparatus used to determine the spectral variation of the mode field diameter can also be used to determine the effective cutoff wavelength λ_{ce}. The cutoff wavelength determined by the offset splice technique depends on the excess loss of both fibers, e.g. that caused by bending [Chapman and Wu 1983]. A comparison of the offset splice and the bending method [Nichols 1982] showed that λ_{ce} determined by the offset splice method is very dependent on bending of the fibers and is smaller by 40–120 nm than λ_{ce} measured by the bending method.

A commercial measuring setup using the offset splice method achieves a resolution of 10 nm and a repeatability (2σ) of the effective cutoff wavelength 20 nm.

From the measured wavelength dependence of the spot size w_a in the single-mode region only, one can also indirectly determine the theoretical cutoff wavelength λ_c [Masuda et al. 1981]. Assuming a step-index profile, the method uses Marcuse's approximate formula (10.5) for the spot size w_G as a function of core radius, a, and normalized frequency, V. By measuring the spot size at two wavelengths λ_1 and λ_2, one can determine the unknown equivalent step index (ESI, Sect. 5.3.2) parameters core radius a_E and numerical aperture A_{NE}, from which one can calculate the theoretical cutoff wavelength:

$$\lambda_c = 2\pi a_E A_{NE}/2.405 \quad . \tag{13.93}$$

However, since the variation of mode field diameter with wavelength is a rather profile-dependent function, fitting of Marcuse's approximate formula for step-index fibers to the measured $w_a(\lambda)$-data can lead to large methodical errors in cutoff wavelength [Streckert 1985].

The spot size method can also be used to determine an ESI core radius a_E and an ESI numerical aperture A_{NE} from the spot size w_a at the cutoff wavelength λ_{ce}, or alternatively from w_a and $dw_a/d\lambda$ at λ_{ce} [Millar 1981b].

13.8.4 Refracted Power Technique

With the so-called refracted power technique [Bhagavatula et al. 1980], light of variable wavelength is produced with the lamp-monochromator combination (Sect. 13.1.1) and focused to a 5–6 μm spot on the end of the fiber to be investigated. The incident light with power P_{inc} launches the two fundamental modes with total power P_{01} and the four LP_{11} modes with total power P_{11}. The rest of the power P_{rad} is radiated. Because of conservation of energy we have

$$P_{inc} = P_{01} + P_{11} + P_{rad} \quad . \tag{13.94}$$

The radiated power is detected by a scattering detector. To this end the fiber is enclosed in an opaque capillary tube with only 1–2 mm of fiber exposed at the beginning. This assembly is placed in a cell containing an immersion fluid with an index slightly higher than the cladding (Fig. 13.21). The radiated

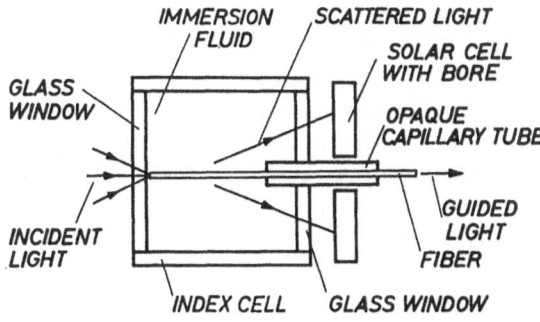

Fig. 13.21. Refracted power technique for measuring the cutoff wavelength of the second order mode: apparatus

power is detected by a solar cell having a hole for the fiber. The method is called the "refracted power technique" because it uses the light refracted from the cladding into the immersion fluid.

In order to become independent of the wavelength dependence of the incident power and of the sensitivity of the photodetector, the spot is focused first on the fiber cladding. No guided modes are excited and the radiated power measured equals the incident power. P_{inc} is then measured as a function of wavelength. Thereafter, the spot is focused on the core. In doing so, it is advantageous to choose an incident light distribution which preferentially excites the LP_{11} mode (Sect. 7.6). Now $P_{\text{rad}}(\lambda)$ is measured and the ratio of the radiated power and the incident power is calculated.

For wavelengths larger than the cutoff wavelength of the LP_{11} mode, this mode is not guided and therefore $P_{11} = 0$, whereas for $\lambda < \lambda_{\text{ce}}$, the LP_{11} mode will transmit a finite amount of power through the hole in the detector. Since the launching efficiency for the fundamental LP_{01} mode changes only slightly around cutoff, the normalized radiated power will decrease when decreasing the wavelength through cutoff.

Near cutoff, the field width of the LP_{11} mode is very large and because of the small spot size of the incident field, the launching efficiency for this mode will be small. Because of the wavelength dependence of coupling efficiency into the LP_{11} mode, one will not observe a sudden step in the radiated power but a transition region [Wang C.C. et al. 1983]. The long wavelength boundary of this transition region is associated with the effective cutoff wavelength.

The main advantage of the refracted power technique is that it determines the cutoff wavelength in a very short section (2–3 mm) of single-mode fiber, so that the measured value approaches the theoretical cutoff wavelength.

13.8.5 Near-Field Method

In the near-field method [Murakami et al. 1979], one uses the change in the near-field intensity distribution for determining the effective cutoff wavelength. As described in Sect. 13.1.1, a white light source and a monochromator are used to produce light of variable wavelength. By means of a microscope objective, this light is coupled into a graded-index multimode fiber which is spliced to the single-mode fiber. The multimode fiber is used as an exciter so that a short length of single-mode fiber can be investigated. The single-mode fiber is immersed in index-matching liquid (Sect. 12.2.8) to strip cladding modes. A magnified image of the fiber output face is detected with a vidicon and is displayed on a TV-monitor (Sect. 13.5). For the scan line through the center of the fiber core, the voltage versus time function is displayed on the screen of an oscilloscope, allowing the radial intensity distribution in the fiber endface to be observed.

For long wavelengths, the near-field pattern is similar to a Gaussian function, whereas for short wavelengths, it has two peaks of equal height with a minimum on the axis. These observations can be explained as follows: For wavelengths greater than the cutoff wavelength, only the fundamental modes will be transmitted by the fiber and one observes their bell-shaped near-field intensity distribution. For wavelengths smaller than the cutoff wavelength of the LP_{11} modes, the multimode fiber couples equal powers into each of the two LP_{01} modes of orthogonal polarization and into each of the four LP_{11} modes of different polarization and orientation (Fig. 4.14). Because of the finite linewidth (> 1 nm) of the source, the fields of the fundamental modes and the LP_{11} modes superimpose incoherently, i.e. their intensities add (Sect. 3.7.2). The intensity is increased in the region of maximum intensity of the LP_{11} modes (Fig. 6.2). Since the intensity of LP_{11} modes is zero on the fiber axis, the resulting near-field pattern shows a dip at this point.

When the wavelength is reduced the top of the near-field intensity pattern first flattens and, at still smaller wavelengths, the dip appears on the axis. The effective cutoff wavelength is that wavelength at which the pattern begins to deviate from the Gaussian form, i.e. where the top of the intensity profile is flat. This wavelength can be determined with an accuracy of ± 10 nm [Franzen 1985b]. But the apparatus for measuring the near field is complex and the fiber end must be carefully imaged.

As with most direct methods, the measured value of the cutoff wavelength is found to decrease with increasing length of the single-mode fiber investigated. The reason is that the attenuation of the LP_{11} modes increases strongly with wavelength (Figs. 5.27, 13.2), so that for a longer fiber, the wavelength must be reduced in order to observe the LP_{11} modes at the fiber end. Murakami et al. [1979] propose to use the mimimum fiber length possible (10–20 mm) for determining the effective cutoff wavelength. It then approaches the theoretical cutoff wavelength.

Instead of the change in the near-field distribution, the change in the far-field radiation pattern has also been used to determine the cutoff wavelength of the second order mode [Worthington 1971].

13.8.6 Pulse Height Method

If the power in the higher order modes at the end of the fiber is very small compared to the power in the fundamental mode, intermodal dispersion will not cause observable pulse distortions. The ratio of the mode powers can also be used to measure the effective cutoff wavelength [Kato et al. 1981].

Using a Raman-laser, short (0.6 ns) light pulses with variable wavelength are produced and focused on the fiber front face exciting the fundamental and the LP_{11} modes with similar powers. The two mode groups propagate with group velocities which in general are slightly different (Sect. 6.7). Therefore, for fibers of sufficient length, the pulses will separate in time and can be observed with a photodetector connected to the fiber output end and a sampling oscilloscope. There is a simple experimental test for deciding which peak in the oscillogram belongs to the LP_{11} mode: just bend the fiber. Since the bend losses of the higher order mode are larger, the peak which becomes smaller is that of the LP_{11} modes.

Kato et al. [1981] define the effective cutoff wavelength as that wavelength for which the power of the higher order modes is 30 dB below the power of the fundamental mode. For a 2 km fiber, this effective cutoff wavelength was found to be larger than the theoretical cutoff wavelength by a factor between 1.18 and 1.29. Thus, using the criterion of the power ratio at the end of a long fiber, the practical normalized cutoff frequency of a step index fiber is about 3 instead of 2.405. Operating a fiber at larger values of V has the advantage that the macro- and microbending losses as well as the absorption losses due to OH ions in the outer cladding are reduced.

The method of determining the cutoff wavelength from the power ratio of the second and first order modes has several disadvantages. One has to use an expensive tunable laser. Long fiber lengths are needed. The cutoff value determined depends on the fiber length used. In fibers which are shorter than the fiber used to measure the effective cutoff wavelength the attenuation of the second order mode is insufficient and, thus, modal noise effects (Sect. 11.6.1) cannot be excluded in short fibers for $\lambda > \lambda_{ce}$.

13.8.7 Polarization Method

As has been discussed in Sect. 6.3.1, the four LP_{11} modes are linear combinations of the exact vector modes HE_{21}, TE_{01}, and TM_{01}. Whereas the LP modes are linearly polarized, the vector modes are not. This fact can be used to determine the cutoff wavelength [Kato et al. 1979; Coppa et al. 1985d; Kato and Miyauchi 1985].

The apparatus needed is very simple: one must simply add a polarizer and an analyzer to the cut-back setup used to measure the attenuation (Sect. 13.3.1).

A beam linearly polarized in the y-direction is focused on the fiber front face with an offset in the x-direction. Considering the transverse field distributions (Fig. 4.14) of the four LP_{11} modes and the overlap integrals with the incident field distribution, one expects that besides the fundamental mode only that LP_{11} mode will be excited which is polarized in the y-direction with the maxima of the intensity oriented along the x-direction. In reality, however, the TM_{01} mode and one of the two HE_{21} modes are launched which because of an amplitude unbalance do not combine to form the pure linearly polarized LP_{11} mode expected. Both the TM_{01} and the HE_{21} modes also have x-components of the electric field vector. Therefore, the analyzer at the end of the fiber oriented in the x-direction will transmit a certain amount of power if the higher order modes are propagated by the fiber. In the single-mode region, almost no power will be detected because the fundamental mode is linearly polarized in the y-direction.

The cutoff wavelength is taken as the longest wavelength for which power transmission through an analyzer oriented in the x-direction is observed.

This polarization method can be extended to measure the fraction of power $P_{01}/(P_{01} + P_{11})$ in the fundamental mode [Coppa et al. 1985d]. Assuming the second order mode field to be completely unpolarized, this power fraction is equal to $(P_{max} - P_{min})/(P_{max} + P_{min})$, where P_{max} and P_{min} are the maximum and minimum powers transmitted through a rotatable analyzer at the fiber end. In the single-mode region, the power ratio is unity and in the two-mode region it is 0.33 to 0.5. The cutoff wavelength is defined as that wavelength at which the curve shows a knee.

For the reasons discussed in Sect. 6.3.3, the cutoff value determined by this direct method decreases with increasing length and curvature of the fiber sample.

13.8.8 Coherence Method

When a fiber only guides the fundamental mode, the fields at two points in a fiber cross section are coherent – we have a spatially coherent wave (Sect. 3.7.5). In the two-mode region, the degree of coherence of the fields at the two different points is smaller than unity. This effect can be used to measure the effective cutoff wavelength [Piazzolla and Spano 1982; Spano et al. 1983].

For a quasi-monochromatic source, because of the different group delays of the fundamental and the second order modes, the fields of the two modes will be incoherent at the end of a fiber of sufficient length. The total field, which is the superposition of the LP_{01} and LP_{11} mode fields, will be partially coherent, with the degree of coherence determined by the relative field strengths of the two modes.

By using a Michelson interferometer with one corner reflector (a reversing wavefront interferometer) it is possible to superimpose the fields at locations (x,y) and $(-x,y)$ of a fiber cross section and to measure the visibility $V(x)$ of the interference fringes (3.23) as a function of the distance x from the center line.

In the single-mode region, the visibility is found to be $V(x) = 1$, whereas in the two-mode region, the visibility decreases from $V(0) = 1$ (since the field of the LP_{11} mode is zero on the fiber axis) to zero at a certain x_0 which depends on the relative power P_{11}/P_{01} in the second order mode. By measuring x_0, one can determine the power ratio.

The power ratio measured as a function of wavelength is found to be zero in the single-mode region and finite in the two-mode region. The cutoff wavelength is determined as the longest wavelength for which the power ratio is different from zero.

The methods for measuring the effective cutoff wavelength described so far are direct methods, since the wavelength is changed until some effect is observed that can be directly attributed to the LP_{11} modes. With the indirect methods to be described in the next sections, other quantities are measured, e.g. the refractive-index profile, and the value of the cutoff wavelength is calculated indirectly from them.

13.8.9 Refractive-Index Profile Method

With this indirect method, the refractive-index profile of the fiber is measured, e.g. by the refracted near-field technique [White K.I. 1979] or by the far-field method [Hotate and Okoshi 1979b] as described in Sect. 13.10.2, and the theoretical cutoff wavelength λ_c is calculated from the index profile by solving the wave equation (6.3) assuming $W = 0$ (6.14).

The disadvantages of this method are that it is difficult to measure the index profile accurately; that one has to use numerical techniques to calculate the theoretical cutoff wavelength; and that the theoretical cutoff wavelength differs considerably from the minimum permissible operating wavelength (Sect. 6.3.3)

13.8.10 Far-Field Method

If the refractive index profile is known to be of the step-index type, the cutoff wavelength can be determined from the far-field radiation pattern [Gambling et al. 1976a, d]. Formulas for the radiation pattern (8.37–39) of a step-index fiber have been given in Sect. 8.1.

From the measured values of the half-intensity angle θ_h and the angle θ_x of the first minimum, the cutoff wavelength can be determined as follows: the ratio $\sin \theta_x / \sin \theta_h$, which is an unambiguous function of the normalized frequency V only, is calculated. Then, from the measured value of $\sin \theta_x / \sin \theta_h$, the value of V can be determined by using a calculated plot of $\sin \theta_x / \sin \theta_h$ versus V. Assuming the relative index difference Δ to be independent of wavelength, the normalized frequency V (5.9) is inversely proportional to wavelength. Thence, from the known cutoff value $V_c = 2.405$ of a step index fiber, the cutoff wavelength is obtained from V and the wavelength λ

$$\lambda_c = \lambda V / 2.405 \quad . \tag{13.95}$$

Besides giving the theoretical cutoff wavelength, this method can also be used to determine the core radius a and the relative refractive index difference Δ (5.2): Since the normalized half-intensity angle $ka \sin \theta_h$ is another unambiguous function of V, the value of the product $ka \sin \theta_h$ can be read from a second calculated diagram. Since the vacuum wave number k and the half-intensity angle θ_h are also known, the core radius a can be determined. Finally, for given values of V, a and λ, the relative index difference Δ can be obtained from the definition of V (5.9).

One drawback of the method, however, is that it assumes that the fiber has a step-index profile, which is usually not the case [Gambling et al. 1978c]. Thus, the far-field pattern method cannot be applied to fibers with an arbitrary refractive-index profile. Moreover, it is no simple task to measure the angle θ_x of the first minimum, since the side lobes of the radiation pattern are extremely small (Sect. 8.1).

For fibers with an arbitrary index profile, a method has been proposed [Pask and Sammut 1980] for determining the V_E value and the core radius a_E of an equivalent step index (ESI) fiber (Sect. 5.3.2).

13.8.11 Transverse Diffraction Method

A technique to determine the most important parameters of a single-mode fiber with power-law index profiles (5.6, 7) consists of two experiments [Brinkmeyer 1979a]: first the fiber is illuminated perpendicularly to its axis by a laser beam.

The fiber is immersed in a liquid the refractive index of which matches the cladding index. The incident light beam is diffracted by the fiber core and the angle $\theta_{\min}$ corresponding to the first minimum of the diffraction pattern is measured. An effective core radius a_e then is defined by (10.7)

$$a_e = 3.832/(k \sin \theta_{\min}) \quad , \tag{13.96}$$

where k is the wavenumber in vacuum.

In the second experiment, the laser beam irradiates the fiber front face to launch the fundamental mode in the fiber, which radiates freely from the other fiber end. In the far field, the angle θ_h is measured, for which the intensity has fallen to one-half of its value at the central maximum. From a_e and the measured angle θ_h, one can then graphically determine the effective normalized frequency V_e (6.39). Finally, the theoretical cutoff wavelength is obtained from the wavelength λ and from V_e according to

$$\lambda_c = \lambda V_e/2.405 \quad . \tag{13.97}$$

If the fiber has a step index profile, the effective core radius a_e equals the true core radius a and the effective normalized frequency V_e equals the normalized frequency V.

When a_e and V_e have been determined, the spot size w_G can be calculated from the simple approximation (10.6).

The disadvantage of the transverse diffraction method is that it can be applied only to power-law refractive-index profiles.

13.8.12 Interference Method

The phase shifts of the fundamental and second order modes in a fiber of length L are $\beta_{01}L$ and $\beta_{11}L$, respectively. Assuming that a coherent source excites both waves in phase at the input end, the phase difference between the two modes at the fiber output is

$$\Delta\Phi = (\beta_{01} - \beta_{11})L \quad . \tag{13.98}$$

This phase difference changes with wavelength causing a varying interference between the fields, which can be utilized for determining the cutoff wavelength [Brinkmeyer and Heckmann 1984].

In order to apply this method, a coherent optical source must be available, which can be tuned in a wavelength range around the cutoff wavelength, e.g. a dye-laser or a fiber Raman laser. Both modes must be launched with comparable powers (Sect. 7.2). Because of the orthogonality of the modes (Sect. 7.1), for a photodiode receiving all photons emitted from the fiber end, the photocurrent is simply proportional to the sum of the two mode powers which does not change for a small change in wavelength. Therefore, the interference between the modes cannot be detected.

A small photodetector must be placed in the radiation field at an offset location with respect to the fiber axis, where the local intensity changes depending on the phase difference between both mode fields.

Because of the interference effect, the photocurrent as a function of wavelength oscillates between maxima and minima. The distance $\delta\lambda$ between two maxima is measured as a function of wavelength and the measured values are approximated by an exponential function plus a constant. From this function, an effective cutoff wavelength can be determined, which is defined as that wavelength, for which the ratio of the normalized group delays of the two modes is $1/e = 0.37$. At this effective cutoff wavelength, more than 63% of the LP_{11} power propagates in the cladding, indicating a large transverse field extent and large macro- and microbending losses. The theoretical cutoff wavelength (assuming an infinite cladding) and this effective cutoff wavelength coincide to within fractions of a percent.

The method is essentially independent of launching conditions and differential mode attenuation, which influence the value of λ_{ce} measured by direct methods.

13.8.13 Miscellaneous Methods

Besides the twelve methods for measuring the cutoff wavelength described so far, several other techniques have been proposed and these will be reviewed briefly.

A sensitive qualitative technique for determining the existence of higher order modes, which utilizes the same effect as the interference method described in Sect. 13.8.12, consists in wrapping the fiber around a piezoelectric transducer acting as a phase modulator and observing the interference between the higher-order modes and the fundamental mode at the fiber end with a small photodetector connected to a lock-in amplifier [Nakajima and Ezekiel 1986]. When the fiber is not single-moded, the demodulated output fluctuates between positive and negative values both as a function of detector position and as a function of time.

The theoretical cutoff wavelength can be determined by measuring the group delay difference

$$\Delta t_{\mathrm{g}} = (\tau_{11} - \tau_{01})L \quad , \tag{13.99}$$

between the second order and the fundamental mode versus wavelength, matching an assumed step index profile with dip to the measured curve and calculating λ_{c} from the index profile [Kato et al. 1981].

In a similar method, the equalization wavelength λ_{equ}, for which the group delay difference Δt_{g} is zero (Sect. 6.7), is measured either by using a pulse method or by observing the interference of the LP_{01} and LP_{11} mode fields [McMillan 1983; McMillan and Robertson 1984]. The equalization wavelength can be measured very accurately and reliably. For a fiber reference length of 10 mm, the effective cutoff wavelength of five different fibers was measured to be 202 to 237 nm longer than the equalization wavelength [McMillan and Robertson 1985].

As has been explained in Sect. 6.3.3, the effective cutoff wavelength can alternatively be defined as that wavelength for which the bending loss of the LP_{11} mode is $22\,000\,dB/km$ for a bend radius of $140\,mm$. Therefore λ_{ce} can be determined indirectly [Nijnuis and van Leeuwen 1984] by measuring the attenuation constant α_{b11} of the second order mode by one of the methods described in Sect. 13.3.6, and by calculating λ_{ce} from the criterion for cutoff (6.50)

$$\alpha_{b11}(\lambda_{ce}, R) = 22\,000\,dB/km \quad . \tag{13.100}$$

In another method, from three wavelength scans of the transmitted power

1. for the unperturbed fiber,
2. for the fiber with one perturbation which attenuates the LP_{11} mode (a bend or a serpentine path), and
3. for the fiber with a second identical perturbation,

the relative power $\varepsilon = P_{01}/(P_{01} + P_{11})$ in the fundamental mode (and the attenuation of the LP_{11} mode caused by the perturbation) can be determined [Coppa et al. 1984b]. The effective cutoff wavelength is defined as that λ value at which the relative power in the fundamental mode falls to a certain predetermined value, e.g. 95%, of its maximum.

As the cutoff of a higher order mode is approached, the evanescent field associated with that mode spreads into the cladding. Consequently, for a fiber that is immersed in index-matching liquid, the leakage rapidly increases as cutoff is approached. This effect has been used [Midwinter and Reeve 1974] to determine λ_{ce} for higher order modes with small values of the mode numbers l and m by observing the light scattered out of the fiber as a function of wavelength.

Periodic microbends cause a coupling between the LP_{01} and LP_{11} modes which is maximum when the inverse of the spatial period equals the difference of the phase constants of the two modes (Sect. 7.5). It has been proposed [Grebel et al. 1986] that this effect be utilized for measuring the cutoff wavelength of the LP_{11} mode.

13.8.14 Comparison of Methods

Since it is very important to know whether a fiber transmits only the fundamental mode or more than one mode, many methods have been proposed for measuring the cutoff wavelength. Because it is too early to predict which methods will ultimately be used, the preceding sections review most of these methods.

In general, different methods will result in different values of the cutoff wavelength. Several groups have compared different measuring methods [Millar 1981b; Nichols 1982; Wang C.C. et al. 1982a,b, 1983; Anderson W.T. and Lenahan 1984; Franzen 1985a,b].

The bending technique (Sect. 13.8.1) can be applied for fiber lengths up to $10\,m$. In longer fiber samples, a different technique must be used, e.g. the spot size method for observing the change in apparent spot size near cutoff (Sect. 13.8.3). For lengths where both methods can be used, they gave values which agree to within ± 10 nm [Anderson W.T. and Lenahan 1984]. In another study [Millar 1981b], the cutoff wavelengths determined from the spectral variation of the apparent spot size (Sect. 13.8.3) agreed to within ± 15 nm with the figures determined by the bending method (Sect. 13.8.1). However, in a third investigation [Nichols 1982], the comparison of the same methods resulted in a discrepancy of between 40 and 120 nm!

C.C. Wang et al. [1982a,b, 1983] compared three experimental methods: the refracted power method (Sect. 13.8.4), the near-field method (Sect. 13.8.5), and the far-field method (Sect. 13.8.10). For the same fiber, differences were found in the measured values of up to 170 nm! The refracted power method gave the longest value (830 nm) for the cutoff wavelength, since it does not suffer from differential mode attenuation. The values determined by the far-field and near-field methods were 660 and 690 nm, respectively. The authors thus conclude that the refracted power measurement technique best gives the intrinsic, i.e. the theoretical, cutoff wavelength.

The National Bureau of Standards (NBS), in cooperation with the Electronic Industries Association (EIA), conducted an interlaboratory comparison among six fiber manufacturers to determine the effective cutoff wavelength of single-mode fibers [Franzen 1985a,b]. The methods employed were the bending method (Sect. 13.8.1), the power step method (Sect. 13.8.2), the near-field method (Sect. 13.8.5), and the multiple bend method (Sect. 13.8.2). Single bend attenuation and the power step method gave an average precision of approximately 8 nm, with the systematic differences between the values determined by the two methods being comparable to the interlaboratory spread in results. There are significant discrepancies (up to 35 nm) between the effective cutoff wavelengths determined by the single bend and the near-field methods.

Since the effective cutoff wavelength depends strongly on the fiber length and the radius of curvature (Sect. 6.3.3), the conditions of measurement are as important as the method used to measure λ_{ce}.

13.9 Measurement of Chromatic Dispersion

The ultimate information capacity of single-mode communication systems is limited by chromatic dispersion and source linewidth (Sects.11.4, 5). Thus chromatic dispersion (5.85)

$$D = d\tau_g/d\lambda \tag{13.101}$$

and dispersion slope (5.93)

$$S = dD/d\lambda = d^2\tau_g/d\lambda^2 \tag{13.102}$$

are essential specification parameters for single-mode fibers.

Pulse broadening in single-mode fibers is too small (down to several ps/km) to measure in kilometer lengths because direct detection receivers cannot resolve pulsewidths narrower than 50 ps. Therefore, most practical measurement techniques measure the specific group delay τ_g as a function of wavelength, obtain the dispersion parameter D and the dispersion slope S by taking the derivatives, and calculate the pulse broadening using the formulas given in Sect. 11.5.

Since one is only interested in the derivatives, it is not necessary to measure the specific group delay absolutely; it is sufficient to measure its change $\tau_g(\lambda) - \tau_g(\lambda_1)$ with wavelength, where λ_1 is a fixed reference wavelength.

Because of random errors in the discrete data points $\tau_g(\lambda_i)$, $i = 1, 2, 3, \ldots$, numerical differentiation is not feasible. A simple analytical function containing only a few parameters is fitted to the delay data. Several fitting functions are used to accurately represent the experimental data, e.g. an n-term polynomial

$$\tau_g(\lambda) = c_1 + c_2(\lambda - \lambda_0)^2 + \ldots + c_n(\lambda - \lambda_0)^n \quad , \tag{13.103}$$

where λ_0 is the wavelength of minimum chromatic dispersion (13.107), a three-term Sellmeier function

$$\tau_g(\lambda) = c_1 \lambda^{-2} + c_2 + c_3 \lambda^2 \quad , \tag{13.104}$$

a five-term Sellmeier function [Payne D.N. and Hartog 1977]

$$\tau_g(\lambda) = c_1 \lambda^{-4} + c_2 \lambda^{-2} + c_3 + c_4 \lambda^2 + c_5 \lambda^4 \quad , \tag{13.105}$$

or a function containing a logarithmic term [Reed and Philen 1986]

$$\tau_g(\lambda) = c_1 + c_2/\lambda + c_3 \ln \lambda \quad . \tag{13.106}$$

In addition to these, other fitting functions for $\tau_g(\lambda)$ have been proposed [Gloge et al. 1980; Pearson et al. 1982].

The function chosen is least-squares fitted to the measured group delay data $\tau_g(\lambda_i)$ and then differentiated to give the dispersion D and the dispersion slope S [Cohen L.G. et al. 1985a]. When D and S are determined, the formulas given in Sect. 11 can be used to calculate the baseband transfer function, the bandwidth, and the pulse broadening.

Two important parameters are the wavelength λ_0 of zero dispersion, where

$$D(\lambda_0) = 0 \quad , \tag{13.107}$$

and the dispersion slope S_0 at λ_0. In terms of these two parameters, the simple three-term Sellmeier function (13.104) can also be written [Cohen L.G. and Lin 1977b; Kapron and Olson 1984]

$$\tau_g(\lambda) = \tau_{g0} + \left(\frac{S_0}{8}\right)(\lambda - \lambda_0^2/\lambda)^2 \quad . \tag{13.108}$$

Here, τ_{g0} is the relative delay minimum at λ_0 (Fig. 5.13b). The dispersion coefficient D is the derivative of (13.108)

$$D(\lambda) = \frac{S_0}{4} (\lambda - \lambda_0^4/\lambda^3) \quad . \tag{13.109}$$

Thus, when using the three-term Sellmeier expansion, only the wavelength λ_0 of zero dispersion and the dispersion slope S_0 at λ_0 need to be known to calculate the dispersion of a fiber at all wavelengths.

A power series expansion of $D(\lambda)$ at $\lambda = \lambda_0$ [$n = 2$ in (13.103)] gives an even simpler formula for the dispersion:

$$D(\lambda) = S_0(\lambda - \lambda_0) \quad . \tag{13.110}$$

Within ± 30 nm of λ_0, this linear approximation lies within 4% of more accurate fits [Kapron and Olson 1984].

Of course, the values of λ_0 and S_0 determined from the experimental data $\tau_g(\lambda_i)$, and the function $D(\lambda)$ depend slightly on the fitting procedure chosen [Kapron and Olson 1984]. For matched-cladding and depressed-cladding fibers, the dispersion curves $D(\lambda)$ determined by the three-term and five-term Sellmeier function differ only very slightly. But for dispersion-shifted fibers, the simple three-term Sellmeier curve for $\tau_g(\lambda)$ (13.104, 108) is not in good agreement with the experimental data, and the five-term function (13.105) must be used to give a better fit [Reed and Philen 1986; Thevenaz et al. 1988].

Methods for measuring the dispersion parameters D and S have been reviewed [Cohen L.G. et al. 1980a, 1985b; Bernard and Visseaux 1985].

The wavelength dependence of the group delay can be determined directly from the pulse delay (Sect. 13.9.1), from the phase shift of a sinusoidal modulation (Sect. 13.9.2), with an optical interferometer (Sect. 13.9.3), from the wavelength dependence of the spot size (Sect. 13.9.4), and by various other methods (Sect. 13.9.5).

13.9.1 Pulse Delay Method

In the time domain, the chromatic dispersion is measured by transmitting short pulses with variable wavelength and measuring the change of the transit time

$$\Delta t_g = [\tau_g(\lambda) - \tau_g(\lambda_1)]L \quad . \tag{13.111}$$

The CCITT [1986] recommends this method as one of two reference test methods for chromatic dispersion.

The measuring setup consists of an optical pulse source emitting short (e.g. 200 ns) pulses, a fast photodiode, and a sampling oscilloscope (Fig. 13.22). The pulse delay technique was first applied to material dispersion measurement in multimode fibers [Lin and Cohen 1978] and in single- and dual-mode fibers [Gloge and Chinnock 1972].

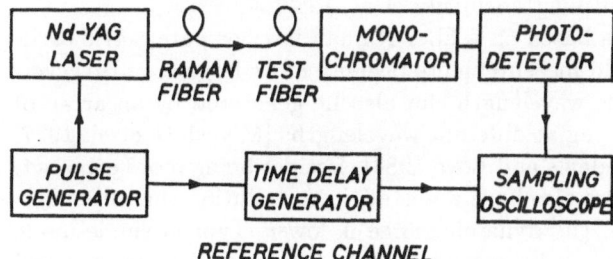

Fig. 13.22. Experimental apparatus for measuring chromatic dispersion by the pulse method

Setups for measuring chromatic dispersion by the pulse delay method have been described in several papers [Cohen L.G. and Lin 1977a,b, 1978; Pocholle et al. 1983a,b; Campos and Srivastava 1984; Srivastava and Franzen 1985].

Various sources can be used for the pulse delay technique. A fiber Raman laser with a silica fiber is commonly used [Cohen L.G. and Lin 1977b, 1978] since it provides multi-wavelength outputs covering the $1.1-1.7\,\mu$m spectrum in the form of subnanosecond pulses (pulsewidth 150 ps). A monochromator is used to sequentially select individual wavelengths. The pulse delay in the fiber is measured by a sampling oscilloscope.

When a GeO_2 fiber instead of a SiO_2 fiber is used in the Raman laser, the dynamic range of the measuring system can easily be improved by more than 10 dB [Chang J. et al. 1985; Takahashi et al. 1986]. It is thus possible to measure the chromatic dispersion of long single-mode links.

When a Raman laser is used, launching can be simplified by butt-jointing the fiber under test directly to the Raman fiber and inserting the monochromator required for wavelength selection between the end of the test fiber and the photodetector. Another advantage of using the monochromator after the test fiber (and not before it, as is done normally) lies in the fact that at the fiber end, the pump pulse with a wavelength of $1.06\,\mu$m can be selected by a dispersive prism and be used for triggering the time sweep generator of the sampling oscilloscope [Campos and Srivastava 1984]. This eliminates the problem of pulse jitter which is observed with digital trigger delay generators and which may be of the order of ± 100 ps.

By reducing pulse jitter, preventing irregularites in the shape of the pulses, and compensation of long term instabilities, the reproducibility error in the measured delay times can be reduced to only 2 ps rms [van Bochove et al. 1985].

An expensive variant of the pulse method uses a parametric oscillator for producing 16 ps pulses with variable wavelength and optical sum frequency mixing in a nonlinear KDP crystal to increase the temporal resolution to better than 30 ps [Mochizuki et al. 1981b].

Another expensive technique for measuring dispersion with high time resolution (13 ps) transmits a $1.06\,\mu$m pulse and a pulse of variable wavelength through the test fiber and uses sum-frequency mixing to detect the coincidence of the pulses at the fiber output [Mochizuki et al. 1982].

A measurement system based on a fiber Raman laser or parametric oscillators is not suitable for field measurements because of its large size. However, the test signal with variable wavelength can also be generated by an array of semiconductor lasers operating at different wavelengths [Miyashita et al. 1977; Lin C. et al. 1982, 1983; Modavis and Love 1984]. Besides being more compact, systems using injection lasers also have a better time resolution than fiber Raman laser setups. However, the dynamic range is lower. Typical single-mode fiber lengths of 5 to 10 km can be measured over the 1.2 to 1.5 mm spectral range, and more than 20 km can be measured in the low-loss regions close to 1.3 and 1.55 mm [Lin C. et al. 1983].

The lasers can be coupled to the test fiber via a wavelength division multiplexer (Sect. 12.3.3) or via a fiber mechanical switch [Lin C. et al. 1983].

In order to reduce the number of laser diodes required, and to increase the number of measured points, one can tune the laser wavelength by changing the laser temperature [Mogensen et al. 1986; Damsgaard et al. 1986]. The tuning range of one laser diode is about 30 nm.

When one is interested only in the dispersion $D(\lambda)$ at a single operating wavelength λ, it can be measured with a single tunable laser diode [Franzen and Kanada 1986]. The differential time delay $\tau_g(\lambda + \Delta\lambda) - \tau_g(\lambda)$ is measured for two laser diode wavelengths separated by $\Delta\lambda = 11$ nm. The dispersion is obtained approximately from the equation

$$D(\lambda) = [\tau_g(\lambda + \Delta\lambda) - \tau_g(\lambda)]/\Delta\lambda \quad . \tag{13.112}$$

By connecting both ends of the test fiber to a directional coupler, a recirculating fiber loop (Sect. 12.3.4) can be made. This method gives a longer effective fiber length and a stable marker which does not depend on electronic trigger or timing [Franzen and Kanada 1986].

The pulse delay versus wavelength can also be measured [Lin C. et al. 1977, 1980a] by applying mirrors at both ends of the fiber under test. When the fiber resonator thus formed is pumped by a pulsed high-power Nd:YAG laser, Raman amplification can cause self-generated optical oscillations. The condition for oscillations in this so-called Raman oscillator is that the frequency of the pump pulses is equal to the round trip frequency of the circulating signal pulses. With this "pulse synchronization technique", the chromatic dispersion can be calculated from the change $\Delta\lambda$ in wavelength caused by a change Δz of the resonator length.

Disadvantages of the conventional pulse delay method for measuring single-mode fiber dispersion are the high cost of the equipment, the requirement for long (more than 0.5 km) lengths of fiber, and hazards due to the high laser power levels.

13.9.2 Phase Shift Method

The group velocity v_g is the signal velocity. When an optical carrier, which is sinusoidally modulated in power with frequency f_m, is transmitted through a single-mode fiber of length L, the modulation envelope is delayed by a time

$$L/v_g = \tau_g L \quad . \tag{13.113}$$

Since a delay by one modulation period $T_m = 1/f_m$ corresponds to a phase shift of 2π, the sinusoidal modulation is phase shifted in the fiber by an angle

$$\phi_m = 2\pi \tau_g L / T_m = 2\pi f_m \tau_g L \quad . \tag{13.114}$$

Thus, the specific group delay

$$\tau_g = \phi_m / (2\pi f_m L) \tag{13.115}$$

can be determined by measuring the phase shift ϕ_m of the modulation either by observing a Lissajous figure on an oscilloscope screen [Imoto et al. 1979] or by using a vector voltmeter [Costa et al. 1982, 1983a,b,c; Srivastava and Franzen 1985].

If one is interested in the chromatic dispersion only, it is sufficient to measure the phase change $\Delta\phi_m$ caused by a wavelength change $\Delta\lambda$. Because of the small bandwidth and high sensitivity of the vector voltmeter as compared to a broadband oscilloscope necessary to measure pulse delay time, LEDs can be used as sources. Because of the large spectral width of LED light, a grating monochromator or a set of interference filters must be used to select a narrow band of wavelengths out of the broad spectrum.

Near the center wavelengths of the LEDs, the optical SNR is larger than 15 dB, and phase differences can be measured with 0.1° resolution. For $f_m = 30\,\text{MHz}$, the equivalent time resolution is 10 ps/km. Typically, the modulation frequency is in the range 30–500 MHz. The selection of the modulation frequency is a compromise between good resolution at high frequencies and phase measurements without ambiguities at low frequencies. By careful attention to experimental technique, it is possible to perform accurate and reproducible chromatic dispersion measurements on fiber lengths of 50 km [Sladen et al. 1986].

Because of the large spectral width of LED light, two or three LED's with different peak wavelengths are sufficient to measure chromatic dispersion in the wavelength range of main interest (1.2–1.6 μm) [Costa et al. 1984].

When the LED's are replaced by semiconductor laser diodes, the dynamic range of the phase shift method can be increased to more than 30 dB so that the chromatic dispersion of fiber links with lengths of up to 120 km can be determined [Tatekura et al. 1984; Vella et al. 1984a,b; Tanaka S. and Kitayama 1984; Horiguchi T. et al. 1985c; Fujise et al. 1986b; Miyajima et al. 1986b].

Chromatic dispersion at a single wavelength only can be measured using a wavelength tunable cleaved-coupled-cavity (C^3) laser [Olsson et al. 1984] or a tunable external-cavity buried-heterostructure laser [Mengel and Gade 1984; Bernard and Visseaux 1985].

Instead of through a single-mode reference fiber, the wave with constant wavelength which serves as a phase reference can also be transmitted through the test fiber itself by utilizing wavelength domain multiplexing [Hatton et al. 1985; Hatton and Nishimura 1985, 1986b]. This reduces the measurement error due to group delay variations (5.83) caused by temperature change during the measurement.

The accuracy and precision of the phase-shift method have been studied several times [Tanaka S. et al. 1984; Horiguchi et al. 1985c; Thomson 1985; Bosselaar et al. 1986].

When the measuring wavelength is shorter than the cutoff wavelength of the second order mode, the LP_{11} mode can cause an error in the measured value of the chromatic dispersion. However, this error rapidly decreases with increasing fiber length and can be minimized by introducing a small loop as a mode filter (Sect. 12.2.14) [Kitayama Y. and Tanaka 1985].

The advantages of the phase-shift method are its very high accuracy and simple instrumentation. The CCITT [1986] recommends the phase shift method as the second of the two reference test methods for chromatic dispersion. Measuring setups based on this method are commercially available.

By periodically changing the optical wavelength between two slightly different values, the output voltage of the vector voltmeter becomes a square wave the amplitude of which is directly proportional to chromatic dispersion D. In this "differential phase shift method", no curve fitting to delay data is required [Barlow A.J. and Mackenzie 1987].

13.9.3 Interferometric Method

Another class of measuring techniques for dispersion uses interferometric procedures to achieve time resolution of 1 ps to 5 fs [Thevenaz et al. 1988], so that the dispersion properties of short lengths (10 cm to some meters) of fiber can be determined.

The method takes advantage of the short coherence time of partially coherent light and was proposed in 1981 [Tateda et al. 1981; Shang 1981]. The interferometric method was originally used to measure the group delay time difference between the LP_{01} and LP_{11} modes in a dual-mode fiber [Shibata N. et al. 1980], and later to measure the change in delay time of the fundamental mode caused by a variation in ambient temperature [Shibata N. et al. 1981a, 1983].

The experimental arrangement, uses e.g. a Mach-Zehnder interferometer (Fig. 13.23) [Saunders and Gardner 1984; Srivastava and Franzen 1985]. Input light is divided into a reference arm containing an air path with a variable delay line (corner cube reflector, roof prism, trombone) and a test arm containing the fiber under test. By means of a half-mirror, the two waves are combined at the output of the interferometer. Since the beams transmitted through the test fiber and the reference path are cross-correlated, the interferometric method is sometimes also called the cross-correlation technique.

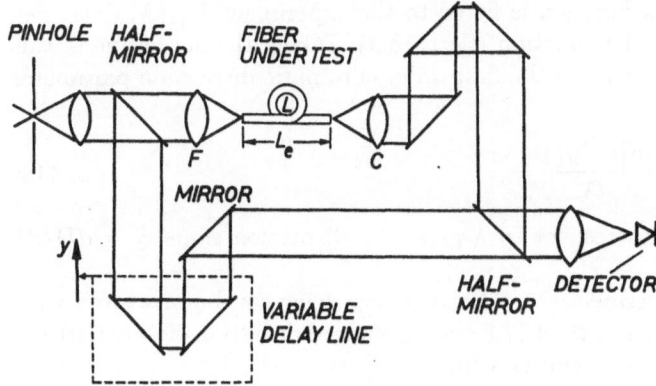

Fig. 13.23. Mach-Zehnder interferometer for measuring chromatic dispersion

Interference maxima and minima can only be observed if the difference in the time delays in the two interferometer arms is smaller than the coherence time of the light. Using a source with a relatively large linewidth (e.g. a 100 W quartz halogen source and a monochromator with a FWHM linewidth of 10 nm) the coherence length for $\lambda = 1.3\,\mu m$ is (3.49)

$$L_c = 0.76\lambda^2/\Delta\lambda_{FWHM} = 0.12\,\text{mm} \tag{13.116}$$

and the coherence time $t_c = L_c/c = 0.4\,\text{ps}$. Thus, interference fringes can only be observed if the group time delays differ by less than about 1 ps. By maximizing the visibility V (3.23) of the interference fringes, the group delays in the two channels can be equalized to within much less than 1 ps.

In order to minimize the time needed to find the interference point, one can modulate the length of the reference path by vibrating the corner cube reflector, and use an ac detection technique [Namihira et al. 1986].

Since the chromatic dispersion parameter $D = d\tau_g/d\lambda$ is the derivative of the group delay per unit length with respect to wavelength, it is not necessary to determine the absolute value of the time delay $\tau_g L$ in the test fiber. For a starting wavelength λ_1, the position y of the variable delay line (Fig. 13.23) is changed continuously, until the fringe visibility is maximum ($y = y_1$). Then the wavelength is changed to λ and the length of the delay line is changed again to find the new position y of maximum fringe visibility. The change in length of the air path in the reference arm is then $2(y - y_1)$, and the change of the time delay in the reference arm is $2(y - y_1)/c$. Since the time delays in the test and the reference arms are equal, the change of the group delay time in the fiber of length L caused by the wavelength change is

$$[\tau_g(\lambda) - \tau_g(\lambda_1)]L = 2(y - y_1)/c \quad . \tag{13.117}$$

This measurement is repeated for many discrete wavelengths λ in the range of interest.

A simple analytical function is fitted to the experimental $\tau_g(\lambda)$ data, e.g. a polynomial or a Sellmeier function (Sect. 13.9). Taking the derivative of this function with respect to λ gives the first order chromatic dispersion parameter [Tateda et al. 1981]:

$$D = d\tau_g/d\lambda = \frac{2}{Lc}\frac{d(y - y_1)}{d\lambda} \quad . \tag{13.118}$$

The derivative of D with respect to λ gives the dispersion slope $S = dD/d\lambda$ (13.102).

With the setup described, the relative group delay can be measured with high precision; for a fiber length of 17.5 cm, a standard deviation of only 0.019 ps has been reported [Saunders and Gardner 1984].

Instead of the Mach-Zehnder interferometer described, one can also use a Michelson interferometer for measuring chromatic dispersion [Bomberger and Burke 1981; Thevenaz et al. 1988]. Since the light travels twice the length of the test fiber, dispersion effects are doubled. In the Michelson interferometer, the classical beam splitter has been replaced by a fiber-optic directional coupler, giving maximum fringe contrast without the need for achromatic optics, careful mirror alignment, and IR viewing equipment [Thevenaz et al. 1988].

The main advantages of the interferometric method are that a white light source can be used, that simple low-frequency detection techniques are adequate, and that very short fiber samples are sufficient. Thus, the interferometric method can be used as a destructive technique for measuring the axial uniformity of the dispersion spectrum in a fiber [Sears et al. 1983, 1984].

On the other hand, the short length of the fiber sample is also a disadvantage since one has to check whether the dispersion value measured for the short fiber is representative for the whole length of the fiber, i.e. whether the fiber is uniform with respect to dispersion [Saunders and Gardner 1985]. The long-length dispersion curve is approximately the average of the dispersion curves obtained from short lengths from both ends of a 1 km fiber [Sears et al. 1983].

If the length L of the fiber under test is larger than some decimeters, the dimensions of the interferometer become unwieldy. It is then possible to use a multi-reflection mirror [Namihira et al. 1986] or a reference fiber with carefully calibrated dispersion spectrum [Cohen L.G. and Stone 1982; Stone and Cohen 1982; Lieber et al. 1986; Thevenaz et al. 1988] in the reference arm in order to equalize the optical path lengths. The variable delay in the reference arm can be obtained by elastically stretching the reference fiber [Thevenaz et al. 1988].

The maximum in the visibility versus path length curve is usually found by visual inspection. However, this method is subject to operator errors. These errors can be eliminated and the resolution enhanced by a simple numerical method, which calculates the cross-correlation of the measured and a theoretical visibility curve [Oksanen and Halme 1984; Lieber et al. 1986]. The amplitude of the intensity oscillation is the Fourier transform of the power spectrum of the source centered at the position where the group delays in both arms are equal [Thevenaz et al. 1988]. For a Gaussian spectrum, the theoretical visibility curve is also a Gaussian function.

Instruments using the interferometic method for measuring chromatic dispersion are commercially available.

There are some variants of the interferometric method. In the method described so far, the length of the path in the reference arm is changed continuously for different fixed wavelengths, and the detector signal is recorded ("path length tuning"). Alternatively, one can measure the detector signal for different fixed time delays in the reference arm, while a monochromator at the fiber output end is scanned through a wavelength range ("λ tuning") [Shang 1981]. With λ tuning, the chromatic dispersion parameter D and the dispersion slope S can be determined directly, i.e. without differentiation, and independently of one another from the wavelengths of the peaks of the detected power versus wavelength curve [Stone J. and Marcuse 1984; Cohen L.G. 1985b].

Chromatic dispersion can also be calculated from the difference in the widths of the interferogram peaks with and without the fiber in the interferometer [Bomberger and Burke 1980]. The experimental setup shown in Fig. 13.23 can also be used to measure the absolute value of the group velocity v_g in very short fiber samples [Saunders and Gardner 1985]. One has to maximize the fringe visibility both with the fiber inserted into the test arm and in its absence. When the fiber is removed, one has to move the fiber collimating lens (C) in the axial direction until its focus coincides with that of the focusing lens (F). This axial movement does not change the time delay in the test channel.

Denoting the positions of the corner reflector with and without fiber by y and y_0 respectively, and the distance between the fiber ends by L_e, the specific group delay time is

$$\tau_g = [2(y - y_0) + L_e]/(Lc) \quad , \tag{13.119}$$

and the group velocity (5.66)

$$v_g = 1/\tau_g \quad . \tag{13.120}$$

With the method described, the group velocity of a fiber of length 34 cm has been measured with an error of 0.03%.

13.9.4 Spot Size Method

The chromatic dispersion D in single-mode fibers is (approximately) the sum of the material dispersion D_m and the waveguide dispersion D_w (5.91). The material dispersion is fairly well known from the properties of fiber materials. The waveguide dispersion is caused by the wavelength dependence of the intensity distribution in the fiber cross section. Since the spot size is the parameter characterizing the radial field distribution, it seems plausible that there is a certain relation between the variation of the spot size and the waveguide dispersion. Indeed (5.89),

$$D_w = \frac{\lambda}{2\pi^2 c n_1} \frac{d}{d\lambda} \left(\frac{\lambda}{w_d^2} \right) \quad , \tag{13.121}$$

where w_d is Petermann's second spot size (10.10). This formula expresses the waveguide dispersion in terms of the spot size only. One does not require knowledge of the refractive index profile. Thus the waveguide dispersion can be determined by measuring the spot size w_d as a function of wavelength and by taking the derivative with respect to wavelength. Using a more general equation similar to (13.121), it is also possible to take into account profile dispersion $(d\Delta/d\lambda \neq 0)$ [Petermann 1983; Srivastava and Franzen 1985].

From the measured waveguide dispersion, the total chromatic dispersion then has to be obtained by the addition of material and profile dispersions (5.91) [Pocholle et al. 1983a,b], the latter often being negligible. This task is by no means easy, since, in principle, knowledge of the refractive-index profile is necessary in order to obtain the properly weighted dispersion [Sharma A.B. et al. 1985].

The method has been applied as follows [Buckland and Nishimura 1985]: The effective far-field width W_e (10.28) is determined from the far-field radiation pattern at just three laser wavelengths. The spot size w_d related to dispersion is calculated from W_e by using $w_d = 1/(\pi W_e)$ (10.31). A Marcuse function (10.5)

$$w_d = A + B\lambda^{3/2} + C\lambda^6 \tag{13.122}$$

is fitted to the w_d versus λ data and inserted into the formula (13.121) for the waveguide dispersion. Adding the material dispersion gives the total chromatic dispersion. The values for the wavelength λ_0 of zero dispersion determined in this manner agree excellently with those obtained from the phase shift technique (Sect. 13.9.2).

In another study [Karstensen and Wetenkamp 1984, 1985], the spot size has been measured by the transverse offset method (Sect. 13.7.2) and the waveguide dispersion obtained from its variation with wavelength. The total dispersion was found to be in good agreement with the results of pulse delay measurements.

In a similar investigation [Smith D.K. and Westwig 1985], the effective far-field width W_e was determined by the variable aperture method (Sect. 13.7.3). The results were shown to compare favorably with the total dispersion measured with the Raman laser method; the difference in λ_0 was about 1 nm.

However, measurements of waveguide dispersion effects in dispersion-shifted and dispersion-flattened fibers have not been reported yet to validate the accuracy of the general procedure [Cohen L.G. 1985b].

The advantages of the spot size method are that temporal detection is not required and that a short fiber sample can be used. The disadvantage is that it measures only waveguide dispersion; material dispersion must be determined independently.

For an ideal dispersion-flattened fiber (Sect. 5.8), the total dispersion D is required to be zero within a given interval of wavelengths. Since the total dispersion is approximately the sum of the material dispersion D_m and the waveguide dispersion D_w (5.91), the waveguide dispersion must be $D_w = -D_m$, where D_m closely resembles the well-known material dispersion of fused silica [Petermann

1983]. Inserting the required waveguide dispersion D_w into (13.121), this equation can be integrated to give the wavelength dependence $w_d(\lambda)$ of the spot size required for an ideal dispersion-flattened fiber. By comparing this function with that measured or calculated from the index profile for a given fiber, one can check whether the fiber is dispersion-flattened or not, without having to calculate the dispersion $D(\lambda)$.

13.9.5 Miscellaneous Methods

Besides the four most important methods described so far, several other methods have been proposed in the literature for measuring chromatic dispersion in single-mode fibers.

Over a very long (20 km) single-mode fiber, the first and second order dispersion parameters D and S can be determined [Kawana et al. 1978; Yamada J. et al. 1980a; Kimura 1980; Cohen L.G. et al. 1981; Miya et al. 1983] by a direct measurement of pulse broadening (for a given source spectral width). The time resolution of the fastest sampling oscilloscopes is limited to 25 ps. Time resolution can be increased by replacing electrical sampling by optical sampling, which is based on phase-matched second-harmonic generation in a $LiIO_3$ crystal [Kanada and Franzen 1986a,b]. Chromatic dispersion has also been determined by measuring the baseband response of very long single-mode fibers [Mori et al. 1983].

The chromatic dispersion can be calculated from the refractive-index profile measured, e.g. by the refracted near-field technique (Sect. 13.10.2). The dispersion of a nearly step-index fiber determined by this indirect method was reported to agree closely with that measured using the pulse delay method [White K.I. et al. 1981].

The dispersion of short lengths of single-mode fibers can be determined with the methods of Fourier transform spectroscopy [Bomberger and Burke 1981; Mengel 1984; Srivastava and Franzen 1985]. The autocorrelation of the input optical field and the cross-correlation of the input and output fields are measured with two interferometers. The fiber transfer function is then determined by dividing the Fourier transform of the cross-correlation by the Fourier transform of the autocorrelation. Since this interferometric technique directly measures the chromatic dispersion, no differentiation of a measured group delay versus wavelength curve is necessary.

Fundamentally, the chromatic dispersion is the second derivative of the phase constant β with respect to angular frequency (5.86). Therefore, if one could measure β as a function of ω, one would be able to determine the dispersion. Since the relative index difference $\Delta = (n_1 - n_2)/n_2$ of single-mode fibers is typically of the order of 0.3%, and since (5.48) $n_2 k < \beta < n_1 k$, the phase constant differs by less than 0.3% from the wavenumbers $n_2 k$ and $n_1 k$ of the cladding and the core, respectively. Thus, one would have to measure the phase constant with extremely high precision for an accurate calculation of its second derivative.

The phase constant β can be determined [Thyagarajan et al. 1986a,b; Sorin et al. 1986a] by a technique which corresponds to the m-line method well-known in integrated optics [Tien and Ulrich 1970]. The fiber is bonded into a fused-quartz block and then polished until only a few micrometers of cladding remain between the fiber core and the polished surface. A high-index prism is placed on the polished region of the fiber. This causes the guided modes to become leaky and radiate energy into the prism (Sect. 7.4). The phase constants of the guided modes are simply related (7.57) to the angle specifying the direction of radiation. Using this technique, the effective refractive indices $n_e = \beta/k$ (5.45) of the LP_{01} and LP_{11} modes have been measured with an accuracy of about 10^{-4}.

Alternatively, the phase constant can be determined indirectly from the field distribution in the outer, homogeneous cladding. In this region, the field decay in the radial direction is described by the modified Hankel function of zero order (5.15) $K_0(W\varrho/a)$, where a/W is the transverse field extent in the cladding. The quantities β and W are related through the separation equation in the cladding (5.17)

$$(W/a)^2 = \beta^2 - n_2^2 k^2 \quad . \tag{13.123}$$

Thus, if one manages to measure the transverse field extent a/W in the cladding, and if one knows the wavelength dependence of the refractive index n_2 of the cladding, one can determine the phase constant as a function of wavelength, the group delay, and the total chromatic dispersion. This scheme was proposed by A.B. Sharma et al. [1985]. In order to measure the field distribution in the outer cladding, a magnified image of the near-field distribution is produced by a microscope objective. A diaphragm with a circular aperture is inserted into the magnified near field. The power passing through the holes is measured as a function of the hole diameter at a given wavelength. From these data, the transverse field extent a/W can be obtained.

The applicability of this new method has been demonstrated via computer simulations and experimental results, which are quite encouraging. The total chromatic dispersion can be obtained without reference to the refractive index profile of the core. It is only necessary to assume a uniform outer cladding with known material dispersion. The main problem of the method is that the field strength in the outer cladding is very small.

The change of $[W/(ak)]^2$ for a wavelength change $\Delta\lambda$ can also be determined from the near-field intensity (NFI) patterns measured at λ and $\lambda + \Delta\lambda$. From this, the change in β and the chromatic dispersion D can be obtained [Coppa et al. 1983a,b]. The chromatic dispersion determined with this "NFI technique" was in excellent agreement with that measured with the phase shift method.

The chromatic dispersion can also be measured with a streak camera [Mochizuki et al. 1985, 1986, 1987]. The source is a pulsed (40 ps) multilongitudinal-mode semiconductor laser. Because of chromatic dispersion, the pulses in the different longitudinal modes experience different time delays. At the output end of a 4.5 km fiber, a diffraction grating (polychromator) produces the spectrum of the transmitted light. The spectrum is detected with a streak camera [Bowley 1986] which displays the time evolution of the spectrum with very high time resolution (about 1 ps). The chromatic dispersion is obtained from the streak image, i.e. from the plot of the relative delays for the successive longitudinal mode pulses versus wavelength.

This method of dispersion measurement has the following advantages:

1. Time delay as a function of wavelength can be measured instantaneously without changing wavelength.
2. It is possible to measure the dispersion using only one semiconductor laser.
3. The measurement can be done in a very short time.

The method is also suitable for link lengths of more than 50 km [Mochizuki et al. 1986, 1987].

In a swept frequency technique [Rao 1984; Srivastava and Franzen 1985], two lasers at different wavelengths λ_1 and λ_2 are sinusoidally intensity modulated in phase and with the same swept frequency. The two waves are combined and transmitted through the test fiber as one composite signal. The photocurrent detected at the fiber end is analyzed by a spectrum analyzer. Minima occur in the amplitude spectrum at frequencies f_n. The differential delay time for the two laser wavelengths can be calculated from the distance between two consecutive minima in the amplitude spectrum:

$$\tau_g(\lambda_1) - \tau_g(\lambda_2) = [L(f_{n+1} - f_n)]^{-1} \ . \tag{13.124}$$

In a similar method [Yamada 1982], a multilongitudinal-mode laser is modulated sinusoidally by a swept frequency signal. The chromatic dispersion can be obtained from the baseband spectrum at the fiber end.

13.9.6 Comparison of Methods

The accuracy, limitations, and ease of implementation of the major dispersion measuring techniques have been compared by L.G. Cohen [1985a,b].

The pulse delay technique requires fiber lengths larger than 0.5 km and either a complicated Raman laser with high peak pulse power or an array of semiconductor lasers at discrete wavelengths. The Raman laser is too bulky and complex for field use.

The phase shift technique also requires long fibers, but can use simple LED's instead of lasers.

Interferometric techniques have a time resolution which can be three orders of magnitude better than that of the pulse delay and the phase shift techniques. Thus, chromatic dispersion can be measured for fiber samples shorter than 1 m.

From the wavelength dependence of the spot size, waveguide dispersion can be determined. However, in order to get the total chromatic dispersion, which determines pulse broadening in single-mode fibers, material dispersion must be added. Usually, the fiber user cannot calculate the material dispersion, since he has no information about the dopants used in the fiber or about the refractive-index profile.

On the same fiber, chromatic dispersion has been determined using four different techniques [Coppa et al. 1985b]: pulse delay, phase shift, wavelength dependence of the spot size, and measured index profile. The measured values agreed very well, thus confirming the reliability of the results obtained with quite different techniques.

In another investigation, the values of the chromatic dispersion measured with the pulse delay and interferometric methods agreed excellently with those calculated from the measured index profile [Eoll et al. 1986].

Both the pulse delay (Sect. 13.9.1) and the phase shift (Sect. 13.9.2) techniques have been accepted by the CCITT [1986] as reference test methods.

13.10 Measurement of the Refractive-Index Profile

The CCITT [1986] characterizes single-mode fibers through propagation parameters (λ_{ce}, w_d, D, S) and not through geometrical and dielectric ones ($2a$, A_N, $n(\varrho)$). Thus, no need arises to specify the refractive-index profile. On the other hand, the refractive-index profile is the basic feature in single-mode fiber design, from which relevant information about fiber characteristics like launching and connection efficiencies, macro- and microbending loss, single-mode condition, and chromatic dispersion can be obtained. Therefore, it is important to have methods for measuring the refractive-index profile.

There are several reviews of methods for measuring the refractive-index profile [Marcuse and Presby 1980; Marcuse 1981a; Stewart W.J. 1982a,b; Okoshi and Hotate 1983].

The refractive-index profile can be measured either along the length of the fiber preform prior to fiber pulling (Sect. 13.10.1) or directly in the drawn fiber (Sect. 13.10.2).

13.10.1 Index Profile of the Preform

Because of the larger diameter of the preform, it is much easier to measure its refractive-index profile than the profile of the drawn single-mode fiber. It is frequently assumed that the fiber profile is the same as the preform profile apart from a scale factor in radial coordinate which equals the ratio of the fiber cladding diameter and that of the preform.

Since the fiber and preform have different thermal histories, and thus different states of strain, it is not obvious that this assumption is warranted [White K.I. et al. 1981]. Sometimes it has been reported that during fiber drawing, due to dopant diffusion or thermally induced stress [Scherer 1980], the index profile can change substantially [Presby et al. 1979; Saunders 1982; Paek et al. 1983; Anderson W.T. et al. 1984].

Many different methods exist for measuring the refractive-index distribution in the preform and these have been reviewed several times [Presby and Marcuse 1980; Marcuse 1981a; Stewart W.J. 1982b; Okoshi and Hotate 1983]. These methods can be applied to the preforms of both multimode and single-mode fibers.

One method for determining the index profile of the preform is to cut from it a thin transverse slab. Using an interference microscope, the phase shift of the light traversing the slab can be measured. From the phase shift, the refractive-index profile can be calculated [Presby et al. 1978; Presby and Marcuse 1980]. The interferometric slab method provides very reliable results. Its main disadvantages are that the preform must be destroyed for the measurement and that the sample preparation is very time consuming.

In a nondestructive ray tracing technique for measuring the refractive-index profile, the preform is illuminated transversely by a very thin laser beam propagating in the x-direction [Chu 1977; Chu and Whitbread 1979a; Watkins 1979; Marcuse 1981a; Stewart W.J. 1982b]. The light beam is deflected by the preform core, the deflection angle θ depending on the transverse distance y of the incident ray from the core axis (z-axis). The deflection function $\theta(y)$ can be measured directly [Peri et al. 1982]. The refractive-index profile of the preform is then reconstructed by taking the inverse Abel transform of the deflection function $\theta(y)$.

Instead of scanning a pencil beam transversely across the preform, the deflection function $\theta(y)$ can also be measured rapidly and highly accurately with a wide parallel beam of light. In this so-called spatial filtering technique [Okoshi et al. 1980; Sasaki I. et al. 1980a, 1981; Okoshi and Hotate 1983; Auge et al. 1983], a triangular mask produces a visual display of the deflection function in the focal plane of a cylindrical lens. In a variant of this method [Sasaki I. et al. 1980b], a rotating chopper converts the light deflection into a measurable time-domain signal.

Another indirect method for obtaining the deflection function $\theta(y)$ is to use an interference microscope to measure the phase change of a wave transmitted transversely through the preform. In terms of geometrical optics, this corresponds to measuring the optical path lengths $\int n(x)dx$ [Born and Wolf 1975] of rays passing through the fiber due to side illumination [Chu 1979; Chu and Whitbread 1979b]. Since there is a simple relation between the path-length difference and the deflection function [Saekeang and Chu 1979], the latter can be calculated from the measured path-length difference.

There are also simple mathematical relations between the moments of the index profile, the deflection function, and the phase function [Chu 1982].

A fourth technique for determining the deflection function $\theta(y)$ is the so-called focusing method [Presby et al. 1978, 1979; Marcuse 1979a; Marcuse and Presby 1979a; Chu and Saekeang 1979; Presby and Marcuse 1979; Saekeang et al. 1980] in which the preform is illuminated transversely with a wide beam of incoherent collimated light. The preform is immersed in index matching liquid to avoid refraction at the outer cladding boundary. The uniform intensity distribution in the beam in front of the preform is modified by the refractive index distribution in the preform. The core of the preform acts as a cylindrical lens and focuses the incident light in an observation plane just outside of the core. The intensity distribution in this image plane is detected with a video camera using a lens with 50 mm focal length. The deflection function is then calculated from the intensity distribution. This method is experimentally more convenient than scanning a thin laser beam transversely through the preform. An analysis of the errors inherent in the focusing method [Irving et al. 1985] predicts that profile errors increase linearly from zero at the outside diameter of the fiber to a maximum value at the center of the core.

The refractive-index profile of single-mode preforms with a Ge-doped core can also be measured by illuminating the preform transversely to its axis by a thin HeCd laser beam with an ultraviolet wavelength of 325 nm [Presby 1981a,b]. The core fluoresces strongly at a wavelength near 420 nm. Since the intensity of the fluorescent light is proportional to the dopant concentration and hence to the index difference, the index profile can be obtained directly by measuring the intensity as a function of the distance ϱ of the beam from the preform axis. The profiles obtained are in excellent agreement with those measured by slab interferometry and by the focusing method.

Most methods described assume from the beginning that the preform is axially symmetric. But since the preforms for polarization maintaining or single polarization fibers are not axially symmetric, a kind of computer tomography has been developed for those preforms [Chu 1979; Chu and Saekeang 1979; Saekeang et al. 1980; Okoshi et al. 1980; Francois et al. 1981, 1982; van Blitterswijk and Smit 1987]. With this method, arbitrary two-dimensional refractive-index distributions can be measured by transversely illuminating the preform from a discrete number of directions ϕ and measuring the deflection angle θ as a function of the distance y of the sampling ray from the preform axis.

The two-dimensional refractive-index distribution $n(x, y)$ can then be computed from the deflection function $\theta(\phi, y)$ [Chu and Saekeang 1979; Francois et al. 1982]. The mathematics of this problem is analogous to that of reconstructing a two-dimensional image from its projections. Various algorithms for this method of computer tomography have been presented in the last decade for medical imaging purposes.

Performing the measurement of deflection function for several preform orientations and several cross sections along the preform permits three-dimensional tomographic reconstruction of the profiles for preforms of arbitrary cross section. The exceedingly large amount of data usually required for such reconstruction can be reduced by applying an interpolation algorithm [Francois et al. 1982] which requires only a few azimuthal projections of the preform. A preform analyzer which measures the deflection function at different rotational orientations and computes the transverse refractive-index distribution using tomographic methods is commercially available. Figure 13.24 shows tomographic representations of refractive-index profiles prepared by means of the low pressure plasma-activated chemical vapour deposition (PCVD) process for the cases of a step-index matched cladding single-mode fiber (a), a depressed cladding single-mode fiber (b), and a dispersion-flattened quadruple-clad single-mode fiber (c) [Bachmann et al. 1983b, 1986a,b; Bachmann 1985, 1986].

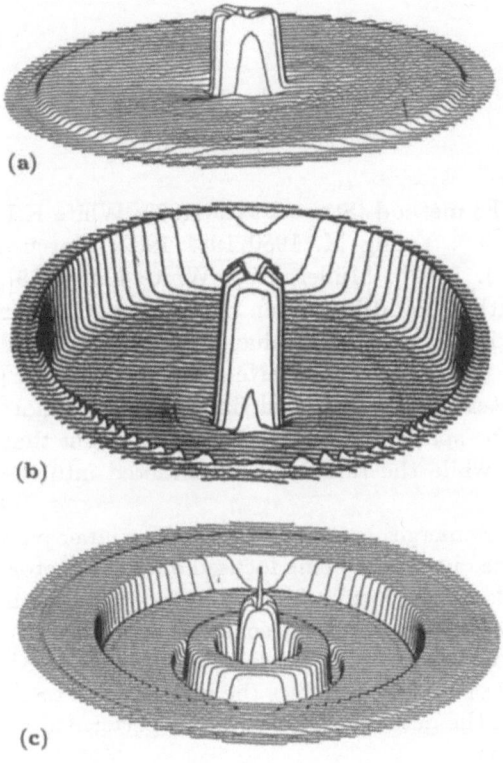

(a)

(b)

(c)

Fig. 13.24a–c. Tomographic representations of the refractive-index profiles of (**a**) a matched-cladding step-index single-mode fiber, (**b**) a depressed-cladding step-index single-mode fiber, and
(**c**) a dispersion-flattened quadruple-clad single-mode fiber. (By courtesy of Philips GmbH, Forschungslaboratorium Aachen)

Fiber preforms are fabricated at high temperatures. Since the thermal expansion coefficient depends on the doping level, residual mechanical stresses are built into the preform when it is cooled down to room temperature. These stresses cause the preform material to become linearly birefringent. Due to this birefringence, a thin laser beam transmitted transversely through the preform, changes its state of polarization. By measuring the retardation as a function of the distance y of the laser beam from the preform axis, one can nondestructively determine the refractive-index profile [Shibata N. et al. 1979].

13.10.2 Index Profile of the Fiber

Conventional techniques for measuring the refractive index profile of multimode fibers cannot be used for single-mode fibers with their smaller core diameter because the spatial resolution is limited by diffraction [Gambling et al. 1978].

Reviews of methods for profiling single-mode fibers have been published in the literature [Marcuse and Presby 1980; Inada 1981; Stewart W.J. 1982b; Okoshi and Hotate 1983]. Several methods for measuring the refractive index profiles of single-mode fibers have been proposed:

—   the refracted near-field technique,
—   the scattering pattern method,

- the near-field method,
- the far-field method,
- the focusing method,
- the reflection method, and
- slab interferometry.

In the refracted near-field (RNF) method [Stewart W.J. 1977; White K.I. 1978, 1979; Marcuse and Presby 1980; Young M. 1980, 1981, 1982; Marcuse 1981a; Saunders 1981; Stewart W.J. 1982b; Müller 1983; White K.I. 1985], the fiber is placed in a cell filled with a liquid having an index slightly higher than the cladding value, i.e. approximately the core value. The fiber is passed through a small hole in an opaque disc. A focused HeNe laser ($\lambda = 633\,\text{nm}$) beam is scanned across the diameter of a flat normal fiber end. The short wavelength of the laser maximizes the spatial resolution. Part of the light that is focused into the fiber is guided, while the remainder is refracted into the cladding (Sect. 13.8.4).

The light not guided by the fiber emerges as a hollow cone. The inner part of this is blocked by the disc, with the outer part being focused onto a detector. The outer angle of this cone depends on the local refractive index at the focus of the beam. The photocurrent varies linearly with the square of the refractive index of the fiber at the point at which the incident light beam is focused. Scanning different points on the end of the fiber enables the index profile to be obtained directly from the output of the detector. Calibration is performed by scanning the disc along the optic axis.

The refracted near-field technique requires no correction factors and gives the relative index differences directly without recourse to external calibration or reference samples. Index resolution is of the order of 1 part in 10^4 with spatial resolution (which improves with increasing power of the focusing lens) of about $0.2\,\mu\text{m}$ [Stewart W.J. and Reid 1982]. The resolution of the RNF technique has been analyzed by W.J. Stewart [1980a].

Since the cladding as well as the core is profiled, all the details of fiber geometry can be obtained, namely, core and cladding diameters, ellipticity, and concentricity. The refracted near-field technique requires minimal sample preparation, no computation, can obtain the absolute index difference, and is applicable to both multimode and single-mode fibers.

The refractive index of the liquid strongly depends on temperature. Therefore, in the RNF technique, the liquid cell temperature is a parameter to be carefully controlled. The influence of temperature on the RNF technique has been studied by Trepat-Marti and Goure [1985].

For five single-mode fibers, the Δ values obtained from the refracted near-field technique have been compared with the Δ values obtained from profiles of the fiber preforms by the focusing method; the average difference being about 6% [Saunders 1981].

Four different techniques for index profiling have been compared [Gauthier et al. 1981]. Among the techniques used, the refracted near-field technique does not suffer from the limitations of the other methods: it can analyze any type of profile with high accuracy. The refracted near-field technique is sensitive to contamination by dust and the apparatus is complex. Nonetheless measuring setups based on this technique are commercially available. They operate at the wavelength 633 nm of a HeNe laser and achieve a spatial resolution of better than 0.5 μm. The resolution and repeatability (2σ) of the refractive index are 10^{-4} and 10^{-3}, respectively.

In the second method for fiber profiling, the scattering method [Okoshi and Hotate 1976a,b, 1983; Brinkmeyer 1977; Hotate and Okoshi 1978a], the fiber is immersed in matching oil and illuminated transversely by a wide laser beam. The fiber core acts as a phase object and diffracts the incident laser beam. The forward diffraction pattern is measured with a very wide dynamic range (60 dB). The refractive-index profile is then calculated by taking the Hankel transform of the diffraction pattern. This "scattering pattern method" has succeeded in achieving a resolution of 0.19 μm [Hotate and Okoshi 1978a] which is the highest resolution ever reported.

The third method, the transmitted or propagation-mode near-field technique [Coppa et al. 1983c,d,e; Morishita 1983a,c, 1985] is based on the relation between the refractive-index profile $n(\varrho)$ and the radial amplitude distribution $E(\varrho)$ of the fundamental fiber mode as described by the scalar wave equation (5.20). The scalar wave equation can be inverted to give the index profile in terms of the derivatives of the radial field distribution:

$$n^2 = \frac{1}{k^2}\left[\beta^2 - \frac{1}{E}\frac{d^2 E}{d\varrho^2} - \frac{1}{\varrho E}\frac{dE}{d\varrho}\right] \ . \tag{13.125}$$

The near-field intensity distribution is measured with the usual apparatus (Sect. 13.5) with an LED as source. The electric field E is proportional to the square root of the measured intensity. When the refractive index is known at a certain point (e.g. in the cladding), the phase constant β can be determined from (13.125), so that $n(\varrho)$ can be determined absolutely everywhere.

Although there are difficulties in taking the first and second derivatives of a measured curve, the index profiles obtained with this technique agree very well with those measured by the refracted near-field technique. The method can also be extended to noncircular optical waveguides, in which the electric field of the fundamental mode depends on the azimuth ϕ in addition to the distance ϱ from the core axis, to give the two-dimensional refractive-index distribution $n(\varrho, \phi)$ [McCaughan and Bergmann 1983; Coppa et al. 1986b,d; Morishita 1986].

In a fourth method for index profiling, one measures the far-field radiation pattern of the fiber end. When the index profile is known to be of the step-index type, the core radius and the refractive-index difference can be determined [Gambling et al. 1976a] by measuring the half-width and the first zero of the exit radiation pattern of the fundamental fiber mode (Sect. 13.8.10). However, the form of the index profile is generally not known in advance. In order to determine the refractive-index profile of an arbitrary fiber from the far field, one has to remember that the near-field and far-field amplitude distributions form a pair of Hankel transforms (8.16, 20). The far-field amplitude is obtained by multiplying the square root of the measured far-field intensity by a factor of $+1$

or -1 corresponding to the polarity of the different radiation lobes. An inverse Hankel transform then gives the radial near-field amplitude distribution $E(\varrho)$. Inserting the function $E(\varrho)$ into the inverted scalar wave equation (13.125) yields the index profile [Hotate and Okoshi 1979b; Alard and Sansonetti 1983; Okoshi and Hotate 1983]. If the far-field radiation pattern is measured over the full interval $0 < \theta < \pi/2$, the spatial resolution of this method is $\lambda/2$.

Calculating the near-field distribution through a Hankel transform of the measured far-field radiation pattern requires an impressive dynamic range of about 90 dB [Coppa et al. 1983e]. The dynamic range required can be reduced to 70 dB if the beam radiated from the fiber end is expanded in a sum of Gaussian beams of different orders, for which far-field and near-field are simply related [Freude and Sharma 1985, 1986; Freude and Richter 1986].

A fifth method is the focusing method mentioned for profiling of preforms, which can also be applied to single-mode fibers [Presby et al. 1979]. The fiber is immersed in matching oil and collimated light is passed transversely through it. The intensity distribution of the transmitted light is observed with a video camera attached to a microscope and processed with a computer-controlled video-analysis system. The profile, measured with a repeatability of better than 1% and with a resolution of better than 1 μm, can be obtained within a few minutes. However, the focusing method requires great care to ensure uniform illumination of the core and cleanliness of the fiber and optics, because the information on the refractive-index profile is expressed as a variable light intensity [Morishita 1983c].

The sixth method uses the fact that the power reflection at a glass-air interface depends on the refractive index of the glass. Conceptually, the reflection method is one of the simplest ways of determining the refractive-index profiles of optical fibers [Eickhoff and Weidel 1975; Ikeda et al. 1975; Tateda 1978]. A focused laser beam scans over the fiber endface and a photodiode detects the light power reflected from the glass-air interface. Fresnel's equation for normal incidence (5.131), gives a simple relation between the local refractive index $n(\varrho)$ and the reflected power $P_r(\varrho)$. For a very small spot size at the focus and small values of the relative refractive-index difference Δ, the difference $n(\varrho) - n_2$ between the index at a point with coordinate ϱ and a point in the cladding is proportional to the difference $P_r(\varrho) - P_2$ of the powers reflected from the respective points.

However, for single-mode fibers, the spot size of the sampling beam is comparable to the core radius and the reflected power distribution does not indicate the refractive-index profile correctly. A correction must be applied to the measured reflectivity profile [Tateda 1978]. The most important disadvantage of the reflection method is that it is particularly sensitive to degradation of the fiber endface by adsorbed foreign material such as dust or water vapor which causes variation in the surface reflectivity. Furthermore, polishing can change the density and the refractive-index of the surface layer [Stone J. and Earl 1976].

The seventh method for fiber profiling is the slab interferometric method described for profiling preforms. It has also been applied both to multimode

[Presby et al. 1976] and to single-mode fibers [Tasker et al. 1978]. While being direct and accurate, thin slice interferometry is destructive and requires a time-consuming sample preparation procedure. Because of its accuracy, axial interferometry is used as a reference method.

This concludes the review of methods for profiling single-mode fibers. These methods usually only measure the index profile at a single wavelength, mostly at the wavelength 0.633 μm of the HeNe laser. However, the refractive-index profile depends on the wavelength. If one knows the dopants used for making the fiber, the refractive-index profile at any other wavelength can be found [Bachmann et al. 1985] by converting the measured refractive-index profile into a dopant concentration profile, from which index profiles at the wavelengths of interest are calculated using a Sellmeier equation with constants given by Hammond [1978].

The methods described assume the fiber material to be isotropic. However, in some types of fibers, anisotropy is introduced intentionally by internal stresses, e.g. for maintaining polarization. The measurement of anisotropic refractive index distributions has been reported neither for a stress-induced birefringent fiber nor for its preform; it is a challenging but difficult task [Okoshi 1981].

For splicing and connecting of single-mode fibers, one needs to know some geometric data of the fiber, such as cladding diameter, cladding ellipticity, and core eccentricity. These parameters can be obtained from the refracted near-field method or can be measured by preparing a short length of fiber (about 1 cm) by scoring and breaking and observing one endface with a microscope while the other is illuminated with a suitable light source making core and cladding visible.

When this measurement is made manually, it is time consuming and subject to errors due to the operator's subjective determination of the fiber boundaries. By automating the procedure [Marcuse and Presby 1979b], the speed and accuracy can be enhanced. The image is processed with a computer-controlled video-analysis system which tracks the core and cladding boundaries, performs various fitting routines, calculates the desired quantities, and produces a plot of fiber boundaries and core offset.

Since core eccentricity causes a large splice loss, the eccentricity can also be determined from the joint loss observed when the claddings are aligned [Tynes et al. 1978]. By rotating one fiber end 360° with a rotational stage and measuring the insertion loss of the butt joint as a function of the rotation angle, core eccentricity can be determined with an accuracy of ± 0.1 μm [Ridgway and Freeman 1982].

13.11 Miscellaneous Measurements

To end this chapter on fiber measurements, some miscellaneous measurement techniques relevant for single-mode fibers will be mentioned.

In single-mode fiber interferometric sensors, it is necessary to measure the optical phase difference between a signal and a reference wave. The principle of measuring optical phase is to make the signal and reference wave interfere.

A very simple technique [Iiyama et al. 1978] for measuring the phase difference between the fundamental modes transmitted by two single-mode fibers uses the principle of Young's two-beam interference experiment. Flat and perpendicular ends are prepared for both fibers. The axes of the two fibers are fixed parallel to each other with the fibers ending in the same transverse plane. The coherent light beams radiated from the two fiber ends interfere. The interference pattern is observed on a distant screen perpendicular to the fiber axes. From the location of the minima of the interference pattern on the screen, the phase difference between the two LP_{01} modes can be calculated.

There are many other intricate methods for measuring optical phase or phase change which are mainly used in interferometric fiber sensors. Phase shifts as small as 10^{-6} rad can be detected by all-fiber interferometers [Jackson D.A. et al. 1980].

If a single-mode fiber is operated above cutoff ($\lambda < \lambda_c$), one is sometimes interested in determining the distribution of the total transmitted power among the different guided modes, i.e. the so-called mode power spectrum. One method for measuring the mode power spectrum [Bartelt and Lohmann 1983] uses computer-generated matched filter holograms for the analysis of arbitrary fiber fields in terms of Gauss-Laguerre beam modes (Sect. 4.8).

A longitudinal stress increases the fiber length and thus the pulse delay time. This effect has been used to determine the residual stress in a jacketed fiber by measuring the increase in pulse delay [Conduit et al. 1979].

List of Symbols

Latin Letters

a	attenuation, level difference, loss, in dB (5.108, 7.10)
a	core radius (Sect. 5.1)
a	profile parameter of a graded-index lens (12.8)
a_{bs}	backscattering loss (13.38, 13.53)
a_c	capture loss (13.36)
a_c	coupling loss of a directional coupler (12.22)
a_c	coupling loss of a fiber tap (12.27)
a_d	directivity of a directional coupler (12.23)
a_e	effective core radius (10.7, 13.96)
a_e	excess loss of a directional coupler (12.24)
$a_{el} = 2a$	electrical level difference (5.116, 5.118)
a_E	core radius of equivalent step-index fiber (Sect. 5.3.2, 13.93)
a_f	far-end crosstalk attenuation of a WDM-coupler (Sect. 12.3.3)
a_f	furcation loss (12.25)
a_i	insertion loss of a fiber tap (12.26, 12.29)
a_n	near-end crosstalk attenuation of a WDM-coupler (Sect. 12.3.3)
a_r	reflection loss in dB (13.19)
a_r	relative optical power level (5.120, 13.4)
a_s	scattering loss in dB (13.33, 13.34)
a_t	transition loss in dB (5.152, 5.153)
a/W	transverse field extent in the cladding (5.16, 10.21)
$a(q)$	complementary circular aperture power transmission function (13.74)
$a(z)$	one-way fiber loss in dB (13.17, 13.54)
A	area
A	element of the $ABCD$-matrix (4.58)
$\underline{A}$	normalized complex amplitude of coupled waves (7.39, 7.40, 7.45, 7.46)
A	Rayleigh scattering coefficient (5.121, 5.122, 5.124)
A_b	beam cross-sectional area (4.29, 7.25, 7.35)
A_e	effective antenna area (7.24, 7.26, 7.27, 7.35)
$A_N = \sqrt{n_1^2 - n_2^2}$	numerical aperture (5.10)
A_{NE}	numerical aperture of an equivalent step-index fiber (13.93)
b	normalized phase constant (5.51, 5.53)
b	slit width (13.80)

B	bandwidth, -3 opt. dB (11.39, 11.43, 11.48, 11.54, 11.95, 11.105, 11.107)
B	element of the $ABCD$-matrix (4.58)
$\underline{B}$	normalized complex amplitude of coupled waves (7.39, 7.40, 7.45, 7.46)
B	radiance (7.27)
$B(z)$	backscattering signature (13.45, 13.47–49, 13.51, 13.52)
$c = 1/\sqrt{\varepsilon_0\mu_0}$	velocity of light in vacuum (4.13)
C	element of the $ABCD$-matrix (4.58)
d	beam offset at a bend (5.151)
d	gap width (5.132)
$D = d\tau_{\mathrm{g}}/d\lambda$	chromatic dispersion parameter (5.85, 11.27, 11.98, 13.109, 13.110, 13.118)
D	dynamic range of an OTDR (13.55)
D	element of the $ABCD$-matrix (4.58)
D_{m}	material dispersion parameter (4.52, 4.53, 5.87, 5.91)
D_{p}	profile dispersion (5.91, 5.92)
D_{w}	waveguide dispersion parameter (5.88, 5.89, 5.91, 13.121)
e	unit vector (3.1)
$\boldsymbol{E}$	electric field vector (3.1)
$E(\varrho)$	radial field amplitude distribution of a LP_{lm} fiber mode (5.11, 6.1)
$E_{\mathrm{a}}(s)$	autocorrelation function of near-field amplitude distribution of fundamental fiber mode (8.46)
$E_{\mathrm{f}}(r,\theta)$	far-field electric field amplitude pattern (8.15, 10.22, 13.58)
$E_{\mathrm{F}}(f_x, f_y)$	Fourier transform of near-field amplitude distribution (8.7)
$E_{\mathrm{G}}(\varrho)$	Gaussian near-field amplitude distribution (8.21)
$E_{\mathrm{GH}}(q)$	Hankel transform of Gaussian near-field amplitude distribution (8.22)
$E_{\mathrm{H}}(q)$	Hankel transform of fundamental mode amplitude distribution (8.16, 10.22, 13.58, 13.60, 13.72)
f	focal length of a lens (12.7)
f	optical frequency (3.44)
f_1	characteristic frequency (11.33, 11.35, 11.48, 11.93)
f_2	characteristic frequency (11.34, 11.36, 11.43, 11.88)
f_{c}	center frequency (11.25)
f_{m}	modulation frequency (11.17)
f_x	spatial frequency (8.4)
f_y	spatial frequency (8.5)
$f(\varrho)$	Petermann's auxiliary microbending function (Sect. 5.9.9)
F	finesse of an interferometer (12.36, 12.37)
$F(\omega_{\mathrm{m}})$	Fourier transform (11.58)
F	parameter in the general joint loss formula (9.16)
g	profile exponent (5.7)
g	transverse gradient of the refractive index (4.54–5)

g_i	antenna gain relative to an isotropic radiator (8.34)
$g(\varrho/a)$	normalized refractive index profile function (5.6)
G	guidance factor (6.34)
G	parameter in the general joint loss formula (9.17)
$h = 6.626\,\mathrm{Js}$	Planck's constant (5.114)
$\boldsymbol{H}$	magnetic field vector (Sect. 4.3)
$H_{\mathrm{m}}(f_{\mathrm{m}})$	baseband transfer function (11.23, 11.32, 11.38, 11.42, 11.47, 11.51, 11.52, 11.57, 11.59)
$H_0(\omega)$	optical transfer function (11.3, 11.5)
i	electrical current (5.113, 11.6)
$\mathrm{j} = \sqrt{-1}$	imaginary unit
I_2	integral (10.4)
I_3	integral (10.26)
J_l	Bessel function of order l (5.21, 6.8)
$k = \omega\sqrt{\varepsilon_0\mu_0}$	wavenumber in vacuum (4.13)
$k_n = nk$	wavenumber in a medium with refractive index n (4.14, 4.36)
K_l	modified Hankel function of order l (5.15, 5.16, 6.4)
l	azimuthal mode number of LP_{lm} modes (6.1)
L	fiber length (5.107)
L_{c}	coherence length (3.36, 3.49)
L_{c}	coupling length (11.107)
L_{p}	beat length (7.51)
m	modulation index (11.17, 11.19)
m	radial mode number of LP_{lm} modes (Sect. 6.1)
m_n	relative nth order harmonic distortion (11.62–63)
$M = V^2/2$	number of modes in a multimode step-index fiber (7.32)
M_n	nth order moment of a spectral line, impulse, or index profile (3.28, 11.64, Sect. 10.1.3)
n	integer number (3.6, 5.57)
n	refractive index (4.37, 4.38)
n_1	maximum refractive index in the core (5.1)
n_1	refractive index of the first medium (5.126)
n_2	refractive index of the homogeneous cladding (5.1)
n_2	refractive index of the second medium (5.126)
n_{b}	optimum refractive index distribution in a bent dielectric waveguide for obtaining low bend losses (5.149)
$n_{\mathrm{e}} = \beta/k$	effective refractive index (5.29, 5.45, 5.47, 5.49, 5.75, 5.76)
n_{eff}	refractive index distribution in the equivalent straight fiber (5.141)
$n_{\mathrm{g}} = d(nk)/dk$	group index of a medium (4.47, 4.50, 5.67)
$n_{\mathrm{ge}} = c/v_{\mathrm{g}}$	effective group index for a fiber mode (5.68, 5.70, 5.75, 5.76)
n_{i}	refractive index of an intermediate medium (9.7)
n_{p}	refractive index of coupling prism (7.57)
n_{s}	refractive index distribution of the straight dielectric waveguide (5.149)

N	number of additions used in OTDR averaging (13.56)
N	number of half wavelengths in a fiber resonator (12.34)
N	number of output ports of a star coupler (12.25)
N	number of wavelengths in a fiber ring resonator (12.38)
p	exponent of the power spectrum of curvature (5.155)
$p = (1/\lambda)\sin\theta_x$	normalized angular position of a slit aperture (13.76) or a knife edge aperture (13.78)
p	parameter in the general joint loss formula (9.18)
p	pitch length of a graded-index lens (12.9)
p_e	bit error probability (Sect. 11.6)
P	optical power, short time average (3.17, 4.27, 5.34, 5.35, 5.37, 5.40, 7.22, 7.34)
$P(\varrho)$	optical power transmitted through a circular area with radius ϱ (4.32, 5.39)
P_1	power transmitted in the core (5.41, 5.42)
P_2	power transmitted in the cladding (5.41, 5.42)
$P_a(z)$	power in the first of two coupled modes (7.47)
$P_b(z)$	power in the second of two coupled modes (7.48, 7.54)
P_i	input power (11.7)
$P_k(x)$	power transmitted past a knife edge (13.79)
P_m	power launched into a multimode fiber (7.30, 7.31)
P_m	power transmitted with mask (13.85, 13.86)
P_o	output power (11.11)
$P_p(\theta)$	power transmitted through a pinhole scanned along a circular path in the far field (13.60)
$P_p'(\varrho)$	power transmitted through a pinhole scanned along a transverse straight line in the far field (13.61)
P_r	power radiated by an incoherent source into air (7.28, 7.29)
P_r	reflected power (13.18)
$P_s(p)$	power transmitted through a slit aperture (13.77, 13.79, 13.80)
P_s	scattered power (13.32)
$P_t(s)$	power transmission function of a joint with transverse offset s (13.63)
$P_v(q)$	power transmitted through variable circular aperture (13.69, 13.72)
$P_\delta(t)$	impulse response of a fiber (11.56, 11.61)
$P_\lambda = dP/d\lambda$	spectral power density (3.37)
$P_\omega = dP/d\omega$	spectral power density (11.80)
$q = 1.602\,\mathrm{As}$	charge of an electron (5.114)
q	complex beam parameter, complex radius of curvature (4.57)
$q = \sin\theta/\lambda$ $= u/(2\pi)$	normalized angular variable for the far-field radiation pattern, spatial frequency (8.14, 8.17, 10.23, 13.59, 13.71)
q	parameter in the general joint loss formula (9.19)

r	field reflection coefficient (5.126)
r	first spherical polar coordinate (Fig. 3.1)
r_c	coupling ratio of a directional coupler (12.21)
R	radius of curvature of the beam axis of a Gaussian beam propagating in an inhomogeneous medium (4.55)
R	radius of curvature of the fiber axis (5.135)
R	radius of curvature of the wavefronts in a Gaussian beam (4.4, 4.17)
$R = \tau_\mathrm{m} v_\mathrm{g}/2$	spatial resolution of an OTDR for locating two discontinuities (13.50)
$R = |r|^2$	power reflectivity (5.128, 5.130–132)
$R = i/P$	responsivity of a photodiode (5.113, 5.114, 11.13)
R_b	bit rate (11.70, 11.106)
$R = \varrho/a$	normalized radial coordinate (6.36)
s	length of arc (5.135)
s	transverse offset (7.17, 7.18, 9.4)
$s_\mathrm{i}(t)$	input signal (11.8)
$s_\mathrm{o}(t)$	output signal (11.14)
S	backscatter capture fraction (5.123, 13.35, 13.37)
S	irradiance, power density, modulus of Poynting's vector (3.16, 3.18, 4.35, 5.36, 5.38)
$\mathbf{S}$	Poynting's vector (3.16)
$S = d^2\tau_\mathrm{g}/d\lambda^2$	dispersion slope, second order dispersion parameter (5.93, 5.94, 11.28, 11.99)
$S_0 = S(\lambda_0)$	zero dispersion slope (5.95)
$S(V, W)$	auxiliary function for calculating pure bend loss (5.143, 5.147)
t	field transmission factor (5.127)
t	time
t_c	center of gravity of a pulse (11.66)
t_c	coherence time (3.35, 3.48)
t_d	delay time (3.35, 7.8)
t_g	group delay time (11.96, 11.102)
$T = 2\pi/\omega$	optical period
T	power transmission factor (5.129, 5.130, 7.15, 7.23)
$T = 1/R_\mathrm{b}$	pulse separation (11.70)
T_b	diffraction tendency of a Gaussian beam (4.21, 5.27)
$-T_\mathrm{f}$	focusing tendency (5.28)
$T_\mathrm{m} = 1/f_\mathrm{m}$	modulation period (13.114)
u	parameter in the general joint loss formula (9.20)
$u = 2\pi q$	spatial angular frequency (8.17, 10.24)
U	radial phase parameter in the core (5.21–24, 5.56, 6.8)
$v_\mathrm{g} = d\omega/d\beta$	group velocity (4.46, 4.47, 5.65, 5.77, 5.81)
$v_\mathrm{p} = \omega/\beta$	phase velocity (3.13, 4.37, 4.45, 5.59–64, 5.81)
V	normalized frequency (5.8, 5.9)

V	visibility of interference fringes (3.23)
V_c	normalized cutoff frequency (6.13, 6.17)
$V_e = 2.405\,V/V_c$	effective normalized frequency (6.39, 10.6, 13.97)
w_0	spot size of a Gaussian beam at the waist (4.1, 4.56)
w_{01}	spot size in the theory of microbending loss (5.156, 5.157)
w_{02}	spot size in the theory of microbending loss (5.156, 5.157)
$w_\infty \approx 2a/W$	parameter in the new microbending loss theory (5.162, 5.163)
w_a	spot size of the fundamental fiber mode, defined from the autocorrelation function (transverse offset method) (Sect. 10.1.5)
w_{aG}	$1/e$ halfwidth of a Gaussian fit to the transverse offset power transmission function (13.65)
$w_{a\,max},\ w_{a\,min}$	spot sizes for noncircular fundamental mode fields (Sect. 10.1.5, 13.67)
w_{ar}	autocorrelation spot size of the reference fiber (13.68)
w_d	spot size of the fundamental fiber mode, defined with the radial derivative of the field distribution, Petermann II spot size (5.73, 5.89, 5.90, 10.10–14, 10.31, 13.122)
w_e	effective spot size of the fundamental fiber mode (10.8, 10.9, 10.29, 13.83, 13.85)
w_e	energy density of electric field (5.80)
w_G	spot size of the fundamental fiber mode, defined as that spot size of a Gaussian beam which maximizes the launching efficiency (5.25, 5.159, 10.5, 10.6)
w_m	energy density of magnetic field (5.80)
w_r	spot size of the mode guided by a graded-index lens (12.10)
w_s	spot size of the fundamental fiber mode, simple $1/e$ definition (Sect. 10.1.1)
$w(p)$	parameter in the new microbending loss theory (5.160, 5.161)
$w(z)$	Gaussian beam $1/e$ halfwidth (4.1, 4.5, 12.1)
W	cladding decay parameter (5.15, 5.17, 5.19, 5.24)
W	energy (5.78, 11.65, 13.3)
W_a	autocorrelation far-field width (Sect. 10.2.5)
W_d	strange definition for the far-field width (10.27, 10.29, 10.33, 13.73, 13.75, 13.77, 13.82, 13.86)
W_e	effective far-field width (10.28, 10.31, 10.33)
W_g	bandgap energy (13.3)
W_G	far-field width related to launching (Sect. 10.2.2, 10.30, 10.33)
W_{IG}	width of a Gaussian function fitted to the far-field intensity pattern (13.65)
W_s	simple definition of the far-field width (Sect. 10.2.1)
W/a	field decay rate in the cladding (5.15–17)
x	first cartesian coordinate (Fig. 3.1)
x_c	center of gravity of a function $f(x)$ (3.29)
y	second cartesian coordinate (Fig. 3.1)

z	third cartesian coordiante (Fig. 3.1)
z	third cylindrical polar coordinate (Fig. 3.1)
z_l	location of the leading edge of a propagating pulse (13.25)
z_{sl}	location at which the leading pulse edge is scattered (13.28)
z_{st}	location at which the trailing pulse edge is scattered (13.29)
z_t	location of the trailing edge of a propagating pulse (13.26)
$z_R = \pi n w_0^2 / \lambda$	Rayleigh distance (4.6, 9.7, 12.2)
z_w	distance between beam waist and fiber end (7.19) or between two fiber ends (9.6)
$Z_0 = 377\,\Omega$	wave impedance of vacuum (4.25)

Greek Letters

$\alpha = 8.69\alpha'$	attenuation coefficient in dB/km (5.110, 13.6, 13.49)
$\alpha = 2\pi a q$	normalized angular variable in the far-field (8.39)
$\alpha = (1/L)dL/dT$	thermal expansion coefficient (5.84)
α_{11}	attenuation coefficient of the LP_{11} mode (13.11, 13.14, 13.15)
α_{mMM}	attenuation coefficient caused by microbending in a multimode fiber (5.159)
α'	total attenuation coefficient in nepers/km (5.97)
α'_a	attenuation coefficient caused by absorption (5.102, 13.9)
α'_b	attenuation coefficient caused by bending (5.104, 5.134, 5.136)
α'_m	attenuation coefficient caused by microbending (5.156)
α'_s	attenuation coefficient caused by scattering (5.103, 5.121, 13.10)
β	phase constant (3.9, 4.36, 5.11, 5.48, 5.54, 6.1)
$\gamma = \alpha' + j\beta$	complex axial propagation coefficient (3.10, 5.106, 11.4)
γ_{12}	complex degree of coherence (Sect. 3.7.3)
$|\gamma_{12}|$	degree of coherence (3.20, 3.27)
δR	critical distance at bends (5.133, 5.137)
$\Delta \approx (n_1 - n_2)/n_1$	relative refractive index difference (5.2, 5.5)
Δa	power level difference (13.90, 13.91)
Δ_E	relative refractive index difference of ESI-fiber (Sect. 5.3.2)
Δf	free spectral range (12.35, 12.39)
Δf	minimum channel separation in a WDM-system (12.33)
Δf_{FWHM}	full width at half maximum of the resonance curve of a fiber resonator (12.36)
$\Delta l = v_g \tau$	length of the pulse emitted by the laser in an OTDR (13.27)
$\Delta n = n_1 - n_2$	refractive index difference (5.1)
Δt_g	group delay time difference (7.8, 13.99, 13.111)
Δt_{37}	$1/e$ halfwidth duration of a pulse (11.69)

Δt_{50}	50% halfwidth duration of a pulse (11.69)
Δx_{37}	$1/e$ halfwidth of a function $f(x)$ (3.34)
Δx_{50}	50% halfwidth of a function $f(x)$ (3.33)
$\Delta\phi$	phase difference (3.18, 13.98)
$\Delta\lambda$	minimum channel separation in a WDM-system (12.33)
$\Delta\lambda_{37}$	$1/e$ half-linewidth (3.42)
$\Delta\lambda_{50}$	50% half-linewidth (3.42)
$\Delta\lambda_{\mathrm{FWHM}}$	full linewidth at half maximum (3.43, 11.37)
$\Delta\tau_{\mathrm{g}} = \tau_{gx} - \tau_{gy}$	polarization-mode dispersion (11.104)
$\varepsilon_0 = 8.854$ $\times 10^{-12}\,\mathrm{As/(Vm)}$	permittivity of free space (4.13)
$\varepsilon = n^2$	relative permittivity (6.35)
η	launching or coupling efficiency (7.9, 9.1)
η	quantum efficiency (5.114)
$\eta(s)$	traversing joint normalized power transmission function (9.4, 10.16, 13.63)
θ	second spherical polar coordinate (Fig. 3.1)
θ_{d}	beam divergence half-angle ($1/e$ amplitude) (4.8, 4.9, 7.20, 7.21, 7.37, 8.30, 9.9)
θ_{h}	radiation angle, half intensity (Sect. 13.8.10)
θ_x	angle characterizing the position of a slit (13.76) or a knife-edge aperture (13.78)
θ_x	radiation angle, first minimum (Sect. 13.8.10)
$\Theta(y)$	deflection function (Sect. 13.10.1)
$\Theta(z)$	phase deviation between a Gaussian beam and a uniform plane wave (4.1, 4.4, 4.16)
κ	coupling coefficient in directional couplers (7.39–42, 7.51)
$\kappa = 1/R$	curvature, inverse of the radius of curvature (4.17, 6.53)
λ	wavelength in vacuum (3.44, 4.15, 4.34)
λ_0	wavelength of minimum dispersion (Sects. 5.5, 8)
λ_{0m}	wavelength of minimum material dispersion (Sect. 4.5, 5.5)
λ_{01}	wavelength of the fundamental fiber mode (Sect. 5.3.1, 5.43, 5.46, 5.50)
λ_{c}	carrier wavelength (3.40)
λ_{c}	theoretical cutoff wavelength of a fiber mode (6.21, 6.32)
λ_{ce}	effective cutoff wavelength of the second order fiber mode as defined by the $CCITT$ (Sect. 6.3.3, Sect. 13.8)
$\lambda_{\mathrm{ce}}(L, R)$	effective cutoff wavelength of the second order fiber mode in a fiber with length L bent with a radius of curvature R (Sect. 6.3.3, 13.92)
λ_{equ}	equalization wavelength, for which $\tau_{01} - \tau_{11} = 0$ (Sect. 6.3.1, Sect. 6.7, Sect. 13.8.13)
λ_i	oscillator wavelength in a Sellmeier expansion (4.38)
λ_l	local wavelength (3.14)
Λ	spatial period in a mode converter (12.30) or a Bragg reflector (12.31)

μ	radial mode number of exact vector modes (Sect. 6.1)
$\mu_0 = 1.257\,\mathrm{Vs/(Am)}$	permeability of free space (4.13)
ν	azimuthal mode number of exact vector modes (Sect. 6.1, 6.58)
$\bar{\nu} = 1/\lambda$	spectroscopic wave number (3.46, 4.15)
$\pi = 3.14159$	
ϱ	first cylindrical polar coordinate (Fig. 3.1)
ϱ_0	radius of a mask (13.84–86)
ϱ_a	radius of a circular aperture (13.70)
σ	parameter in the general joint loss formula (9.14)
σ_f	RMS-linewidth in the frequency domain (3.45)
σ_t	RMS-duration of a pulse (11.67–71)
σ_x	RMS-width of a function $f(x)$ (3.30)
σ_λ	RMS-linewidth in the wavelength domain (3.38, 11.37)
$\sigma_\varrho = 0.707\,w_\mathrm{G}$	intensity-based spot size (4.30)
σ_ω	RMS-linewidth in the radian frequency domain (3.45)
τ	pulse duration in an OTDR (13.27)
$\tau_0 = 1/c$ $= 3.33\,\mu\mathrm{s/km}$	delay time per unit length in vacuum (4.43)
$\tau_\mathrm{g} = 1/v_\mathrm{g}$	group delay time per unit length (4.48, 4.49, 5.66)
τ_m	measured FWHM pulse width of an OTDR (13.50)
$\tau_\mathrm{p} = 1/v_\mathrm{p}$	phase delay time per unit length (4.40–42, 4.44, 5.63)
ϕ	second cylindrical polar coordinate (Fig. 3.1)
ϕ	third spherical polar coordinate (Fig. 3.1)
Φ	phase angle (3.2, 4.4, 4.12, 5.13, 11.20, 13.114)
$\Phi(\Omega)$	power spectrum of curvature (5.155–156)
$\omega = 2\pi f$	optical angular frequency (3.4)
$\omega_\mathrm{m} = 2\pi f_\mathrm{m}$	modulation angular frequency (11.17)
Ω	solid angle (7.27, 7.36)
Ω	spatial frequency of microbends (5.155)
$\Omega(z)$	local taper angle (12.19, 12.20)

References

1986 Abdula R.M., Saleh B.E.A.: Dynamic spectra of pulsed laser diodes and propagation in single-mode fibers. IEEE J. QE-**22**, 2123–2130

1987 Abe K., Thomson E., Leith G.: Fiber strength degradation induced by local injection/detection block. Conf. Opt. Fiber Commun. (OFC'87), TUQ5

1986 Aberson J.A., White I.A.: Low-stress high-efficiency local-power injection into single-mode fibers. Intern. Wire & Cable Sympos. Proceed. 1986, 334–337

1987 Aberson J.A., White I.A.: Local power injection into single-mode fibers using periodic microbends: a low-stress high-efficiency injector. Conf. Opt. Fiber Commun. (OFC'87), TUQ12

1965 Abramowitz M., Stegun I.A.: *Handbook of Mathematical Functions.* (Dover Publ., New York)

1979 Adams M.J., Payne D.N., Ragdale C.M.: Birefringence in optical fibers with elliptical cross-section. Electron. Lett. **15**, 298–299

1981 Adams M.J.: *An Introduction to Optical Waveguides.* (Wiley & Sons, Chichester, New York)

1986 Agrawal G.P., Potasek M.J: Effect of frequency chirping on the performance of optical communication systems. Opt. Lett. **11**, 318–320

1981 Ainslie B.J., Beales K.J., Day C.R., Rush J.D.: The reproducible fabrication of ultra low loss single-mode fiber. 7th Europ. Conf. Opt. Commun. 1981, 2.5.1–5.4

1985 Ainslie B.J., Craig S.P., Cooper D.M., Day C.R.: Low loss dual window single-mode fibers with very low bending sensitivity. 11th Europ. Conf. Opt. Commun. 1985, 317–320

1986 Ainslie B.J., Day C.R.: A review of single-mode fibers with modified dispersion characteristics. J. Lightwave Tech. LT-**4**, 967–979

1981 Alard F., Jeunhomme L., Sansonetti P.: Fundamental mode spot-size measurement in single-mode optical fibers. Electron. Lett. **17**, 958–960

1982a Alard F., Jeunhomme L., Monerie M., Sansonetti P., Vassallo C.: Reply: Fundamental mode spot-size measurements in single-mode optical fibers. Electron. Lett. **18**, 693–694

1982b Alard F., Lamouler P., Moutonnet D., Sansonetti P.: The mode spot size: a universal parameter for single-mode fiber properties. 8th Europ. Conf. Opt. Commun. 1982, 89–92

1982c Alard F., Sansonetti P., Jeunhomme L.: Fundamental-mode spot size measurement in single-mode optical fibers. Conf. Opt. Fiber Commun. (OFC'82), 62–63

1983 Alard F., Sansonetti P.: Derivation of fundamental-mode electric field and modal properties of single-mode fibers from variable-aperture launch method. Electron. Lett. **19**, 313–314

1979 Albrecht J., Neumann E.-G.: Simulation of the near field of single-mode fibers by means of a microwave model. Microwaves, Optics and Acoustics **3**, 109–114

1979 Alferness R.C.: Optical directional couplers with weighted coupling. Appl. Phys. Lett. **35**(3), 260–262

1981 Alferness R.C.: Guided-wave devices for optical communication. IEEE J. QE-**17**, 946–959

1982 Alferness R.C., Ramaswamy V.R., Korotky S.K., Divino M.D., Buhl L.L.: Efficient single-mode fiber to titanium diffused lithium niobate waveguide coupling for $\lambda = 1.32\,\mu$m. IEEE J. QE-**18**, 1807–1813

1984 Alferness R.C., Divino M.D.: Efficient fiber to x-cut Ti:LiNbO$_2$ waveguide coupling for $\lambda = 1.32\,\mu$m. Electron. Lett. **20**, 465–466

1987 Alferness R.C.: Integrated-optic devices for lightwave communication systems. Conf. Opt. Fiber Commun. (OFC'87), paper TUJ3

1985   Alphones A.: Comment: Fast method for calculating cutoff frequencies in single-mode fibers with arbitrary index profiles. Electron. Lett. **21**, 966

1984   Althouse E.L., Williams J.C., May R.G.: A low-loss, bidirectional optical rotary joint for fiber-optic applications. Proc. SPIE **479**, 117–120

1986   Amitay N., Presby H.M., Dimarcello F.V., Nelson K.T.: Single-mode optical fiber tapers for self-aligned beam expansion. Electron. Lett. **22**, 702–703

1987   Amitay N., Presby H.M., Dimarcello F.V., Nelson K.T.: Optical fiber tapers – a novel approach to self-aligned beam expansion and single-mode hardware. J. Lightwave Tech. LT-**5**, 70–76

1986   Anderson D., Lisak M.: Propagation characteristics of frequency-chirped super-Gaussian optical pulses. Opt. Lett. **11**, 569–571

1983a Anderson W.T., Philen D.L.: Spot size measurements for single-mode fibers – a comparison of four techniques. J. Lightwave Tech. LT-**1**, 20–26

1983b Anderson W.T., Philen D.L.: Spot size measurements in single-mode fibers. Conf. Opt. Fiber Commun. (OFC'83), 24–25

1984   Anderson W.T.: Consistency of measurement methods for the mode field radius in a single-mode fiber. J. Lightwave Tech., LT-**2**, 191–197

1984   Anderson W.T., Lenahan T.A.: Length dependence of the effective cutoff wavelength in single-mode fibers. J. Lightwave Tech. LT-**2**, 238–242

1984   Anderson W.T., Glodis P.F., Philen D.L.: Thermally induced refractive-index changes in a single-mode optical-fiber preform. Conf. Opt. Fiber Commun. (OFC'84), 78–79

1986   Anderson W.T.: Status of single-mode fiber measurements. Techn. Dig. Sympos. Opt. Fiber Measur. 1986, NBS spec.publ. **720**, 1–6

1986   Anderson W.T., Kilmer J.P.: Consistency of mode field diameter definitions for fibers with Gaussian and non-Gaussian field profiles in different measurement domains. Techn. Dig. Sympos. Opt. Fiber Measur. 1986, NBS spec. publ. **720**, 49–52

1986   Anderson W.T., Johnson A.J., Kilmer J.P., Thomas E.A.: Fusion splicing of dissimilar fibers – a comparison of mode field diameter and cross-correlation loss predictions with experimental results. Techn. Dig. Sympos. Opt. Fiber Measur. 1986, NBS spec. publ. **720**, 65–68

1987   Anderson W.T., Shah V., Curtis L., Johnson A.J., Kilmer J.P.: Mode-field diameter measurements for single-mode fibers with non-Gaussian field profiles. J. Lightwave Tech. LT-**5**, 211–217

1986a Andreasen S.B.: private communication

1986b Andreasen S.B.: Combined numerical and analytical method of calculating microbending losses in single-mode fibers with arbitrary index profiles. J. Lightwave Tech. LT-**4**, 596–600

1985   Anelli P., Grasso G., Modone E., Sordo B., Esposto F.: Investigation on hydrogen induced effects on optical cables and possible countermeasures. 11th Europ. Conf. Opt. Commun. 1985, 511–514

1986   Anelli P., Grasso G.: A new high capacity hydrogen absorber. 12th Europ. Conf. Opt. Commun. 1986, 15–18

1986   Ankiewicz A., Snyder A.W., Zheng X.: Coupling between parallel optical fiber cores – critical examination. J. Lightwave Tech. LT-**4**, 1317–1323

1984   Anslow P.J., Farrington J.G., Goddard I.J., Throssell W.R.: System penalty effects caused by spectral variations and chromatic dispersion in single-mode fiber-optic systems. J. Lightwave Tech. LT-**2**, 960–967

1986   Anslow P.J., Goddard I.J.: Modulation-induced spectral penalties in high bit-rate single-mode systems. J. Lightwave Tech. LT-**4**, 751–754

1981   Aoyama K., Nakagawa K., Itoh T.: Optical time domain reflectometry in a single-mode fiber. IEEE J. QE-**17**, 862–868

1985   Aoyama T.: Optical passive devices for longwave fiber-optic communications. Conf. Opt. Fiber Commun. (OFC'85), 108–109

1969   Arnaud J.A.: Degenerate optical cavities: II: Effect of misalignments. Appl. Opt. **8**, 1909–1917

1974a Arnaud J.A.: Transverse coupling in fiber optics. Part I: Coupling between trapped modes. Bell Syst. Tech. J. **53**, 217–224

1974b Arnaud J.A.: Transverse coupling in fiber optics, part III: Bending losses. Bell Syst. Tech. J. **53**, 1379–1394

1975 Arnaud J.A.: Comments on "Radiation from curved dielectric slabs and fibers". IEEE Trans. MTT-**23**, 935–936

1976 Arnaud J.A.: *Beam and Fiber Optics* (Academic Press, New York, San Francisco, London)

1980 Arnold G., Petermann K.: Intrinsic noise of semiconductor lasers in optical communication systems. Opt. Quantum Electron. **12**, 207–219

1985a Arnold G., Krumpholz O.: Coupling of monomode fibers to edge-emitting diodes. Conf. Opt. Fiber Commun. (OFC'85), 48–50

1985b Arnold G., Krumpholz O.: Optical transmission with single-mode fibers and edge-emitting diodes. Electron. Lett. **21**, 390–392

1985 Arnold G., Gottsmann H., Krumpholz O., Schlosser E., Schurr E.-A.: 1.3 μm edge-emitting diodes launching 250 μW into a single-mode fiber at 100 mA. Electron. Lett. **21**, 993–994

1977a Arnold J.M.: Attenuation of an optical fiber immersed in a high-index surrounding medium. Microwaves, Opt., Acoust. **1**, 93–102

1977b Arnold J.M.: Asymptotic evaluation of the normalized cut-off frequencies of an optical waveguide with quadratic index-variation. Microw., Opt., Acoustics **1**, 203–208

1983 Arnold J.M., Allen R.: Microbending loss in optical waveguides. Proc. Inst. Electr. Eng. **130**, Pt.H, 331–339

1984 Arnold J.M., Felsen L.B.: Ray invariants, plane wave spectra, and adiabatic modes for tapered dielectric waveguides. Radio Science **19**, 1256–1264

1986a Artiglia M., Coppa G., Di Vita P.: Simple and accurate microbending loss evaluation in generic single-mode fibers. 12th Europ. Conf. Opt. Commun. 1986, 341–344

1986b Artiglia M., Coppa G., Di Vita P.: New analysis of microbending losses in single-mode fibers. Electron. Lett. **22**, 623–625

1980 Arvidsson G., Thylen L.: Novel method for coupling between single-mode fibers and integrated optical components and its possible applications. Proc. Inst. Electr. Eng. **127**, Pt.H, 37–40

1986 Asano Y., Suzuki, S., Yokota H., Tanaka G.: Splicing of silica core single-mode fibers. Conf. Opt. Fiber Commun. (OFC'86), 74–75

1983 Augé J., Valentin M., Turpin M., Dubos J., Pocholle J.P.; Raffy J., Bourbin Y.: Caractérisations dans les fibers optiques monomodes. Revue Techn. Thomson-CSF **15**, 1013–1048

1984 Augé J., Dupont P., Jeunhomme L.B.: Bending and microbending loss sensitivity of step index single-mode fibers. Symp. Opt. Fiber Meas. 1984, NBS spec. publ. **683**, 25–28

1985 Augé J., Jeunhomme L.B., Kelekis A.: The prediction of single-mode fiber transmission characteristics from mode field diameter. 11th Europ. Conf. Opt. Commun. 1985, 333–335

1979 Azzam R.M.A., Bashara N.M.: *Ellipsometry and Polarized Light*. (North Holland, Amsterdam, New York, Oxford)

1980 Baack C., Elze G., Enning B., Walf G.: Modal noise and optical feedback in high-speed optical systems at 0.85 μm. Electron. Lett. **16**, 592–593

1984 Baba K., Shiraishi K., Hanaizumi O., Kawakami S.: Buried ion-exchanged optical waveguides with refractive-index profiles controlled by rediffusion. Appl. Phys. Lett. **45**(8), 815–817

1983a Bachmann P., Geittner P., Hübner H., Leers D., Lennartz M.: Material dispersion characteristics of optical fibers prepared by the PCVD process. Electron. Lett. **19**, 765–767

1983b Bachmann P., Leers D., Lennartz M., Wehr H.: Preparation of single-mode fibers by the low pressure PCVD process. 9th Europ. Conf. Opt. Commun. 1983, 5–7

1985 Bachmann P.: Review of plasma deposition applications: preparation of optical waveguides. Pure & Appl. Chem. **57**, 1299–1310

1985 Bachmann P., Leers D., Wehr H., Weirich F., Wiechert D., van Steenwijk J.A., Tjaden D.L.A., Werhahn E.: PCVD DFSM-fibers: Performance, limitations, design optimization. 11th Europ. Conf. Opt. Commun. 1985, 197–200

1986 Bachmann P.: Dispersion flattened and dispersion shifted single-mode fibers: worldwide status. 12th Europ. Conf. Opt. Commun. 1986, Vol. II, 17–25

1986a Bachmann P., Geittner P., Lydtin H.: Progress in the PCVD process. Conf. Opt. Fiber Commun. (OFC'86), 76–78

1986b Bachmann P., Leers D., Wehr H., Wiechert D.U., van Steenwijk J.A., Tjaden D.L.A., Wehrhahn E.R.: Dispersion-flattened single-mode fibers prepared with PCVD: performance, limitations, design optimization. J. Lightwave Tech. LT-4, 858–863

1987 Bachmann P., Leers D., Wiechert D.U.: The bending performance of matched cladding, depressed cladding, and dispersion flattened single-mode fibers. Conf. Opt. Fiber Commun. (OFC'87), post deadline paper

1983 Bachus E.-J., Braun R.-P., Strebel B.: Polarization-maintaining single-mode fiber resonator. Electron. Lett. 19, 1027–1028

1986 Baden J.L., Aberson, J.A., Swiderski M.J.: Mass splicing of single-mode fibers. Conf. Opt. Fiber Commun. (OFC'86), 52–53

1982 Baets R., Lagasse P.E.: Calculation of radiation loss in integrated-optic tapers and Y-junctions. Appl. Opt. 21, 1972–1978

1983 Baets R., Lagasse P.E.: Loss calculation and design of arbitrary curved integrated-optic waveguides. J. Opt. Soc. Am. 73, 177–182

1987 Barlow A.J., Mackenzie I.: Direct measurement of chromatic dispersion by the differential phase technique. Conf. Opt. Fiber Commun. (OFC'87), TUQ1

1953 Barlow H.M., Cullen A.L.: Surface waves. Proc. Inst. Electr. Eng. 100, Pt. III, 329–341

1976 Barnoski M.K.; Jensen S.M.: Fiber waveguides: a novel technique for investigating attenuation characteristics. Appl. Opt. 15, 2112–2115

1976 Barnoski M.K.; Rourke M.D., Jensen S.M.: A novel technique for investigating attenuation of fiber waveguides. 2nd Europ. Conf. Opt. Commun. 1976, 75–79

1977 Barnoski M.K.; Rourke M.D., Jensen S.M.; Melville R.T.: Optical time domain reflectometer. Appl. Opt. 16, 2375–2379

1981 Barnoski M.K. (ed.): Fundamentals of Optical Fiber Communications, 2nd ed. (Academic Press, New York)

1978 Barrel K.F., Carpenter D.J., Pask C.: Interpretation of optical-fiber baseband frequency measurement. Microwave, Optics and Acoustics 2, 41–44

1979 Barrell K.F., Pask C.: Optical fiber excitation by lenses. Optica acta 26, 91–108

1983 Bartelt H.O., Lohmann A.W.: Mode analysis of optical fibers using computer-generated matched filters. Electron. Lett. 19, 247–249

1984 Bayer-Helms F.: Coupling coefficients of an incident wave and the modes of a spherical optical resonator in the case of mismatching and misalignment. Appl. Opt. 23, 1369–1380

1984 Beales K.J., Cooper D.M., Duncan W.J., Rush J.D.: Practical barrier to hydrogen diffusion into optical fibers. Electron. Lett. 20, 159–161

1980 Bear P.D.: Microlenses for coupling single-mode fibers to single-mode thin-film waveguides. Appl. Opt. 19, 2906–2909

1983 Beasley J.D., Moore D.R., Stowe D.W.: Evanescent wave fiber optic couplers: three methods. Proc. SPIE 417, 36–43

1985 Becker J.A.; Zell W.: Entwicklungstendenzen bei Monomodeschweißgeräten. NTG-Fachberichte 89, 169–172

1983 Bennett M.J.: Dispersion characteristics of monomode optical-fiber systems. Proc.IEE (London) 130, Pt. H, 309–314

1986 Bennion I., Reid D.C.J., Rowe C.J., Stewart W.J.: High-reflectivity monomode-fiber grating filters. Electron. Lett 22, 341–343

1980a Bergh R.A., Kotler G., Shaw H.J.: Single-mode fiber optic directional coupler. Electron. Lett. 16, 260–261

1980b Bergh R.A., Lefevre H.C., Shaw H.J.: Single-mode fiber-optic polarizer. Opt. Lett. 5, 479–481

1982 Bergh R.A., Digonnet M.J.F., Lefèvre H.C., Newton S.A., Shaw H.J.: "Single-mode fiber optic components" in Fiber-Optic Rotation Sensors, ed. by S. Ezekiel and H.J. Arditty, (Springer, Berlin, Heidelberg, New York), pp. 136–143

1984a Bernard J.J., Depresles E., Jeunhomme L., Moncelet J.L., Carratt M.: 1.3 µm portable reflectometer for the field test of single-mode fiber cables. Symp. Opt. Fiber Meas. 1984, NBS spec. publ. 683, 95–98

1984b Bernard J.J., Ducate J., Gausson Y., Guillon J., Le Blevennec G.: Field portable reflectometer for single-mode fiber cables. 10th Europ. Conf. Opt. Commun. 1984, 84–85

444

1985   Bernard J.J., Visseaux B.: High resolution chromatic dispersion measurements for single-mode fibers. Proc. SPIE **584**, 215–222

1986   Bernard J.J., Guillon, J.: Transmission capacity improvement of a 1.55 μm single-mode fiber link by chromatic dispersion equalization. Conf. Opt. Fiber Commun. (OFC'86), 108–110

1984   Berthou H., Cochet F., Jaccard. P., Parriaux O.: Single to multimode fiber coupler design and technology. 10th Europ. Conf. Opt. Commun. 1984, 184–185

1986   Bethea C.G., Levine B.F., Marchut L., Mattera V.D., Peticolas L.J.: Photon-counting optical time-domain reflectometer using a planar InGaAsP avalanche detector. Electron. Lett. **22**, 302–303

1980   Bhagavatula V.A., Love W.F., Keck D.B., Westwig R.A.: Refracted power technique for cutoff wavelength measurement in single-mode waveguides. Electron. Lett. **16**, 695–696

1985   Bhagavatula V.A.: Recent designs of single-mode fibers for telecommunications. 11th Europ. Conf. Opt. Commun. 1985, Vol. II, 47–53

1986   Bhagavatula V.A.: Dispersion-shifted and dispersion-flattened single-mode designs. Conf. Opt. Fiber Commun. (OFC'86), 94–96

1987   Bhagavatula V.A., Emig K.A.: Dispersion transformer for fiber-optic systems. Conf. Opt. Fiber Commun. (OFC'87), THB6

1977   Bianciardi E., Rizzoli V.: Propagation in graded-core fibers: a unified numerical description. Opt. Quantum Electron. **9**, 121–133

1965   Biernson G., Kinsley D.J.: Generalized plots of mode patterns in a cylindrical dielectric waveguide applied to retinal cones. IEEE Trans. MTT-**13**, 345–356

1983   Biet M., Pocholle J.P.: Backscattering analysis of optical fibers in the dual mode regime. J. Opt. Commun. **4**, 42–46

1986   Bilodeau F., Faucher S., Hill K.O., Johnson D.C.: Wavelength, polarization, and mechanical properties of compact, low-loss fused fiber beamsplitters: fabrication and overcoupled operation in many orders. 12th Europ. Conf. Opt. Commun. 1986, 129–132

1971a Bisbee D.L.: Optical fiber joining technique. Bell Syst. Tech. J. **50**, 3153–3158

1971b Bisbee D.L.: Measurements of loss due to offsets and end separations of optical fibers. Bell Syst. Tech. J. **50**, 3159–3168

1976   Bisbee D.L.: Splicing silica fibers with an electric arc. Appl. Opt. **15**, 796–798

1986a Bjarklev A.: Microdeformation losses of single-mode fibers with step-index profiles. J. Lightwave Tech. LT-**4**, 341–346

1986b Bjarklev A.: Different microdeformation performances of dispersion-shifted and dispersion-flattened single-mode fibers. 12th Europ. Conf. Opt. Commun. 1986, 357–360

1986   Blake J.N., Kim B.Y., Shaw H.J.: Fiber-optic modal coupler using periodic microbending. Opt. Lett. **11**, 177–179

1979   Bloom D.M., Mollenauer L.F., Lin C., Taylor D.W., DelGaudio A.M.: Direct demonstration of distortionless picosecond-pulse propagation in kilometer-length optical fibers. Opt. Lett. **4**, 297–299

1982   Blow K.J., Doran N.J., Hornung S.: Power spectrum of microbends in monomode optical fibers. Electron. Lett. **18**, 448–450

1982   Bludau W., Rossberg R.: Characterization of laser-to-fiber coupling techniques by their optical feedback. Appl. Opt. **21**, 1933–1939

1985   Bludau W., Rossberg R.H.: Low-loss laser-to-fiber coupling with negligible optical feedback. J. Lightwave Tech. LT-**3**, 294–302

1986   Böck G.: Wellenselektiver koaxialer Lichtwellenleiter-Richtkoppler. Siemens Forsch.-u. Entwickl.-Ber. **15**, 32–39

1963   Böhme G.: Die Strahlungsverluste bei der Wellenausbreitung längs gekrümmter dielektrischer Leitungen. Nachrichtentechnik **13**, 46–49

1971   Börner M.: Ein optisches Nachrichtenübertragungssystem mit Glasfaser-Wellenleitern. Wiss. Ber. AEG-Telefunken **44**, 41–45

1972   Börner M., Gruchmann D., Guttmann J., Krumpholz O., Löffler W.: Lösbare Steckverbindung für Ein-Mode-Glasfaserlichtwellenleiter. AEÜ, Arch. für Elektron. und Übertragungstech.: Electron. and Commun. **26**, 288–289

1976   Börner M., Maslowski S.: Single-mode transmission systems for civil telecommunication. Proc. Inst. Electr. Eng. **123**, 627–632

1980   Börner M.: Zur Geschichte der Optischen Nachrichtentechnik im Hause AEG-Telefunken. Wiss. Ber. AEG-Telefunken **53**, 2–4

1986 Bogush A.J., Elkins R.E.: Gaussian field expansions for large aperture antennas. IEEE Trans. AP-**34**, 228–243

1980 Bomberger W.D., Burke J.J.: Interferometric technique for the determination of dispersion in a short length of single-mode optical fiber. Techn. Dig. Symp. Opt. Fiber Measur. 1980, NBS spec. publ. **597**, 101–104

1981 Bomberger W.D., Burke J.J.: Interferometric measurement of dispersion of a single-mode optical fiber. Electron. Lett. **17**, 495–496

1984 Bonaventura G., Rossi U.: Standardization within CCITT of optical fibers for telecommunication systems. Globe Com. '84, 1–6

1985 Bonaventura G.: CCITT standardization of optical fibers for telecommunication systems. Proc. SPIE **584**, 235–242

1975 Born M., Wolf E.: *Principles of Optics*, 5th edition. (Pergamon Press, Oxford, New York)

1985 Boscher D.: Cables et composants passifs pour liaisons monomodes. L'Onde Electrique **65**, 59–69

1986 Bosselaar L., Kuijt G., van Luijk J.F., Matthijsse P.: Accuracy of phase shift technique with LED's for measuring total dispersion in single-mode fibers. Techn. Dig. Sympos. Opt. Fiber Measur. 1986, NBS spec. publ. **720**, 15–18

1980 Botez D., Herskowitz G.J.: Components for optical communications systems: a review. Proc. IEEE **68**, 689–731

1983 Boucher D. (ed.): *Optical fibers in Adverse Environments*. Proc. of the Soc. of Photo-Optical Instrum. Eng., Vol. 404 (SPIE, Bellingham)

1982 Boucouvalas A.C., Papageorgiou C.D.: Cutoff frequencies in optical fibers of arbitrary refractive index profile using the resonance technique. IEEE J. QE-**18**, 2027–2031

1985a Boucouvalas A.C.: Mode-cutoff frequencies of coaxial optical couplers. Opt. Lett. **10**, 95–97

1985b Boucouvalas A.C.: Coaxial optical fiber coupling. J. Lightwave Tech. LT-**3**, 1151–1158

1985a Boucouvalas A.C., Georgiou G.: Tapering of single mode-optical fibers. 11th Europ. Conf. Opt. Commun. 1985, 575–578

1985b Boucouvalas A.C., Georgiou G.: Biconical taper coaxial optical fiber coupler. Electron. Lett. **21**, 864–865

1985c Boucouvalas A.C., Georgiou G.: Biconical taper coaxial coupler filter. Electron. Lett. **21**, 1033–1034

1986a Boucouvalas A.C., Georgiou G.: External refractive-index response of tapered coaxial couplers. Opt. Lett. **11**, 257–259

1986b Boucouvalas A.C., Georgiou G.: A method of beam forming and fabricating optical fiber gap devices. 12th Europ. Conf. Opt. Commun. 1986, 361–364

1982 Bowers J.E., Newton S.A., Sorin W.V., Shaw H.J.: Filter response of single-mode fiber recirculating delay lines. Electron. Lett. **18**, 110–111

1986 Bowers J.E.: Optical transmission using PSK-modulated subcarriers at frequencies to 16 GHz. Electron. Lett. **22**, 1119–1121

1986 Bowley D.J.: Streak recording of picosecond pulses. Laser Focus **22**(4), 110–115

1961 Boyd G.D., Gordon J.P.: Confocal multimode resonator for millimeter through optical wavelength masers. Bell Syst. Tech. J. **40**, 489–508

1980 Boyd R.W.: Intuitive explanation of the phase anomaly of focused light beams. J. Opt. Soc. Am. **70**, 877–880

1956 Bracewell R.N.: Strip integration in radio astronomy. Austral. J. Phys. **9**, 198–217

1959 Bracey M.F., Cullen A.L., Gillespie E.F.F., Staniforth J.A.: Surface-wave research in Sheffield. IRE Trans. AP-**7**, S219–S225

1986 Brain M.C., Cottrell E.A.: Practical one-way-loss range for OTDR's. Techn. Dig. Sympos. Opt. Fiber Measur. 1986, NBS spec. publ. **720**, 73–76

1986 Brenner B.: Kokak's asphere capability faciliates move into single-mode connectors. Laser Focus **22**(1), 68–70

1958 Bresler A.D., Joshi G.H., Marcuvitz N.: Orthogonality properties for modes in passive and active uniform waveguides. J. Appl. Phys. **29**, 794–799

1984 Bricheno T., Fielding A.: Stable low-loss single-mode couplers. Electron. Lett. **20**, 230–232

1986 Brierley M.C., Millar C.A., Ainslie B.J.: Fundamental characterization of polished fiber half-couplers in contact with variable refractive-index media. Conf. Opt. Fiber Commun. (OFC'86), 66–67

1977 Brinkmeyer E.: Refractive-index profile determination from the diffraction pattern. Appl. Opt. **16**, 2802–2803

1979a Brinkmeyer E.: Profile-independent representation of near- and far-field characteristics of single-mode fibers and its use for the determination of fiber parameters. 5th Europ. Conf. Opt. Commun. 1979, 17.2-1-2-4

1979b Brinkmeyer E.: Spot size of graded-index single-mode fibers: profile independent representation and new determination method. Appl. Opt. **18**, 932–937

1980a Brinkmeyer E.: Backscattering in single-mode fibers. Electron. Lett. **16**, 329–330

1980b Brinkmeyer E.: Analysis of the backscattering method for single-mode optical fibers. J. Opt. Soc. Am. **70**, 1010–1012

1981 Brinkmeyer E.: Forward-backward transmission in birefringent single-mode fibers: interpretation of polarization-sensitive measurements. Opt. Lett. **6**, 575–577

1984 Brinkmeyer E., Heckmann S.: Cutoff wavelength determination in single-mode fibers by mode interference. Opt. Lett. **9**, 28–30

1984 Brinkmeyer E., Neumann E.-G.: Comments on "Single-mode fiber OTDR: experiment and theory". IEEE J. QE-**20**, 1293–1294

1986 Brinkmeyer E., Streckert J.: Reduction of polarization sensitivity of optical-time domain reflectometers for single-mode fibers. J. Lightwave Tech. LT-**4**, 513–515

1966 Brown J.: Electromagnetic momentum associated with waveguide modes. Proc. Inst. Electr. Eng. **113**, 27–34

1985 Buckland E.L., Nishimura M.: Measurement of wavelength variation of mode radius using far-field pattern method. Electron. Lett. **21**, 1149–1151

1980 Bulmer C.H., Sheem S.K., Moeller R.P., Burns W.K.: High-efficiency flip-chip coupling between single-mode fibers and LiNbO₃ channel waveguides. Appl. Phys. Lett. **37**(4), 351–353

1981 Bulmer C.H., Sheem S.K., Moeller R.P., Burns W.K.: Fabrication of flip-chip optical couplers between single-mode fibers and LiNbO₃ channel waveguides. IEEE Trans. CHMT-**4**, 350–355

1983 Bures J., Lacroix S., Lapierre J.: Analyse d'un coupleur bidirectionnel à fibres optiques monomodes fusionnées. Appl. Opt. **22**, 1918–1922

1984 Bures J., Lacroix S., Veilleux C., Lapierre J.: Some particular properties of monomode fused fiber couplers. Appl. Opt. **23**, 968–969

1977 Burns W.K., Hocker G.B.: End fire coupling between optical fibers and diffused channel waveguides. Appl. Opt. **16**, 2048–2050

1982 Burns W.K., Moeller R.P., Villaruel C.A.: Observation of low noise in a passive fibre gyroscope. Electron. Lett. **18**, 648–650

1983 Burns W.K., Moeller R.P.: Measurement of polarization mode dispersion in high-birefringence fibers. Opt. Lett. **8**, 195–197

1983 Burns W.K., Moeller R.P., Chen C.: Depolarization in a single-mode optical fiber. J. Lightwave Tech. LT-**1**, 44–50

1985 Burns W.K., Abebe M., Villaruel C.A.: Parabolic model for shape of fiber taper. Appl. Opt. **24**, 2753–2755

1986 Burns W.K., Abebe M., Villaruel C.A., Moeller R.P.: Loss mechanisms in single-mode fiber tapers. J. Lightwave Tech. LT-**4**, 608–613

1972 Burrus C.A.: Radiance of small-area high-current-density electroluminescent diodes. Proc. IEEE **58**, 231–232

1985 Buus J.: Propagation of chirped semiconductor laser pulses in monomode fibers. Appl. Opt. **24**, 4196–4198

1985 Byron K.C., Pitt G.D.: Limits to power transmission in optical fibers. Electron. Lett. **21**, 850–852

1986 Calvo F., Marqués J.B., Villuendas F.: A new method for non-Gaussian mode field radius measurement in axially non-symmetrical single-mode optical fibers. 12th Europ. Conf. Opt. Commun. 1986, 75–78

1985 Calzavara M., Coppa G., Di Vita P.: Self-imaging offset measurements in single-mode optical fibers. 11th Europ. Conf. Opt. Commun. 1985, 329–332

1986a Calzavara M., Coppa G., Di Vita P.: A new technique for mode field diameter determination by transverse offset measurements. Techn. Dig. Sympos. Opt. Fiber Measur. 1986, NBS spec. publ. **720**, 53–56

1986b Calzavara M., Coppa G., Di Vita P.: Self-imaging transverse offset measurements in single-mode optical fibers. Electron. Lett. **22**, 144–146

1986c Calzavara M., Coppa G., di Vita P., Potenza M.: Mode-field diameter measurements in polarization-maintaining optical fibers. Conf. Opt. Fiber Commun. (OFC'86), 104–106

1984 Cameron K.H.: Simple and practical technique for attaching single-mode fibers to lithium-niobate waveguides. Electron. Lett. **20**, 974–976

1979 Campbell J.C.: Tapered waveguides for guided wave optics. Appl. Opt. **18**, 900–902

1984 Campos A.C., Srivastava R.: Wavelength-dependent differential modal delay in graded-index fibers. Electron. Lett. **20**, 20–22

1984 Cancellieri G., Ravaioli U.: *Measurements of Optical Fibers and Devices: Theory and Experiments.* (Artech House, Dedham)

1985 Cancellieri G., Orfei A.: Discussion on the effective cut-off wavelength of the LP_{11}-mode in single-mode optical fibers. Opt. Commun. **55**, 311–315

1985 Cancellieri G., Fantini P., Pesciarelli U.: Effects of joints on single-mode single-polarization optical fiber links. Appl. Opt. **24**, 964–969

1986 Cannon T.C.: Single-mode lightguide connectors. Conf. Opt. Fiber Commun. (OFC'86), 118–120

1984 Caponi R., Coppa G., Di Vita P., Rossi U.: Spot-size measurements in single-mode fibers. Symp. Opt. Fiber Meas. 1984, NBS spec. publ. **683**, 37–40

1985 Caponi R., Coppa G., Di Vita P., Rossi U.: Optical processing technique for spot-size measurements in single-mode fibers. Electron. Lett. **21**, 56–57

1975 Carlson A.Br.: *Communication systems,* 2nd ed. (McGraw-Hill, New-York)

1976 Carpenter D.J., Pask C.: Optical fiber excitation by partially coherent sources. Opt. Quantum Electron. **8**, 545–556

1977 Carpenter D.J., Pask C.: Optical fiber excitation by polychromatic partially coherent sources. Opt. Commun. **20**, 262–264

1982 Carter W.H.: Focal shift and concept of effective Fresnel number for a Gaussiaan laser beam. Appl. Opt. **21**, 1989–1994

1972 Case K.M.: On wave propagation in inhomogeneous media. J. Math. Phys. **13**, 360.

1976 Caspers Fr., Neumann E.-G.: Optical-fiber end preparation by spark erosion. Electron. Lett. **12**, 443–444

1973 Casperson L.W.: Gaussian light beams in inhomgeneous media. Appl. Opt. **12**, 2434–2441

1976 Casperson L.W.: Beam modes in complex lenslike media and resonators. J. Opt. Soc. Am. **66**, 1373–1379

1981 Casperson L.W.: Synthesis of Gaussian beam optical systems. Appl. Opt. **20**, 2243–2249

1985 Casperson L.W., Kirkwood J.L.: Beam propagation in tapered quadratic index waveguides: numerical solutions. J. Lightwave Tech. LT-**3**, 256–263

1985 Cassidy D.T., Johnson D.C., Hill K.O.: Wavelength-dependent transmission of monomode fiber tapers. Appl. Opt. **24**, 945–950

1986 Cassidy D.T., Johnson D.C., Hill K.O.: Wavelength-dependent transmission of monomode optical fiber tapers: errata. Appl. Opt. **25**, 328.

1983 CCITT: "Optical fiber terms and definitions", CCITT Temporary Document No. **40-E**, 1983

1984 CCITT: "Revised version of Recommendation G.652 – Characteristics of a single-mode fiber cable", CCITT document COM.XV/TD **46-E**, May 1984.

1986 CCITT: "Recommendation G.652 – Characteristics of a single-mode fiber cable", CCITT document Fascicle III.2, 272–291

1978 Chandra R., Thyagarayan K., Ghatak A.K.: Mode excitation by tilted and offset Gaussian beams in W-type fibers. Appl. Opt. **17**, 2842–2847

1979a Chang C.T.: Minimum dispersion at 1.55 µm for single-mode step-index fibers. Electron. Lett. **15**, 765–767

1979b Chang C.T.: Minimum dispersion in a single-mode step-index optical fiber. Appl. Opt. **18**, 2516–2522

1976 Chang D.C., Kuester E.F.: Radiation and propagation of a surface-wave mode on a curved open waveguide of arbitrary cross section. Radio Science **11**, 449–457

1986 Chang H., Huang H.S., Wu.J.: Wave coupling between parallel single-mode and multimode optical fibers. IEEE Trans. MTT-**34**, 1337–1343

448

1985 Chang J., Takahashi H., Sugimoto I., Oyobe A.: Measurement of chromatic dispersion with a GeO_2 fiber Raman laser. Conf. Opt. Fiber Commun. (OFC'85), 34–35

1983 Chapman J.E., Wu D.: Single-mode fiber cutoff measurements using offset splice technique: excess loss effects. Electron. Lett. **19**, 290–291

1982 Chen P.Y.P.: Fast method for calculating cutoff frequencies in single-mode fibers with arbitrary index profiles. Electron. Lett. **18**, 1048-1049

1976 Cherin A.H., Rich P.J.: An injection-molded plastic connector for splicing optical cables. Bell Syst. Tech. J. **55**, 1057–1067

1976 Chesler R.B., Dabby F.W.: Simple testing methods give users a feel for cable parameters. Electronics, **49**(August 5), 90–92

1980 Cheung N.K., Denkin N.M.: An automatic inspection system for single fiber connector plugs. Symp. Opt. Fiber Meas. 1980, NBS spec. publ. **597**, 45–47

1981 Cheung N.K.: Transfer-molded biconical connector for single-mode fiber interconnections. Conf. Opt. Fiber Commun. (OFC'81), 98–99

1983 Cheung N.K.: Dispersion penalties for 432 Mb/s single-mode fiber transmission systems in the 1.3 μm wavelength region. 9th Europ. Conf. Opt. Commun. 1983, 271–274

1983 Cheung N.K., Sandahl C.R.: Effect of reflections from single-mode connectors on InGaAsP/InP lasers in high-bit-rate transmission systems. Conf. Opt. Fiber Commun. (OFC'83), 66–69

1984 Cheung N.K.: Reflection and modal noise associated with connectors in single-mode fibers. Proc. SPIE **479**, 56–59

1984a Cheung N.K., Kaiser P.: Cutoff wavelength and modal noise in single-mode fiber systems. Symp. Opt. Fiber Meas. 1984, NBS spec. publ. **683**, 15–18

1984b Cheung N.K., Kaiser P.: Modal noise in single-mode fiber transmission systems. 10th Europ. Conf. Opt. Commun. 1984, 242–243

1985 Cheung N.K., Tomita A., Glodis P.F.: Observation of modal noise in single-mode fiber transmission systems. Electron. Lett. **21**, 5–7

1986 Chiang K.S.: Analysis of fused couplers by the effective-index method. Electron. Lett. **22**, 1221–1222

1984 Chraplyvy A.R.: Optical power limits in multichannel wavelength-division-multiplexed systems due to stimulated Raman scattering. Electron. Lett. **20**, 58–59

1986 Chraplyvy A.R., Marcuse D., Tkach R.W.: Effect of Rayleigh backscattering from optical fibers on DFB laser wavelength. J. Lightwave Tech. LT-**4**, 555–559

1986a Christodoulides D.N., Reith L.A., Saifi M.A.: On the numerical aperture of a single-mode fiber. Techn. Dig. Sympos. Opt. Fiber Measur. 1986, NBS spec. publ. **720**, 133–136

1986b Christodoulides D.N., Reith L.A., Saifi M.A.: Coupling efficiency and sensitivity of an LED to a single-mode fiber. Electron. Lett. **22**, 1110–1111

1977 Chu P.L.: Nondestructive measurement of index profile of an optical-fiber preform. Electron. Lett. **13**, 736–738

1979 Chu P.L.: Nondestructive refractive-index profile measurement of elliptical optical fiber or preform. Electron. Lett. **15**, 357–358

1979 Chu P.L., Saekeang C.: Nondestructive determination of refractive-index profile and cross-sectional geometry of optical-fiber preform. Electron. Lett. **15**, 635–637

1979a Chu P.L., Whitbread T.: Measurement of refractive-index profile of optical-fiber preform. Electron. Lett. **15**, 295–296

1979b Chu P.L., Whitbread T.: Nondestructive determination of refractive index profile of an optical fiber: fast Fourier transform method. Appl. Opt. **18**, 1117–1122

1982 Chu P.L.: Relations between moments of index profile and moments of deflection function and phase function of optical fiber or preform. Electron. Lett. **18**, 832–833

1987 Chuang S.-L.: A coupled mode formulation by reciprocity and a variational principle. J. Lightwave Tech. LT-**5**, 5–15

1985 Cisternino F., Costa B., Rao M.M., Sordo B.: Tunable backscattering for single-mode optical fibers in 1.1–1.6 μm spectral region. 11th Europ. Conf. Opt. Commun. 1985, 263–266

1986 Cisternino F., Costa B., Rao M.M., Sordo B.: Tunable backscattering for single-mode optical fibers in 1.1–1.6 μm spectral region. J. Lightwave Tech. LT-**4**, 884–888

1970 Clarricoats P.J.B., Chan K.B.: Electromagnetic wave propagation along radially inhomogeneous dielectric cylinders. Electron. Lett. **6**, 694–695

1972 Clarricoats P.J.B., Sharpe A.B.: Modal matching applied to a discontinuity in a planar surface waveguide. Electron. Lett. **8**, 28–29

1973 Clarricoats P.J.B., Chan K.B.: Propagation behaviour of cylindrical-dielectric-rod waveguides. Proc. Inst. Electr. Eng. **120**, 1371–1378

1976 Clarricoats P.J.B.: "Optical fiber waveguides – a review", in *Progress in Optics XIV*, ed. by E.Wolf (North-Holland), Chap. VII, pp. 329–402

1986 Cochrane P., Hall R.D., Moss J.P., Betts R.A., Bickers L.: Local-line single-mode optics – viable options for today and tomorrow. IEEE J. Selec. Areas Comm. SAC-**4**, 1438–1445

1972 Cohen L.G.: Power coupling from GaAs injection lasers into optical fibers. Bell Syst. Tech. J. **51**, 573–594

1974 Cohen L.G., Schneider M.V.: Microlenses for coupling junction lasers to optical fibers. Appl. Opt. **13**, 89–94

1977a Cohen L.G., Lin Ch.: Transmission measurements of zero material dispersion in optical fibers. IEEE J. QE-**13**, 91D–92D

1977b Cohen L.G., Lin Ch.: Pulse delay measurements in the zero material dispersion wavelength for optical fibers. Appl. Opt. **16**, 3136–3139

1978 Cohen L.G., Lin Ch.: A universal fiber-optic (UFO) measurement system based on a near-IR fiber Raman laser. IEEE J. QE-**14**, 855–859

1979 Cohen L.G., Fleming J.W.: Effect of temperature on transmission in lightguides. Bell Syst. Tech. J. **58**, 945–951

1979 Cohen L.G., Lin Ch., French W.G.: Tailoring zero chromatic dispersion into the 1.5–1.6 μm low-loss spectral region of single-mode fibers. Electron. Lett. **15**, 334–335

1980a Cohen L.G., Kaiser P., Lin Ch.: Experimental techniques for evaluation of fiber transmission loss and dispersion. Proc. IEEE **68**, 1203–1209

1980b Cohen L.G., Mammel W.L., Lin C., French W.G.: Propagation characteristics of double-mode fibers. Bell Syst. Tech. J. **59**, 1061–1072

1981 Cohen L.G., Lumish S.: Effects of water absorption peaks on transmission characteristics of LED-based lightwave systems operating near 1.3 μm wavelength. IEEE J. QE-**17**, 1270–1276

1981 Cohen L.G., Mammel W.L., Stone J., Pearson A.D.: Transmission studies of a long single-mode fiber – measurements and considerations for bandwidth optimization. Bell Syst. Tech. J. **60**, 1713–1725

1982 Cohen L.G., Stone J.: Interferometric measurements of minimum dispersion spectra in short lengths of single-mode fiber. Electron. Lett. **18**, 564–566

1982a Cohen L.G., Mammel W.L., Lumish St.: Dispersion and bandwidth spectra in single-mode fibers. IEEE J. QE-**18**, 49–53

1982b Cohen L.G., Mammel W.L., Lumish St.: Tailoring the shapes of dispersion spectra to control bandwidths in single-mode fibers. Opt. Lett. **7**, 183–185

1982c Cohen L.G., Marcuse D., Mammel W.L.: Radiating leaky-mode losses in single-mode lightguides with depressed-index claddings. IEEE J. QE-**18**, 1467–1472

1983a Cohen L.G., Mammel W.L., Jang S.J.: Ultrabroadband single-mode fibers. Conf. Opt. Fiber Commun. (OFC'83), 10–11

1983b Cohen L.G., Mammel W.L., Jang S.J., Pearson A.D.: High-bandwidth single-mode fibers. 9th Europ. Conf. Opt. Commun. 285 and in post deadline papers

1985a Cohen L.G.: Comparison of single-mode fiber dispersion measurement techniques. Conf. Opt. Fiber Commun. (OFC'85), 36–37

1985b Cohen L.G.: Comparison of single-mode fiber dispersion measurement techniques. J. Lightwave Tech. LT-**3**, 958–966

1974 Cohen R.L.: Loss measurements in optical fibers. 1: sensitivity limit of bolometric techniques. Appl. Opt. **13**, 2518–2521

1974 Cohen R.L., West K.W., Lazay P.D., Simpson J.: Loss measurements in optical fibers. 2: Bolometric measuring instrumentation. Appl. Opt. **13**, 2522–2524

1960 Collin R.E.: *Field theory of guided waves* (McGraw-Hill, New York)

1979 Conduit A.J., Hartog A.H., Payne D.N.: Residual stress diagnosis in jacketed optical fibers by a pulse delay technique. 5th Europ. Conf. Opt. Commun. 1979, 8.2-1–2-4

1980a Conduit A.J., Hartog A.H., Payne D.N.: Spectral- and length-dependent losses in optical fibers investigated by a two-channel backscatter technique. Electron. Lett. **16**, 77–78

1980b Conduit A.J., Hullett J.L., Hartog A.H., Payne D.N.: An optimized technique for backscatter attenuation measurements in optical fibers. Opt. Quantum Electron. **12**, 169–178

1980c Conduit A.J., Payne D.N., Hartog A.H.: Optical fiber backscatter-loss signatures: identification of features and correlation with known defects using the two-channel technique. 6th Europ. Conf. Opt. Commun. 1980, 152–155

1973 Cook J.S., Mammel W.L., Grow R.J.: Effect of misalignments on coupling efficiency of single-mode optical fiber butt joints. Bell Syst. Tech. J. **52**, 1439–1448

1982 Cooper P.R.: Refractive-index measurements of paraffin, a silicon elastomer, and an epoxy resin over the 500-1500 nm spectral range. Appl. Opt. **21**, 3413–3415

1983 Cooper P.R.: Refractive-index measurements of liquids used in conjunction with optical fibers. Appl. Opt. **22**, 3070–3072

1983a Coppa G., Di Vita P., Potenza M., Rossi U.: A new technique for chromatic dispersion measurement in monomode fibers. 9th Europ. Conf. Opt. Commun. 1983, 189–192

1983b Coppa G., Di Vita P., Potenza M., Rossi U.: Near-field measurements in monomode fibers: determination of chromatic dispersion. Electron. Lett. **19**, 731–733

1983c Coppa G., Di Vita P., Rossi U.: A simple technique for the measurement of the refractive-index profile in monomode fibers. Techn. Digest Int. Conf. Integrated Optics and Optic. Fiber Commun. (IOOC'83), 38–39

1983d Coppa G., Di Vita P., Rossi U.: "Processing of near-field intensity measurements in optical fibers", in *Optical waveguide science*, Proc. of the Intern. Sympos. in Kweilin, China, June 20–23, 1983, ed. by Hung-chia H., Snyder A.W. (Martinus Nijhoff, The Hague) pp. 207–212

1983e Coppa G., Di Vita P., Rossi U.: Characterization of single-mode fibers by near-field measurement. Electron. Lett. **19**, 293–294

1984a Coppa G., Costa B., Di Vita P., Rossi U.: Characterization of single-mode optical fibers. 28th SPIE annual meeting, San Diego, Aug. 19.–24. 1984, Proc. SPIE **500**, paper 500–02

1984b Coppa G., Di Vita P., Rossi U.: A new method for cut-off measurement in monomode fibers. 10th Europ. Conf. Opt. Commun. 1984, 120–121

1985a Coppa G., Costa B., di Vita P., Rossi U.: Cut-off wavelength and mode-field diameter measurements in single-mode fibers. Proc. SPIE **584**, 210–214

1985b Coppa G., Costa B., Di Vita P., Rossi U.: Single-mode optical fiber characterization. Optical Engineering **24**, 676–680

1985c Coppa G., Di Vita P., Potenza M.: Measurement of mode field radius in single-mode fibers by means of harmonic analysis. 11th Europ. Conf. Opt. Commun. 1985, 325–328

1985d Coppa G., di Vita P., Rossi U.: Measurement of cut-off wavelength in monomode fibers by a polarization method. Opt. Quantum Electron. **17**, 41–45

1986a Coppa G., Di Vita P., Potenza M.: New method for mode-field radius measurements in single-mode fibers based on harmonics detection. Electron. Lett. **22**, 4–5

1986b Coppa G., Di Vita P., Potenza M.: Determination of refractive-index distribution from near-field measurements in polarization-maintaining fibers. Techn. Dig. Sympos. Opt. Fiber Measur. 1986, NBS spec. publ. **720**, 35–38

1986c Coppa G., Di Vita P., Potenza M.: Application of harmonics detection to the measurement of mode field radius in single-mode optical fibers. J. Lightwave Tech. LT-**4**, 889–893

1986d Coppa G., Di Vita P., Potenza M.: Two-dimensional index distribution determination from near-field measurements in single-mode fibers. Electron. Lett. **22**, 1038–1040

1983 Cornbleet S.: Geometrical optics reviewed, a new light on an old subject. Proc. IEEE **71**, 471–502

1982 Costa B., Mazzoni D., Puleo M., Vezzoni E.: Phase shift technique for the measurement of chromatic dispersion in optical fibers using LED's. IEEE J. QE-18, 1509–1515

1983a Costa B., Puleo M., Vezzoni E.: "Measurement of chromatic dispersion in single-mode fibers by incoherent sources" in *Optical waveguide science*, Proc. of the Intern. Sympos. in Kweilin, China, June 20–23, 1983, ed. by Hung-chia H., Snyder A.W. (Martinus Nijhoff, The Hague) pp. 85–90

1983b Costa B., Puleo M., Vezzoni E.: Measurement of chromatic dispersion characteristics in single-mode fibers by incoherent sources. 9th Europ. Conf. Opt. Commun. 1983, 385–388

1983c Costa B., Puleo M., Vezzoni E.: Phase-shift technique for the measurement of chromatic dispersion in single-mode optical fibers using LED's. Electron. Lett. **19**, 1074–1076

1984 Costa B., Puleo M., Vezzoni E.: High dynamic chromatic dispersion measurement in single-mode fibers. 10th Europ. Conf. Opt. Commun. 1984, 72–73

1985 Costa B.: Introduction to technical program. 11th Europ. Conf. Opt. Commun. 1985, 23–28

1987 Costa B., Di Vita P.: Single-mode fiber measurements. Conf. Opt. Fiber Commun. (OFC'87), THD1

1983a Cotter D.: Stimulated Brillouin scattering in monomode optical fiber. J. Opt. Commun. **4**, 10–19

1983b Cotter D.: "Optical nonlinearity in fibers: a new factor in systems design", in *Optical Waveguide Science*, Proc. of the Intern. Sympos. in Kweilin, China, June 20–23, 1983, ed. by Hung-chia H., Snyder A.W. (Martinus Nijhoff, The Hague), pp. 145–149

1986 Cotter D.: Nonlinearity limits of power transmission through optical fibers. 12th Europ. Conf. Opt. Commun. 1986 Vol. II, 107–110

1986 Cottrell E.A., Brain M.C.: Long-reach single-mode fiber OTDR using a 1.5 μm semiconductor laser. Electron. Lett. **22**, 443–445

1982 Cozens J.R., Boucouvalas A.C.: Coaxial optical coupler. Electron. Lett. **18**, 138–140

1982 Cozens J.R., Boucouvalas A.C., Al-Assam A., Lee M.J., Morris D.G.: Optical coupling in coaxial fibers. Electron. Lett. **18**, 679–681

1986 Crosignani B., Yariv A., Di Porto P.: Time-dependent analysis of a fiber-optic passive-loop resonator. Opt. Lett. **11**, 251–253

1980 CSELT, technical staff of: *Optical fiber communication* (Centro Studi e Laboratori Telecommunicazioni, Torino)

1984 Csencsits R., Lemaire P.J., Reed W.A., Shenk D.S., Walker K.L.: Fabrication of low-loss single-mode fibers. Conf. Opt. Fiber Commun. (OFC'84), 54–55

1975 Cullen A.L., Eng, C., Ozkan O.: Coupled parallel rectangular dielectric waveguides. Proc. Inst. Electr. Eng. **122**, 593–599

1984 Cvijetic M.M.: Dual-mode optical fibers with zero intermodal dispersion. Opt. Quantum Electron. **16**, 307–317

1984a Daino B., Spano P.: Fault location in inaccessible cables. 10th Europ. Conf. Opt. Commun. 1984, postdeadline paper

1984b Daino B., Spano P.: New method of fault detection in optical fiber links. Electron. Lett. **20**, 781–783

1972 Dakin J.P., Gambling W.A., Payne D.N., Sunak H.R.D.: Launching into glass-fiber optical waveguides. Opt. Commun. **4**, 354–357

1974 Dakin J.P.: A simplified photometer for rapid measurement of total scattering attenuation of fiber optical waveguides. Opt. Commun. **12**, 83–88

1980 Dalgleish J.F.: Splices, connectors, and power couplers for field and office use. Proc. IEEE **68**, 1226–1232

1986 Damsgaard H., Hansen O., Mogensen F.: Chromatic dispersion measurement of single-mode fibers by wavelength temperature tuning of laser diodes. 12th Europ. Conf. Opt. Commun. 1986, 353–356

1981 Dandridge A., Miles R.O.: Spectral characteristics of semiconductor laser diodes coupled to optical fibers. Electron. Lett. **17**, 273–274

1983 Danielsen P.: Simple power spectrum of microbendings in single-mode fibers. Electron. Lett. **19**, 318–320

1983 Danielsen P., Yevick D.: Propagation beam analysis of bent optical waveguides. J. Opt. Commun. **4**, 94–98

1980 Danielson B.L.: An assessment of the backscatter technique as a means for estimating loss in optical waveguides. National Bureau of Standards, Techn. Note **1018** (Feb.), 1–76

1981 Danielson B.L.: Backscatter measurements on optical fibers. NBS Techn. Note **1034**, 1–46

1982 Danielson B.L.: Backscatter measurements on optical fibers. NBS spec. publ. **637**, Vol. 1, 1–46

1985 Danielson B.L.: Optical time-domain reflectometer specifications and performance testing. Appl. Opt. **24**, 2313–2322

1984 Das A.K., Bhattacharyya S.: Optimum conditions for fusion splicing of optical fiber. Proc. IEEE **72**, 983–984

1985 Das A.K., Bhattacharyya S.: Low-loss fusion splices of optical fibers. J. Lightwave Tech. LT-**3**, 83–92

1985 Das S., Goud P.A., Englefield C.G.: Microbending dependence of phase in single-mode fibers. Opt. Lett. **10**, 294–296

1983 D'Auria L., Combemale Y., Muller B.: Connecteur à double excentrique pur fibre optique monomode. Revue Technique Thomson-CSF **15**, 711–729

1984 Davis A.W., Allsop B.E.: A 1.3 μm single-mode fiber optical system for long distance repeaterless FM TV transmission. 7th Europ. Conf. Opt. Commun. 1981, 14.3-1–3-4

1983 Day G.W.: Birefringence measurements in single-mode optical fiber. Proc. SPIE-**425**, 72–79

1984 de Blok C.M., Matthijsse P.: Core alignment procedure for single-mode fiber jointing. Electron. Lett. **20**, 109–110

1986 de Bortoli, M., Moncalvo A.: Economic alternatives in single-mode optical subscriber links: lasers vs. LED's. 12th Europ. Conf. Opt. Commun. 1986, 469–472

1984 de Fornel F., Ragdale C.M., Mears R.J.: Analysis of single-mode fused tapered fiber couplers. Proc. Inst. Electr. Eng. **131**, Pt. H, 221–228

1964 Deschamps G.A., Mast P.E.: Beam tracing and applications. Proc. Symp. on Quasi-Optics, New York 1964, 379–395

1971 Deschamps G.A.: Gaussian beam as a bundle of complex rays. Electron. Lett. **7**, 684–685

1983 Desurvire E., Pocholle J.P., Raffy J., Papuchon M.: Deux applications de l'optique non linéaire guidée. Revue Techn. Thomson-CSF **15**, 809–863

1983 Deveau G.F., Miller C.M., Smith M.Y.: Low loss single-mode fiber splices using ultraviolet curable cement. Conf. Opt. Fiber Commun. (OFC'83), PD Sect. 6-1–6-4

1984 Dick J.M., Modavis R.A., Racki J.G., Westwig R.A.: Automated-mode radius measurement using the variable aperture method in the far-field. Conf. Opt. Fiber Commun. (OFC'84), 90–91

1986 Dick J.M., Shaar C.: Mode field diameter: toward a standard definition. Lasers & Applications 1986(5), 91–94

1979 Dickson L.D.: Ronchi ruling method for measuring Gaussian beam diameter. Optical Engin. **18**(1), 70–75

1977 Diebold G.J.: Effects of A/D converter resolution in signal averaging. Rev. Sci. Instrum. **48**, 1689–1694

1982a Digonnet M.J.F., Shaw H.J.: Single-mode fiber-optic wavelength multiplexer. Conf. Opt. Fiber Commun. (OFC'82), 36–37

1982b Digonnet M.J.F., Shaw H.J.: Analysis of a tunable single-mode optical fiber coupler. IEEE Trans. MTT-**30**, 592–600, and IEEE J. QE-**18**, 746–752

1983 Digonnet M.J.F., Shaw H.J.: Wavelength multiplexing in single-mode fiber couplers. Appl. Opt. **22**, 484–491

1985 Digonnet M.J.F., Feth J.R., Stokes L.F., Shaw H.J.: Measurement of the core proximity in polished fiber substrates and couplers. Opt. Lett. **10**, 463–465

1973 Dil J.G., Blok H.: Propagation of electromagnetic surface waves in a radially inhomogeneous optical waveguide. Opto-electronics **5**, 415–428

1975 di Vita P., Vannucci R.: The "radiance law" in radiation transfer processes. Appl. Phys. **7**, 249–255

1979 di Vita P., Rossi U.: Backscattering measurements in optical fibers: separation of power decay from imperfection contribution. Electron. Lett. **15**, 467–469

1980 di Vita P., Rossi U.: The backscattering technique: its field of applicability in fiber diagnostics and attenuation measurements. Opt. Quantum Electron. **11**, 17–22

1984 di Vita P., Coppa G., Rossi U.: Characterization methods for single-mode fibers. 10th Europ. Conf. Opt. Commun. 1984, 48–49

1985 Döldissen W., Heidrich H., Hoffmann D.: "Reduction of bend-losses in integrated optics devices", in *Integrated Optics*, ed. by Nolting H.P. and Ulrich R., Springer Series in Optical Sciences (Springer, Berlin, Heidelberg, New York), pp. 210–214

1981 Domergue J.-P., Richin P., Pocholle J.-P.: Méthode de clivage d'une fibre optique: définition et expérimentation. Revue Technique Thomson-CSF **13**, 1085–1105

1986 Dreyer D.R., Emig K.A., Minthorn A.R.: Characterization of splice loss as a function of wavelength in step-index, single-mode, optical waveguide fibers. Internat. Wire & Cable Sympos. Proceed. 1986, 220–223

1985 Duff D.G., Stone F.T., Wu J.: Measurements of modal noise in single-mode lightwave systems. Conf. Opt. Fiber Commun. (OFC'85), 52–53

1986 DuPuy R.E.: OTDR's meet the challenge of single-mode technology. Laser Focus **22** (3), 120–130

1970 Dyott R.B., Stern J.R.: Group delay in glass fiber waveguide. Conf. Trunk Telecommunications by Guided Waves, London 1970, IEE publ. **71**, 176–181

1971 Dyott R.B., Stern J.R.: Effects of multiple scattering in optical-fiber transmission line. Electron. Lett. **7**, 624–625

1972 Dyott R.B., Stern J.R., Stewart J.H.: Fusion junctions for glass-fiber waveguides. Electron. Lett. **8**, 290–292

1982 Dyott R.B., Schrank P.F.: Self-locating elliptically cored fiber with an accessible guiding region. Electron. Lett. **18**, 980–981

1973 Eaves R.E.: On the phase velocity and group velocity of guided waves. J. Math. Phys. **14**, 432–433

1983 Eccleston D.J., Dick J.M.: Industrialized measurement system for single-mode fibers. Proc. SPIE **425**, 49–55

1977 Egashira E., Kobayashi M.: Optical fiber splicing with a low-power CO_2 laser. Appl. Opt. **16**, 1636–1638

1975 Eickhoff W., Weidel E.: Measuring method for the refractive index profile of optical glass fibers. Opt. Quantum Electron. **7**, 109–113

1981 Eickhoff W.: Multiple-scattering noise in single-mode fiber systems. 7th Europ. Conf. Opt. Commun. 1981, P4-1–P4-4

1981a Eickhoff W., Ulrich R.: Statistics of backscattering in single-mode fibers. 3rd Intern. Conf. Integrat. Opt. Opt. Commun. (IOOC 81), 76–78

1981b Eickhoff W., Ulrich R.: Optical frequency-domain reflectometry in single-mode fibers. 3rd Intern. Conf. Integrat. Opt. Opt. Commun. (IOOC 81), 106–107

1981c Eickhoff W., Ulrich R.: Optical frequency-domain reflectometry in single-mode fiber. Appl. Phys. Lett. **39**(9), 693–695

1982 Eisenstein G., Vitello D.: Chemically etched conical microlenses for coupling single-mode lasers into single-mode fibers. Appl. Opt. **21**, 3470–3474

1986 Engel R.: Analysis of SMF splices through evaluation of their spectral loss characteristics. 12th Europ. Conf. Opt. Commun. 1986, 145–148

1986 Engel R., McNair E.: Characterization of optical fiber splices in the field. Techn. Dig. Sympos. Opt. Fiber Measur. 1986, NBS spec. publ. **720**, 43–48

1981 Engelmann R.W.H.: Bandwidth formulas for chromatic limitation in optical fibers. Electron. Lett. **17**, 333–334

1983 Enning B., Wenke G.: Demonstration of coloured noise and signal distortions in baseband spectra of broadband optical transmission systems. ntz Archiv **5**, 301–303

1986 Enochs S.: Optical fiber interconnect to a single-mode laser diode. Conf. Opt. Fiber Commun. (OFC'86), 58–60

1986 Eoll C., Lieber W., Loch M., Etzkorn H., Heinlein W.: Advances in the dispersion characterization of different single-mode fibers. 12th Europ. Conf. Opt. Commun. 1986, 71–74

1978 Epworth R.E.: The phenomenon of modal noise in analogue and digital optical fiber systems. 4th Europ. Conf. Opt. Commun. 1978, 492–501

1981 Epworth R.E., Pettitt M.J.: Polarization modal noise and fiber birefringence in single-mode fiber systems. Conf. Opt. Fiber Commun. (OFC'81), 58–60

1984 Epworth R.E., Smith D.F., Wright S.: A practical $1.3\,\mu m$ semiconductor source with significantly better short term coherence than a gas laser. 10th Europ. Conf. Opt. Commun. 1984, 132–133

1979 Esposito J.J.: Evanescent wave coupling in fiber optic connectors: an observation. Appl. Opt. **18**, 1292–1293

1976 Ettenberg M., Kressel H., Wittke J.P.: Very high radiance edge-emitting LED. IEEE J. QE-**12**, 360–364

1981 Eyges L., Wintersteiner P.: Modes of an array of dielectric waveguides. J. Opt. Soc. Am. **71**, 1351–1360

1984 Facq P., De Fornel F., Jean F.: Tunable single-mode excitation in multimode fibers. Electron. Lett. **20**, 613–614

1986   Farries M.C., Townsend J.E., Poole S.B.: Very high-rejection optical fiber filters. Electron. Lett. **22**, 1126–1128

1981   Farrington J.G.: Reflection noise effects on error rate in a 35 km single-mode transmission system. 7th Europ. Conf. Opt. Commun. 1981, 14.2-1–2-4

1983   Favre F., Le Guen D.: Emission frequency stability in single-mode fiber optical feedback controlled semiconductor lasers. Electron. Lett. **19**, 663–665

1985   Favre F., Le Guen D.: Spectral properties of a semiconductor laser coupled to a single-mode fiber resonator. IEEE J. QE-**21**, 1937–1946

1981   Feit M.D., Fleck J.A.: Propagating beam theory of optical fiber cross coupling. J. Opt. Soc. Am. **71**, 1361–1372

1985   Felsen L.B.: Novel ways for tracking rays. J. Opt. Soc. Am. A **2**, 954–963

1986   Felsen L.B.: Adiabatic spectra for tapered dielectric waveguides. AEÜ, Arch. für Elektron. und Übertragungstech.: Electron. and Commun. **40**, 259–262

1986   Fermann M.E., Poole S.B., Payne D.N., Martinez F.: A new technique for the relative measurement of scatter levels in single-mode fibers. Techn. Dig. Sympos. Opt. Fiber Measur. 1986, NBS spec. publ. **720**, 77–80

1978   Findakly T., Chen C.: Optical directional couplers with variable spacing. Appl. Opt. **17**, 769–773

1985   Finegan T.: An investigation into approximate analytical methods for describing planar optical waveguide junctions. Opt. Quantum Electron. **17**, 109–118

1986   Finvers I., Clegg D.D., So V., Vella P.J.: Live fiber identifier. Intern. Wire & Cable Sympos. Proceed. 1986, 440–442

1976   Fleming J.W.: Material and mode dispersion in $GeO \cdot B_2O_3 \cdot SiO_2$ glasses. J. Amer. Ceramic Soc. **59**, 503–507

1978   Fleming J.W.: Material dispersion in lightguide glasses. Electron. Lett. **14**, 326–328

1983   Fleming J.W., Wood D.L.: Refractive index dispersion and related properties in fluorine doped silica. Appl. Opt. **22**, 3102–3104

1985   Fleming S.C., Gale R.W., Nelson J.C.C.: Characterization of optical fibers by measurement of the transmitted near field. Proc. SPIE **584**, 189–194

1982   Föllinger O.: *Laplace- und Fourier-Transformation* (AEG-Telefunken, Berlin, Frankfurt)

1980   Fox M., Dennis M.R.: Comment: Index matching fluids for long wavelength (1.2–1.6 μm) fiber-optic applications. Electron. Lett. **16**, 651–652

1986   Fox M.: Comment: New approach for determining non-Gaussian mode fields of single-mode fibers from measurements in far-field. Electron. Lett. **22**, 109–110

1981   Francois P.L., Sasaki I., Adams M.J.: Three-dimensional fiber preform profiling. Electron. Lett. **17**, 876–878

1982   Francois P.L.: Dispersion-free single-mode doubly clad fibers with small pure bend loss. Electron. Lett. **18**, 818–819

1982   Francois P.L., Sasaki I., Adams M.J.: Practical three-dimensional profiling of optical fiber preforms. IEEE Trans. MTT-**30**, 370–380 and IEEE J. QE-**18**, 524–534

1984   Francois P.L., Alard F., Bayon J.F., Rose B.: Multimode nature of quadruple-clad fibers. Electron. Lett. **20**, 37–38

1986   Francois P.L., Vassallo C.: Comparison between pseudomode and radiation mode methods for deriving microbending losses. Electron. Lett. **22**, 261–262

1979   Franken A.J.J., Khoe G.D., Renkens J., Verwer C.J.G.: Versuchsgerät zum halbautomatischen Verschweißen von Glasfasern für die optische Nachrichtenübertragung. Philips techn. Rdsch. **38**, 176–177

1986   Frantsesson A.V., Yang R., Unger H.-G.: Macrobending loss of triple-clad single-mode fibers. AEÜ, Arch. für Elektron. und Übertragungstech.: Electron. and Commun. **40**, 132–133

1981   Franzen D.L., Kim E.M.: Long optical-fiber Fabry-Perot interferometers. Appl. Opt. **20**, 3991–3992

1985a   Franzen D.L.: Interlaboratory measurement comparison among fiber manufacturers to determine the effective cutoff wavelength and mode field diameter of single-mode fiber. Conf. Opt. Fiber Commun. (OFC'85), 36–37

1985b   Franzen D.L.: Determining the effective cutoff wavelength of single-mode fibers: an interlaboratory comparison. J. Lightwave Tech. LT-**3**, 128–134

1985   Franzen D.L., Srivastava R.: Determining the mode-field diameter of single-mode optical fiber: an interlaboratory comparison. J. Lightwave Tech. LT-**3**, 1073–1077

1986 Franzen D.L., Kanada T.: High-resolution dispersion measurements in single-mode fiber using a tunable laser diode and recirculating coupler. Conf. Opt. Fiber Commun. (OFC'86), 102–103

1984 Freude W., Sharma A.: Refractive-index profile and modal dispersion prediction for a single-mode optical waveguide from its far-field radiation pattern. Symp. Opt. Fiber Meas. 1984, NBS spec. publ. **683**, 29–32

1985 Freude W., Sharma A.: Refractive-index profile and modal dispersion prediction for a single-mode optical waveguide from its far-field radiation pattern. J. Lightwave Tech. LT-3, 628–634

1986 Freude W., Sharma A.: Errata: "Refractive-index profile and modal dispersion prediction for a single-mode optical waveguide from its far-field radiation pattern". J. Lightwave Tech. LT-4, 375

1986 Freude W., Richter H.: Refractive-index profile determination of single-mode fibers by far-field power measurements at 1300 nm. Electron. Lett. **22**, 945–947

1986 Friberg A.T.: Energy transport in optical systems with partially coherent light. Appl. Opt. **25**, 4547–4556

1986 Frigo N.J.: A generalized geometrical representation of coupled mode theory. IEEE J. QE-**22**, 2131–2140

1984 Frisch D.A., Henning I.D.: Effects of laser chirp on optical systems – initial tests using a 1480 nm DFB laser. Electron. Lett. **20**, 631–633

1984 Fuhr P.L.: Direct-current polarization characteristics of various AlGaAs laser diodes. Opt. Lett. **9**, 438–440

1980 Fujii Y., Aoyama K., Minowa J.: Optical demultiplexer using a silicon Echelette grating. IEEE J. QE-**16**, 165–169

1984 Fujise M., Kuwazuru M.: Rayleigh scatter measurements of a 42 km single-mode fiber at 1.55 μm wavelength using an LD and an LN_2 cooled Ge-PIN detector. Electron. Lett. **20**, 232–233

1984 Fujise M., Iwamoto Y.: Core alignment by a simple local monitoring method. Appl. Opt. **23**, 2643–2648

1986a Fujise M., Iwamoto Y., Takei S.: Self-core alignment arc-fusion splicer based on a simple local monitoring method. J. Lightwave Tech. LT-4, 1211–1218

1986b Fujise M., Kuwazuru M., Nunokawa M., Iwamoto Y.: Chromatic dispersion measurement over a 100 km dispersion-shifted single-mode fiber by a new phase-shift technique. Electron. Lett. **22**, 570–572

1976 Fujita H., Suzaki Y., Tachibana A.: Optical fiber splicing technique with a CO_2 laser. Appl. Opt. **15**, 320–321

1980 Fukuma M., Noda J.: Optical properties of titanium-diffused $LiNbO_3$ strip waveguides and their coupling-to-a-fiber characteristics. Appl. Opt. **19**, 591–597

1982 Furakawa S., Murakami Y., Ishihara K., Hoshino K.: Non-monitored splicing of single-mode optical fiber. Trans. IECE Jap. **J65-B**, 662–663 (in Japanese)

1986 Furukawa S.-I., Koyamada Y.: New simple one-end splice loss measurement method using an OTDR for single-mode optical fiber. Electron. Lett. **22**, 535–537

1978 Furuya K., Suematsu Y.: Random bend losses in single-mode optical fiber cables: power-spectrum estimation from spectral losses. Electron. Lett. **14**, 653–654

1979 Furuya K., Suematsu Y.: Random-bend loss of single-mode and graded-multimode optical fiber cables. 5th Europ. Conf. Opt. Commun. 1979, 17.4-1–4-4

1979a Furuya K., Miyamoto M., Suematsu Y.: Bandwidths of single-mode optical fibers. Trans. IECE of Japan E **62**, 305–310

1979b Furuya K., Miyamoto M., Suematsu Y.: Bandwidth limitations due to harmonic distortion in single-mode optical fibers. Proc. IEEE **67**, 694

1980 Furuya K., Suematsu Y.: Random-bend loss in single-mode and parabolic-index multimode optical fiber cables. Appl. Opt. **19**, 1493–1500

1986 Fye D.M., Olshansky R., Lacourse J., Powazinik W., Lauer R.B.: Low-current 1.3 μm edge-emitting LED for single-mode fiber subscriber loop applications. Electron. Lett. **22**, 87–88

1986 Gallawa R.L., Yang S.: Optical fiber power meters: a round robin test of uncertainty. Appl.Opt. **25**, 1066–1068

1973 Gambling W.A., Payne D.N., Matsumura H.: Mode excitation in a multimode optical fiber waveguide. Electron. Lett. **9**, 412–414

1976a Gambling W.A., Payne D.N., Matsumura H.: Propagation studies on single-mode phosphosilicate fibers. 2nd Europ. Conf. Opt. Commun. 1976, 95–100

1976b Gambling W.A., Payne D.N., Matsumura H.: Routine characterization of single-mode fibers. Electron. Lett. **12**, 546–547

1976c Gambling W.A., Payne D.N., Matsumura H.: Radiation from curved single-mode fibers. Electron. Lett. **12**, 567–569

1976d Gambling W.A., Payne D.N., Matsumura H., Dyott R.B.: Determination of core diameter and refractive-index difference of single-mode fibers by observation of the far-field pattern. Microw., Optics and Acoust. **1**, 13–17

1977 Gambling W.A., Matsumura H.: Simple characterization factor for practical single-mode fibers. Electron. Lett. **13**, 691–693

1977a Gambling W.A., Matsumura H., Sammut R.A.: Mode shift at bends in single-mode fibers. Electron. Lett. **13**, 695–697

1977b Gambling W.A., Payne D.N., Matsumura H.: Cut-off frequency in radially inhomogeneous single-mode fiber. Electron. Lett. **13**, 139–140

1977c Gambling W.A., Payne D.N., Matsumura H.: Effect of dip in the refractive index on the cut-off frequency of a single-mode fiber. Electron. Lett. **13**, 174–175

1977d Gambling W.A., Payne D.N., Matsumura H., Norman S.R.: Measurement of normalized frequency in single-mode optical fibers. Electron. Lett. **13**, 133–136

1978a Gambling W.A., Matsumura H.: Propagation in radially inhomogeneous single-mode fibers. Opt. Quantum Electron. **10**, 31–40

1978b Gambling W.A., Matsumura H.: Propagation characteristics of curved optical fibers. Trans. IECE Japan, **E 61**, 196–201

1978a Gambling W.A., Matsumura H., Cowley A.G.: Jointing loss in single-mode fibers. Electron. Lett. **14**, 54–55

1978b Gambling W.A., Matsumura H., Ragdale C.M.: Field deformation in a curved single-mode fiber. Electron. Lett. **14**, 130–132

1978c Gambling W.A., Matsumura H., Ragdale C.M.: Loss mechanisms in practical single-mode fibers. 4th Europ. Conf. Opt. Commun. 1978, 260–269

1978d Gambling W.A., Matsumura H., Ragdale C.M.: Wave propagation in a single-mode fiber with dip in the refractive index. Opt. Quantum Electron. **10**, 301–309

1978e Gambling W.A., Matsumura H., Ragdale C.M.: Joint loss in single-mode fibers. Electron. Lett. **14**, 491–493

1978f Gambling W.A., Matsumura H., Ragdale C.M.: Zero-mode dispersion in single-mode fibers. Electron. Lett. **14**, 618–620

1978g Gambling W.A., Matsumura H., Ragdale C.M., Sammut R.A.: Measurement of radiation loss in curved single-mode fibers. Microwaves, Optics and Acoustics **2**, 134–140

1979 Gambling W.A., Matsumura H.: "A comparison of single-mode and multimode fibers for long-distance telecommunications", in *Fiber and Integrated Optics*, ed. by D.B. Ostrowski (Plenum, New York), pp. 333–343

1979a Gambling W.A., Matsumura H., Ragdale C.M.: Curvature and microbending losses in single-mode optical fibers. Opt. Quantum Electron. **11**, 43–59

1979b Gambling W.A., Matsumura H., Ragdale C.M.: Mode dispersion, material dispersion and profile dispersion in graded-index single-mode fibers. Microwaves, Optics and Acoustics **3**, 239–246

1979c Gambling W.A., Matsumara H., Ragdale C.M.: Zero total dispersion in graded-index single-mode fibers. Electron. Lett. **15**, 474–476

1975 Gardner W.B.: Microbending loss in optical fibers. Bell Syst. Tech. J. **54**, 457–465

1975 Gardner W.B., Gloge D.: Microbending loss in coated and uncoated optical fibers. Topical Meeting on Optical Fiber Transm. 1975, WA3-1-3-4

1982 Gardner W.B.: Single-mode fiber loss round robin. Techn. Dig. Sympos. Opt. Fiber Measur. 1982, NBS Publ. **641**, 85–87

1986a Garg A.O., Eoll C.K.: New measurement technique for measurement of microbending losses in single-mode fibers. Techn. Dig. Sympos. Opt. Fiber Measur. 1986, NBS spec. publ. **720**, 125–128

1986b Garg A.O., Eoll C.K.: Experimental characterization of bend and microbend losses in single-mode fibers. Intern. Wire & Cable Sympos. Proceed. 1986, 406–414

1981 Garlichs G.: Polarization behaviour fluctuations of a single-mode fiber. Electron. Lett. **17**, 894–895

1982 Garrett I., Todd C.J.: Review: Components and systems for long-wavelength monomode fiber transmission. Opt. Quantum Electron. **14**, 95–143

1983 Garrett I.: Towards the fundamental limits of optical-fiber communications. J. Lightwave Tech. LT-**1**, 131–138

1980 Garside B.K., Lim T.K., Marton J.P.: Propagation characteristics of parabolic-index fiber modes: linearly polarized approximation. J. Opt. Soc. Am. **70**, 395–400

1985 Gartside C.H., Baden J.L.: Single-mode ribbon cable and array splicing. Conf. Opt. Fiber Commun. (OFC'85), 106–107

1981 Gauthier F., Auge J., Gallou D., Wehr M., Blaison S.: Consistent refractive-index profile measurements of a step-index monomode optical fiber attained by several techniques. IEEE J. QE-**17**, 885–889

1986a Geckeler S.: *Lichtwellenleiter für die optische Nachrichtentechnik* (Springer, Berlin, Heidelberg)

1986b Geckeler S.: Einfluß von chromatischer Dispersion und Modenverteilungsrauschen auf die nutzbare Bandbreite optischer Übertragungssysteme. Siemens Forsch.-u. Entwickl.-Ber. **15**, 23–31

1985 Georgiou G., Boucouvalas A.C.: Low-loss single-mode optical couplers. Proc. Inst. Electr. Eng. **132**, Pt. J, 297–302

1986 Georgiou G., Boucouvalas A.C.: High-isolation single-mode wavelength-division multiplexer/demultiplexer. Electron. Lett. **22**, 62–63

1985 Ghafoori-Shiraz H., Okoshi T.: Optical-fiber diagnosis using optical-frequency-domain reflectometry. Opt. Lett. **10**, 160–162

1986a Ghafoori-Shiraz H., Okoshi T.: Optical frequency-domain reflectometry. Opt. Quantum Electron. **18**, 265–272

1986b Ghafoori-Shiraz H., Okoshi T.: Fault location in optical fibers using optical frequency domain reflectometry. J. Lightwave Tech. LT-**4**, 316–322

1986 Ghafoori-Shiraz H., Asano T.: Microlens for coupling a semiconductor laser to a single-mode fiber. Opt. Lett. **11**, 537–539

1983 Ghatak A.K., Srivastava R., Faria I.F., Thyagarajan K., Tiwari R.: Accurate method for characterizing single-mode fibers: theory and experiment. Electron. Lett. **19**, 97–99

1985a Gimlett J.L., Stern M., Curtis L., Young W.C., Cheung P.W., Shumate P.W.: Dispersion penalties for single-mode-fiber transmission using 1.3 and 1.5 μm LED's. Electron. Lett. **21**, 668–670

1985b Gimlett J.L., Stern M., Vodhanel R.S., Cheung N.K., Chang G.K., Leblanc H.P., Shumate, P.W., Suzuki A.: Transmission experiments at 560 Mb/s and 140 Mb/s using single-mode fiber and 1300 nm LEDs. 11th Europ. Conf. Opt. Commun. 1985, postdeadline papers, 53–56

1964 Gloge D.: Bündelung kohärenter Lichtstrahlen durch ein ortsabhängiges Dielektrikum. AEU, Arch. für Elektron. und Übertragungstech.: Electron. and Commun. **18**, 451–452

1971a Gloge D.: Weakly guiding fibers. Appl. Opt. **10**, 2252–2258

1971b Gloge D.: Dispersion in weakly guiding fibers. Appl. Opt. **10**, 2442–2445

1972 Gloge D., Chinnock E.L.: Fiber-dispersion measurements using a mode-locked Krypton laser. IEEE J. QE-**8**, 852–854

1973 Gloge D., Smith P.W., Bisbee D.L., Chinnock E.L.: Optical fiber end preparation for low-loss splices. Bell Syst. Tech. J. **52**, 1579–1588

1975 Gloge D.: Propagation effects in optical fibers. IEEE Trans. MTT-**23**, 106–120

1979a Gloge D.: Effect of chromatic dispersion on pulses of arbitrary coherence. Electron. Lett. **15**, 686–687

1979b Gloge D.: The optical fiber as a transmission medium. Reports on Progress in Physics **42**, 1779–1824

1980 Gloge D., Ogawa K., Cohen L.G.: Baseband characteristics of long-wavelength LED systems. Electron. Lett. **16**, 366–367

1981 Gold M.P., Hartog A.H.: Measurement of backscatter factor in single-mode fibers. Electron. Lett. **17**, 965–966

1982a Gold M.P., Hartog A.H.: Determination of structural parameter variations in single-mode optical fibers by time-domain reflectometry. Electron. Lett. **18**, 489–490

1982b Gold M.P., Hartog A.H.: Analysis of backscatter waveforms from single-mode fibers. 8th Europ. Conf. Opt. Commun. 1982, 633–638

1983a Gold M.P., Hartog A.H.: A practical high-performance single-mode OTDR system for the long-wavelength region. 9th Europ. Conf. Opt. Commun. 1983, 181–184

1983b Gold M.P., Hartog A.H.: Ultra-long-range OTDR in single-mode fibers at $1.3\,\mu$m. Electron. Lett. **19**, 463–464

1984a Gold M.P., Hartog A.H.: Long-range single-mode OTDR: ultimate performance and potential uses. 10th Europ. Conf. Opt. Commun. 1984, 128–129

1984b Gold M.P., Hartog A.H.: Improved-dynamic-range single-mode OTDR at $1.3\,\mu$m. Electron. Lett. **20**, 285–287

1984 Gold M.P., Hartog A.H., Payne D.N.: New approach to splice-loss monitoring using long-range OTDR. Electron. Lett. **20**, 338–340

1985 Gold M.P.: Design of a long-range single-mode OTDR. J. Lightwave Tech. LT-**3**, 39–46

1982 Goldberg L., Taylor H.F., Dandridge A., Weller J.F., Miles R.O.: Spectral characteristics of semiconductor lasers with optical feedback. IEEE Trans. MTT-**30**, 401–409

1968 Goodman J.W.: *Introduction to Fourier Optics* (McGraw-Hill, New York)

1985 Goodman J.W.: Fan-in and fan-out with optical interconnections. Optica Acta **32**, 1489–1496

1966 Gordon E.I.: A review of acoustooptical deflection and modulation devices. Proc. IEEE **54**, 1391–1401

1977 Gordon K.S., Rawson E.G., Nafarrate A.B.: Fiber-break testing by interferometry: a comparison of two breaking methods. Appl. Opt. **16**, 818–819

1952 Goubau G.: On the excitation of surfaces waves. Proc. IRE, **40**, 865–868

1961 Goubau G., Schwering F.: On the guided propagation of electromagnetic wave beams. IEEE Trans. AP-**9**, 248–256

1986 Goulley B., Clapeau M., Facq P.: Mode spot size measurements in noncircular optical waveguides. Electron. Lett. **22**, 942–943

1965 Gradshteyn I.S., Ryzhik I.M.: *Tables of Integrals, Series, and Products* (Academic Press, New York)

1978 Grau G.: *Quantenelektronik: Optik und Laser* (Vieweg, Braunschweig)

1978 Grau G.: Anregung und Kopplung von Normalmoden in Wellenleitern. AEÜ, Arch. für Elektron. und Übertragungstech.: Electron. and Commun. **32**, 195–199

1980 Grau G., Leminger O.G., Sauter E.G.: Mode excitation in parabolic index fibers by Gaussian beams. AEÜ, Arch. für Elektron. und Übertragungstech.: Electron. and Commun. **34**, 259–265

1981 Grau G.: *Optische Nachrichtentechnik* (Springer, Berlin, Heidelberg, New York)

1986 Grebel H., Herskowitz G.J.: Mode coupling in single-mode fiber: the advantage of using higher-order spatial-distortion-function harmonics. Opt. Lett. **11**, 674–676

1986 Grebel H., Herskowitz G.J., Mezhoudi M.: Measurement of cutoff wavelength of single-mode fiber with periodic perturbation of fiber axis. Electron. Lett. **22**, 1135–1136

1982 Grosskopf G., Küller L., Patzak E.: Laser mode partition noise in optical wideband transmission links. Electron. Lett. **18**, 493–494

1973 Guttmann J., Krumpholz O.: Theoretische und experimentelle Untersuchungen zur Verkopplung zweier Glasfaser-Lichtwellenleiter. Wiss. Ber. AEG-Telefunken **46**, 8–15

1975 Guttmann J., Krumpholz O.: Location of imperfections in optical glass-fiber waveguides. Electron. Lett. **11**, 216–217

1975a Guttmann J., Krumpholz O., Pfeiffer E.: A simple connector for glass fiber optical waveguides. AEÜ, Arch. für Elektron. und Übertragungstech.: Electron. and Commun. **29**, 50–52

1975b Guttmann J., Krumpholz O., Pfeiffer E.: Optical fiber-stripline-coupler. Appl. Opt. **14**, 1225–1227

1983 Haibara T., Matsumoto M., Tanifuji T., Tokuda M.: Monitoring method for axis alignment of single-mode optical fiber and splice-loss estimation. Opt. Lett. **8**, 235–237

1986 Haibara T., Matsumoto M., Miyauchi M.: Design and development of an automatic cutting tool for optical fibers. J. Lightwave Tech. LT-**4**, 1434–1439

1977 Hammond C.R., Norman S.R.: Silica based binary glass systems – refractive index behaviour and composition in optical fibers. Opt. Quantum Electron. **9**, 399–409

1978 Hammond C.R.: Silica-based binary glass systems: wavelength dispersive properties and composition in optical fibers. Opt. Quantum Electron. **10**, 163–170

1985 Hanaizumi O., Miyagi M., Kawakami S.: Wide Y-junctions with low-losses in three-dimensional dielectric optical waveguides. IEEE J. QE-**21**, 168–173

1976 Hannay J.H.: Mode coupling in an elastically deformed optical fiber. Electron. Lett. **12**, 173–174

459

1985 Hardwick N.E., Reynolds M.R.: Finite-element analysis of thermal stresses in the single-mode bonded splice. Opt. Lett. **10**, 241–243

1986 Hardwick N.E., Davies S.T.: Single-mode rapid ribbon splice. Internat. Wire & Cable Sympos. Proceed. 1986, 208–212

1985 Hardy A., Streifer W.: Coupled mode theory of parallel waveguides. J. Lightwave Tech. LT-**3**, 1135–1146

1986 Hardy A., Streifer W.: Coupled modes of multiwaveguide systems and phased arrays. J. Lightwave Tech. LT-**4**, 90–99

1986a Hardy A., Osinski M., Streifer W.: Application of coupled-mode theory to nearly parallel waveguide systems. Electron. Lett. **22**, 1249–1250

1986b Hardy A., Shakir S., Streifer W.: : Coupled-mode equations for two weakly guiding single-mode fibers. Opt. Lett. **11**, 324–326

1982 Harmon R.A.: Polarization stability in long lengths of monomode fiber. Electron. Lett. **18**, 1058–1060

1986 Harris A.J., Castle P.F.: Bend loss measurements on high numerical aperture single-mode fibers as a function of wavelength and bend radius. J. Lightwave Tech. LT-**4**, 34–40

1986 Harstead E.E., Shi Y.C., Klafter L.: Low-loss multifiber optical rotary joint. Conf. Opt. Fiber Commun. (OFC'86), 70–72

1981 Hartog A.H., Payne D.N., Conduit A.J.: Polarization measurements on monomode fibers using optical time-domain reflectometry. Proc. Inst. Electr. Eng. **128**, Pt.H. 168–170

1984 Hartog A.H.: Advances in optical time-domain reflectometry. Symp. Opt. Fiber Measurem. 1984, NBS spec. publ. **683**, 89–94

1984 Hartog A.H., Gold M.P.: On the theory of backscattering in single-mode optical fibers. J. Lightwave Tech. LT-**2**, 76–82

1980 Hashimoto M.: Cutoff frequencies of vector wave modes in cladded inhomogeneous optical fiber. Electron. Lett. **16**, 806–808

1978a Hatakeyama I., Tsuchiya H.: Fusion splices for single-mode optical fibers. IEEE J. QE-**14**, 614–619

1978b Hatakeyama I., Tsuchiya H.: Fusion splices for optical fibers by discharge heating. Appl. Opt. **17**, 1959–1964

1979 Hatakeyama I., Tsuchiya H.: Fusion splices for single-mode optical fibers by discharge heating. Rev. Electr. Commun. Lab. NTT, Japan **27**, 532–542

1985 Hatton W.H., Nishimura M.: New field measurement system for single-mode fiber dispersion utilizing wavelength division multiplexing technique. Electron. Lett. **21**, 1072–1073

1985 Hatton W.H., Nishimura M., Haltiwanger W.: New field measurement system for single-mode fiber dispersion utilizing wavelength division multiplexing technique. Proc. Intern. Wire & Cable Symp. 1985, 142–149

1986a Hatton W.H., Nishimura M.: Temperature dependence of chromatic dispersion of single-mode optical fiber. Conf. Opt. Fiber Commun. (OFC'86), 104–105

1986b Hatton W.H., Nishimura M.: New method for measuring the chromatic dispersion of installed single-mode fibers utilizing wavelength division multiplexing techniques. J. Lightwave Tech. LT-**4**, 1116–1119

1986c Hatton W.H., Nishimura M.: Temperature dependence of chromatic dispersion in single-mode fibers. J. Lightwave Tech. LT-**4**, 1552–1555

1987 Haus H.A., Huang W.P., Kawakami S., Whitaker N.A.: Couple-mode theory of optical waveguides. J. Lightwave Tech. LT-**5**, 16–23

1986 Hayata K., Koshiba M., Suzuki M.: Modal spot size of axially nonsymmetrical fibers. Electron. Lett. **22**, 127–129

1980 Healey P.: Optical time domain reflectometry by photon counting. 6th Europ. Conf. Opt. Commun. 1980, 156–159

1980 Healey P., Hensel P.: Optical time domain reflectometry by photon counting. Electron. Lett. **16**, 631–633

1981a Healey P.: Pulse compression coding in optical time domain reflectometry. 7th Europ. Conf. Opt. Commun. 1981, 5.2-1–2-4

1981b Healey P.: OTDR in monomode fibers at 1.3 μm using a semiconductor laser. Electron. Lett. **17**, 62–64

1981c Healey P.: Multichannel photon-counting backscatter measurements on monomode fiber. Electron. Lett. **17**, 751–752

1982 Healey P., Malyon D.J.: OTDR in single-mode fiber at 1.5 μm using heterodyne detection. Electron. Lett. **18**, 862–863

1982 Healey P., Smith D.R.: OTDR in single-mode fiber at 1.55 μm using a semiconductor laser and PINFET receiver. Electron. Lett. **18**, 959–961

1984a Healey P.: Fading in heterodyne OTDR. Electron. Lett. **20**, 30–32

1984b Healey P.: Optical time domain reflectometry – performance comparison of the analogue and photon counting techniques. Opt. Quantum Electron. **16**, 267–276

1984c Healey P.: Fading rates in coherent OTDR. Electron. Lett. **20**, 443–444

1984 Healey P., Booth R.C., Daymond-John B.E., Nayar B.K.: OTDR in single-mode fiber at 1.5 μm using homodyne detection. Electron. Lett. **20**, 360–362

1985a Healey P.: Statistics of Rayleigh backscatter from a single-mode optical fiber. Electron. Lett. **21**, 226–228

1985b Healey P.: Review of long wavelength single-mode optical fiber reflectometry techniques. J. Lightwave Tech. LT-3, 876–886

1985 Healey P., Faulkner D.W.: New transmissive star couplers for full duplex channels. 11th Europ. Conf. Opt. Commun. 1985, 589–592

1986 Healey P.: Instrumentation principles for optical time domain reflectometry. J. Phys. E.: Sci. Instrum., **19**, 334–341

1986 Healey P., Faulkner D.W.: New transmissive star couplers for full duplex channels. Electron. Lett. **22**, 332–334

1981a Heckmann S.: Modal noise in single-mode fibers. Opt. Lett. **6**, 201–203

1981b Heckmann S.: Modal noise in single-mode fibers operated slightly above cutoff. Electron. Lett. **17**, 499–500

1981 Heckmann S., Brinkmeyer E., Streckert J.: Long-range backscattering experiments in single-mode fibers. Electron. Lett. **6**, 634–635

1983 Heckmann S.: *Analyse des Transmissionsverhaltens optischer Nachrichtensysteme mit Grundmodenfasern* (Dissertation, Universität Wuppertal)

1984 Heckmann S.: Deterioration of the signal quality in realistic single-mode fiber systems. Proc. Int. Wire & Cable Symp. 1984, 266–275

1985 Heckmann S., Rybach J., Brinkmeyer E., Knoechel R.: The future of optical time domain reflectometry: utilization of the coherent detection technique. 34th Int. Wire a. Cable Symp. Cherry Hill 1985, 125–134

1986 Heffner B.L., Kino G.S., Khuri-Yakub B.T., Risk W.P.: Switchable fiber-optic tap using the acousto-optic Bragg interaction. Opt. Lett. **11**, 476–478

1984 Hegarty J., Poulsen S.D., Jackson K.A., Kaminow I.P.: Low-loss single-mode wavelength-division multiplexing with etched fiber arrays. Electron. Lett. **20**, 685–686

1975 Heiblum M., Harris J.H.: Analysis of curved optical waveguides by conformal transformation. IEEE J. QE-11, 75–83

1980 Heitmann W., Day C.R., Zwick U.: Broadband spectral attenuation measurements on optical fibers – an interlaboratory comparison by members of COST 208. 6th Europ. Conf. Opt. Commun. 1980, 148–151

1981 Heitmann W., Day C.R., Zwick U., Mengel F.: Broadband spectral attenuation measurements on optical fibers: an interlaboratory comparison by members of COST 208. Opt. Quantum Electron. **13**, 47–54

1985 Henry, P.S.: Lightwave primer. IEEE J. QE-21, 1862–1879

1985 Henry W.M., Love J.D.: Variational approximations for couplers and Y-junctions. Opt. Quantum Electron. **17**, 359–370

1975 Hensel P.: Simplified optical-fiber breaking machine. Electron. Lett. **11**, 581–582

1986 Hentschel C.: Making accurate fiber-optic power measurement. Laser Focus/Electro-Optics 22(9), 108–119

1983a Herloski R., Marshall S., Antos R.: Gaussian beam ray-equivalent modeling and optical design. Appl. Opt. **22**, 1168–1174

1983b Herloski R., Marshall S., Antos R.: Gaussian beam ray-equivalent modeling and optical design: erratum. Appl. Opt. **22**, 3151

1984 Hevey L.M., Saikkonen S.L., Taylor D.H.: Theoretical splice loss study of single-mode fibers. Proc. SPIE **479**, 48–52

1970 Heyke H.J.: Launching of fiber modes by Gaussian beams. AEÜ, Arch. für Elektron. und Übertragungstech.: Electron. and Commun. **24**, 521–522

461

1984   Heyman E.: On the tunneling hypothesis for ray reflection and transmission at a concave dielectric boundary. IEEE Trans. MTT-**32**, 978–986

1984   Hill K.O., Johnson D.C., Lamont R.G.: Efficient coupling-ratio control in single-mode-fiber biconical-taper couplers. Conf. Opt. Fiber Commun. (OFC'84), 98–99

1985   Hill K.O., Johnson D.C., Lamont R.G.: Wavelength dependence in fused biconical taper splitters: measurement and control. 11th Europ. Conf. Opt. Commun. 1985, 567–570

1987   Hill K.O., Johnson D.C., Bilodeau F., Faucher S.: Fuse-pull-and-taper monomode-fiber directional-couplers: coupling mechanisms and pull signature diagnostics. Conf. Opt. Fiber Commun. (OFC'87), TUQ22

1982   Hillerich B., Weidel E.: Polarization noise in single-mode fiber WDM-systems and its reduction by depolarizing elements. 8th Europ. Conf. Opt. Commun. 1982, 149–153

1983   Hillerich B., Weidel E.: Polarization noise in single-mode fibers and its reduction by depolarizers. Opt. Quantum Electron. **15**, 281–287

1986   Hillerich B.: New analysis of LED to single-mode fiber coupling. Electron. Lett. **22**, 1176–1177

1987   Hillerich B.: Theory of light emitting diode to single-mode fiber coupling. Conf. Opt. Fiber Commun. (OFC'87), MD5

1977   Hirai M., Uchida N.: Melt splice of multimode optical fiber with an electric arc. Electron. Lett. **13**, 123–125

1981   Hirota O., Suematsu Y., Kwok K.: Properties of intensity noises of laser diodes due to reflected waves from single-mode optical fibers and its reduction. IEEE J. QE-**17**, 1014–1020

1986   Hoffe R., Chrostowski J.: Optical pulse compression and breaking in nonlinear fiber couplers. Opt. Commun. **57**, 34–38

1910   Hondros D., Debye P.: Elektromagnetische Wellen an dielektrischen Drähten. Annal. Physik, 4.Folge, **32**, 465–476

1986   Honmou H., Ishikawa R., Ueno H., Kobayashi M.: 1.0 dB low-loss coupling of laser diode to single-mode fiber using a planoconvex graded-index rod lens. Electron. Lett. **22**, 1122–1123

1983   Hooper R.C., Midwinter J.E., Smith D.W., Stanley I.W.: Progress in monomode transmission techniques in the United Kingdom. J. Lightwave Tech. LT-**1**, 596–610

1985   Hooper R.C., Payne D.B., Reeve M.H.: The development of single-mode fiber transmission systems at BTRL. Part 1 – early developments. British Telecommun. Engin. **4**, 74–84

1986   Hooper R.C., Smith D.W.: The development of single-mode fiber transmission systems at BTRL. Part 2 – recent developments. British Telecommun. Engin. **4**, 193–198

1985   Hordvik A., Eriksrud M.: Loss mechanisms in monomode fibers with a loose tube, jelly-filled jacket. Proc. SPIE **584**, 79–85

1986   Hordvik A., Eriksrud M.: Loss mechanism in single-mode fibers jacketed with a loose jelly-filled tube. J. Lightwave Tech. LT-**4**, 1178–1182

1985   Horgan J.: Thwarting the information thieves. IEEE spectrum **22**(7), 30–41

1984   Horiguchi T., Tokuda M.: Optical time domain reflectometer for single-mode fibers. Trans. IECE Japan, E**67**, 509–515

1984   Horiguchi T., Nakazawa M., Tokuda M., Uchida N.: An acoustooptical directional coupler for an optical time-domain reflectometer. J. Lightwave Tech. LT-**2**, 108–115

1985a   Horiguchi T., Suzuki K., Shibata N., Nakazawa M., Seikai S.: A novel technique for reducing polarization noise in optical time domain reflectometers for single-mode fibers. J. Lightwave Tech. LT-**3**, 901–908

1985b   Horiguchi T., Suzuki K., Shibata N., Seikai S.: : Birefringent launching fibers for reducing backscattered power fluctuations in polarization-sensitive optical-time-domain reflectometers. J. Opt. Soc. Am. A **2**, 1698–1704

1985c   Horiguchi T., Tokuda M., Negishi Y.: Chromatic dispersion measurements over a 50-km single-mode fiber. J. Lightwave Tech. LT-**3**, 51–54

1981   Hornung S., Reeve M.H.: Single-mode optical fiber microbending loss in a loose tube coating. Electron. Lett. **17**, 774–775

1982   Hornung S., Doran N.J., Allen R.: Monomode fiber microbending loss measurements and their interpretation. Opt. Quantum Electron. **14**, 359–362

1982   Hosain S.I., Sharma A., Ghatak A.K.: Splice-loss evaluation for single-mode graded-index fibers. Appl. Opt. **21**, 2716–2720

1983 Hosain S.I., Goyal I.C., Ghatak A.K.: Accuracy of scalar approximation for single-mode fibers. Opt. Commun. **47**, 313–316

1981 Hosaka T., Okamoto K., Miya T., Sasaki Y., Edahiro T.: Low-loss single polarization fibers with asymmetrical strain birefringence. Electron. Lett. **17**, 530–531

1983 Hoshikawa M., Toda Y.: "Optical fiber splicing", in *Japan Annual Review in Electronics, Computers, & Telecommunications*, Vol. 5, *Optical Fibers & Devices*, ed. by Y. Suematsu (Ohm, Tokyo, Osaka, Kyoto; North-Holland, Amsterdam, New York, Oxford 1983), pp. 209–218

1978a Hotate K., Okoshi T.: Semiautomatic measurement of refractive-index profiles of single-mode fibers by scattering-pattern method. Trans. IECE Japan **E61**, 202–205

1978b Hotate K., Okoshi T.: Formula giving single-mode limit of optical fiber having arbitrary refractive-index profile. Electron. Lett. **14**, 246–248

1979a Hotate K., Okoshi T.: A formula giving cutoff frequencies of modes in an optical fiber having arbitrary refractive-index profile. Trans. IECE of Japan, **E62**, 1–6

1979b Hotate K., Okoshi T.: Measurement of refractive-index profile and transmission characteristics of a single-mode optical fiber from its exit-radiation pattern. Appl. Opt. **18**, 3265–3271

1983 Hotate K., Kozasa M., Higashiguchi M.: Direct measurement of semiconductor-laser spectrum by stabilized optical-fiber Fabry-Perot interferometers. 9th Europ. Conf. Opt. Commun. 1983, 147–150

1977 Howard A.Q.: Bend radiation in optical fibers. Fiber a. Integr. Optics **1**, 181–196

1986 Howard R.M., Hullett J.L., Jeffery R.D.: Range and accuracy in backscatter measurements. Opt. Quantum Electron. **18**, 291–308

1976 Hsu H.P., Milton A.F.: Flip-chip approach to endfire coupling between single-mode optical fibers and channel waveguides. Electron. Lett. **12**, 404–405

1974 Hudson M.C.: Calculation of the maximum coupling efficiency into multimode optical waveguides. Appl. Opt. **13**, 1029–1033

1985 Hughes R., So V., Vella P.J.: Local launch and monitor for single-mode splicing. Proc. Intern. Wire & Cable Symp. 1985, 397–401

1981a Hullett J.L., Jeffery R.D.: Long-range optical fiber backscatter loss signatures using two-point processing. Conf. Opt. Fiber Commun. (OFC'81), 106–108

1981b Hullett J.L., Jeffery R.D.: Noise in optical fiber backscatter measurement. Opt. Quantum Electron. **13**, 117–124

1982 Hullett J.L., Jeffery R.D.: Long-range optical fiber backscatter loss signatures using two-point processing. Opt. Quantum Electron. **14**, 41–49

1981 Hung-chia H., Zi-Hua W.: Analytical approach to prediction of dispersion properties of step-index single mode optical fibers. Electron. Lett. **17**, 202–204

1983 Hung-chia H.: "Microwave approach to optical waveguides", in *Optical Waveguide Science*, Proc. of the Intern. Sympos. in Kweilin, China, June 20–23, 1983, ed. by H. Hung-chia, A.W. Snyder (Martinus Nijhoff, The Hague) pp. 35–43

1984 Hung-chia H.: *Coupled Mode Theory as Applied to Microwave and Optical Transmission* (VNU Science Press, Utrecht)

1984 Hunsperger R.G.: *Integrated Optics: Theory and Technology*, 2nd ed. Springer Series in Optical Sciences, Vol. 33 (Springer, Berlin, Heidelberg, New York, Tokyo)

1982a Hussey C.D., Pask C.: Theory of the profile-moments description of single-mode fibers. Proc. Inst. Electr. Eng. **129**, Pt. H. 123–134

1982b Hussey C.D., Pask C.: Characterization and design of single-mode optical fibers. Opt. Quantum Electron. **14**, 347–358

1984 Hussey C.D.: Field to dispersion relationships in single-mode fibers. Electron. Lett. **20**, 1051–1052

1985 Hussey C.D., Martinez F.: Approximate analytic forms for the propagation characteristics of single-mode optical fibers. Electron. Lett. **21**, 1103–1104

1984 IEEE Std 812–1984: IEEE Standard definitions of terms relating to fiber optics.

1984 Iga K., Kokubun Y., Oikawa M.: *Fundamental of Microoptics, Distributed-Index, Microlens, and Stacked Planar Optics* (Academic Press, Tokyo, Ohm, Tokyo)

1985 Ihee Y.K., Liou K.-Y., Burrus C.A., Hall K.L.: Linewidth reduction of cleaved-coupled-cavity lasers by optical feedback from a single-mode polarization-preserving fiber external cavity. Electron. Lett. **21**, 1146–1148

1978 Iiyama M., Kamiya T., Yanai H.: Optical field mapping using single-mode optical fibers. Appl. Opt. **17**, 1965–1971

1985 Iizuka K.: *Engineering Optics* (Springer, Berlin, Heidelberg, New York, Tokyo)

1975 Ikeda M., Tateda M., Yoshikiyo H.: Refractive index profile of a graded index fiber: measurement by a reflection method. Appl. Opt. **14**, 814–815

1974 Imai M., Hara E.H.: Excitation of fundamental and low-order modes of optical fiber waveguides by Gaussian beams. 1: Tilted beams. Appl. Opt. **13**, 1893–1899

1975 Imai M., Hara E.H.: Excitation of the fundamental and low-order modes of optical fiber waveguides with Gaussian beams. 2: Offset beams. Appl. Opt. **14**, 169–173

1983 Imon K., Tokuda M.: Axis-alignment method for arc-fusion splice of single-mode fiber using a beam splitter. Opt. Lett. **8**, 502–503

1979 Imoto N., Sugimura A., Daikoku K., Miya T.: Dispersion characteristics of single-mode optical fibers. Rev. Electr. Commun. Lab. NTT, Japan **27**, 515–531

1981 Imoto N., Ikeda M.: Polarization dispersion measurement in long single-mode fibers with zero dispersion wavelength at $1.5\,\mu m$. IEEE J. QE-**17**, 542–545

1982 Imoto N., Tsuchiya H.: "Wavelength and polarization dispersion in single-mode optical fibers", in *Japan Annual Review in Electronics, Computers, & Telecommunications*, Vol. 1, *Optical Fibers & Devices*, ed. by Y. Suematsu (Ohm, Tokyo, Osaka, Kyoto; North-Holland, Amsterdam, New York, Oxford), pp. 274–291

1976 Inada K.: A new graphical method relating to optical fiber attenuation. Opt. Commun. **19**, 437–439

1981 Inada K.: Single-mode fiber measurements. 7th Europ. Conf. Opt. Commun. 1981, 5.1-1-1-8

1982 Inada K., Yamauchi R., Miyamoto M.: Wavelength dependence of geometrical imperfection losses in single-mode fibers. 8th Europ. Conf. Opt. Commun. 1982, 596–600

1984 Inada K.: "Low-loss single-mode fiber and its splicing", in *Japan Annual Review in Electronics, Computers, & Telecommunications*, Vol. 11, *Optical Fibers & Devices*, ed. by Y. Suematsu (Ohm, Tokyo, Osaka, Kyoto; North-Holland, Amsterdam, New York, Oxford), pp. 195–207

1986 Inada K., Watanabe O., Taya H.: Splicing of fibers by the fusion method. IEEE J. Select. Areas Commun. SAC-**4**, 706–713

1985 Inoue K., Toba H., Nosu K.: Tunable optical multi/demultiplexer for optical FDM transmission system. Electron. Lett. **21**, 387–389

1985 Irving D.H., Whitbread T.W., Phillips C.J.E.: Error analysis of the focusing method for optical fiber and preform refractive-index profile measurement. Appl. Opt. **24**, 3128–3133

1983 Ishihara H., Hirooka A., Sato K.: Commercialized single-mode optical fiber cable technique. 9th Europ. Conf. Opt. Commun. 1983, postdeadline paper

1986 Ishihara S., Mitsuhashi Y., Tagawa M., Yamazaki H.: Simple fabrication of an optical fiber mirror. Appl. Opt. **25**, 3982–3983

1985 Ishikawa R., Honmou H., Ueno H., Kobayashi M.: Marked low-loss coupling of laser diode to single-mode fiber by a plan-convex graded-index rod lens. 11th Europ. Conf. Opt. Commun. 1985, 637–640

1986a Ishikura A., Kato Y., Abe T., Miyauchi M.: Optimum fusion splice method for polarization-preserving fibers. Appl. Opt. **25**, 3455–3459

1986b Ishikura A., Kato Y., Miyauchi M.: Taper splice method for single-mode fibers. Appl. Opt. **25**, 3460–3465

1984 Ishio H., Minowa J., Nosu K.: Review and status of wavelength-division multiplexing technology and its application. J. Lightwave Tech. LT-**2**, 448–463

1985 Ito N., Numazaki T.: Optical two-way communication system using a rotary coupler. Appl. Opt. **24**, 2221–2224

1986 Ito T.: Precise measurement of the change in the optical length of a fiber-optic Fabry-Perot interferometer. Appl. Opt. **25**, 1072–1075

1985 Itoh H., Ohmori Y., Nakahara N.: Loss increase due to chemical reactions of hydrogen in silica glass optical fibers. J. Lightwave Tech. LT-**3**, 1100–1104

1981 Iwahashi E.: Trends in long-wavelength single-mode transmission systems and demonstrations in Japan. IEEE J. QE-**17**, 890–896

1982 Iwashita K., Nakagawa K., Nakano Y., Suzuki Y.: Chirp pulse transmission through a single-mode fiber. Electron. Lett. **18**, 873–874

1977   Izawa T., Shibata N., Takeda A.: Optical attenuation in pure and doped fused silica in the IR wavelength region. Appl. Phys. Lett. **31**(1), 33–35

1982   Izutsu M., Nakai Y., Sueta T.: Operation mechanism of the single-mode optical waveguide Y junction. Opt. Lett. **7**, 136–138

1983   Jaccard P., Scheja B., Berthou H., Parriaux O.: A batch fabrication technique for single-mode fiber couplers. 9th Europ. Conf. Opt. Commun. 1983, 409–412

1984   Jaccard P., Scheja B., Berthou H., Cochet F., Parriaux O., Brugger A.: A new technique for low-cost all-fiber device fabrication. Proc. SPIE **479**, 16–19

1980   Jackson D.A., Dandridge A., Sheem S.K.: Measurement of small phase shifts using a single-mode optical fiber interferometer. Opt. Lett. **5**, 139–141

1975   Jackson J.D.: *Classical Electrodynamics*, 2nd ed. (Wiley, New York, London, Sydney, Toronto)

1985   Jackson K.P., Newton S.A., Moslehi B., Tur M., Cutler C.C., Goodman J.W., Shaw H.J.: Optical fiber delay-line signal processing. IEEE Trans. MTT-**33**, 193–210

1986   Jauncey I.M., Reekie L., Mears R.J., Payne D.N., Rowe C.J., Reid D.C.J., Bennion I., Edge C.: Narrow-linewidth fiber laser with integral fiber grating. Electron. Lett. **22**, 987–988

1986   Jedrzejewski K.P., Martinez F., Minelly J.D., Hussey C.D., Payne F.P.: Tapered-beam expander for single-mode optical fiber gap devices. Electron. Lett. **22**, 105–106

1980   Jeffery R.D., Hullett J.L.: N-point processing of optical fiber backscatter signals. Electron. Lett. **16**, 822–823

1986   Jennings D.A., Evenson K.M., Knight D.J.E.: Optical frequency measurements. Proc. IEEE **74**, 168–179

1979   Jeunhomme L.: Dispersion minimization in single-mode fibers between $1.3\,\mu$m and $1.7\,\mu$m. Electron. Lett. **15**, 478–479

1982   Jeunhomme L.: Single-mode fiber design for long haul transmission. IEEE Trans. MTT-**30**, 573–578 and IEEE J. QE-**18**, 727–732

1983   Jeunhomme L.: *Single-mode Fiber Optics. Principles and Applications* (Dekker, New York, Basel)

1984   Jeunhomme L.: Les fibres optiques monomodes: des télécommunications à l'électronique. L'Ondes Electrique **64**(5/6), 68–77

1986   Jeunhomme L.: Les utilisations des fibres monomodes en transmissions à grandes distances. L'Onde Electrique **66**, 103–108

1976   Jocteur R., Tardy A.: Optical fibers splicing with plasma torch and oxhydric microburner. 2nd Europ. Conf. Opt. Commun. 1976, 261–266

1986   Johnson D.C., Hill K.O.: Control of wavelength selectivity of power transfer in fused biconical monomode directional couplers. Appl. Opt. **25**, 3800–3803

1983   Johnson L.M., Leonberger F.J.: Low-loss $LiNbO_3$ waveguide bends with coherent coupling. Opt. Lett. **8**, 111–113

1984   Johnson L.M., Yap D.: Theoretical analysis of coherently coupled optical waveguide bends. Appl. Opt. **23**, 2988–2990

1965   Jones A.L.: Coupling of optical fibers and scattering in fibers. J. Opt. Soc. Am. **55**, 261–271

1985   Joyce W.B., DeLoach, B.C.: Alignment-tolerant optical-fiber tips for laser transmitters. J. Lightwave Tech. LT-**3**, 755–757

1975   Jürgensen K.: Dispersion-optimized optical single-mode glass fiber waveguides. Appl. Opt. **14**, 163–168

1977   Jürgensen K.: Transmission of Gaussian pulses through monomode dielectric optical waveguides. Appl. Opt. **16**, 22–23

1978   Jürgensen K.: Gaussian pulse transmission through monomode fibers, accounting for source linewidth. Appl. Opt. **17**, 2412–2415

1979   Kaiser P., French W.G., Bisbee D.L., Shiever J.W.: Cabling of single-mode fibers. 5th Europ. Conf. Opt. Commun. 1979, 7.4-1–4-4

1982   Kaiser P., Young W.C., Curtis L.: Optical connector measurement aspects, including single-mode connectors. Techn. Dig. Sympos. Opt. Fiber Measur. 1982, NBS Publ. **641**, 123–126

1983   Kaiser P. Optical technology used in the Atlanta single-mode experiment. Proc. SPIE **425**, 127–133

1985 Kaiser P.: Single-mode fiber technology for the subscriber loop. 11th Europ. Conf. Opt. Commun. 1985, invited papers, 125–131

1986 Kaiser P.: Network architecture and systems technology for future broadband ISDN systems. 12th Europ. Conf. Opt. Commun. 1986 Vol. III, 81–85

1983 Kajfez D.: Modal field patterns in dielectric rod waveguide. Microwave Journ. 26(5), 181–192

1986 Kamikawa N., Chang C.: Predicting microbending losses in single-mode fibers. Techn. Dig. Sympos. Opt. Fiber Measur. 1986, NBS spec. publ. 720, 129–132

1981 Kaminow I.P.: Polarization in optical fibers. IEEE J. QE-17, 15–22

1984 Kaminow I.P.: Polarization-maintaining fibers. Appl. Scient. Research 41, 257–270

1986a Kanada T., Franzen D.L.: Optical waveform measurement by optical sampling with a mode-loked laser diode. Opt. Lett. 11, 4–6

1986b Kanada T., Franzen D.L.: Single-mode fiber dispersion measurements using optical sampling with a mode-locked laser diode. Opt. Lett. 11, 330–332

1986 Kanamori H., Yokota H., Tanaka G., Watanabe M., Ishiguro Y., Yoshida I., Kakii T., Itoh S., Asano Y., Tanaka S.: Transmission characteristics and reliability of pure-silica-core single-mode fibers. J. Lightwave Tech. LT-4, 1144–1150

1966 Kao K.C., Hockham G.A.: Dielectric-fiber surface waveguides for optical frequencies. Proc. Inst. Electr. Eng. 113, 1151–1158

1970 Kao K.C., Dyott R.B., Snyder A.W.: Design and analysis of an optical fiber waveguide for communication. Conf. Trunk Telecommunications by Guided Waves, London 1970, IEE publ. 71, 211–217

1961 Kapany N.S., Burke J.J.: Fiber optics. IX. Waveguide effects. J. Opt. Soc. Am. 51, 1067–1078

1965 Kapany N.S., Burke J.J., Frame K.: Radiation characteristics of circular dielectric waveguides. Appl. Opt. 4, 1534–1543

1967 Kapany N.S.: *Fiber Optics, Principles and Applications* (Academic Press, New York, San Francisco, London)

1970 Kapany N.S., Burke J.J., Sawatari T.: Fiber optics. XII. A technique for launching an arbitrary mode on an optical dielectric waveguide. J. Opt. Soc. Am. 60, 1178–1185

1972 Kapany N.S., Burke J.J.: *Optical Waveguides* (Academic Press, New York and London)

1970a Kapron F.P., Keck D.B., Maurer R.D.: Radiation losses in glass optical waveguides. Conf. Trunk Telecommunications by Guided Waves, London 1970, IEE publ. 71, 148–153

1970b Kapron F.P., Keck D.B., Maurer R.D.: Radiation losses in glass optical waveguides. Appl. Phys. Lett. 17, 423–425

1971 Kapron F.P., Keck D.B.: Pulse transmission through a dielectric optical waveguide. Appl. Opt. 10, 1519–1523

1972 Kapron F.P., Maurer R.D., Teter M.P.: Theory of backscattering effects in waveguides. Appl. Opt. 11, 1352–1356

1977 Kapron F.P.: Maximum information capacity of fiber-optic waveguides. Electron. Lett. 13, 96–97

1977 Kapron F.P., Lukowski T.I.: Monomode fiber design: power containment. Appl. Opt. 16, 1465–1466

1979 Kapron F.P.: Baseband response function of monomode fibers. Conf. Opt. Fiber Commun. (OFC'79), 104

1980 Kapron F.P.: Source and modulation effects in monomode fibers. 6th Europ. Conf. Opt. Commun. 1980, 129–132

1981 Kapron F.P., Kneller D.G., Garel-Jones P.M.: Aspects of optical frequency-domain reflectometry. Intern. Conf. Integrat. Opt. Opt. Commun. (IOOC 81), 106–107

1983 Kapron F.P., Lazay P.D.: Monomode fiber measurement techniques and standards. Proc. SPIE 425, 40–48

1984a Kapron F.P.: Critical review of fiber-optic communication technology: optical fibers. Proc. SPIE 512, 2–16

1984b Kapron F.P.: Dispersion-slope parameter for monomode fiber bandwidth. Conf. Opt. Fiber Commun. (OFC'84), 90–92

1984 Kapron F.P., Olson T.C.: Accurate specification of single-mode dispersion measurements. Symp. Opt. Fiber Meas. 1984, NBS spec. publ. 683, 111–114

1985 Kapron F.P.: The evolution of optical fibers. Microwave Journ. 28(4), 111–120

1986 Kapron F.P., Kozikowski C., Olson T.: Novel OTDR effects in determining losses of single-mode fibers and splices. Intern. Wire & Cable Sympos. Proceed. 1986, 338–343

1987 Kapron F.P.: Chromatic dispersion format for single-mode and multimode fibers. Conf. Opt. Fiber Commun. (OFC'87), TUQ2

1978 Karr M.A., Rich T.C., DiDomenico M.: Lightwave fiber tap. Appl. Opt. **17**, 2215–2218

1984 Karstensen H., Wetenkamp L.: Comparison of chromatic dispersion measurements of single-mode optical fibers by spot-size and pulse delay method. Symp. Opt. Fiber Meas. 1984, NBS spec. publ. **683**, 131–134

1985 Karstensen H., Wetenkamp L.: Comparison of chromatic dispersion measurements of single-mode optical fibers by spot-size and pulse delay method. AEÜ, Arch. für Elektron. und Übertragungstech.: Electron. and Commun. **39**, 65–68

1982 Kashyap R., Pantelis P.: Measurement of optical fiber absorption loss: a novel technique. Techn. Dig. Sympos. Opt. Fiber Measur. 1982, NBS Publ. **641**, 67–70

1984 Katagiri T., Tachikura M., Sankawa I.: Optical microscope observation method of a single-mode optical fiber core for precise core-axis alignment. J. Lightwave Tech. LT-**2**, 277–283

1979 Kato Y., Kitayama K., Seikai Sh., Uchida N.: Novel method for measuring cutoff wavelengths of the HE_{21}-, TE_{01}- and TM_{01}-modes. Electron. Lett. **15**, 410–411

1981 Kato Y., Kitayama K., Seikai Sh., Uchida N.: Effective cutoff-wavelength of the LP_{11}-mode in single-mode fiber cables. IEEE J. QE-**17**, 35–39

1982a Kato Y., Kitayama K., Seikai Sh.: Design consideration on broad-band W-type two-mode optical fibers. IEEE Trans. MTT-**30**, 1–5

1982b Kato Y., Seikai S., Shibata N., Tachigami S., Toda Y., Watanabe O.: Arc-fusion splicing of single-mode fibers. 2: A practical splice machine. Appl. Opt. **21**, 1916–1921

1982c Kato Y., Seikai S., Tateda M.: Arc-fusion splicing of single-mode fibers. 1: Optimum splice conditions. Appl. Opt. **21**, 1332–1336

1982d Kato Y., Tanifuji T., Tokuda M., Uchida N.: New optical monitoring method for arc-fusion splice of single-mode fibers and high-precision estimation of splice loss. Electron. Lett. **18**, 972–973

1983 Kato Y., Seikai S., Tanifuji T.: Arc-fusion splice of single-mode fiber cables. Rev. Electr. Commun. Lab. NTT, Japan **31**, 282–289

1984a Kato Y., Seikai S., Tanifuji T.: Arc-fusion splicing of single-mode fibers: an apparatus with an automatic core-axis alignment mechanism and its field trial results. J. Lightwave Tech. LT-**2**, 442–447

1984b Kato Y., Tanifuji T., Kashima N., Arioka R.: Arc-fusion splicing of single-mode fibers. 3: A highly efficient splicing technique. Appl. Opt. **23**, 2654–2659

1985 Kato Y.: Fusion splicing of polarization preserving fibers. Appl. Opt. **24**, 2346–2350

1985 Kato Y., Miyauchi M.: Measuring cutoff wavelength of HE_{21}, TE_{01}, and TM_{01} modes: a novel method using single-polarization launching. Appl. Opt. **24**, 2351–2354

1976 Katsuyama Y., Tokuda M., Uchida N., Nakahara M.: New method for measuring V-value of a single-mode optical fiber. Electron. Lett. **12**, 669–670

1979 Katsuyama Y: Single-mode propagation in 2-mode region of optical fiber by using mode filter. Electron. Lett. **15**, 442–444

1979 Katsuyama Y., Ishida Y., Ishihara K., Miyashita T., Tsuchiya H.: Suitable parameters of single-mode optical fiber. Electron. Lett. **15**, 94–95

1980 Katsuyama Y., Mitsunaga Y., Ishida Y., Ishihara K.: Transmission loss of coated single-mode fiber at low temperatures. Appl. Opt. **19**, 4200–4205

1985 Katsuyama Y., Hatano S., Hogari K., Matsumoto T., Kokubun T.: Single-mode optical fiber ribbon cable. Electron. Lett. **21**, 134–135

1985 Katsuyama Y., Hatano S., Hogari K., Matsumoto T., Maekawa E.: Single-mode optical fiber ribbon cable for subscriber line. 11th Europ. Conf. Opt. Commun. 1985, 383–386

1986 Kaufmann H., Buchmann P., Hirter R., Melchior H., Guekos G.: Self-adjusted permanent attachment of fibers to GaAs waveguide components. Electron. Lett. **22**, 642–644

1982a Kawachi M., Edahiro T., Toba H.: Microlens formation on VAD single-mode fiber ends. Electron. Lett. **18**, 71–72

1982b Kawachi M., Kawasaki B.S., Hill K.O.: Fabrication of single-polarization single-mode-fiber couplers. Electron. Lett. **18**, 962–964

1975 Kawakami S.: Relation between dispersion and power-flow distribution in a dielectric waveguide. J. Opt. Soc. Am. **65**, 41–45

1975 Kawakami S., Miyagi M., Nishida S.: Bending losses of dielectric slab optical waveguide with double or multiple claddings: theory. Appl. Opt. **14**, 2588–2597

1976   Kawakami S.: Mode conversion losses of randomly bent, singly and doubly clad wave-guides for single-mode transmission. Appl. Opt. **15**, 2778–2784

1976   Kawakami S., Nishida S., Sumi M.: Transmission characteristics of W-type optical fibers. Proc. Inst. Electr. Eng. **123**, 586–590

1985   Kawakami S., Baba K.: Field distribution near an abrupt bend in single-mode wave-guides: a simple model. Appl. Opt. **24**, 3643–3647

1978   Kawana A., Kawachi M., Miyashita T., Saruwatari M., Asatani K., Yamada J., Oe K.: Pulse broadening in long-span single-mode fibers around a material dispersion free wavelength. Opt. Lett. **2**, 106–108

1980   Kawana A., Miya T., Imoto N., Tsuchiya H.: Pulse broadening in long-span dispersion-free single-mode fibers at 1.5 μm. Electron. Lett. **16**, 188–189

1982   Kawana A., Hosaka T., Miya T.: Uniform bending losses of single-mode fibers. Trans. IECE of Japan **E65**, 331–336

1985a   Kawano K., Mitomi O., Saruwatari M.: Combination lens method for coupling a laser diode to a single-mode fiber. Appl. Opt. **24**, 984–989

1985b   Kawano K., Saruwatari M., Mitomi O.: A new confocal combination lens method for a laser-diode module using a single-mode fiber. J. Lightwave Tech. **LT-3**, 739–745

1986   Kawano K.: Coupling characteristics of lens systems for laser diode modules using single-mode fiber. Appl. Opt. **25**, 2600–2605

1986   Kawano K., Mitomi O., Saruwatari M.: Laser diode module for single-mode fiber based on new confocal combination lens method. J. Lightwave Tech. **LT-4**, 1407–1413

1978   Kawasaki B.S., Hill K.O., Johnson D.C., Fujii Y.: Narrow-band Bragg reflectors in optical fibers. Opt. Lett. **3**, 66–68

1981a   Kawasaki B.S., Hill K.O., Johnson D.C.: Optical time domain reflectometer for single-mode fiber at selectable wavelengths. Appl. Phys. Lett. **38**(10), 740–742

1981b   Kawasaki B.S., Hill K.O., Lamont R.G.: Biconical-taper single-mode fiber coupler. Opt. Lett. **6**, 327–328

1983a   Kawasaki B.S., Johnson D.C., Hill K.O.: Configurations, performance, and applications of biconical taper optical fiber coupling structures. Canadian J. Phys. **61**, 352–360

1983b   Kawasaki B.S., Kawachi M., Hill K.O., Johnson D.C.: A single-mode fiber coupler with a variable coupling ratio. J. Lightwave Tech. **LT-1**, 176–178

1983   Kawata O., Hoshino K., Ishihara K.: Low-loss single-mode fiber splicing technique using core direct monitoring. Electron. Lett. **19**, 1048–1049

1984   Kawata O.: Statistical analysis of fusion splice losses for single-mode fibers. Appl. Opt. **23**, 3289–3293

1984   Kawata O., Hoshino K., Miyajima Y., Ohnishi M., Ishihara K.: A splicing and inspection technique for single-mode fibers using direct core monitoring. J. Lightwave Tech. **LT-2**, 185–191

1981   Kayoun P., Puech C., Papuchon M., Arditty H.J.: Improved coupling between laser diode and single-mode fiber tipped with a chemically etched self-centred diffracting element. Electron. Lett. **17**, 400–402

1972   Keck D.B., Tynes A.R.: Spectral response of low-loss optical waveguides. Appl. Opt. **11**, 1502–1506

1972   Keck D.B., Schultz P.C., Zimar F.: Attenuation of multimode glass optical waveguides. Appl. Phys. Lett. **21**(5), 215–217

1973   Keck D.B., Maurer R.D., Schultz P.C.: On the ultimate lower limit of attenuation in glass optical waveguides. Appl. Phys. Lett. **22**, 307–309

1983a   Keck D.B.: Single-mode fibers outperform multimode cables. IEEE spectrum **20**(3), 30–37

1983b   Keck D.B.: Fibre monomodali per telecomunicazioni ottiche. L'Elettrotecnica **70**, 911–919

1984a   Keil R., Klement E., Mathyssek K., Wittmann J.: Experimental investigation of the beam spot size radius in single-mode fiber tapers. Electron. Lett. **20**, 621–622

1984b   Keil R., Mathyssek K., Klement E., Auracher F.: Coupling between semiconductor laser diodes and single-mode optical fibers. Siemens Forsch.- u. Entwickl.-Ber. **13**, 284–288

1983   Kersten R.Th.: *Einführung in die Optische Nachrichtentechnik* (Springer, Berlin, Heidelberg, New York)

1979a   Khoe G.D.: New coupling techniques for single-mode optical fiber transmission systems. 5th Europ. Conf. Opt. Commun. 1979, 6.1.1–1.4

1979b Khoe G.D.: Practical machine for electric arc splicing of optical fibers in the field. Electron. Lett. **15**, 152–153

1980 Khoe G.D., Kuyt G., Luijendijk J.A: Automatic fabrication of optical fiber ends: perpendicular fracture mirrors in glass fibers, coated glass fibers, and plastic-clad fibers. 6th Europ. Conf. Opt. Commun. 1980, 286–289

1981a Khoe G.D., Kuyt G., Luijendijk J.A: Optical fiber end preparation: a new method for producing perpendicular fractures in glass fibers, coated-glass fibers, and plastic-clad fibers. Appl. Opt. **20**, 707–714

1981b Khoe G.D., Kuyt G., van Leest J.H.F.M., Luijendijk J.A:: A simple plug-in single-mode fiber connector consisting of low precision parts. 7th Europ. Conf. Opt. Commun. 1981, 7.4.1–4.4

1982a Khoe G.D., van Leest J.H.F.M., Rittich D., Schmidt B.: Compact single-mode fiber connector with an average loss of 0.3 dB without index match. 8th Europ. Conf. Opt. Commun. 1982, 341–343

1982b Khoe G.D., van Leest J.H.F.M., Luijendijk J.A.: Single-mode fiber connectors using core-centered ferrules. IEEE J. QE-**18**, 1573–1580

1983a Khoe G.D., Luijendijk J.A., Jacobs A.C.: Application of UV-excited fluorescence for the preparation of single-mode fiber connectors and splices. 9th Europ. Conf. Opt. Commun. 1983, 413–416

1983b Khoe G.D., Poulissen J., de Vrieze H.M.: Efficient coupling of laser diodes to tapered monomode fibers with high-index end. Electron. Lett. **19**, 205–207

1984a Khoe G.D., Kock H.G., Küppers D., Poulissen J.H.F.M., de Vrieze H.M.: Progress in monomode optical-fiber interconnection devices. J. Lightwave Tech. LT-**2**, 217–227

1984b Khoe G.D., Meuleman L.J., Poulissen J.: Laser diode devices for coherent fiber optics and special applications. 10th Europ. Conf. Opt. Commun. 1984, 134–135

1985 Khoe G.D., Dieleman A.H.: TTOSS, a subscriber network for direct detection and coherent systems. 11th Europ. Conf. Opt. Commun. 1985, 479–482

1985 Khoe G.D., Kock H.G.: Laser-to-monomode-fiber coupling and encapsulation in a modified TO-5 package. J. Lightwave Tech. LT-**3**, 1315–1320

1985 Khoe G.D., Valster A., Bachmann P.: Coupling of laser diodes and side emitting LED's to flat dispersion quadruple clad monomode fibers at 1300 and 1530 nm. 11th Europ. Conf. Opt. Commun. 1985, 641–644

1986 Khoe G.D., Dieleman A.H.: TTOSS, an integrated subscriber system for direct and coherent detection. J. Lightwave Tech. LT-**4**, 778–784

1986 Khoe G.D., Luijendijk J.A., Vroomen L.J.C.: Arc-welded monomode fiber splices made with the aid of local injection and detection of blue light. J. Lightwave Tech. LT-**4**, 1219–1222

1981 Kim E.M., Franzen D.L.: Measurement of far-field and near-field radiation patterns from optical fibers. Nat. Bureau of Standards Spec. Publ. **1032**, 1–40

1983 Kim E.M., Franzen D.L.: Measurement of near-field radiation patterns from optical fibers. Nat. Bureau of Standards Spec. Publ. **637**, Vol. 2, 101–186

1977 Kimura T., Daikoku K.: A proposal on optical fiber transmission systems in a low-loss 1.0–1.4 μm wavelength region. Opt. Quantum Electron. **9**, 33–42

1978 Kimura T., Saruwatari M., Yamada J., Uehara S., Miyashita T.: Optical fiber (800 Mb/s) transmission experiment at 1.05 μm. Appl. Opt. **17**, 2420–2426

1979 Kimura T.: Single-mode systems and components for longer wavelengths. IEEE Trans. CAS-**26**, 987–1010

1980 Kimura T.: Single-mode digital transmission technology. Proc. IEEE **68**, 1263–1268

1982 Kimura T., Yamada J.: "Single-mode digital transmission technology", in *Japan Annual Review in Electronics, Computers, & Telecommunications*, Vol. 1, *Optical Fibers & Devices*, ed. by Y. Suematsu (Ohm, Tokyo, Osaka, Kyoto; North-Holland, Amsterdam, New York, Oxford), pp. 332–350

1985a Kimura T., Noda K., Shimizu H.: Distance resolution characterization of fiber reflectometry using sampling phase shifting digital wave memory. J. Opt. Commun. **6**, 2–7

1985b Kimura T., Shimizu H., Miyagi H., Noda K.: Time division multiplexed OTDR system. Electron. Lett. **21**, 1070–1071

1978 Kinoshita K., Egashira K.: Optical fiber end preparation using a CO_2 laser. Appl. Opt. **17**, 1210–1212

1979 Kinoshita K., Kobayashi M.: End preparation and fusion splicing of an optical fiber array with a CO_2 laser. Appl. Opt. **18**, 3256–3260

1986 Kinoshita T., Sano K., Yoneda E.: Tunable 8-channel wavelength demultiplexer using an acousto-optic light deflector. Electron. Lett. **22**, 669–670

1983 Kist R., Sohler W.: Fiber-optic spectrum analyzer. J. Lightwave Tech. LT-**1**, 105–110

1986 Kitano I., Ueno H., Toyama M.: Gradient-index lens for low-loss coupling of a laser diode to single-mode fiber. Appl. Opt. **25**, 3336–3339

1979 Kitayama K., Kato Y., Seikai S., Uchida N., Ikeda M.: Experimental verification of modal dispersion free characteristics in a two-mode optical fiber. IEEE J. QE-**15**, 6–8

1980 Kitayama K., Kato Y., Seikai S., Uchida N., Akiyama M., Fukuda O.: Transmission characteristic measurement of two-mode optical fiber with a nearly optimum index-profile. IEEE Trans. MTT-**28**, 604–608

1981 Kitayama K., Kato Y., Seikai S., Uchida N.: Structural optimization for two-mode fiber: theory and experiment. IEEE J. QE-**17**, 1057–1063

1982 Kitayama K., Kato Y., Seikai S., Uchida N., Akiyama M.: Transmission bandwidth of the two-mode fiber link. IEEE J. QE-**18**, 1871–1876

1984 Kitayama K., Ohashi M., Ishida Y.: Length dependence of effective cutoff wavelength in single-mode fibers. J. Lightwave Tech. LT-**2**, 629–634

1985 Kitayama K., Ishida Y.: Wavelength-selective coupling of two-core optical fiber: application and design. J. Opt. Soc. Am. A **1**, 90–94

1985 Kitayama K., Shibata N., Ohashi M.: Two-core optical fibers: experiment. J. Opt. Soc. Am. A **2**, 84–89

1984 Kitayama Y., Tanaka S.: Effective cutoff-wavelength of single-mode fiber. Conf. Opt. Fiber Commun. (OFC'84), 80–82

1985 Kitayama Y., Tanaka S.: Length dependence of LP_{11}-mode cutoff and its influence on the chromatic dispersion measurement by the phase shift method. Proc. SPIE **584**, 229–234

1982 Klein K.-F., Heinlein W.E.: Orientation and polarization dependent cutoff wavelengths in elliptical-core single-mode fibers. Electron. Lett **18**, 640–641

1985a Klemas A.T., Reed W.A., Shenk D.S., Saifi M.: Analysis of mode-field radius calculated from far-field radiation patterns. Conf. Opt. Fiber Commun. (OFC' 85), 30–31

1985b Klemas A.T., Shenk D.S., Reed W.A., Saifi M.A.: Analysis of mode field radius calculated from single-mode fiber far-field radiation patterns. J. Lightwave Tech. LT-**3**, 967–970

1979 Kobayashi K., Ishikawa R., Minemura K., Sugimoto S.: Micro-optic devices for fiber-optic communications. Fiber a. Integr. Optics **2**, 1–17

1984 Koch T.L., Bowers J.E.: Nature of wavelength chirping in directly modulated semiconductor lasers. Electron. Lett. **20**, 1038–1040

1986 Koch T.L., Corvini P.J.: Studies of high-bit-rate dispersive optical-fiber transmission using single-frequency lasers. Conf. Opt. Fiber Commun. (OFC'86), 32–34

1984 Köster W.: Wellenlängenmultiplex mit Monomodefasern durch Energietransfer zwischen den Mantelfeldern. Frequenz **38**, 273–277

1964a Kogelnik H.: Coupling and conversion coefficients for optical modes. Symp. on Quasi-Optics, Brooklyn, 333–347

1964b Kogelnik H.: Matching of optical modes. Bell Syst. Tech. J. **43**, 334–337

1965a Kogelnik H.: Imaging of optical modes – resonators with internal lenses. Bell Syst. Tech. J. **44**, 455–494

1965b Kogelnik H.: On the propagation of Gaussian beams of light through lenslike media including those with a loss or gain variation. Appl. Opt. **4**, 1562–1569

1966 Kogelnik H., Li T.: Laser beams and resonators. Proc. IEEE **54**, 1312–1329 and Appl. Opt. **5**, 1550–1567

1974 Kogelnik H., Weber H.P.: Rays, stored energy, and power flow in dielectric waveguides. J. Opt. Soc. Am. **64**, 174–185

1975 Kogelnik H.: "Theory of dielectric waveguides", in *Integrated Optics*, ed. by T. Tamir, Topics in Applied Physics, Vol. 7 (Springer, Berlin, Heidelberg, New York), pp. 13–81

1976 Kohanzadeh Y.: Hot splices of optical waveguide fibers. Appl. Opt. **15**, 793–795

1980 Kokubun Y., Iga K.: Formulas for TE_{01} cutoff in optical fibers with arbitrary index profile. J. Opt. Soc. Am. **79**, 36–40

1982 Kokubun Y., Iga K.: Single-mode condition of optical fibers with axially symmetric refractive-index distribution. Radio Science **17**, 43–49

1986 Komatsu K., Kondo M., Ohta Y.: Titanium/magnesium double diffusion method for efficient fiber-$LiNbO_3$ waveguide coupling. Electron. Lett. **22**, 881–882

1984 Kopera P.M., Krueger H.H.A., Tekippe V.J., Wuensch D.L.: Performance evaluation of single-mode couplers. Proc. SPIE **479**, 9–15

1985 Korotky S.K., Marcatili E.A.J., Veselka J.J., Bosworth R.H.: "Greatly reduced losses for small-radius bends in Ti:LiNbO$_3$ waveguides", in *Integrated Optics*, ed. by H.P. Nolting and R. Ulrich, Springer Series in Optical Sciences (Springer, Berlin, Heidelberg, New York), pp. 207–209

1986 Korotky S.K.: Three-space representation of phase-mismatch switching in coupled two-state optical systems. IEEE J. QE-**22**, 952–958

1986 Korotky S.K., Marcatili E.A.J., Veselka J.J., Bosworth R.H.: Greatly reduced losses for small-radius bends in Ti:LiNbO$_3$ waveguides. Appl. Phys. Lett. **48**(2), 92–94

1986 Kotrotsios G., Parriaux O., Cochet F.: Geometric representation of degenerate coupling between two guided waves. Optik **72**, 82–84

1985 Koyama F., Suematsu Y.: Analysis of dynamic spectral width of dynamic-single-mode (DSM) lasers and related transmission bandwidth of single-mode fibers. IEEE J. QE-**21**, 292–297

1985 Krause D., Paquet V., Siefert W.: A novel plasma-impulse-CVD process for the preparation of fiber preforms. 11th Europ. Conf. Opt. Commun. 1985, 7–9

1985 Krause J.T., Kurkjian C.R.: Fiber splices with 'perfect fiber' strengths of 5.5 GPa, $\nu<0.01$. Electron. Lett. **21**, 533–535

1985 Krause J.T., Reed W.A., Walker K.L.: Splice loss of single-mode fibers as related to fusion time, temperature, and index profile alteration. 11th Europ. Conf. Opt. Commun. 1985, 629–632

1986 Krause J.T.: Ultrahigh-strength fiber splices by modified H$_2$/O$_2$ flame fusion. Electron. Lett. **22**, 1075–1077

1986 Krause J.T., Reed W.A., Walker K.L.: Splice loss of single-mode fiber as related to fusion time, temperature, and index profile alteration. J. Lightwave Tech. LT-**4**, 837–840

1983 Krivoshlykov S.G., Petrov N.I., Sissakian I.N.: Mode energy transformation between two connected multimode general square-law-index optical waveguides. Opt. Quantum Electron. **15**, 193–207

1983 Krüger R.: Optische Justiereinrichtung zum Einkoppeln eines Laserstrahles in eine Monomode-Faser. Feinwerktechnik & Messtechnik **91**, 184–186

1980 Krumbholz D., Brinkmeyer E., Neumann E.-G.: Core/cladding power distribution, propagation constant, and group delay: simple relation for power-law graded-index fibers. J. Opt. Soc. Am. **70**, 179–183

1970 Krumpholz O.: Mode propagation in fibers: discrepancies between theory and experiment. Conf. Trunk Telecommunications by Guided Waves, London 1970, IEE publ. **71**, 56–61

1971 Krumpholz O.: Modenreine Glasfaser-Lichtwellenleiter. Wiss. Ber. AEG-Telefunken **44**, 64–70

1975 Krumpholz O.: Light coupling into monomode fibers. Wiss. Ber. AEG-Telefunken **48**, 90–94

1981 Krumpholz O., Westermann F.: Power coupling between monomode fibers and semiconductor lasers with strong astigmatism. 7th Europ. Conf. Opt. Commun. 1981, 7.7-1–7-4

1985a Krumpholz O.: Teilnehmeranschluß mit Monomodefaser und LED. NTG-Fachberichte **89**, 70–73

1985b Krumpholz O.: Subscriber links using single-mode fibers and LEDs. 11th Europ. Conf. Opt. Commun. 1985, invited papers, 133–139

1982 Ku R.T., Dufft W.H.: Hemispherical microlens coupling of semiconductor laser to single-mode fiber. Conf. Opt. Fiber Commun. (OFC'82), 60–62

1984 Ku R.T.: Progress in efficient/reliable semiconductor laser-to-single-mode fiber coupler development. Conf. Opt. Fiber Commun. (OFC'84), 4–6

1985 Kühne R., Petermann K., Majewski A.: Limits for microbending loss in multiple clad fibers. 11th Europ. Conf. Opt. Commun. 1985, 313–316

1975a Kuester E.F., Chang D.C.: Nondegenerate surface-wave mode coupling between dielectric waveguides. IEEE Trans. MTT-**23**, 877–882

1975b Kuester E.F., Chang D.C.: Surface-wave radiation loss from curved dielectric slabs and fibers. IEEE J. QE-**11**, 903–907

1977 Kuester E.F.: An alternative expression for the curvature loss of a dielectric wave guide and its application to the rectangular dielectric channel. Radio Science **12**, 573–578

1984   Kuester E.F.: Generalization of the partial-power law (Brown's identity) to waveguides with lossy media. Electron. Lett. **20**, 456–457

1974   Kuhn M.H.: The influence of the refractive index step due to the finite cladding of homogeneous fibers on the hybrid properties of modes. AEÜ, Arch. für Elektron. und Übertragungstech.: Electron. and Commun. **28**, 393–401

1975a Kuhn M.H.: Optimum attenuation of cladding modes in homogeneous single-mode fibers. AEÜ, Arch. für Elektron. und Übertragungstech.: Electron. and Commun. **29**, 201–204

1975b Kuhn M.H.: Curvature loss in single-mode fibers with lossy jacket. AEÜ, Arch. für Elektron. und Übertragungstech.: Electron. and Commun. **29**, 400–402

1976   Kuhn M.H.: Erratum: Curvature loss in single-mode fibers with lossy jacket. AEÜ, Arch. für Elektron. und Übertragungstech.: Electron. and Commun. **30**, 124

1978   Kumar A., Chandra R., Sammut R.A., Ghatak A.K.: Cutoff frequencies of a parabolic-core W-type fiber. Electron. Lett. **14**, 676–678

1982   Kummer R.B., Fleming St.R.: Monomode optical fiber splice loss: combined effects of misalignments and spot size mismatch. Conf. Opt. Fiber Commun. (OFC'82), 44–45

1977   Kurochi N., Ushirogawa A., Morimoto Y., Shimozone M.: A development study on design and fabrication of an optical fiber connector. 3rd Europ. Conf. Opt. Commun. 1977, 97–99

1980a Kuwahara H., Sasaki M., Tokoyo N.: Efficient coupling from semiconductor lasers into single-mode fibers with tapered hemispherical ends. Appl. Opt. **19**, 2578–2583

1980b Kuwahara H., Sasaki M., Tokoyo N., Saruwatari M., Nakagawa K.: Efficient and reflection insensitive coupling from semiconductor lasers into tapered hemispherical-end single-mode fibers. 6th Europ. Conf. Opt. Commun. 1980, 191–194

1981   Kuwahara H., Goto M.: Generation of harmonic distortion at fiber connectors. Electron. Lett. **17**, 626–627

1983   Kuwahara H., Onoda Y., Goto M., Nakagami T.: Reflected light in the coupling of semiconductor lasers with tapered hemispherical end fibers. Appl. Opt. **22**, 2732–2738

1981   Kuyt G., Luijendijk J.A., Khoe G.D., Kock H.G.: Metal sealed hermetic laser diode package suitable for direct single-mode fiber coupling. 7th Europ. Conf. Opt. Commun. 1981, P13.1–13.4

1983   Kuznetsov M., Haus H.A.: Radiation loss in dielectric waveguide structures by the volume current method. IEEE J. QE-19, 1505–1514

1986a Lacroix S., Gonthier F., Bourbonnais R., Black R.J., Bures J., Lapierre J.: Abruptly tapered fibers. 12th Europ. Conf. Opt. Commun. 1986, 191–194

1986b Lacroix S., Gonthier F., Bures J.: All-fiber wavelength filter from succesive biconical tapers. Opt. Lett. **11**, 671–673

1986c Lacroix S., Black R.J., Veilleux C., Lapierre J.: Tapered single-mode fibers: external refractive-index dependence. Appl. Opt. **25**, 2468–2469

1986d Lacroix S., Bourbonnais R., Gonthier F., Bures J.: Tapered monomode optical fibers: understanding large power transfer. Appl. Opt. **25**, 4421–4425

1981   Lam D.K.W., Garside B.K.: Characterization of single-mode optical fiber filters. Appl. Opt. **20**, 440–445

1985a Lamont R.G., Hill K.O., Johnson D.C.: Tuned-port twin biconical-taper fiber splitters: fabrication from dissimilar low-mode-number fibers. Opt. Lett. **10**, 46–48

1985b Lamont R.G., Johnson D.C., Hill K.O.: Power transfer in fused biconical-taper single-mode fiber couplers: dependence on external refractive index. Appl. Opt. **24**, 327–332

1980   Lang R., Kobayashi K.: External optical feedback effects on semiconductor injection laser properties. IEEE J. QE-16, 347–355

1985   Larner D.S.: Effects of dispersion on a 565 Mb/s optical fiber laboratory system. Electron. Lett. **21**, 187–189

1985   Larner D.S., Bhagavatula V.A.: Dispersion reduction in single-mode fiber links. Electron. Lett. **21**, 1171–1172

1965   Larsen H.: Dielektrische Wellenleiter bei optischen Frequenzen. AEÜ, Arch. für Elektron. und Übertragungstech.: Electron. and Commun. **19**, 535–540

1983   Laude J.P., Flamand J., Gautherin J.C., Lepere D., Gacoin P., Bos F., Lerner J.: Stimax, a grating multiplexer for monomode or multimode fibers. 9th Europ. Conf. Opt. Commun. 1983, 417–420

1984 Laude J.P.: Les multiplexeurs de longeur d'onde en télécommunication optique. J. Optics (Paris) **15**, 419–423

1984 Lawson C.M., Kopera P.M., Hsu T.Y., Tekippe V.J.: In-line single-mode wavelength division multiplexer/demultiplexer. Electron. Lett. **20**, 963–964

1975 Lax M., Louisell W.H., McKnight W.B.: From Maxwell to paraxial wave optics. Phys. Review A **11**, 1365–1370

1980 Lazay P.D.: Effect of curvature on the cutoff wavelength of single-mode fibers. NBS spec. publ. **597**, 93–95

1982 Lazay P.D., Pearson A.D.: Developments in single-mode fiber design, materials, and performance at Bell Laboratories. IEEE Trans. MTT-**30**, 350–356, and IEEE J. QE-**18**, 504–510

1982 Leach J.S., Cannell G.J., Robertson A.J., Gurton P.: Low-loss splicing of a 62.4 km single-mode fiber link. Electron. Lett. **18**, 697–698

1982 Lecoy P., Richter H.: Optimization of the coupling between an astigmatic laser-diode and a single-mode fiber with an elliptical microlens. Opto'82, 124–126

1985 Lee K.S., Barnes F.S.: Microlenses on the end of single-mode optical fibers for laser applications. Appl. Opt. **24**, 3134–3139

1985 Leminger O.G., Zengerle R.: Determination of single-mode fiber coupler design parameters from loss measurements. J. Lightwave Tech. LT-**3**, 864–867

1983 Lenahan T.A.: Calculation of modes in an optical fiber using the finite element method and EISPACK. Bell Syst. Tech. J. **62**, 2663–2694

1985 Le Noane G., Bizeul J.C., Mehadji K., Le Marer R., Calevo R.: Successful mass splicing of single-mode fibers: the flat design. 11th Europ. Conf. Opt. Commun. 1985, 633–636

1985a Levine B.F., Bethea C.G., Campbell J.C.: 1.52 μm room-temperature photon-counting optical time domain reflectometer. Electron. Lett. **21**, 194–196

1985b Levine B.F., Bethea C.G., Campbell J.C.: Room-temperature 1.3-μm optical time domain reflectometer using a photon counting InGaAs/InP avalanche detector. Appl. Phys. Lett. **46**(4), 333–335

1985c Levine B.F., Bethea C.G., Cohen L.G., Campbell J.C., Morris G.D.: Optical time reflectometer using a photon-counting InGaAs/InP avalanche photodiode at 1.3 μm. Electron. Lett. **21**, 83–84

1973 Lewin L.: Local form of the radiation condition: application to curved dielectric structures. Electron. Lett. **9**, 468–469

1974 Lewin L.: Radiation from curved dielectric slabs and fibers. IEEE Trans. MTT-**22**, 718–727

1975 Lewin L.: Correction to "Radiation from curved dielectric slabs and fibers". IEEE Trans. MTT-**23**, 779

1976 Lewin L.: Polarization degeneracy in bent optical fiber. Electron. Lett. **12**, 465–466

1977 Lewin L., Chang D.C., Kuester E.F.: *Electromagnetic waves and curved structures* (Peregrinus, Stevenage)

1980 Li T.: Structures, parameters, and transmission properties of optical fibers. Proc. IEEE **68**, 1175–1180

1983 Li T.: Advances in optical fiber communications: an historical perspective. IEEE J. SAC-**1**, 356–372

1981 Liao F.J., Boyd J.T.: Single-mode fiber coupler. Appl. Opt. **20**, 2731–2734

1986 Lieber W., Düppre Th., Fassian B.: Hochauflösendes Meßsystem zur interferometrischen Bestimmung der chromatischen Dispersion von Einmodenfaser. Opto Elektronik Magazin **2**(6), 532–539

1982 Lilly C.J.: The state of the art and application of optical fiber systems operating at the longer wavelengths of 1300-1600 nm. 8th Europ. Conf. Opt. Commun. 1982, 17–24

1984 Lilly C.J., Walker S.D.: The design and performance of digital optical fiber systems. The Radio a. Electron. Eng. **54**, 179–191

1979 Lim T.K., Garside B.K., Marton J.P.: Guided modes in fibers with parabolic-index core and homogeneous cladding. Opt. Quantum Electron. **11**, 329–344

1977 Lin C., Stolen R.H., Cohen L.G.: A tunable 1.1-μm fiber Raman oscillator. Appl. Phys. Lett. **31**, 97–99

1978 Lin C., Cohen L.G.: Pulse delay measurements in the zero-material-dispersion region for germanium- and phosporus-doped silica fibers. Electron. Lett. **14**, 170–172

1980a Lin C., Cohen L.G., French W.G., Presby H.M.: Measuring dispersion in single-mode fibers in the 1.1–1.3 μm spectral region – a pulse synchronization technique. IEEE J. QE-16, 33–36

1980b Lin C., Kogelnik H., Cohen L.G.: Optical pulse equalization and low dispersion transmission in single-mode fibers in the 1.3–1.7 μm spectral region. 6th Europ. Conf. Opt. Commun. 1980, 91–94

1980c Lin C., Kogelnik H., Cohen L.G.: Optical-pulse equalization of low-dispersion transmission in single-mode fibers in the 1.3–1.7 μm spectral region. Opt. Lett. 5, 476–478

1981 Lin C., Marcuse D.: Optimum optical pulse width for high bandwidth single-mode fiber transmission. Electron. Lett. 17, 54–55

1982 Lin C., Tynes A.R., Tomita A., Liu P.L.: Pulse delay measurements in single-mode fibers using picosecond InGaAsP injections lasers in the 1.3-μm spectral region. Conf. Opt. Fiber Commun. (OFC'82), 28–30

1983 Lin C., Tynes A.R., Tomita A., Liu P.L., Philen D.L.: Chromatic dispersion measurements in single-mode fibers using picosecond InGaAsP injection lasers in the 1.2–1.5 μm spectral region. Bell Syst. Tech. J. 62, 457–462

1986 Lin C.: Nonlinear optics in fibers for fiber measurements and special device functions. J. Lightwave Tech. LT-4, 1103–1115

1985 Lin S.H., Wang S.Y., Newton S.A., Houng Y.M.: Low-loss GaAs/GaAlAs strip-loaded waveguides with high coupling efficiency to single-mode fibers. Electron. Lett. 21, 597–598

1984 Linke R.A.: Transient chirping in single-frequency lasers: lightwave system consequences. Electron. Lett. 20, 472–474

1983 Lipson J., Harvey G.T.: Low-loss wavelength-division multiplexing (WDM) devices for single-mode systems. J. Lightwave Tech. LT-1, 387–390

1985 Lipson J., Minford W.J., Murphy E.J., Rice T.C., Linke R.A., Harvey G.T.: A six-channel wavelength multiplexer and demultiplexer for single-mode systems. J. Lightwave Tech. LT-3, 1159–1163

1985 Lissberger P.H., Roy A.K.: Narrowband position-tuned multilayer interference filter for use in single-mode fiber systems. Electron. Lett. 21, 798–799

1984 Liu P., Ogawa K.: Statistical measurements as a way to study mode partition in injection lasers. J. Lightwave Tech. LT-2, 44–48

1977 Loewen E.G., Nevière, M., Maystre D.: Grating efficiency theory as it applies to blazed and holographic gratings. Appl. Opt. 16, 2711–2721

1955 Louisell W.H.: Analysis of the single tapered mode coupler. Bell Syst. Tech. J. 34, 853–870

1979 Love J.D.: Power series solution of the scalar wave equation for cladded, power-law profiles of arbitrary exponent. Opt. Quantum Electron. 11, 464–466

1984 Love J.D., Hussey C.D.: Variational approximations for higher-order modes of weakly guiding fibers. Opt. Quantum Electron. 16, 41–48

1984a Love J.D., Ankiewicz A.: Cutoff in single-mode optical fiber couplers. Electron. Lett. 20, 362–363

1984b Love J.D., Ankiewicz A.: Fundamental mode cutoff in depressed cladding multiple-fiber couplers. Electron. Lett. 20, 767–768

1985 Love J.D.: Explicit formulae for cutoff values on fibers and couplers. Opt. Quantum Electron. 17, 139–147

1985 Love J.D., Ankiewicz A.: Modal cutoffs in single- and few-mode fiber couplers. J. Lightwave Tech. LT-3, 100–110

1986 Love J.D., Henry W.M.: Quantifiying loss minimization in single-mode fiber tapers. Electron. Lett. 22, 912–914

1985 Lowe R.S.: Industrial measurements for single-mode fibers. Proc. SPIE 584, 172–175

1977 Lukowski T.I., Kapron F.P.: Parabolic fiber cutoffs: A comparison of theories. J. Opt. Soc. Am. 67, 1185–1187

1981 MacDonald R.I.: Frequency domain optical reflectometer. Appl. Opt. 20, 1840–1844

1979 Machida S., Kawana A., Ishihara K., Tsuchiya H.: Interference of an AlGaAs laser diode using a 4.15 km single-mode fiber cable. IEEE J. QE-15, 535–537

1983 Mahlein H.F.: Fiber-optic communication in the wavelength-division multiplex mode. Fiber a. Integr. Optics 4, 339–371

1985 Mahr H.: Ein Plädoyer für den Begriff Frequenzmultiplex in der optischen Nachrichtentechnik. Frequenz 39, 314-319

1965　Malitson I.H.: Interspecimen comparison of the refractive index of fused silica. J. Opt. Soc. Am. **55**, 1205–1209

1983　Mallinson S.R.: A monomode connector using shape memory effect metal. 9th Europ. Conf. Opt. Commun. 1983, 105–108

1985　Mallinson S.R.: Crosstalk limits of Fabry-Perot demultiplexers. Electron. Lett. **21**, 759–760

1985　Mallinson S.R., Warnes G.: Optimization of thick lenses for single-mode optical-fiber microcomponents. Opt. Lett. **10**, 238–240

1986　Mantica P.G., Montrosset I., Tascone R., Zich R.: Source-field representation in terms of Gaussian beams. J. Opt. Soc. Am. A **3**, 497–507

1986　Mao Z., Fang X., Li B.: Mode excitation theory and experiment of single-mode fiber directional coupler with strong coupling. J. Lightwave Tech. LT-**4**, 466–472

1969a　Marcatili E.A.J.: Dielectric rectangular waveguide and directional coupler for integrated optics. Bell Syst. Tech. J. **48**, 2071–2102

1969b　Marcatili E.A.J.: Bends in optical dielectric guides. Bell Syst. Tech. J. **48**, 2103–2132

1969　Marcatili E.A.J., Miller S.E.: Improved relations describing directional control in electromagnetic wave guidance. Bell Syst. Tech. J. **48**, 2161–2188

1980　Marcatili E.A.J.: The directional coupler chart. 6th Europ. Conf. Opt. Commun. 1980, 322–325

1985　Marcatili E.A.J.: Dielectric tapers with curved axes and no loss. IEEE J. QE-**21**, 307–314.

1986　Marcatili E.A.J.: Improved coupled-mode equations for dielectric guides. J. Lightwave Tech. LT-**4**, 988–993

1987　Marcatili E.A.J., Buhl L.L., Alferness R.C.: Improved coupled-mode equations: experimental verification. Conf. Opt. Fiber Commun. (OFC'87), TUO3

1982　Marciniak H.: Elastooptic effect in a bent slab waveguide. Optica Applicata **12**, 165–172

1970a　Marcuse D.: Radiation losses of the dominant mode of round optical fibers. Conf. Trunk Telecommunications by Guided Waves, London 1970, IEE publ. **71**, 89–94

1970b　Marcuse D.: Radiation losses of tapered dielectric slab waveguides. Bell Syst. Tech. J. **49**, 273–290

1970c　Marcuse D.: Radiation losses of the dominant mode in round dielectric waveguides. Bell Syst. Tech. J. **49**, 1665–1693

1970d　Marcuse D.: Excitation of the dominant mode of a round fiber by a Gaussian beam. Bell Syst. Tech. J. **49**, 1695–1703

1971a　Marcuse D.: The coupling of degenerate modes in two parallel dielectric waveguides. Bell Syst. Tech. J. **50**, 1791–1816

1971b　Marcuse D.: Attenuation of unwanted cladding modes. Bell Syst. Tech. J. **50**, 2565–2583

1973a　Marcuse D. (Ed.): Integrated Optics. (IEEE Press, New York)

1973b　Marcuse D.: Coupled mode theory of round optical fibers. Bell Syst. Tech. J. **52**, 817–842

1974　Marcuse, D.: Theory of Dielectric Optical Waveguides. (Academic Press, New York, London)

1975a　Marcuse D.: Coupled-mode theory for anisotropic optical waveguides. Bell Syst. Tech. J. **54**, 985–995

1975b　Marcuse D.: Excitation of parabolic-index fibers with incoherent sources. Bell Syst. Tech. J. **54**, 1507–1530

1975c　Marcuse D.: Reflection losses from imperfectly broken fiber ends. Appl. Opt. **14**, 3016–3020

1975d　Marcuse D.: Radiation losses of the HE_{11}-mode of a fiber with sinusoidal perturbed core boundary. Appl. Opt. **14**, 3021–3025

1976a　Marcuse D.: Curvature loss formula for optical fibers. J. Opt. Soc. Am. **66**, 216–220

1976b　Marcuse D.: Field deformation and loss caused by curvature of optical fibers. J. Opt. Soc. Am. **66**, 311–320

1976c　Marcuse D.: Microbending losses of single-mode, step-index and multimode, parabolic-index fibers. Bell Syst. Tech. J. **55**, 937–955

1977a　Marcuse D.: Review of monomode fibers. 3rd Europ. Conf. Opt. Commun. 1977, 60–65

1977b　Marcuse D.: Loss analysis of single-mode fiber splices. Bell Syst. Tech. J. **56**, 703–718

475

1978a Marcuse D.: Gaussian approximation of the fundamental modes of graded-index fibers. J. Opt. Soc. Am. **68**, 103–109

1978b Marcuse D.: Radiation losses of parabolic-index slabs and fibers with bent axes. Appl. Opt. **17**, 755–762

1979a Marcuse D.: Refractive index determination by the focusing method. Appl. Opt. **18**, 9–13

1979b Marcuse D.: Interdependence of waveguide and material dispersion. Appl. Opt. **18**, 2930–2932

1979a Marcuse D., Presby H.M.: Focusing method for nondestructive measurement of optical fiber index profiles. Appl. Opt. **18**, 14–22

1979b Marcuse D., Presby H.M.: Automatic geometric measurements of single-mode and multimode optical fibers. Appl. Opt. **18**, 402–408

1980 Marcuse D.: Pulse distortion in single-mode fibers. Part 1. Appl. Opt. **19**, 1653–1660

1980 Marcuse D., Presby H.M.: Index profile measurements of fibers and their evaluation. Proc. IEEE **68**, 666–688

1981a Marcuse D.: Principles of Optical Fiber Measurements. (Academic Press, New York, San Francisco, London)

1981b Marcuse D.: Pulse distortion in single-mode fibers. Part 2. Appl. Opt. **20**, 2969–2974

1981c Marcuse D.: Pulse distortion in single-mode fibers. 3: Chirped pulses. Appl. Opt. **20**, 3573–3579

1981 Marcuse D., Lin Ch.: Low dispersion single-mode fiber transmission – the question of practical versus theoretical maximum transmission bandwidth. IEEE J. QE-**17**, 869–878

1982a Marcuse D.: Light transmission Optics, 2nd edition (van Nostrand Reinhold, New York)

1982b Marcuse D.: Influence of curvature on the losses of doubly clad fibers. 8th Europ. Conf. Opt. Commun. 1982, 583–586

1982c Marcuse D.: Influence of curvature on the losses of doubly clad fibers. Appl. Opt. **21**, 4208–4213

1984 Marcuse D.: Microdeformation losses of single-mode fibers. Appl. Opt. **23**, 1082–1091

1985 Marcuse D.: Directional-coupler filter using dissimilar optical fibers. Electron. Lett. **21**, 726–727

1985 Marcuse D., Stone J.: Experimental comparison of the bandwidths of standard and dispersion-shifted fibers near their "zero-dispersion" wavelengths. Opt. Lett. **10**, 163–165

1986 Marcuse D., Stone J.: Coupling efficiency of front surface and multilayer mirrors as fiber-end reflectors. J. Lightwave Tech. LT-**4**, 377–381

1987 Marcuse D.: Mode conversion in optical fibers with monotonically increasing core radius. J. Lightwave Tech. LT-**5**, 125–133

1984 Marhic M.E.: Hierarchic and combinatorial star couplers. Opt. Lett. **9**, 368–370

1985a Mark J., Bódtker E., Tromborg B.: Measurement of Rayleigh backscatter-induced linewidth reduction. Electron. Lett. **21**, 1008–1009

1985b Mark J., Bódtker E., Tromborg B.: Statistical characteristics of a laser diode exposed to Rayleigh backscatter from a single-mode fiber. Electron. Lett. **21**, 1010–1011

1984 Marom E., Ramer O.G., Ruschin Shl.: Relation between normal-mode and coupled-mode analyses of parallel waveguides. IEEE J. QE-**20**, 1311–1319

1984 Martinez O.E., Gordon J.P., Fork R.L.: Negative group-velocity dispersion using refraction. J. Opt. Soc. Am. A**1**, 1003–1006

1987 Martinez O.E.: 3000 times grating compressor with positive group velocity dispersion: Application to fiber compensation in $1.3–1.6\,\mu m$ region. IEEE J. QE-**23**, 59–64

1980 Masuda S.: Variable attenuator for use in single-mode fiber transmission systems. Appl. Opt. **19**, 2435–2438

1981 Masuda S., Iwama T., Daido Y.: Nondestructive measurement of core radius, numerical aperture, and cutoff wavelength for single-mode fibers. Appl. Opt. **20**, 4035–4038

1982 Masuda S., Iwama T.: Low-loss lens connector for single-mode fibers. Appl. Opt. **21**, 3475–3483

1984 Mathyssek K., Keil R., Klement E.: New coupling arrangement between laser diode and single-mode fiber with high coupling efficiency and particularly low feedback effect. 10th Europ. Conf. Opt. Commun. 1984, 186–187

1985 Mathyssek K., Wittmann J., Keil R.: Fabrication and investigation of drawn fiber tapers with spherical microlenses. J. Opt. Commun. **6**, 142–146

1974 Matsuhara M., Kumagai N.: Theory of coupled open transmission lines and its applications. IEEE Trans. MTT-**22**, 378–382

1975 Matsuhara M., Watanabe A.: Coupling of curved transmission lines, and application to optical directional couplers. J. Opt. Soc. Am. **65**, 163–168

1980 Matsumura H., Suganuma T.: Normalization of single-mode fibers having an arbitrary index profile. Appl. Opt. **19**, 3151–3158

1980a Matsumura H., Katsuyama T., Suganuma T.: Fundamental study of single polarization fibers. 6th Europ. Conf. Opt. Commun. 1980, 49–52

1980b Matsumura H., Suganuma T., Katsuyama T.: Simple normalization of single-mode fibers with arbitrary index profile. 6th Europ. Conf. Opt. Commun. 1980, 103–106

1979a Matthijsse P., de Blok C.M.: Measurement of splice insertion loss using the backscattering method. 5th Europ. Conf. Opt. Commun. 1979, 9.5-1–5-4

1979b Matthijsse P., de Blok C.M.: Field measurement of splice loss applying the backscattering method. Electron. Lett. **15**, 795–797

1981 Mazurczyk V.J.: Sensitivity of single-mode buried heterostructure lasers to reflected power at 274 Mb/s. Electron. Lett. **17**, 143–144

1985 McCartney D.J., Payne D.B., Duncan S.S.: Position-tunable holographic filters in dichromated gelatin for use in single-mode-fiber demultiplexers. Opt. Lett. **10**, 303–305

1986 McCartney D.J.: A review of the performance of two types of position tuned filters for use in the 1200–1600 nm range. 12th Europ. Conf. Opt. Commun. 1986, 133–136

1983 McCaughan L., Bergmann E.E.: Index distribution of optical waveguides from their mode profile. J. Lightwave Tech. LT-**1**, 241–244

1984 McHenry M.A., Chang D.C.: Coupled mode theory of two nonparallel dielectric waveguides. IEEE Trans. MTT-**32**, 1469–1475

1973 McIntyre P.D., Snyder A.W.: Power transfer between optical fibers. J. Opt. Soc. Am. **63**, 1518–1527

1974 McIntyre P.D., Snyder A.W.: Power transfer between nonparallel and tapered optical fibers. J. Opt. Soc. Am. **64**, 285–288

1975 McMahon D.H.: Efficiency limitations imposed by thermodynamics on optical coupling in fiber-optic data links. J. Opt. Soc. Am. **65**, 1479–1482

1983 McMillan J.L.: Novel technique for measurement of wavelength of zero relative delay between modes of a dual-mode optical fiber. Electron. Lett. **19**, 240–242

1984 McMillan J.L., Robertson S.C.: Cutoff wavelength determination in single-mode optical fibers by measurement of equalization wavelength. Electron. Lett. **20**, 698–699

1985 McMillan J.L., Robertson S.C.: Characterization of single-mode optical fibers by measurement of equalization wavelength. Electron. Lett. **21**, 295–296

1980 Melliar-Smith C.M., Lazay P.D., Pasteur G.A., Chandross F.A., Wood D.L.: Index matching fluids for long wavelength (1.2–1.6 μm) fiber-optic applications. Electron. Lett. **16**, 403–404

1985 Melman P., Davies R.W.: Application of the Clausius-Mossotti equation to dispersion calculations in optical fibers. J. Lightwave Tech. LT-**3**, 1123–1124

1984 Mengel F.: Interferometric monomode fiber measurements: influence of source spectrum and second-order dispersion. Electron. Lett. **20**, 66–67

1984 Mengel F., Gade N.: Monomode fiber dispersion measurement with a tunable 1.3 μm external cavity laser. 10th Europ. Conf. Opt. Commun. 1984, 126–127

1982 Meslener G.J.: Dispersion-induced harmonic distortion in fiber systems using coherent light sources. Conf. Opt. Fiber Commun. (OFC'82), 68–71

1984 Meslener G.J.: Chromatic dispersion induced distortion of modulated monochromatic light employing direct detection. IEEE J. QE-**20**, 1208–1216

1985 Meunier J.-M., Maystre F., Robert Ph.: Optical fibers far-field measurement using an integrating sphere. Proc. SPIE **584**, 264–268

1980a Meunier J.P., Pigeon J., Massot J.N.: Perturbation theory for the evaluation of the normalized cutoff frequencies in radially inhomogeneous fibers. Electron. Lett. **16**, 27–29

1980b Meunier J.P., Pigeon, J., Massot J.N.: Analyse perturbative des caractéristiques de propagation des fibres optiques à gradient d'indice quasi-parabolique. Opt. Quantum Electron. **12**, 41–49

477

1982 Meunier J.P., Pigeon J., Massot J.N.: Comments on "A simple numerical method for the cutoff frequency of a single-mode fiber with an arbitrary index profile". IEEE Trans. MTT-**30**, 108–109

1984a Meunier J.P., Pigeon J., Massot J.N.: An efficient method for calculating cutoff frequencies in optical fibers with arbitrary index profiles. J. Lightwave Tech. LT-**2**, 171–175

1984b Meunier J.P., Pigeon J., Massot J.N.: LP_{11}-mode cutoff: a simplified numerical approach. Opt. Quantum Electron. **16**, 327–330

1974 Midwinter J.E., Reeve M.H.: A technique for the study of mode cut-offs in multimode optical fibers. Opto-electron. **6**, 411–416

1975 Midwinter J.E.: The prism-taper coupler for the excitation of single modes in optical transmission lines. Opt. Quantum Electron. **7**, 297–303

1981 Midwinter J.E.: Studies of monomode long wavelength fiber systems at the British Telecom Research Laboratories. IEEE J. QE-**17**, 911–918

1983a Midwinter J.E.: Monomode fibers for long haul transmission systems. Br. Telecom. Technol. J. **1**, 5–11

1983b Midwinter J.E.: "Monomode fibers for long haul transmission systems", in Optical waveguide science, Proc. of the Intern. Sympos. in Kweilin, China, June 20–23, 1983, ed. by Hung-chia H., Snyder A.W. (Martinus Nijhoff, The Hague), pp. 69–76

1985 Midwinter J.E.: Current status of optical communications technology. J. Lightwave Tech. LT-**3**, 927–930

1980 Millar C.A., Payne D.B.: Monomode fiber connector using fiber bead location. 6th Europ. Conf. Opt. Commun. 1980, 306–309

1981a Millar C.A.: A measurement technique for optical fiber break angles. Opt. Quantum Electron. **13**, 125–131

1981b Millar C.A.: Direct method of determining equivalent-step-index profiles for monomode fibers. Electron. Lett. **17**, 458–460

1982a Millar C.A.: Application of automated equivalent step-index profiling to a 31.6 km monomode fiber system. Conf. Opt. Fiber Commun. (OFC'82), 62–64

1982b Millar C.A.: Comment: fundamental mode spot-size measurement in single-mode optical fibers. Electron. Lett. **18**, 395–396

1983 Millar C.A.: A device to measure the coaxial alignment of butted optical fibers. 9th Europ. Conf. Opt. Commun. 1983, 121–124

1983 Millar C.A., Gooch D.J.: Optimizing cleaving tools by end-angle measurement. Conf. Opt. Fiber Commun. (OFC'83), 12–13

1984a Millar C.A.: A simple near-field scannning system for refractive index profiles and mode spot shape. Symp. Opt. Fiber Meas. 1984, NBS spec. publ. **683**, 33–36

1984b Millar C.A.: Simplified optical fiber end-cleave measurement device. Electron. Lett. **20**, 528–530

1985 Millar C.A., Mallison S.R., Warnes G.C.: Cladding alignment of butted optical fibers using a diffraction alignment device (DAD) and its application to mode-spot concentricity-error measurement. J. Lightwave Tech. LT-**3**, 686–875

1986 Millar C.A., Ainslie B.J., Brierley M.C., Craig S.P.: Fabrication and characterization of D-fibers with a range of accurately controlled core/flat distances. Electron. Lett. **22**, 322–324

1975 Miller C.M.: A fiber-optic cable connector. Bell Syst. Tech. J. **54**, 1547–1555

1977 Miller C.M.: Optical fiber splicing. Techn. Digest Opt. Fiber Transmission II 1977, WA3-1-3-6

1978 Miller C.M.: Fiber-optic array splicing with etched silicon chips. Bell Syst. Tech. J. **57**, 75–90

1982a Miller C.M.: Local detection device for single-mode fiber splicing. Conf. Opt. Fiber Commun. (OFC'82), 44–45

1982b Miller C.M.: Connectorless taps for coated single-mode or multimode optical fibers. 8th Europ. Conf. Opt. Commun. 1983, 344–347

1984 Miller C.M.: Single-mode fiber splicing. Conf. Opt. Fiber Commun. (OFC'84), 70–71

1985 Miller C.M., DeVeau G.F.: Simple high-performance mechanical splice for single-mode fibers. Conf. Opt. Fiber Commun. (OFC'85), 26–27

1986a Miller C.M.: Optical Fiber Splices and Connectors. (Dekker, New York, Basel)

1986b Miller C.M.: Mechanical optical fiber splices. J. Lightwave Tech. LT-**4**, 1228–1231

1954 Miller S.E.: Coupled wave theory and waveguide applications. Bell Syst. Tech. J. **33**, 661–719

1965 Miller S.E.: Light propagation in generalized lens-like media. Bell Syst. Tech. J. **44**, 2017–2064

1966 Miller S.E., Tillotson L.C.: Optical transmission research. Proc. IEEE **54**, 1300–1311

1973a Miller S.E., Marcatili E.A., Li T.: Research toward optical-fiber transmission systems. Part I: The transmission medium. Proc. IEEE, **61**, 1703–1726

1973b Miller S.E., Li T., Marcatili E.A.: Research toward optical-fiber transmission systems. Part II: Devices and sytems considerations. Proc. IEEE, **61**, 1726–1751

1979 Miller S.E., Chynoweth A.G. (eds.): Optical fiber telecommunications (Academic Press, New York, San Francisco, London)

1982 Minowa J., Saruwatari M., Suzuki N.: Optical componentry utilized in field trial of single-mode fiber longhaul transmission. IEEE Trans. MTT-**30**, 551–563 and IEEE J. QE-**18**, 705–717

1983 Minowa J., Fujii Y.: Dielectric multilayer thin-film filters for WDM transmission systems. J. Lightwave Tech. LT-**1**, 116–121

1984 Mitsunaga Y., Katsuyama Y., Ishida Y.: Thermal characteristics of jacketed optical fibers with initial imperfections. J. Lightwave Tech. LT-**2**, 18–24

1981 Miya T., Nakahara M., Yoshioka N., Watanabe M., Furui Y., Fukuda O.: Transmission characteristics of VAD single-mode fiber. 7th Europ. Conf. Opt. Commun. 1981, postdeadl. papers, 22–25

1981 Miya T., Nakahara M., Inagaki N.: Fabrication of dispersion-free VAD single-mode fibers in the 1.5 μm wavelength region. J. Lightwave Tech. LT-**1**, 14–19

1976 Miyagi M., Yip G.L.: Field deformation and polarization change in a step-index optical fiber due to bending. Opt. Quantum Electron. **8**, 335–341

1977 Miyagi M., Yip G.L.: Mode conversion and radiation losses in a step-index optical fiber due to bending. Opt. Quantum Electron. **9**, 51–60

1977a Miyagi M., Nishida Sh.: Transmission characteristics of a dielectric-tube waveguide with an outer higher-index cladding. Electron. Lett. **13**, 274–275

1977b Miyagi M., Nishida Sh.: Peculiar properties of cutoff frequencies in a dielectric optical tube waveguide. Proc. IEEE **65**, 1411–1412

1978 Miyagi M., Nishida Sh.: Bending losses of dielectric rectangular waveguides for integrated optics. J. Opt. Soc. Am. **68**, 316–319

1979a Miyagi M., Nishida Sh.: Pulse spreading in a single-mode optical fiber due to third-order dispersion. Appl. Opt. **18**, 678–682

1979b Miyagi M., Nishida Sh.: Pulse spreading in a single-mode optical fiber due to third-order dispersion: effect of optical source bandwidth. Appl. Opt. **18**, 2237–2240

1986a Miyajima Y., Amaike M., Kawata O., Negishi Y.: Acoustic emission in optical fiber cutting and its application for fiber cut-end inspection method. 12th Europ. Conf. Opt. Commun. 1986, 79–82

1986b Miyajima Y., Ohnishi M., Negishi Y.: Chromatic dispersion measurement over a 120 km dispersion-shifted single-mode fiber in the 1.5 μm wavelength region. Electron. Lett. **22**, 1185–1186

1977 Miyashita T., Horiguchi M., Kawana A.: Wavelength dispersion in a single-mode fiber. Electron. Lett. **13**, 227–228

1981a Mochizuki K., Namihira Y., Wakabayashi H.: Polarization mode dispersion measurements in long single mode fibers. Electron. Lett. **17**, 153–154

1981b Mochizuki K., Namihira Y., Wakabayashi H.: Dispersion measurements in single-mode fibers using sum-frequency mixing as a picosecond optical shutter. Electron. Lett. **17**, 646–648

1982 Mochizuki K., Namihira Y., Wakabayashi H.: Pulse delay measurements in single-mode fibers using sum-frequency mixing as a picosecond shutter. IEEE J. QE-**18**, 278–282

1983 Mochizuki K., Namihira Y., Yamamoto H.: Transmission los increase in optical fibers due to hydrogen permeation. Electron. Lett. **19**, 743–745

1984a Mochizuki K., Namihira Y., Kuwazura M., Iwamoto Y.: Behavior of hydrogen molecules adsorbed on silica in optical fibers. IEEE J. QE-**20**, 694–697

1984b Mochizuki K., Namihira Y., Kuwazura M., Nunokawa M.: Influence of hydrogen on optical fiber loss in submarine cables. J. Lightwave Tech. LT-**2**, 802–807

1985 Mochizuki K., Fujise M., Suzuki H., Watanabe M., Koishi M., Tsuchiya T.: Direct measurement of chromatic dispersion in single-mode fibers using a streak camera. Electron. Lett. **21**, 524–525

1986 Mochizuki K., Fujise M., Kuwazuru S., Nunokawa M., Iwamoto Y.: Dispersion measurement of a long-length concatenated single-mode fiber with a streak camera. Conf. Opt. Fiber Commun. (OFC'86), 102–103

1987 Mochizuki K., Fujise M., Kuwazuru S., Nunokawa M., Iwamoto Y.: Optical fiber dispersion measurement technique using a streak camera. J. Lightwave Tech. LT-5, 119–124

1984 Modavis R.A., Love W.F.: Multiple-wavelength system for characterizing dispersion in single-mode optical fibers. Symp. Opt. Fiber Meas. 1984, NBS spec. publ. **683**, 115–118

1986 Mogensen F., Damsgaard H., Hansen O.: Chromatic dispersion measurements of single-mode fibers in the 1.3 μm and 1.55 μm wavelength range by the wavelength temperature tuning technique. Techn. Dig. Sympos. Opt. Fiber Measur. 1986, NBS spec. publ. **720**, 23–26

1985 Mollenauer L.F.: Solitons in optical fibers and the soliton laser. Phil. Trans. R. Soc. London **A315**, 437–450

1980 Monerie M., Jeunhomme L.: Polarization mode coupling in long single-mode fibers. Opt. Quantum Electron. **12**, 449–461

1982 Monerie M.: Fundamental-mode cutoff in depressed inner cladding fibers. Electron. Lett. **18**, 642–644

1985 Monerie M.: Théorie de la transmission sur fibre mononomde. L'Onde Electrique **65**, 44–58

1986 Moore W.M., Carter S.F., France P.W., Williams J.R.: Absorption and scattering measurements on a low loss ZrF_4-based fiber. 12th Europ. Conf. Opt. Commun. 1986, 299–302

1983 Mori M., Soeda K., Nishikawa T.: Measurement of chromatic dispersion in a long single-mode fiber by the baseband response. 9th Europ. Conf. Opt. Commun. 1983, 381–383

1984 Morimoto Y., Komatsu K., Ushirogawa A.: Design and fabrication of low loss single-mode optical fiber connectors. Proc. SPIE **479**, 36–41

1979 Morishita K., Inagaki S., Kumagai N.: Analysis of discontinuities in dielectric waveguides by means of the least squares boundary residual method. IEEE Trans. MTT-**27**, 310–315

1980 Morishita K., Kondoh Y., Kumagai N.: On the accuracy of scalar approximation technique in optical fiber analysis. IEEE Trans. MTT-**28**, 33–36

1983a Morishita K.: "Determination of single-mode fiber refractive-index profiles by a propagation-mode near-field scanning technique", in Optical waveguide science, Proc. of the Intern. Sympos. in Kweilin, China, June 20–23, 1983, ed. by Hung-chia H., Snyder A.W. (Martinus Nijhoff, The Hague), pp. 101–197

1983b Morishita K.: Hybrid modes in circular cylindrical optical fibers. IEEE Trans. MTT-**31**, 344–350

1983c Morishita K.: Refractive-index profile determination of single-mode optical fibers by a propagation-mode near-field scanning technique. J. Lightwave Tech. LT-1, 445–449

1985 Morishita K.: Measurement of refractive-index profile of single-mode optical fibers by the propagation-mode near-field method J. Lightwave Tech. LT-3, 244–247

1986 Morishita K.: Index profiling of three-dimensional optical waveguides by the propagation-mode near-field method. J. Lightwave Tech. LT-4, 1120–1124

1985a Mortimore D.B.: Low-loss 8 × 8 single-mode star coupler. Electron. Lett. **21**, 502–504

1985b Mortimore D.B.: Wavelength-flattened fused couplers. Electron. Lett. **21**, 742–743

1986a Mortimore D.B.: A wavelength flattened 8 × 8 single-mode star coupler. 12th Europ. Conf. Opt. Commun. 1986, 459–462

1986b Mortimore D.B.: Wavelength-flattened 8 × 8 single-mode star coupler. Electron. Lett. **22**, 1205–1206

1986 Mortimore D.B., Wright J.V.: Low-loss joints between dissimilar fibers by tapering fusion splices. Electron. Lett. **22**, 318–319

1984 Mossman P.: Components for optical fiber systems. GEC Journ. of Res. 112–118

1986 Mozer A.P.: Einfluß der optischen Pulsform auf das Bitraten-Längenprodukt. AEÜ, Arch. für Elektron. und Übertragungstech.: Electron. and Commun. **40**, 203–207

1983 Müller T.: Resolution improvement in refracted near-field index measurement by a lens-shaped liquid cell. Electron. Lett. **19**, 580–582

1985 Müller T.: LP_{11}-Grenzwellenlänge von Einmodenfasern – Abhängigkeit von Probenlänge und Krümmungsradius bei der Biegemethode. NTG-Fachberichte **89**, 155–158

1981 Mukai T., Yamamoto Y.: Gain, frequency bandwidth, and saturation output power of AlGaAs DH laser amplifiers. IEEE J. QE-**17**, 1028–1034

1978 Murakami Y., Tsuchiya,H.: Bending losses of coated single-mode optical fibers. IEEE J. QE-**14**, 495–501

1978 Murakami Y., Hatakeyama I., Tsuchiya H.: Normalized frequency dependence of splice losses in single-mode optical fibers. Electron. Lett. **14**, 277–278

1979 Murakami Y., Kawana A., Tsuchiya H.: Cut-off wavelength measurements for single-mode optical fibers. Appl. Opt. **18**, 1101–1105

1980 Murakami Y.: Coupling between curved dielectric waveguides. Appl. Opt. **19**, 398–403

1980 Murakami Y., Yamada J., Sakai J., Kimura T.: Microlens tipped on a single-mode fiber end for InGaAsP laser coupling improvement. Electron. Lett. **16**, 321–322

1981 Murakami Y., Sudo S.: Coupling characteristics measurements between curved waveguides using a two-core fiber coupler. Appl. Opt. **20**, 417–422

1982 Murakami Y., Noguchi K., Ashiya F., Negishi Y., Kojima N.: Maximum measurable distances for a single-mode optical fiber fault locator using the stimulated Raman scattering (SRS) effect. IEEE J. QE-**18**, 1473–1477

1975 Murata H., Inao S., Matsuda Y., Takahashi T.: Splicing of optical fiber cable on site. 1st Europ. Conf. Opt. Commun. 1975, 93–95

1981 Murata H., Inagaki N.: Low-loss single-mode fiber development and splicing research in Japan. IEEE J. QE-**17**, 835–849

1983 Murayama Y., Yamauchi R., Sugawara Y., Inada K., Ohashi M., Kato Y.: Monomode operation of high V-value quasi-monomode fibers. 9th Europ. Conf. Opt. Commun. 1983, 69–72

1983 Murphy E.J., Rice T.C.: Low-loss coupling of multiple fiber arrays to single-mode waveguides. J. Lightwave Tech. LT-**1**, 479–482

1985 Murphy E.J., Rice T.C., McCaughan L., Harvey G.T., Read P.H.: Permanent attachment of single-mode fiber arrays to waveguides. J. Lightwave Tech. LT-**3**, 795–799

1986 Murphy E.J., Rice T.C.: Self-alignment technique for fiber attachment to guided wave devices. IEEE J. QE-**22**, 928–932

1985 Myogadani T., Tanaka Sh.: Mode field diameter – measurement in far-field pattern method and characteristics. Proc. SPIE **584**, 257–263

1978 Nagano K., Kawakami S., Nishida S.: Change of the refractive index in an optical fiber due to external forces. Appl. Opt. **17**, 2080–2085

1984 Nagel S.R.: Review of the depressed cladding single-mode fiber design and performance for the SL undersea system application. J. Lightwave Tech. LT-**2**, 792–801

1984 Nakagawa K.: "1.5 μm single-mode transmission" in Japan Annual Review in Electronics, Computers, & Telecommunications, Vol. 11, Optical Fibers and Devices, ed. by Y. Suematsu (Ohm, Tokyo, Osaka, Kyoto; North-Holland, Amsterdam, New York, Oxford), pp. 230–242

1983 Nakahara T., Hoshikawa M., Tanaka S.: Characteristics of VAD single-mode fiber with depressed cladding layer. Proc. SPIE **425**, 2–8

1986 Nakajima Y., Ezekiel S.: Sensitive determination of the existence of higher-order modes in a quasi-single-mode fiber. Opt. Lett. **11**, 815–817

1978 Nakamura M., Aiki K., Chinone N., Ito R., Umeda J.: Longitudinal-mode behaviors of mode-stabilized $Al_x Ga_{1-x} As$ injection lasers. J. Appl. Phys. **49**(9), 4644–4648

1981 Nakazawa M., Tokuda M.: Marked extension of diagnosis length in optical time domain reflectometry using 1.32 μm YAG laser. Electron. Lett. **17**, 783–785

1981 Nakazawa M., Tanifuji T., Tokuda M., Uchida N.: Photon probe fault locator for single-mode optical fiber using an acoustooptical light deflector. IEEE J. QE-**17**, 1264–1269

1983a Nakazawa M.: Theory of backward Rayleigh scattering in polarization-maintaining single-mode fibers and its application to polarization optical time domain reflectometry. IEEE J. QE-**19**, 854–861

1983b Nakazawa M.: Rayleigh backscattering theory for single-mode optical fibers. J. Opt. Soc. Am. **73**, 1175–1180

1983 Nakazawa M., Aoyama K.: Measurement technique for single-mode optical fiber. Rev. Electr. Commun. Lab. NTT, Japan **31**, 290–298

1983 Nakazawa M., Tokuda M.: Measurement of the fiber loss spectrum using fiber Raman optical-time-domain reflectometry. Appl. Opt. **22**, 1910–1914

1983 Nakazawa M., Tokuda M., Uchida N.: Analyses of optical time-domain reflectometry for single-mode fibers and of polarization optical time-domain reflectometry for polarization-maintaining fibers. Opt. Lett. **8**, 130–134

1984a Nakazawa M., Tokuda M., Morishige Y., Toratani H.: 1.55 µm OTDR for single-mode optical fiber longer than 110 km. Electron. Lett. **20**, 323–324

1984b Nakazawa M., Tokuda M., Washio K., Asahara Y.: 130-km-long fault location for single-mode optical fiber using 1.55 µm Q-switched Er^{3+}: glass laser. Opt. Lett. **9**, 312–314

1985 Nakazawa M., Nakashima T., Seikai S., Ikeda M.: Self-detecting optical time domain reflectometer for single-mode fibers. Opt. Lett. **10**, 157–159

1981 Namihira Y., Mochizuki K., Tatekura K.: Effects of thermal stress on group delay in jacketed single-mode fibers. Electron. Lett. **17**, 813–815

1982 Namihira Y., Wakabayashi H., Yamamoto H.: High-stability measuring equipment for very small variations of optical-fiber loss. Electron. Lett. **18**, 124–126

1986 Namihira Y., Iwamoto Y., Murakami Y., An S.: High speed and high accuracy single-mode fiber dispersion measuring equipment using modified interferometric method. Techn. Dig. Sympos. Opt. Fiber Measur. 1986, NBS spec. publ. **720**, 19–22

1984 Narasimhan M.S., Karthikeyan M.: Evaluation of Fourier transform integrals using FFT with improved accuracy and its applications. IEEE Trans. AP-**32**, 404–408

1979 Nawata K., Iwahara Y., Suzuki N.: Ceramic capillary splices for optical fibers. Electron. Lett. **15**, 470–472

1980 Nawata K.: Multimode and single-mode fiber connectors technology. IEEE J. QE-**16**, 618–627

1982 Nawata K., Suzuki N.: "Multimode and single-mode fiber connectors technology", in Japan Annual Review in Electronics, Computers, & Telecommunications, Vol. 1, Optical Fibers & Devices, ed. by Y.Suematsu (Ohm, Tokyo, Osaka, Kyoto; North-Holland, Amsterdam, New York, Oxford), pp. 152–171

1983 Nayar B.K., Smith D.R.: Monomode-polarization-maintaining fiber directional couplers. Opt. Lett. **8**, 543–545

1986 Nazarathy M., Newton S.A.: Rayleigh backscattering in optical fiber recirculating delay lines. Appl. Opt. **25**, 1051–1055

1975 Nelson A.R.: Coupling optical waveguides by tapers. Appl. Opt. **14**, 3012–3015

1983 Nelson B.P., Wright J.V.: Problems in the use of ESI parameters in specifying monomode fibers. Proc. SPIE **425**, 9–14

1984 Nelson B.P., Wright J.V.: Problems in the use of ESI parameters in specifying monomode fibers. Brit. Telecom. Techn. J. **2**, 81–85

1976 Nemoto S., Makimoto T.: A relationship between phase and group indices of guided modes in dielectric waveguides. Int. J. Electronics **40**, 187–190

1979 Nemoto S., Makimoto T.: Analysis of splice loss in single-mode fibers using a Gaussian field approximation. Opt. Quantum Electron. **11**, 447–457

1964a Neumann E.-G.: Über das elektromagnetische Feld der schwach geführten Dipolwelle. Z. angew. Phys. **16**, 452–460

1964b Neumann E.-G.: Ein Perot-Fabry-Interferometer für Millimeterwellen mit freien und geführten Wellen. Z. angew. Phys. **17**, 304–308

1964c Neumann E.-G.: Ein neues Prinzip zur Umlenkung elektromagnetischer Wellen. Z. angew. Phys. **18**, 71–76

1967a Neumann E.-G.: Über das elektromagnetische Feld am freien Ende einer dielektrischen Leitung: I. Abstrahlung. Z. angew. Physik **24**, 1–11

1967b Neumann E.-G.: Über das elektromagnetische Feld am freien Ende einer dielektrischen Leitung: II. Empfang. Z. angew. Physik **24**, 12–18

1974 Neumann E.-G., Rudolph H.-D.: Poynting's vector and the wavefronts near a plane conductor. Electron. Lett. **10**, 446–447

1975 Neumann E.-G.: Inhomogeneities in monomode optical waveguides. Nouv. Rev. Optique **6**, 263–271

1975a Neumann E.-G., Rudolph H.-D.: Losses from corners in dielectric-rod or optical-fiber waveguides. Appl. Phys. **8**, 107–116

1975b Neumann E.-G., Rudolph H.-D.: Radiation from bends in dielectric rod transmission lines. IEEE Trans. MTT-**23**, 142–149

1977 Neumann E.-G., Opielka D.: Scattering matrix and radiation characteristics of the junction between two different monomode microwave or optical dielectric waveguides. Opt. Quantum Electron. **9**, 209–222

1978 Neumann E.-G.: Optical time domain reflectometer: comment. Appl. Opt. **17**, 1675

1980 Neumann E.-G.: Analysis of the backscattering method for testing optical fiber cables. AEÜ, Arch. für Elektron. und Übertragungstech.: Electron. and Commun. **34**, 157–160

1981 Neumann E.-G.: Reducing radiation loss of tilts in dielectric optical waveguides. Electron. Lett. **17**, 369–371

1982a Neumann E.-G.: Low loss dielectric optical waveguide bends. Fiber and Integr. Optics **4**, 203–211

1982b Neumann E.-G.: Curved dielectric optical waveguides with reduced transition losses. Proc. Inst. Electr. Eng. **129**, Pt.H, 278–280

1983 Neumann E.-G., Richter W.: Sharp bends with low losses in dielectric optical waveguides. Appl. Opt. **22**, 1016–1022

1987 Neumann E.-G.: Beam transformers for obtaining low-loss splices between dissimilar single-mode fibers. J. Opt. Soc. Am. A**4**, 1021–1029

1982 Newton S.A., Bowers J.E., Shaw H.J.: Single-mode fiber recirculating delay line. Proc. SPIE **326**, 108–115

1983 Newton S.A., Cross P.S.: Microwave-frequency response of an optical-fiber delay-line filter. Electron. Lett. **19**, 480–481

1983 Newton S.A., Howland R.S., Jackson K.P., Shaw H.J.: High-speed pulse-train generation using single-mode fiber recirculating delay lines. Electron. Lett. **19**, 756–758

1986 Newton S.A., Nazarathy M., Trutna W.R.: Measured backscatter signature of a fiber recirculating delay line. Appl. Opt. **25**, 1879–1881

1985 Nicholls S.T.: Automatic manufacture of polished single-mode fiber directional coupler. Electron. Lett. **21**, 825–826

1982 Nichols P.D.: Comparison between techniques for measuring the LP_{11}-mode cutoff wavelength in monomode fibers. Electron. Lett. **18**, 1008–1009

1984 Nicholson G.: Effect of fiber endface angle on the measurement of the mode spot size from the far-field pattern. Opt. Quantum Electron. **16**, 405–408

1978 Nicia A.: Practical low-loss lens connector for optical fibers. Electron. Lett. **14**, 511–512

1981 Nicia A.: Wavelength multiplexing and demultiplexing systems for single-mode and multimode fibers. 7th Europ. Conf. Opt. Commun. 1981, 8.1-1–1-7

1981 Nicia A., Tholen A.: High-efficient ball-lens connector and related functional devices for single-mode fibers. 7th Europ. Conf. Opt. Commun. 1981, 7.5.1–5.4

1981 Nicia A., Rittich D.: Ball lens connector system for optical fibers and cables. 30th Intern. Cable a. Wire Sympos., Cherry Hill, 341–351

1982 Nicia A.: Loss analysis of laser-fiber coupling and fiber combiner, and its application to wavelength division multiplexing. Appl. Opt. **21**, 4280–4289

1983 Nicolaisen E., Danielsen P.: Calculated and measured spot size of equivalent Gaussian field in single-mode optical fibers. Electron. Lett. **19**, 27–29

1982 Nielsen C.J.: Influence of polarization-mode coupling on the transmission bandwidth of single-mode fibers. J. Opt. Soc. Am. **72**, 1142–1146

1983 Nielsen C.J.: Impulse response of single-mode fibers with polarization-mode coupling. J. Opt. Soc. Am. **73**, 1603–1611

1984 Nielsen C.J.: Bandwidth-limiting effects in coherent optical communications systems based on polarization-preserving fibers. Electron. Lett. **20**, 403–405

1984 Niizeki N.: "Development of optical fiber" in Japan Annual Review in Electronics, Computers, & Telecommunications, Vol. 11, Optical Fibers and Devices, ed. by Y. Suematsu (Ohm, Tokyo, Osaka, Kyoto; North-Holland, Amsterdam, New York, Oxford), pp. 168–178

1984 Nijnuis H.T., van Leeuwen K.A.H.: Length and curvature dependence of effective cutoff-wavelength and LP_{11}-mode attenuation in single-mode fibers. Symp. Opt. Fiber Meas. 1984, NBS spec. publ. **683**, 11–14

1984 Nishimura M., Suzuki S.: Measurement of mode field radius by far-field pattern method. 10th Europ. Conf. Opt. Commun. 1984, 118–119

1978 Noda J., Mikami O., Minakata M., Fukuma M.: Single-mode optical-waveguide fiber coupler. Appl. Opt. **17**, 2092–2096

1983 Noda J., Shibata N., Edahiro T., Sasaki Y.: Splicing of single polarization-maintaining fibers. J. Lightwave Tech. LT-1, 61–66

1986 Noda J., Okamoto K., Sasaki Y.: Polarization-maintaining fibers and their applications. J. Lightwave Tech. LT-4, 1071–1089

1982 Noguchi K., Murakami Y., Yamashita K., Ashiya F.: 52 km-long single-mode optical fiber fault location using the stimulated Raman scattering effect. Electron. Lett. **18**, 41–42

1984 Noguchi K.: A 100-km-long single-mode optical fiber fault location. J. Lightwave Tech. LT-**2**, 1–6

1985 Noguchi K., Shibata N., Uesugi N., Negishi Y.: Loss increase for optical fibers exposed to hydrogen atmosphere. J. Lightwave Tech. LT-**3**, 236–243

1985 Nolting H.-P., Ulrich R. (Eds.): Integrated Optics, Springer Series in Optical Sciences, Vol. 48 (Springer, Berlin, Heidelberg, New York, Tokyo)

1982 Nosu K.: Single-mode fiber measurement in Japan. Techn. Dig. Sympos. Opt. Fiber Measur. 1982, NBS Publ. **641**, 71–77

1983 Nosu K.: "Fiber-optic wavelength-division-multiplexing technology and its application", in Japan Annual Review in Electronics, Computers, & Telecommunications, Vol. 5, Optical Fibers & Devices, ed. by Y. Suematsu (Ohm, Tokyo, Osaka, Kyoto; North-Holland, Amsterdam, New York, Oxford), pp. 268–290

1981 Nyquist D.P., Johnson D.R., Hsu S.V.: Orthogonality and amplitude spectrum of radiation modes along open boundary waveguides. J. Opt. Soc. Am. **71**, 49–54

1978 O'Connor P., Tauc J.: Light scattering in optical waveguides. Appl. Opt. **17**, 3226–3231

1977 Odagiri Y., Shikada M., Kobayashi K.: High-efficiency laser-to-fiber coupling circuit using a combination of a cylindrical lens and a Selfoc lens. Electron. Lett. **13**, 395–396

1980 Odagiri Y., Seki M., Nomura H., Sugimoto M., Kobayashi K.: Practical $1.5\,\mu$m LD-isolator-single-mode fiber module using a V-grooved diamond heatsink. 6th Europ. Conf. Opt. Commun. 1980, 282–285

1982a Ogawa K.: Analysis of mode partition noise in laser transmission systems. IEEE J. QE-**18**, 849–855

1982b Ogawa K.: Considerations for single-mode fiber systems. Bell Syst. Tech. J. **61**, 1919–1931

1982 Ogawa K., Vodhanel R.S.: Measurements of mode partition noise of laser diodes. IEEE J. QE-**18**, 1090–1093

1984 Ohashi M., Kitayama K., Kobayashi T., Ishida Y.: LP_{11}-mode loss measurements in the two-mode propagation region of optical fibers. Opt. Lett. **9**, 303–305

1985 Ohashi M., Kitayama K., Ishida Y., Negishi Y.: Simple approximations for chromatic dispersion in single-mode fibers with various index profiles. J. Lightwave Tech. LT-**3**, 110–115

1986 Ohashi M., Kitayama K., Seikai S.: Mode field diameter measurement conditions for fibers by transmitted field pattern methods. J. Lightwave Tech. LT-**4**, 109–115

1985 Ohtsuka Y.: Analysis of a fiber-optic passive loop-resonator gyroscope: dependence on resonator parameters and light-source coherence. J. Lightwave Tech. LT-**3**, 378–384

1980 Okada K., Hashimoto K., Shibata T., Nagaki Y.: Optical cable fault location using correlation technique. Electron. Lett. **16**, 629–630

1982 Okada K.: "Backscattering measurement", in Japan Annual Review in Electronics, Computers & Telecommunications 1982, ed. by. Y.Suematsu (North-Holland, Amsterdam), pp. 299-313

1976 Okamoto K., Okoshi T.: Analysis of wave propagation in optical fibers having core with α-power refractive-index distribution and uniform cladding. IEEE Trans. MTT-**24**, 416–421

1980 Okano Y., Nakagawa K., Ito T.: Laser mode partition noise evaluation for optical fiber transmission. IEEE Trans. COM-**28**, 238–243

1976a Okoshi T., Hotate K.: Computation of the refractive-index distribution in an optical fiber from its scattering pattern for a normally incident laser beam. Opt. Quantum Electron. **8**, 78–79

1976b Okoshi T., Hotate K.: Refractive-index profile of an optical fiber: its measurement by the scattering pattern method. Appl. Opt. **15**, 2756–2764

1980 Okoshi T., Nishimura M., Kosuge M.: Nondestructive measurement of axially non-symmetric refractive-index distribution of optical fiber preforms. Electron. Lett. **16**, 722–724

1981 Okoshi T.: Single-polarization single-mode optical fibers. IEEE J. QE-**17**, 879–884

1982 Okoshi T.: Optical Fibers. (Academic Press, New York, London)

1983 Okoshi T.: Single-polarization single-mode optical waveguiding schemes. in Optical Waveguide Science, Proc. of the Intern. Sympos. in Kweilin, China, June 20–23, 1983, ed. by Hung-chia H., Snyder A.W. (Martinus Nijhoff, The Hague), pp. 77–84

1983 Okoshi T., Hotate K.: "Measurement of refractive-index profiles of optical fibers and preforms", in Japan Annual Review in Electronics, Computers, & Telecommunications, Vol. 5, Optical Fibers & Devices, ed. by Y. Suematsu (Ohm, Tokyo, Osaka, Kyoto; North-Holland, Amsterdam, New York, Oxford), pp. 219–234

1984 Oksanen L., Halme S.J.: Interferometric dispersion measurement in single-mode fibers with a numerical method to extract the group delays from the measured visibility curves. Symp. Opt. Fiber Meas. 1984, NBS spec. publ. 683, 127–130

1984 Oliner A.A.: Historical perspectives on microwave field theory. IEEE Trans. MTT-32, 1022–1045

1971 Oliver B.M., Cage J.M.: Electronic Measurements and Instrumentation. (McGraw-Hill, New York)

1976 Olshanski R.: Microbending loss of single-mode fibers. 2nd Europ. Conf. Opt. Commun. 1976, 101–103

1985a Olshansky R., Fye D.M:, Manning J., Stern M.B., Meland E., Powazinik W., Ulbricht L.W., Lauer R.B.: High power, high speed edge-emitting LEDs for optical communication systems. 11th Europ. Conf. Opt. Commun. 1985, 433–436

1985b Olshansky R., Fye D., Manning J., Stern M., Meland E., Powazinik W., Ulbricht L., Lauer R.: High-power InGaAsP edge-emitting LED's for single-mode optical communication systems. Electron. Lett. 21, 730–731

1984 Olson T.C., Kapron F.P., Geyer T.W.: Describing dispersion in concatenated single-mode fiber cables. Intern. Wire & Cable Symposium Proc. 1984, 276–282

1984 Olsson N.A., Dutta N.K., Logan R.A., Besomi P.: Fiber-dispersion and propagation-delay measurements with frequency- and amplitude-modulated cleaved-coupled-cavity semiconductor lasers. Opt. Lett. 9, 180–182

1986 Osaka K., Yanagi T., Asano Y.: Mass fusion splicing machine for ribbon-type optical fibers. Internat. Wire & Cable Sympos. Proceed. 1986, 90–94

1959 Osterberg H., Snitzer E., Polanyi M., Hilberg R., Hicks J.W.: Optical waveguide modes in small glass fibers. II Experimental. J. Opt. Soc. Am. 49, 1128.

1974 Ostermayer F.W., Benson W.W.: Integrating sphere for measuring scattering loss in optical fiber waveguides. Appl. Opt. 13, 1900–1902

1979 Ostrowski D.B. (Ed.): Fiber and Integrated Optics. (Plenum Press, New York and London)

1986 Otten W.G., Kapron F.P.: Mode-field diameter by knife-edge scanning. Conf. Opt. Fiber Commun. (OFC'86), 74–75

1975 Ozeki T., Ito T., Tamura T.: Tapered section of multimode cladded fibers as mode filters and mode analyzers. Appl. Phys. Lett. 26(7), 386–388

1979 Pacey G.K., Dalgleish J.F.: Fusion splicing of optical fibers. Electron. Lett. 15, 32–34

1975 Paek U.C., Weaver A.L.: Formation of a spherical lens at optical fiber ends with a CO_2 laser. Appl. Opt. 14, 294–298

1982 Paek U.C., Peterson G.E., Carnevale A.: Parametric effects on the bandwidth of a single-mode fiber with experimental verification. Appl. Opt. 21, 704–709

1983 Paek U.C., Potasek M.J., Watros T.L.: Comparison of the wavelengths for zero total dispersion calculated from measured index profiles of depressed clad single-mode preforms and drawn fibers. Appl. Opt. 22, 1758–1762

1981 Pal B.P., Kumar A., Ghatak A.K.: Predicting dispersion minimum in a step-index monomode fiber: a comparison of the theoretical approaches. J. Opt. Commun. 2, 97–100

1985 Parent M., Bures J., Lacroix S., Lapierre J.: Propriétés de polarisation des réflecteurs de Bragg induits par photosensibilité dans les fibres optiques monomodes. Appl. Opt. 24, 354–357

1981a Parriaux O., Bernoux F., Chartier G.: Wavelength selective distributed coupling between single-mode optical fibers for multiplexing. J. Opt. Commun. 2, 105–109

1981b Parriaux O., Gidon S., Kuznetsov A.A.: Distributed coupling on polished single-mode optical fibers. Appl. Opt. 20, 2420–2423

1982 Parriaux O., Chartier G., Bernoux F.: "Coupling and multiplexing between single-mode optical fibers", in *Fiber-Optic Rotation Sensors*, ed. by S. Ezekiel and H.J. Arditty, (Springer, Berlin, Heidelberg, New York), pp. 144–148

1984   Parriaux O., Cochet F., Berthou H., Brügger A.: Loss analysis in polished single-mode fiber couplers. 10th Europ. Conf. Opt. Commun. 1984, 90–91

1975   Pask C., Barrell K.: Optical-fiber excitation by finite beams. Electron. Lett. **11**, 270–272

1978   Pask C.: Equal excitation of all modes on an optical fiber. J. Opt. Soc. Am. **68**, 572–576

1980   Pask C., Sammut R.A.: Experimental characterization of graded-index single-mode fibers. Electron. Lett. **16**, 310–311

1984   Pask C.: Physical interpretation of Petermann's strange spot size for single-mode fibers. Electron. Lett. **20**, 144–145

1980   Payne D.B., Millar C.A.: Triple-ball connector using fiber-bead location. Electron. Lett. **16**, 11–12

1982   Payne D.B., McCartney D.J., Healey P.: Fusion splicing of a 31.6 km monomode optical fiber system. Electron. Lett. **18**, 82–84

1984   Payne D.B., Reeve M.H., Millar C.A., Todd C.J.: Single-mode fiber specification and system performance. Symp. Opt. Fiber Meas. 1984, NBS spec. publ. **683**, 1–5

1975   Payne D.N., Gambling W.A.: Zero material dispersion in optical fibers. Electron. Lett. **11**, 176–178

1977   Payne D.N., Hartog A.H.: Determination of the wavelength of zero material dispersion in optical fibers by pulse delay measurements. Electron. Lett. **13**, 627–629

1982   Payne D.N., Barlow A.J., Ramskov Hansen J.J.: Development of low- and high-birefringence optical fibers. IEEE Trans MTT-30, 323–333, and IEEE J. QE-18, 477–478

1985   Payne F.P., Hussey C.D., Yataki M.S.: Modelling fused single-mode fiber couplers. Electron. Lett. **21**, 461–462

1986   Payne F.P., Finegan T., Yataki M.S., Mears R.J., Hussey C.D.: Dependence of fused taper couplers on external refractive index. Electron. Lett. **22**, 1207–1209

1982   Pearson A.D., Lazay P.D., Reed W.A., Saunders M.J.: Bandwidth optimization of depressed index single-mode fiber by means of a parametric study. 8th Europ. Conf. Opt. Commun. 1982, 93–97

1982   Peri D., Chu P.L., Whitbread T.: Direct display of the deflection function of optical fiber preforms. Appl. Opt. **21**, 809–814

1985   Perina J.: Coherence of Light, 2nd ed. (Reidel, Dordrecht, Boston, Lancaster)

1971   Personick S.D.: Time dispersion in dielectric waveguides. Bell Syst. Tech. J. **50**, 843–859

1973a Personick S.D.: Receiver design for digital fiber optic communication systems, I. Bell Syst. Tech. J. **52**, 843–874

1973b Personick S.D.: Baseband linearity and equalization in fiber optic digital communication systems. Bell Syst. Tech. J. **52**, 1175–1194

1974   Personick S.D., Hubbard W.M., Holden W.S.: Measurement of the baseband frequency response of a 1-km fiber. Appl. Opt. **13**, 266–268

1977   Personick S.D.: Photon probe – an optical-fiber time-domain reflectometer. Bell Syst. Tech. J. **56**, 355–366

1981   Personick S.D.: "Design of repeaters for fiber systems" in Fundamentals of Optical Fiber Communications, 2nd ed. ed. by M.K. Barnoski (Academic Press, New York, San Francisco, London), pp. 183–204

1976a Petermann K.: Microbending loss in monomode fibers. Electron. Lett. **12**, 107–109

1976b Petermann K.: Theory of microbending loss in monomode fibers with arbitrary refractive index profile. AEÜ, Arch. für Elektron. und Übertragungstech.: Electron. and Commun. **30**, 337–342

1976   Petermann K., Storm H.: Microbending loss in single-mode W-fibers. Electron. Lett. **12**, 537–538

1977a Petermann K.: Fundamental mode microbending loss in graded-index and W-fibers. Opt. Quantum Electron. **9**, 167–175

1977b Petermann K.: Uncertainties of the leaky mode correction for near-square-law optical fibers. Electron. Lett. **13**, 513–514

1981a Petermann K.: Transmission characteristics of a single-mode fiber transmission line with polarization coupling. 7th Europ. Conf. Opt. Commun. 1981, 3.2-1-2-4

1981b Petermann K.: Nonlinear transmission behaviour of a single-mode fiber transmission line due to polarization coupling. J. Opt. Commun. **2**, 59–64

1982   Petermann K., Arnold G.: Noise and distortion characteristics of semiconductor lasers in optical fiber communication systems. IEEE Trans. MTT-**30**, 389–400

1983   Petermann K.: Constraints for fundamental-mode spot size for broadband dispersion-compensated single-mode fibers. Electron. Lett. **19**, 712–714

1986   Petermann K., Kühne R.: Upper and lower limits for the microbending loss in arbitrary single-mode fibers. J. Lightwave Tech. LT-**4**, 2–7

1986   Petermann K., Krüger U.: Chirp reduction in intensity-modulated semiconductor lasers for maximum transmission capacity of single-mode fibers. AEÜ, Arch. für Elektron. und Übertragungstech.: Electron. and Commun. **40**, 283–288

1987   Petermann K.: Mode-field characteristics of single-mode fiber designs. Conf. Opt. Fiber Commun. (OFC'87), TUA1

1986   Petersen R.C., Philen D.L.: Safety issues concerning diode lasers used in telecommunications systems. Proc. Intern. Wire & Cable Symp. 1986, 426–432

1980   Philen D.L.: Optical time domain reflectometry on single mode fibers using a Q-switched Nd:YAG laser. Techn. Dig. Symp. Opt. Fiber Measur. 1980, NBS Spec. Publ. **597**, 97–100

1982   Philen D.L., White I.A., Kuhl J.F., Mettler S.C.: Single-mode fiber OTDR: Experiment and theory. IEEE J. QE-**18**, 1499–1508

1984   Philen D.L., White I.A., Kuhl J.F., Mettler S.C.: Authors' reply to "Comments on 'Single-mode fiber OTDR: experiment and theory'". IEEE J. QE-**20**, 1293–1294

1980   Piazzolla S., De Marchis G.: Spatial coherence in optical fibers. Opt. Commun. **32**, 380–382

1982   Piazzolla S., Spano P.: Spatial coherence in incoherently excited optical fibers. Opt. Commun. **43**, 175–179

1973   Pinnow D.A., Rich T.C.: Development of a calorimetric method for making precision optical absorption measurements. Appl. Opt. **12**, 984–992

1975   Pinnow D.A., Rich T.C.: Measurements of the absorption coefficient in fiber optical waveguides using a calorimetric technique. Appl. Opt. **14**, 1258–1259

1985   Plastow R., Monham K.L., Carter A.C., Ritter J.E., Croft T.D., Gibson M.: Transmission over 107 km of dispersion-shifted fiber at 16 Mbit/s using a 1.55 μm edge-emitting source. Electron. Lett. **21**, 369–370

1983   Pleibel W., Stolen R.H., Rashleigh S.C.: Polarization-preserving coupler with self aligning birefringent fibers. Electron. Lett. **19**, 825–826

1979   Pocholle J.P.: Single mode optical fiber characterization by the LP_{11}-mode radiation pattern. Opt. Commun. **31**, 143–147

1983   Pocholle J.P.: Caractéristiques de la propagation guidée dans les fibres optiques monomodes. Revue Techn. Thomson-CSF **15**, 881–976

1983   Pocholle J.P., Auge J.: New simple method for measuring the mode spot size in monomode fibers. Electron. Lett. **19**, 191–193

1983a   Pocholle J.P., Raffy J., Auge J., Papuchon M.: Determination of modal dispersion in monomode fibers from wavelength dependence of the mode spot size. Electron. Lett. **19**, 1093–1094

1983b   Pocholle J.P., Raffy J., Papuchon M.: Analyse de la dispersion chromatique dans les fibres optiques. Revue Techn. Thomson-CSF **15**, 1073–1131

1984   Pophillat L.: Video transmission using a 1.3 μm LED and monomode fiber. 10th Europ. Conf. Opt. Commun. 1984, 238–239

1982   Popov M.M.: A new method of computation of wave fields using Gaussian beams. Wave motion **4**, 85–97

1984   Povlsen J.H.: Characterization of graded single-mode fibers by fundamental-mode spot size variation. Electron. Lett. **20**, 543–545

1985a   Povlsen J.H.: Analysis on splice-, microbending-, and Rayleigh losses in GeO_2 doped, dispersion-shifted, single-mode fibers. 11th Europ. Conf. Opt. Commun. 1985, 321–324

1985b   Povlsen J.H.: Two-parameter characterization of single-cladded, dispersion-shifted single-mode fibers. Electron. Lett. **21**, 322–324

1986a   Povlsen J.H., Andreasen S.B.: Microbending theory and its implications. 12th Europ. Conf. Opt. Commun. 1986, 369–372

1986b   Povlsen J.H., Andreasen S.B.: Analysis on splice-, microbending-, and Rayleigh losses in GeO_2 doped dispersion-shifted single-mode fibers. J. Lightwave Tech. LT-**4**, 706–710

1976   Presby H.M., Mammel W., Derosier R.M.: Refractive-index profiling of graded-index optical fibers. Rev. Sci. Instrum. **47**, 348–352

1978 Presby H.M., Marcuse D., Astle H.W.: Automatic refractive-index profiling of optical fibers. Appl. Opt. **17**, 2209–2214

1979 Presby H.M.: Axial refractive-index depressions in preforms and fibers. Fibers Integr. Opt. **2**, 111–126

1979 Presby H.M., Marcuse D.: Preform index profiling (PIP). Appl. Opt. **18**, 671–677

1979 Presby H.M., Marcuse D., French W.G.: Refractive-index profiling of single-mode optical fibers and preforms. Appl. Opt. **18**, 4006–4011

1980 Presby H.M., Marcuse D.: The index-profile characterization of fiber preforms and drawn fibers. Proc. IEEE **68**, 1198–1203

1981a Presby H.M.: Fluorescence profiling of single-mode optical fiber preforms. Appl. Opt. **20**, 446–450

1981b Presby H.M.: Ultraviolet-excited fluorescence in optical fibers and preforms. Appl. Opt. **20**, 701–706

1987 Presby H.M., Amitay N., Dimarcello F.V., Nelson K.T.: Optical fiber up-tapers at 1.3 μm for self-aligning beam expansion and single-mode hardware. Conf. Opt. Fiber Commun. (OFC'87), paper TUQ8

1959 Primak W, Post D.: Photoelastic constants of vitreous silica and its elastic coefficient of refractive index. J. Appl. Phys. **30**, 779–788

1986 Qian J.: Generalized coupled-mode equations and their applications to fiber couplers. Electron. Lett. **22**, 304–306

1986a Qian J., Huang W.: Coupled-mode theory for LP-modes. J. Lightwave Tech. LT-**4**, 619–625

1986b Qian J., Huang W.: LP-modes and ideal modes on optical fibers. J. Lightwave Tech. LT-**4**, 626–630

1984 Ragdale C.M., Slonecker M.H., Williams J.C.: Review of fused single-mode coupler technology. Proc. SPIE **479**, 2–8

1982 Ramaswamy V., Alferness R.C., Divino M.: High efficiency single-mode fiber to Ti: LiNbO₃ waveguide coupling. Electron. Lett. **18**, 30–31

1981 Ramer O.G.: Single-mode fiber-to-channel waveguide coupling. J. Opt. Commun. **2**, 122–127

1965 Ramo S., Whinnery J.R., van Duzer Th.: Fields and Waves in Communication Electronics (Wiley, New York)

1975 Ramsay M.M., Hockham G.A., Kao K.C.: Ausbreitung von Lichtwellen in Lichtleitfasern. Elektr. Nachrichtenwesen **50**, 168–176

1984 Rao R.: Field dispersion measurements – a swept frequency technique. Symp. Opt. Fiber Meas. 1984, NBS spec. publ. **683**, 135–138

1986 Rao, R., Cook J.S.: High return loss connector design without using fiber contact or index matching. Electron. Lett. **22**, 731–732

1978 Rashleigh S.C., Ulrich R.: Polarization mode dispersion in single-mode fibers. Opt. Lett. **3**, 60–62

1982 Rashleigh S.C., Marrone M.J.: Polarization-holding in a high-birefringence fiber. Electron. Lett. **18**, 326–327

1983 Rashleigh S.C.: Origins and control of polarization effects in single-mode fibers. J. Lightwave Tech. LT-**1**, 312–331

1983 Rashleigh, S.C., Stolen R.H.: Preservation of polarization in single-mode fibers. Fiberoptic Technology, (5), 155–161

1974 Rawson E.G.: Analysis of scattering from fiber waveguides with irregular core surfaces. Appl. Opt. **13**, 2370–2377

1982 Rediker R.H., Leonberger F.J.: Analysis of integrated-optics near 3 dB coupler and Mach-Zehnder interferometric modulator using four-port scattering matrix. IEEE J. QE-**18**, 1813–1816

1986 Reed W.A., Philen D.L.: Study of algorithms used to fit group delay data for single-mode optical fibers. Techn. Dig. Sympos. Opt. Fiber Measur. 1986, NBS spec. publ. **720**, 7–10

1984 Reichelt A., Michel H., Rauscher W.: Wavelength-division multi-demultiplexers for two-channel single-mode transmission systems. J. Lightwave Tech. LT-**2**, 675–681

1986 Reith L.A., Shumate P.W., Koga Y.: Laser coupling to single-mode fiber using graded-index lenses and compact disc 1.3 μm laser package. Electron. Lett. **22**, 836–838

1987 Reith L.A., Shumate P.W.: Coupling sensitivity of an edge-emitting LED to single-mode fiber. J. Lightwave Tech. LT-5, 29–34

1979 Reitz P.R.: A quality check for fiber end faces. Optical Spectra, 39–40

1980a Reitz P.R.: Préparation et controle des extrémités de fibres optiques. L'Onde Electrique 60(10), 63–64

1980b Reitz P.R.: Fiber end preparation and inspection techniques. Electron. Engin. (Mid-April), 86

1983 Richin P., Moronvalle C.: L'épissurage par fusion entre fibres monomodes. Revue Techn. Thomson-CSF 15, 731–747

1985 Richter H.: Two-dimensional optical far-field power measurements of high angular resolution and/or high dynamic range. Proc. SPIE 584, 249–256

1982 Ridgway D.N., Freeman L.J.: A simple technique for high accuracy core-cladding concentricity measurement of single-mode fibers. Techn. Dig. Sympos. Opt. Fiber Measur. 1982, NBS Publ. 641, 139–142

1986 Ripamonti G., Cova S.: Optical time-domain reflectometry with centimeter resolution at 10^{-15} W sensitivity. Electron. Lett. 22, 818–819

1985 Ritger A.J.: Bandwidth improvement in MCVD multimode fibers by fluorine etching to reduce the center dip. 11th Europ. Conf. Opt. Commun. 1985, 913–916

1981 Rittich D.: Low loss optical couplers. Advances in Ceramics. Vol. 2. Physics of Fiber Optics, 451–462

1983 Rivoallan L, Guilloux J.Y., Lamouler P.: Monomode fiber fusion splicing with CO_2 laser. Electron. Lett. 19, 54–55

1985 Rocher E.Y.: Applications of monomode fiber to local networks. IEEE J. Select. Areas Commun. SAC-3, 897–907

1986 Rodrigues J.M.P., MacLean T.S.M., Gazey B.K., Miller J.F.: Completely fused tapered couplers: comparison of theoretical and experimental results. Electron. Lett. 22, 402–404

1981 Rogers A.J.: Polarization-optical time domain reflectometry: a technique for the measurement of field distributions. Appl. Opt. 20, 1060–1074

1980 Ronchi L., Scheggi A.M.: Beam waveguides and guided propagation. Advances in Electronics and Electron Physics 51, 63–136

1982 Ross J.N.: Birefringence measurement in optical fibers by polarization-optical time-domain reflectometry. Appl. Opt. 21, 3489–3495

1984 Ross J.N.: The rotation of the polarization in low birefringence monomode optical fibers due to geometric effects. Opt. Quantum Electron. 16, 455–461

1985 Roßberg R.: Optische Stecker für Lichtwellenleiter. NTG-Fachberichte 89, 91–95

1976 Rudolph H.-D., Neumann E.-G.: Approximations for the eigenvalues of the fundamental mode of a step-index glass fiber waveguide. Electron. Lett. 29, 328–329

1978 Runge P.K., Cheng S.S.: Demountable single-fiber optic connectors and their measurement on location. Bell Syst. Tech. J. 57, 1771–1790

1986 Runge P.K., Trischitta P.R. (Eds).: Undersea Lightwave Communications. (IEEE, New York)

1985 Russell P.St.J., Ulrich R.: Grating-fiber coupler as a high-resolution spectrometer. Opt. Lett. 10, 291–293

1986 Rybach J., Heckmann S.: Ein Rückstreumeßgerät für die Lichtwellenleiter der Zukunft. PKI Tech. Mitt. (2), 55–56

1985 Saad S.M.: On the higher order modes of elliptical optical fibers. IEEE Trans. MTT-33, 1110–1113

1979 Saekeang C., Chu P.L.: Nondestructive determination of refractive index profile of an optical fiber: backward light scattering method. Appl. Opt. 18, 1110–1116

1980 Saekeang C., Chu P.L., Whitbread T.W.: Nondestructive measurement of refractive-index profile and cross-sectional geometry of optical fiber preforms. Appl.Opt. 19, 2025–2030

1980 Saijonmaa J., Sharma A.B., Halme S.J.: Selective excitation of parabolic-index optical fibers by Gaussian beams. Appl. Opt. 19, 2442–2452

1983 Saijonmaa J., Yevick D.: Beam-propagation analysis of loss in bent optical waveguides and fibers. J. Opt. Soc. Am. 73, 1785–1791

1983 Saijonmaa J., Yevick D., Hermansson B.: An analysis of single-mode fiber bending loss. Proc. IOOC'83, Tokyo, 404–405

1981 Sakaguchi H., Seki N., Yamamoto S.: Power coupling from laser diodes into single-mode fibers with quadrangular pyramid-shaped hemiellipsoidal ends. Electron. Lett. **17**, 425–426

1977 Sakai J., Kimura T.: Large-core, broadband optical fiber. Opt. Lett. **1**, 169–171

1978a Sakai J., Kimura T.: Bending loss of propagation modes in arbitrary-index profile optical fibers. Appl. Opt. **17**, 1499–1506

1978b Sakai J., Kimura T.: Splice loss evaluation for optical fibers with arbitrary index profile. Appl. Opt. **17**, 2848–2853

1978c Sakai J., Kimura T.: Splicing and bending losses of single-mode optical fibers. Appl. Opt. **22**, 3653–3659

1978 Sakai J., Kitayama K., Ikeda M., Kato Y., Kimura T.: Design considerations of broad-band dual-mode optical fibers. IEEE Trans. MTT-**26**, 658–665

1979 Sakai J.: Simplified bending loss formula for single-mode optical fibers. Appl. Opt. **18**, 951–952

1979 Sakai J., Kimura T.: Practical microbending loss formula for single-mode optical fibers. IEEE J. QE-**15**, 497–500

1980a Sakai J.: Microbending loss evaluation in arbitrary-index single-mode optical fibers. Part I: Formulation and general properties. IEEE J. QE-**16**, 36–44

1980b Sakai J.: Microbending loss evaluation in arbitrary-index single-mode optical fibers. Part II: Effects of core index profiles. IEEE J. QE-**16**, 44–49

1980 Sakai J., Kimura T.: Design of a miniature lens for semiconductor laser to single-mode fiber coupling. IEEE J. QE-**16**, 1059–1067

1981 Sakai J., Kimura T.: Fields in a curved optical fiber. IEEE J. QE-**17**, 29–34

1982 Sakai J., Kimura T.: Analytical bending loss formula of optical fibers with field deformation. Radio Science **17**, 21–29

1984 Sakai J., Shimizu M.: Power spectrum of fiber output fluctuation in polarization-preserving optical fibers. Electron. Lett. **20**, 923–925

1978 Sakamoto K., Miyajiri T., Kakuzen H., Hirai M., Uchida N.: The automatic splicing machine employing electric arc fusion. 4th Europ. Conf. Opt. Commun. 1978, 296–303

1975 Sammut R.A., Pask C., Snyder A.W.: Excitation and power of the unbound modes within a circular dielectric waveguide. Proc. Inst. Electr. Eng. **122**, 25–33

1976a Sammut R.A., Snyder A.W.: Leaky modes on a dielectric waveguide: orthogonality and excitation. Appl. Opt. **15**, 1040–1044

1976b Sammut R.A., Snyder A.W.: Leaky modes on a dielectric waveguide: orthogonality and excitation: erratum. Appl. Opt. **15**, 2953.

1977 Sammut R.A.: Some integrals in the theory of curved optical fibers. Opt. Quantum Electron. **9**, 539.

1978 Sammut R.A.: Range of monomode operation of W-fibers. Opt. Quantum Electron. **10**, 509–514

1978 Sammut R.A., Ghatak A.K.: Perturbation theory of optical fibers with power-law core profile. Opt. Quantum Electron. **10**, 475–482

1979 Sammut R.A.: Analysis of approximations for the mode dispersion in monomode fibers. Electron. Lett. **15**, 590–591

1984 Samson P.J.: Computation of fundamental-mode spot size from far-field pattern of single-mode optical fibers. Electron. Lett. **20**, 680–681

1985a Samson P.J.: Usage-based comparison of ESI-techniques. J. Lightwave Tech. LT-**3**, 165–175

1985b Samson P.J.: Measurement of wavelength dependence of the far-field pattern of single-mode optical fibers. Opt. Lett. **10**, 306–307

1985c Samson P.J.: Measurement of single-mode optical-fiber spot size using strip integration of far-field. Electron. Lett. **21**, 589–591

1986 Samson P.J.: Far-field techniques for the characterization of single-mode fibers. Opt. Quantum Electron. **18**, 5–22

1986a Sankawa I., Nagasawa S., Satake T., Ishida M., Arioka R.: Low-loss small-size optical-fiber connector with a precision plastic-molded ferrule. J. Lightwave Tech. LT-**4**, 1237–1242

1986b Sankawa I., Satake T., Kashima N.: Methods for reducing optical fiber connector reflection through use of index matching materials. Rev. Electr. Commun. Lab. NTT, Japan **34**, 703–709

1982a Sansonetti P.: Prediction of modal dispersion in single-mode fibers from spectral behaviour of mode spot size. Electron. Lett. **18**, 136–138

1982b Sansonetti P.: Modal dispersion is single-mode fibers: simple approximation issued from mode spot size spectral behaviour. Electron. Lett. **18**, 647–648

1985 Saravanos S., Lowe R.S.: New approach for determining non-Gaussian mode fields of single-mode fibers from measurements in far-field. Electron. Lett. **21**, 898–899

1986a Saravanos S., Lowe R.S.: Measurement of non-Gaussian mode fields by the far-field axial scanning technique. Conf. Opt. Fiber Commun. (OFC'86), 104–105

1986b Saravanos S., Lowe R.S.: New approach for determining non-Gaussian mode fields of single-mode fibers from measurements in far-field: Reply. Electron. Lett. **22**, 110.

1986c Saravanos S., Lowe R.S.: The measurement of non-Gaussian mode fields by the far-field axial scanning technique. J. Lightwave Tech. LT-4, 1563–1566

1984 Sarkar S., Thyagarajan K., Kumar A.: Gaussian approximation of the fundamental mode in single-mode elliptic core fibers. Opt. Commun. **49**, 178–183

1985 Sarkar S., Kumar A., Goyal I.C., Sharma E.K., Tewari R.: On microbending loss calculations in single-mode fibers with arbitrary index profiles. Opt. Commun. **53**, 91–94

1979 Saruwatari M., Nawata K.: Semiconductor to single-mode fiber coupler. Appl. Opt. **18**, 1847–1856

1980 Saruwatari M., Sugie T.: Efficient laser diode – single-mode fiber coupling using two confocal lenses. Electron. Lett. **16**, 955–956

1981 Saruwatari M, Sugie T.: Efficient laser diode to single-mode fiber coupling using a combination of two lenses in confocal condition. IEEE J. QE-17, 1021–1027

1983 Saruwatari M., Suzuki N., Sugie T.: Active and passive optical devices for use in single-mode fiber transmission systems. Rev. Electr. Commun. Lab. NTT, Japan **31**, 299–309

1984 Saruwatari M.: "Laser diode module for single-mode optical fiber", in Japan Annual Review in Electronics, Computers, & Telecommunications, Vol. 11, Optical Fibers & Devices, ed. by Y. Suematsu (Ohm, Tokyo, Osaka, Kyoto; North-Holland, Amsterdam, New York, Oxford), pp. 129–140

1980a Sasaki I., Payne D.N., Adams M.J.: Measurement of refractive-index profiles in optical-fiber preforms by spatial-filtering technique. Electron. Lett. **16**, 219–221

1980b Sasaki I., Payne D.N., Mansfield R.J., Adams M.J.: Variation of refractive-index profiles in single-mode fiber preforms measured using an improved high-resolution spatial-filtering technique. 6th Europ. Conf. Opt. Commun. 1980, 140–143

1981 Sasaki I., Francois P.L., Payne D.N.: Accuracy and resolution of preform index-profiling by the spatial-filtering method. 7th Europ. Conf. Opt. Commun. 1981, 6.4-1...6.4-4

1982 Sasaki Y., Shibata N., Noda J.: Splicing of single-polarization fibers by an optical short-pulse method. Electron. Lett. **18**, 997–999

1984 Sasaki Y., Hosaka T., Noda J.: Fabrication of polarization-maintaining optical fibers with stress-induced birefringence. Rev. Electr. Commun. Lab. NTT, Japan **32**, 452–460

1986 Satake T., Nagasawa S., Arioka R.: A new type of demountable plastic-molded single-mode multifiber connector. J. Lightwave Tech. LT-4, 1232–1236

1985 Saul R.H., King W.C., Olson N.A., Zipfel C.L., Chin B.H., Chin A.K., Camlibet I., Minneci G.: 180 Mb/s, 35 km transmission over single-mode fiber using 1.3 μm edge-emitting LED's. Electron. Lett. **21**, 773–775

1979 Saunders M.J.: Torsion effects on fractured fiber ends. Appl. Opt. **18**, 1480–1481

1981 Saunders M.J.: Optical fiber profiles using the refracted near-field technique: a comparison with other methods. Appl. Opt. **20**, 1645–1651

1982 Saunders M.J.: A comparison of single-mode refractive index profiles in preforms and fibers. 8th Europ. Conf. Opt. Commun. 1982, 587–589

1984 Saunders M.J., Gardner W.B.: Precision interferometric measurement of dispersion in short single-mode fibers. Symp. Opt. Fiber Meas. 1984, NBS spec. publ. **683**, 123–126

1985 Saunders M.J., Gardner W.B.: Interferometric determination of dispersion variations in single-mode fibers. 11th Europ. Conf. Opt. Commun. 1985, 893–896

1980 Scherer W.G.: Stress-induced index profile distortion in optical waveguides. Appl. Opt. **19**, 2000–2006

1980 Schiffner G., Schneider H., Schöner G.: Double-core single-mode optical fiber as directional coupler. Appl. Phys. **23**, 41–45

1981 Schöner G., Schiffner G.: Coupling properties of a double-core single-mode optical fiber. Siemens Forsch.-u. Entwickl.-Ber. **10**, 172–178

1983 Schulte H.J.: Transmission tests during the SL lightwave submarine cable system sea trial. Proc. SPIE **425**, 142–148

1985 Schwander T., Schwaderer B.: Ankopplung von Monomodefasern an Halbleiterlaser mit hohem Wirkungsgrad bei geringen Rückwirkungen. Frequenz **39**, 139–143

1985 Schwander T., Schwaderer B., Storm H.: Coupling of lasers to single-mode fibers with high efficiency and low optical feedback. Electron. Lett. **21**, 287–289

1980 M.Schwartz: Information Transmission, Modulation and Noise, 3nd ed. (McGraw-Hill, New York)

1986 Schwierz: private communication.

1983 Sears F.M., Cohen L.G., Stone J.: Measurements of the axial uniformity of dispersion spectra in single-mode fibers. Proc. SPIE **425**, 56–62

1984 Sears F.M., Cohen L.G., Stone J.: Interferometric measurements of dispersion-spectra variations in a single-mode fiber. J. Lightwave Tech. LT-**2**, 181–184

1985 Sears F.M., White I.A., Kummer R.B., Stone F.T.: Probability of modal noise in single-mode lightguide systems. 11th Europ. Conf. Opt. Commun. 1985, 823–826

1986a Sears F.M., White I.A., Gentry S.P.: Cable cutoff wavelength and attenuation of the LP_{11}-mode. Conf. Opt. Fiber Commun. (OFC'86), 66–67

1986b Sears F.M., White I.A., Kummer R.B., Stone F.T.: Probability of modal noise in single-mode lightguide systems. J. Lightwave Tech. LT-**4**, 652–655

1985 Seikai A., Ohashi M., Shibata N., Uesugi N.: Development of single-mode fiber cables in NTT. Proc. SPIE **584**, 202–209

1982 Seki M., Kobayashi K., Odagiri Y., Shikada M., Tanigawa T., Ishikawa R.: 20-channel micro-optic grating demultiplexer for 1.1-1.6 μm band using a small focusing parameter graded-index rod lens. Electron. Lett. **18**, 257–258

1983 Self S.A.: Focusing of spherical Gaussian beams. Appl. Opt. **22**, 658–661

1987 Shaar C.S., Kritler D.A., Lovely P.S.: Errors in single-mode cutoff wavelength measurements due to periodic features in the transmission spectrum of a short multimode reference fiber. Opt. Lett. **12**, 132–134

1984 Shah V.: Effective cut-off wavelength for single-mode fibers: the combined effect of curvature and index profile. Symp. Opt. Fiber Meas. 1984, NBS spec. publ. **683**, 7–10

1986 Shah V., Curtis L., Young W.C.: Joint losses as a function of offset and mode field diameters for dispersion-shifted and standard fibers and fiber combinations. Techn. Dig. Sympos. Opt. Fiber Measur. 1986, NBS spec. publ. **720**, 57–60

1987 Shah V.: Curvature dependence of the effective cutoff wavelength in single-mode fibers. J. Lightwave Tech. LT-**5**, 35–43

1987 Shah V., Young W.C., Curtis L.: Large fluctuations in transmitted power at fiber joints with polished endfaces. Conf. Opt. Fiber Commun. (OFC'87), TUF4

1981a Shang H.-T.: Chromatic dispersion measurement by white-light interferometry on meter-length single-mode optical fibers. Electron. Lett. **17**, 603–605

1982 Shank C.V., Fork R.L., Yen R., Stolen R.H., Tomlinson W.J.: Compression of femtosecond optical pulses. Appl. Phys. Lett. **40**(9), 761–763

1981a Sharma A., Ghatak A.K.: A variational analysis of single-mode graded-index fibers. Opt. Commun. **36**, 22–24

1981b Sharma A., Ghatak A.K.: A simple numerical method for the cutoff frequency of a single-mode fiber with an arbitrary index-profile. IEEE Trans. MTT-**29**, 607–610

1982 Sharma A., Hosain S.I., Ghatak A.K.: The fundamental mode of graded-index fibers: simple and accurate variational methods. Opt. Quantum Electron. **14**, 7–15

1981 Sharma A.B., Halme S.J., Butusov M.M.: Optical Fiber Systems and Their Components. (Springer, Berlin, Heidelberg, New York)

1984a Sharma A.B., Al-Ani A.-H., Halme S.J.: Experimental modification of the bend-loss formula of monomode fibers. 10th Europ. Conf. Opt. Commun. 1984, 210–211

1984b Sharma A.B., Al-Ani A.-H., Halme S.J.: Constant-curvature loss in monomode fibers: an experimental investigation. Appl. Opt. **23**, 3297–3301

1985 Sharma A.B., Vanninen M., Oksanen L.: Chromatic dispersion in single-mode fibers: a new theoretical and experimental approach. 11th Europ. Conf. Opt. Commun. 1985, 667–670

1981 Sharma E.K., Goyal I.C., Ghatak A.K.: Calculation of cutoff frequencies in optical fibers for arbitrary profiles using the matrix method. IEEE J. QE-**17**, 2317–2321

1984 Sharma E.K., Tewari R.: Accurate estimation of single-mode fiber characteristics from near-field measurements. Electron. Lett. **20**, 805–806

1979a Sheem S.K., Giallorenzi T.G.: Single-mode fiber-optical power divider: encapsulated etching technique. Opt. Lett. **4**, 29–31

1979b Sheem S.K., Giallorenzi T.G.: Single-mode fiber multiterminal star directional coupler. Appl. Phys. Lett. **35**(2), 131–133

1979 Sheem S.K., Cole J.H.: Acoustic sensitivity of single-mode optical power dividers. Opt. Lett. **4**, 322–324

1980 Sheem S.K.: Fiber-optic gyroscope with [3 × 3] directional coupler. Appl. Phys. Lett. **37**(10), 869–871

1984 Shen Y.R.: The Principles of Nonlinear Optics. (Wiley, New York)

1971 Shevchenko V.V.: Radiation losses in bent waveguides for surface waves. Radiophys. Quantum Electron. **14**, 607–614

1985 Shi Y.C., Klafter L., Harstead E.E.: A dual-fiber optical rotary joint. J. Lightwave Tech. LT-**3**, 999–1004

1979 Shibata N., Jinguji K., Kawachi M., Edahiro T.: Nondestructive structure measurement of optical-fiber preforms with photoelastic effect. Japanese J. of Appl. Phys. **18**, 1267–1273

1980 Shibata N., Tateda M., Seikai S., Uchida N.: Spatial technique for measuring modal delay differences in a dual-mode fiber. Appl. Opt. **19**, 1489–1492

1981a Shibata N., Katsuyama Y., Tateda M., Seikai S.: Thermal characteristic measurement in jacketed and unjacketed fibers by a spatial technique. Electron. Lett. **17**, 345–347

1981b Shibata N., Shibata S., Edahiro T.: Refractive index dispersion of lightguide glasses at high temperature. Electron. Lett. **17**, 310–311

1981c Shibata N., Tateda M., Seikai S., Uchida N.: Wavelength dependence of polarization mode dispersion in elliptical-core single-mode fibers. Electron. Lett. **17**, 564-565

1982 Shibata N., Tateda M., Seikai S.: Polarization mode dispersion measurement in elliptical core single-mode fibers by a spatial technique. IEEE J. QE-**18**, 53–58

1983 Shibata N., Katsuyama Y., Mitsunaga Y., Tateda M., Seikai S.: Thermal characteristics of optical pulse transit time delay and fiber strain in a single-mode optical fiber cable. Appl. Opt. **22**, 979–984

1984 Shibata N., Tsubokawa M., Seikai S.: Measurements of polarization mode dispersion by optical heterodyne detection. Electron. Lett. **20**, 1055–1057

1985 Shibata N., Tsubokawa M., Ohashi M., Seikai S.: Bending loss measurement of LP_{11}-mode in quasi-single-mode operation region. Electron. Lett. **21**, 1042–1043

1981 Shibata S., Horiguchi M., Jinguij K., Mitachi S., Kanamori T., Manabe T.: Prediction of loss minima in infra-red optical fibers. Electron. Lett. **17**, 775–777

1986 Shigihara K., Shiraishi K., Kawakami S.: Modal field transformer between dissimilar waveguides. 12th Europ. Conf. Opt. Commun. 1986, 185–188

1986 Shiina T., Shiraishi K., Kawakami S.: Waveguide-bend configuration with low-loss characteristics. Opt. Lett. **11**, 736–738

1978 Shimizu N., Tsuchiya H.: Single-mode fiber connectors. Electron. Lett. **14**, 611–613

1979 Shimizu N., Tsuchiya H.: Single-mode optical fiber connectors. Trans. IECE Japan, E**62**, 276–278

1979 Shimizu N., Tsuchiya H., Izawa T.: Low-loss single-mode fiber connectors. Electron. Lett. **15**, 28–29

1984 Shinohara S.: "Single-mode system" in Japan Annual Review in Electronics, Computers, & Telecommunications, Vol. 11, Optical Fibers and Devices, ed. by Y. Suematsu (Ohm, Tokyo, Osaka, Kyoto; North-Holland, Amsterdam, New York, Oxford), pp. 254–263

1986 Shirasaki M., Nakajima H., Fukushima N., Asama K.: Broadening of bandwidths in grating multiplexer by original dispersion-dividing prism. Electron. Lett. **22**, 764–765

1985 Shumate P.W., Gimlett J.L., Stern M., Romeiser M.B., Cheung N.K.: Transmission of 140 Mbit/s signals over single-mode fiber using surface- and edge-emitting 1.3 μm LED's. Electron. Lett. **21**, 522–524

1971 Siegman A.E.: An Introduction to Lasers and Masers. (McGraw-Hill, New York)

1986 Simon R., Sudarshan E.C.G., Mukunda N.: Gaussian-Maxwell beams. J. Opt. Soc. Am. A **3**, 536–540

1975 Sklyarov O.K.: An electric-spark method of treating the ends of an optical fiber. Sov. J. Opt. Technol. **42**, 606–607

1986 Sladen F.M.E., Reichard H.S., Uveges S.: Chromatic dispersion measurements on long fiber lengths using LED's. Electron. Lett. **22**, 841–842

1982 Slonecker M.H.: Single-mode fused biconical taper coupler. Conf. Opt. Fiber Commun. (OFC'82), 36–37

1983 Slonecker M.H., Williams J.C.: Recent advances in single-mode fused taper coupler technology. Proc. SPIE **412**, 50–53

1984 Slonecker M.H., Goodman S.E., Williams J.C.: Low-loss environmentally stable single-mode optical coupler for 1300 nm operation. Conf. Opt. Fiber Commun. (OFC'84), 98–99

1985 Smith D.K., Westwig R.A.: Non-Gaussian fitting of variable aperture far-field data and waveguide dispersion from multiwavelength measurements of mode radius. Conf. Opt. Fiber Commun. (OFC'85), 28–30

1972 Smith R.G.: Optical power handling capacity of low loss optical fibers as determined by stimulated Raman and Brillouin scattering. Appl. Opt. **11**, 2489–2494

1959 Snitzer E., Hicks J.W.: Optical waveguide modes in small glass fibers. I. Theoretical. J. Opt. Soc. Am. **49**, 1128.

1961 Snitzer E.: Cylindrical dielectric waveguide modes. J. Opt. Soc. Am. **51**, 491–498

1961 Snitzer E., Osterberg H.: Observed dielectric waveguide modes in the visible spectrum. J. Opt. Soc. Am. **51**, 499–505

1969 Snitzer E.: "Optical dielectric waveguides" in Advances in Quantum Electronics (Columbia University Press), pp. 348–369

1965 Snyder A.W.: Surface mode coupling along a tapered dielectric rod. IEEE Trans. AP-**13**, 821–822

1966 Snyder A.W.: Surface waveguide modes along a semi-infinite dielectric fiber excited by a plane wave. J. Opt. Soc. Am. **56**, 601–606

1969a Snyder A.W.: Asymptotic expressions for eigenfunctions and eigenvalues of a dielectric or optical waveguide. IEEE Trans. MTT-**17**, 1130–1138

1969b Snyder A.W.: Excitation and scattering of modes on a dielectric or optical fiber. IEEE Trans. MTT-**17**, 1138–1144

1970a Snyder A.W.: Coupling of modes on a tapered dielectric cylinder. IEEE Trans. MTT-**18**, 383–392

1970b Snyder A.W.: Radiation losses due to variations of radius on dielectric or optical fibers. IEEE Trans. MTT-**18**, 608–615

1972 Snyder A.W.: Coupled-mode theory for optical fibers. J. Opt. Soc. Am. **62**, 1267–1277

1973 Snyder A.W., Mitchell D.J.: Leaky rays cause failure of geometric optics on optical fibers. Electron. Lett. **9**, 437–438

1973 Snyder A.W., Pask C.: Incoherent illumination of an optical fiber. J. Opt. Soc. Am. **63**, 806–812

1974 Snyder A.W.: Leaky-ray theory of optical waveguides of circular cross section. Appl. Phys. **4**, 273–298

1974 Snyder A.W., Mitchell D.J.: Leaky-mode analysis of circular optical waveguides. Opto-electronics **6**, 287–296

1974 Snyder A.W., Mitchell D.J., Pask C.: Failure of geometric optics for analysis of circular optical fibers. J. Opt. Soc. Am. **64**, 608–614

1975 Snyder A.W., White I., Mitchell D.J.: Radiation from bent optical waveguides. Electron. Lett. **11**, 332–333

1978 Snyder A.W., Young W.R.: Modes of optical waveguides. J. Opt. Soc. Am. **68**, 297–309

1979 Snyder A.W., Sammut R.A.: Fundamental (HE$_{11}$) modes of graded optical fibers. J. Opt. Soc. Am. **69**, 1663–1671

1980 Snyder A.W.: Weakly guiding optical fibers. J. Opt. Soc. Am. **70**, 405–411

1981 Snyder A.W.: Understanding momomode optical fibers. Proc. IEEE **69**, 6–12

1983 Snyder A.W., Love J.D.: Optical Waveguide Theory. (Chapman and Hall, London)

1985 Snyder A.W., Rühl F.F.: Radiation losses from couplers. J. Lightwave Tech. LT-**3**, 31–36

1985 Snyder A.W., Zheng X.: Fused couplers of arbitrary cross-section. Electron. Lett. **21**, 1079–1080

1986 Snyder A.W., Ankiewicz A.: Fiber couplers composed of unequal cores. Electron. Lett. **22**, 1237–1238

1986 So, V., Lamont J., Vella P.J.: Fault-locating accuracy with an OTDR: reflective versus nonreflective faults. IEEE J. Select. Areas Commun. SAC-**4**, 737–740

1973   Someda C.G.: Simple, low-loss joints between single-mode optical fibers. Bell Syst. Tech. J. **52**, 583–596

1975   Someda C.G., Zoboli M.: Cutoff wavelengths of guided modes in optical fibers with square-law core profile. Electron. Lett. **11**, 602–603

1984   Someda C.G.: Simple way to understand the behaviour of an optical Y-junction. Electron. Lett. **20**, 349–350

1983   Sorin W.V., Jackson K.P., Shaw H.J.: Evanescent amplification in a single-mode optical fiber. Electron. Lett. **19**, 820–822

1985a Sorin W.V., Shaw H.J.: Single-mode fiber evanescent grating reflector. Conf. Opt. Fiber Commun. (OFC'85), 132–133

1985b Sorin W.V., Shaw H.J.: A single-mode fiber evanescent grating reflector. J. Lightwave Tech. LT-3, 1041–1043

1986a Sorin W.V., Kim B.Y., Shaw H.J.: Phase-velocity measurements using prism output coupling for single- and few-mode optical fibers. Opt. Lett. **11**, 106–108

1986b Sorin W.V., Kim B.Y., Shaw H.J.: Highly selective evanescent modal filter for two-mode optical fibers. Opt. Lett. **11**, 581–583

1987   Sorin W.V., Zorabedian P., Newton S.A.: Tunable single-mode fiber reflective grating filter. Conf. Opt. Fiber Commun. (OFC'87), THB4

1983   Spano P., De Marchis G., Grosso G.: Coherence properties and cutoff wavelength determination in dielectric waveguides. Appl. Opt. **22**, 1915–1917

1979   Sporleder F., Unger H.-G.: Waveguide Tapers Transitions and Couplers. (Peregrinus, Stevenage and New York)

1985   Srivastava R., Franzen D.L.: Single-mode optical fiber characterization. National Bureau of Standards Report July 1985, 1–101

1986   Stallard W.A., Beaumont A.R., Booth R.C.: Integrated optic devices for coherent transmission. J. Lightwave Tech. LT-4, 852–857

1981   Stensland L., Borak G.: Raman time-domain reflectometry. Conf. Opt. Fiber Commun. (OFC'81), 104–105

1970   Stern J.R., Dyott R.B.: Launching into single-mode optical fiber waveguide. Conf. Trunk Telecommunications by Guided Waves, London 1970, IEE publ. **71**, 191–196

1970   Stern J.R., Peace M., Dyott R.B.: Launching into optical-fiber waveguide. Electron. Lett. **6**, 160–162

1971   Stern J.R., Dyott R.B.: Off-axis launching into a fiber-optical waveguide. Electron. Lett. **7**, 52–53

1982   Stern J.R., Payne D.B., Wood T.D.S., Todd C.J.: The characterization of monomode fiber links installed in operational duct. Techn. Dig. Sympos. Opt. Fiber Measur. 1982, NBS Publ. **641**, 79–84

1976   Stewart J.H., Hensel P.: Technique for jointing small-core optical fibers. Electron. Lett. **12**, 570.

1975   Stewart W.J.: A new technique for determining the V values and refractive index profiles of optical fibers. Digest of Techn. Papers Topical Meeting on Optical Fiber Transmiss., Williamsburg 1975, TuD8-1–TuD8-4

1977   Stewart W.J.: A new technique for measuring the refractive index profiles of graded optical fibers. Tech. Dig. Intern. Conf. Integr. Optics Opt. Fiber Commun. IOOC'77, 395–398

1980a Stewart W.J.: "Resolution of near-field optical fiber refractive index profiling methods", in Scanned Image Microscopy, ed. by Ash E.A. (Academic Press, New York), pp. 233–239

1980b Stewart W.J.: Simplified parameter-based analysis of single-mode optical guides. Electron. Lett. **16**, 380–382

1982a Stewart W.J.: Index profile measurements. Techn. Dig. Sympos. Opt. Fiber Measur. 1982, NBS Publ. **641**, 135–138

1982b Stewart W.J.: Optical fiber and preform profiling technology. IEEE J. QE-18, 1451–1466

1982   Stewart W.J., Reid D.C.J.: High resolution optical fiber index profiling. 8th Europ. Conf. Opt. Commun. 1982, 193–196

1983   Stewart W.J., Reid D.C.J., Rees G.: Spot size and concentricity measurements of monomode fibers: improved techniques. Conf. Opt. Fiber Commun. (OFC'83), 24–25

1984   Stewart W.J., Rees G.J., Reid D.C.J.: Waveguide dispersion measurement in monomode fibers from spot size. 10th Europ. Conf. Optic. Commun. 1984, 122–123

1985   Stewart W.J., Love J.D.: Design limitations on tapers and couplers in single-mode fibers. 11th Europ. Conf. Opt. Commun. 1985, 559–562

1982   Stokes L.F., Chodorow M., Shaw H.J.: All-single-mode fiber resonator. Opt. Lett. **7**, 288–290

1983   Stokes L.F., Chodorow M., Shaw H.J.: Sensitive all-single-mode fiber resonant ring interferometer. J. Lightwave Tech. LT-1, 110–115

1979a Stolen R.H.: "Nonlinear properties of optical fibers" in Optical Fiber Telecommunications, ed. by Miller S.E. and Chynoweth A.G. (Academic Press, New York, San Francisco, London), pp. 125–150

1979b Stolen R.H.: "Fiber Raman lasers", in Fiber and Integrated Optics, ed. by Ostrowsky D.B. (Plenum Press, New York, London), pp. 157–182

1980   Stolen R.H.: Nonlinearity in fiber transmission. Proc. IEEE **68**, 1232–1236

1978   Stone F.T., Gardner W.B., Lovelace C.R.: Calorimetric measurement of absorption and scattering losses in optical fibers. Opt. Lett. **2**, 48–50

1982   Stone F.T.: Separation of total-loss data into its absorption and scattering components: a more accurate model for fiber loss. Appl. Opt. **21**, 2721–2726

1984   Stone F.T.: The relation between source spectral width and coherence length in single-mode fibers: the effect on modal noise. 10th Europ. Conf. Opt. Commun. 1984, 100–101

1976   Stone J., Earl H.E.: Surface effects and reflection refractometry of optical fibers. Opt. Quantum Electron. **8**, 459–463

1982   Stone J., Cohen L.G.: Minimum-dispersion spectra of single-mode fibers measured with subpicosecond resolution by white-light crosscorrelation. Electron. Lett. **18**, 716–718

1984   Stone J., Marcuse D.: Direct measurement of second-order dispersion in short optical fibers using white-light inferferometry. Electron. Lett. **20**, 715–752

1984   Stone J., Chraplyvy A.R., Wiesenfeld J.M., Burrus C.A.: Overtone absorption and Raman spectra of H_2 and D_2 in silica optical fibers. AT&T Bell Lab. Techn. J. **63**, 991–999

1985a Stone J.: Optical-fiber Fabry-Perot interferometer with finesse of 300. Electron. Lett. **21**, 504–505

1985b Stone J.: An optical fiber Fabry-Perot inteferometer with finesse of 300. 11th Europ. Conf. Opt. Commun. 1985, 855–858

1985   Stone J., Chraplyvy A.R., Kasper B.L.: Long-range $1.5\,\mu$m OTDR in a single-mode fiber using a D_2 gas-in-glass laser (100 km) or a semiconductor laser (60 km). Electron. Lett. **21**, 541–542

1986   Stone J., Marcuse D: Ultrahigh finesse fiber Fabry-Perot interferometers. J. Lightwave Tech. LT-4, 382–385

1941   Stratton J.A.: Electromagnetic Theory. (McGraw-Hill, New York)

1980   Streckert J.: New method for measuring the spot size of single-mode fibers. Opt. Lett. **5**, 505–506

1982   Streckert J., Brinkmeyer E.: Characteristic parameters of monomode fibers. Appl. Opt. **21**, 1910–1915

1984   Streckert J.: Verfahren zum Erfassen des Felddurchmessers eines einwelligen Lichtwellenleiters. German patent pending, No. P 34 24 844.7, 6. July 1984

1985   Streckert J.: A new fundamental mode spot size definition usable for non-Gaussian and noncircular field distributions. J. Lightwave Tech. LT-3, 328–331

1986   Streckert J.: Private communication.

1967   Streifer W., Kurtz C.N.: Scalar analysis of radially inhomogeneous guiding media. J. Opt. Soc. Am. **57**, 779–786

1982   Stueflotten S.: Low temperature excess loss of loose tube fiber cables. Appl. Opt. **21**, 4300–4307

1984   Sucha G.D., Carter W.H.: Focal shift for a Gaussian beam: an experimental study. Appl. Opt. **23**, 4345–4347

1986   Suchoski P.G., Ramaswamy V.: Exact numerical technique for the analysis of step discontinuities and tapers in optical dielectric waveguides. J. Opt. Soc. Am. A **3**, 194–203

1983   Sudbø A.S.: An optical time domain reflectometer with low-power InGaAsP diode lasers. J. Lightwave Tech. LT-1, 616–618

1978   Sudo S., Kawachi M., Edahiro T., Izawa T., Shioda T., Gotoh H.: Low-OH-content optical fiber fabricated by vapour-phase axial-deposition method. Electron. Lett. **14**, 534–535

1975 Suematsu Y., Furuya K.: Quasi-guided modes and related radiation losses in optical dielectric waveguides with external higher index surroundings. IEEE Trans. MTT-23, 170–175

1977 Suematsu Y., Kishino K.: Coupling coefficient in strongly coupled dielectric waveguides. Radio Science 12, 587–592

1982 Suematsu Y., Hirota O., Toda J., Tejima S.: Property of modal noise in single-mode optical fibers. 8th Europ. Conf. Opt. Commun. 1982, 154–158

1983 Suematsu Y.: Long-wavelength optical fiber communication. Proc. IEEE 71, 692–721

1982 Sugie T., Saruwatari M.: Nonreciprocal circuit for laser diode to single-mode fiber coupling employing a YIG sphere. Electron. Lett. 18, 1026–1028

1983 Sugie T., Saruwatari M.: An effective nonreciprocal circuit for semiconductor laser to optical fiber coupling using a YIG sphere. J. Lightwave Tech. LT-1, 121–130

1986 Sugie T., Saruwatari M.: Distributed feedback laser diode (DFB-LD) to single-mode fiber coupling module with optical isolator for high bit rate modulation. J. Lightwave Tech. LT-4, 236–245

1980 Sugimura A., Daikoku K., Imoto N., Miya T.: Wavelength dispersion characteristics of single-mode fibers in low-loss region. IEEE J. QE-16, 215–225

1986 Sugita E., Iwasa K., Shintaku T.: Design for push-pull coupling single fiber connectors featuring zirconia (ZrO_2) ceramic ferrules. 12th Europ. Conf. Opt. Commun. 1986, 141–144

1982 Sumida M., Takemoto K.: Lens aberration effect on a laser diode to single-mode fiber coupler. Electron. Lett. 18, 586–587

1984 Sumida M., Takemoto K.: Lens coupling of laser diodes to single-mode fibers. J. Lightwave Tech. LT-2, 305–311

1983 Suzuki K., Shibata N., Ishida Y.: Polarization-mode dispersion as a bandwidth-limiting factor in a long-haul single-mode optical-transmission system. Electron. Lett. 19, 689–691

1984 Suzuki K., Horiguchi T., Seikai S.: Optical time-domain reflectometer with a semiconductor laser amplifier. Electron. Lett. 20, 714–716

1986 Suzuki K., Noguchi K., Uesugi N.: Long-range OTDR for single-mode optical fiber using a P_2O_5 highly doped fiber Raman laser. Electron. Lett. 22, 1273–1274

1983 Suzuki M., Kikuchi Y., Yamada T., Watanabe O.: Arc fusion splicing machine for single polarization single-mode optical fibers. 9th Europ. Conf. Opt. Commun. 1983, 117–120

1979 Suzuki N., Iwahara Y., Saruwatari M., Nawata K.: Ceramic capillary connector for 1.3 μm single-mode fibers. Electron. Lett. 15, 809–810

1986 Suzuki N., Saruwatari M., Okuyama M.: Low insertion and high return loss optical connectors with spherically convex polished end. Electron. Lett. 22, 110–112

1978 Szczepanek P.S., Berthold J.W.: Side launch excitation of selected modes in graded-index optical fibers. Appl. Opt. 17, 3245–3247

1984 Tachikura M., Kashima N.: Fusion mass-splices for optical fibers using high-frequency discharge. J. Lightwave Tech. LT-2, 25–31

1985 Tai S., Kyuma K., Nakayama T.: Novel measuring method for spectral linewidth of laser diodes using fiber-optic ring resonator. Electron. Lett. 21, 91–93

1985 Tajima K.: Self amplitude modulation in PSK coherent optical transmission systems. 11th Europ. Conf. Opt. Commun. 1985, 351–354

1985 Tajima K., Washio K.: Generalized view of Gaussian pulse-transmission characteristics in single-mode optical fibers. Opt. Lett. 10, 460–462

1985 Takada A., Sugie T., Saruwatari M.: Picosecond optical pulse compression from gain-switched 1.3 μm distributed-feedback laser diode through highly dispersive single-mode fiber. Electron. Lett. 21, 969–971

1986 Takahashi H., Chang J., Nakamura K., Sugimoto I., Takabayashi T., Oyobe A.: Efficient single-pass Raman generation in a GeO_2 optical fiber and its application to measurement of chromatic dispersion. Opt. Lett. 11, 383–385

1987 Takahashi S.: Low-loss fluoride fiber for mid-infrared optical communication. Conf. Opt. Fiber Commun. (OFC'87), WH1

1986   Takato N., Yasu M., Kawachi M.: Low-loss high-silica single-mode channel waveguides. Electron. Lett. **22**, 321–322

1975a Tamir T. (ed.): Integrated Optics, Topics in Applied Physics, Vol. 7 (Springer, Berlin, Heidelberg, New York)

1975b Tamir T.: "Beam and waveguide couplers", in Integrated Optics, ed. by T.Tamir, Topics in Applied Physics, Vol. 7, (Springer, Berlin, Heidelberg, New York), pp. 83–137

1986   Tamir T., Blok H.: Preface to the Feature Issue on Propagation and Scattering of Beam Fields. J. Opt. Soc. Am. A **3**, 462–464

1985   Tamura Y., Maeda H., Shikii S., Yokoyama B.: Single-mode fiber WDM in the 1.2/1.3 micron wavelength region. 11th Europ. Conf. Opt. Commun. 1985, 579–582

1984   Tanaka K.: Analysis and synthesis of Gaussian beam optical systems. Optik **67**, 345–361

1984   Tanaka S., Ono T.: Reduced thermal coefficient of transmission delay time in optical fiber. 10th Europ. Conf. Opt. Commun. 1984, 284–285

1984   Tanaka S., Kitayama Y.: Measurement accuracy of chromatic dispersion by the modulation phase technique. J. Lightwave Tech. LT-**2**, 1040–1044

1977   Tanaka T., Yamada S., Sumi M., Mikoshiba K.: Microbending losses of doubly clad (W-type) optical fibers. Appl. Opt. **16**, 2391–2394

1983   Tanifuji T., Kato Y., Seikai S.: Realization of a low-loss splice for single-mode fibers in the field using an automatic arc-fusion splicing machine. Conf. Opt. Fiber Commun. (OFC'83), 14–15

1985   Tanifuji T.: Modal noise arising from slightly nonuniform efficiency of a detector. J. Opt. Soc. Am. A**2**, 1852–1856

1983   Tardy A., Jureczyczyn M., Jeunhomme L., Carratt M., Hakoun R., Mignien M., Trezeguet J.P.: Automatic single-mode fiber splicing machine without power monitoring. 9th Europ. Conf. Opt. Commun. 1983, 113–116

1978   Tasker G.W., French W.G., Simpson J.R., Kaiser P., Presby H.M.: Low-loss single-mode fibers with different B_2O_3-SiO_2 compositions. Appl. Opt. **17**, 1836–1842

1978   Tateda M.: Single-mode fiber refractive-index profile measurement by reflection method. Appl. Opt. **17**, 475–478

1980   Tateda M., Tanaka S., Sugawara Y.: Thermal characteristics of phase shift in jacketed optical fibers. Appl. Opt. **19**, 770–773

1981   Tateda M., Shibata N., Seikai Sh.: Interferometric method for chromatic dispersion measurement in a single-mode optical fiber. IEEE J. QE-**17**, 404–407

1982   Tatekura K., Yamamoto H., Nunokawa M.: Novel core alignment method for low-loss splicing of single-mode fibers utilizing UV-excited fluorescence of Ge-doped silica core. Electron. Lett. **18**, 712–713

1984   Tatekura K., Nishikawa H., Fujise M., Wakabayashi H.: High accurate automatic measurement equipment for chromatic dispersion making use of the phase-shift technique with LD's. Symp. Opt. Fiber Meas. 1984, NBS spec. publ. **683**, 119–122

1973   Taylor H.F.: Frequency-selective coupling in parallel dielectric waveguides. Opt. Commun. **8**, 421–425

1974   Taylor H.F., Yariv A.: Guided wave optics. Proc. IEEE **62**, 1044–1060

1984   Taylor H.F.: Bending effects in optical fibers. J. Lightwave Tech. LT-**2**, 617–628

1981   Tekippe V.J.: Evanescent wave coupling of optical fibers. Advances in Ceramics **2**, 482–490

1985   Tekippe V.J., Willson W.R.: Single-mode directional couplers. Laser Focus, **21**(5), 132–144

1986   Tewari R., Thyagarajan K.: Analysis of tunable single-mode directional couplers using simple and accurate relations. J. Lightwave Tech. LT-**4**, 386–390

1986   Tewari R., Thyagarajan K., Ghatak A.K.: Novel method for characterization of single-mode fibers and prediction of crossover wavelength and bandpass in nonidentical fiber directional couplers. Electron. Lett. **22**, 792–794

1988   Thevenaz L., Pellaux J., von der Weid J.: All-fiber interferometer for chromatic dispersion measurements. J. Lightwave Tech. LT-**5**, to be published

1985   Thomson-CSF, groupe d'ingénieurs: L'Optique Guideé Monomode et ses Applications. (Masson, Paris)

1985   Thomson J.E.: Precision of the phase shift method for measuring chromatic dispersion in single-mode optical fibers. Proc. SPIE **584**, 223–228

1982 Thyagarajan K., Gupta A., Pal B.P., Ghatak A.K.: An analytical model for studying two-mode fibers: splice loss prediction. Opt. Commun. **42**, 92–96

1985 Thyagarajan K., Tewari R.: Accurate analysis of single-mode graded-index fiber directional couplers. J. Lightwave Tech. LT-**3**, 59–62

1986a Thyagarajan K., Shenoy M.R., Ramadas M.R.: Prism coupling technique: a novel method for measurement of propagation constant and beat length in single-mode fibers. Techn. Dig. Sympos. Opt. Fiber Measur. 1986, NBS spec. publ. **720**, 39–42

1986b Thyagarajan K., Shenoy M.R., Ramadas M.R.: Prism coupling technique: a method for measurement of propagation constant and beat length in single-mode fibers. Electron. Lett. **22**, 832–833

1980 Thylen L.: Bend radiation of optical fibers. Opt. Quantum Electron. **12**, 1–7

1965 Tien P.K., Gordon J.P., Whinnery J.R.: Focusing of a light beam of Gaussian field distribution in continuous and periodic lens-like media. Proc. IEEE **53**, 129–136

1969 Tien P.K., Ulrich R., Martin R.J.: Modes of propagating light waves in thin deposited semiconductor films. Appl. Phys. Lett. **14**, 291–294

1970 Tien P.K., Ulrich R.: Theory of prism-film coupler and thin-film light guides. J. Opt. Soc. Am. **60**, 1325–1337

1971 Tien P.K.: Light waves in thin films and integrated optics. Appl. Opt. **10**, 2395–2413

1977 Timmermann C.C.: Asymptotic expression for the far-field distribution of step-index single-mode fibers. Appl. Opt. **16**, 1793–1794

1978 Tjaden D.L.A.: First-order correction to "weak-guidance" approximation in fiber optics theory. Philips J. Res. **33**, 103–112

1984 Tokuda M., Nakazawa M., Horiguchi T.: "Optical time domain reflectometer at a 1.3–1.7 μm wavelength region" in Japan Annual Review in Electronics, Computers, & Telecommunications, Vol. 11, Optical Fibers and Devices, ed. by Y. Suematsu (Ohm, Tokyo, Osaka, Kyoto; North-Holland, Amsterdam, New York, Oxford), pp. 179–194

1982 Tomita A.: Interferometric technique for measuring small gaps in single- and multimode fiber connectors. Appl. Opt. **21**, 2655–2656

1982 Tomita A., Glodis P.F., Kalish D., Kaiser P.: Characterization of the bend sensitivity of single-mode fibers using the basket-weave test. Techn. Dig. Sympos. Opt. Fiber Measur. 1982, NBS Publ. **641**, 89–92

1983 Tomita A., Duff D.G., Cheung N.K.: Mode partition noise caused by wavelength-dependent attenuation in lightwave systems. Electron. Lett. **19**, 1079–1080

1984 Tomita A., Cohen L.G.: Effective cutoff and modal noise in single-mode fiber systems. 10th Europ. Conf. Opt. Commun. 1984, 212–213

1984 Tomita A., Lemaire P.J.: Observation of a short wavelength loss edge caused by hydrogen in optical fibers. 10th Europ. Conf. Opt. Commun. 1984, postdeadline paper

1985 Tomita A., Lemaire P.J.: Hydrogen-induced loss increases in germanium-doped single-mode optical fibers: long term predictions. Electron. Lett. **21**, 71–72

1985 Tomita A. Cohen L.G.: Leaky-mode loss of the second propagating mode in single-mode fibers with index well profiles. Appl. Opt. 1704–1707

1977 Tomlinson W.J.: Wavelength multiplexing in multimode optical fibers. Appl. Opt. **16**, 2180–2194

1980 Tomlinson W.J.: Applications of GRIN-rod lenses in optical fiber communication systems. Appl. Opt. **19**, 1127–1138

1981 Tran D.C., Koo K.P.: Stabilizing single-mode fiber couplers by using gel glass. Electron. Lett. **17**, 187–188

1981 Tran D.C., Koo K.P., Sheem S.K.: Single-mode fiber directional couplers fabricated by twist-etching techniques (stabilization). IEEE J. QE-**17**, 988–991

1985 Travis A.R.L., Carroll J.E.: Possible fused fiber in-phase/quadrature measuring multiport. Electron. Lett. **21**, 954–955

1985 Trepat-Marti J., Goure J.P.: Thermal effect on the refracted near-field optical fiber profiling technique. Proc. SPIE **584**, 243–248

1985 Trommer G.: Unterdrückung von HE_{21}-, E_{01}-, H_{01}-Moden in Glasfasern mit gut geführtem HE_{11}-Mode bei normierten Frequenzen von $V > 2.4$. ntz-Archiv **7**, 215–218

1986 Truesdale C.M., Nolan D.A.: Core/clad mode coupling in a new multiple-index waveguide structure. 12th Europ. Conf. Opt. Commun. 1986, 181–184

1986 Truesdale C.M., Nolan D.A., Keck D.B., Berkey G.E.: Multiple-index waveguide coupler. Conf. Opt. Fiber Commun. (OFC'86), 60–61

1986   Tsubokawa M., Shibata N., Ohashi M., Seikai S.: Length dependence of LP_{11} mode loss in the quasi-single-mode operating region. Opt. Lett. **11**, 327–329

1977   Tsuchiya H., Nakagome H., Shimizu N., Ohara S.: Double eccentric connectors for optical fibers. Appl. Opt. **16**, 1323–1331

1979   Tsuchiya H., Shimizu N., Izawa T.: Single mode fiber connectors. Rev. Electr. Commun. Lab. NTT, Japan **27**, 543–554

1983   Tur M., Moslehi B.: Laser phase noise effects in fiber-optic signal processors with recirculating loops. Opt. Lett. **8**, 229–231

1985   Tur M., Moslehi B., Goodman J.W.: Theory of laser phase noise in recirculating fiber-optic delay lines. J. Lightwave Tech. LT-3, 20–31

1970   Tynes A.R.: Integrating cube scattering detector. Appl. Opt. **9**, 2706–2710

1971   Tynes A.R., Pearson A.D., Bisbee D.L.: Loss mechanisms and measurements in clad glass fibers and bulk glass. J. Opt. Soc. Am. **61**, 143–153

1977   Tynes A.R., Derosier R.M.: Low-loss splices for single-mode fibers. Electron. Lett. **13**, 673–674

1978   Tynes A.R., Derosier R.M., French W.G.: Measurement of single-mode fiber core-cladding concentricity. Electron. Lett. **14**, 113–115

1979   Tynes A.R., Derosier R.M., French W.G.: Low V-number optical fibers: Secondary maxima in the far-field radiation pattern. J. Opt. Soc. Am. **69**, 1587–1596

1981   Tzoar N., Pascone R.: Radiation loss in tapered waveguides. J. Opt. Soc. Am. **71**, 1107–1114

1973   Uchida N., Niizeki N.: Acoustooptic deflection materials and techniques. Proc. IEEE **61**, 1073–1092

1986   Uchida N., Uesugi N.: Infrared optical loss increase in silica fibers due to hydrogen. J. Lightwave Tech. LT-4, 1132–1138

1970   Uchida T., Furukawa M., Kitano I., Koizumi K., Matsumura H.: Optical characteristics of a light-focusing fiber guide and its applications. IEEE J. QE-6, 606–612

1978   Uchida T., Sugimoto S.: Micro-optic devices for optical communications. 4th Europ. Conf. Opt. Commun. 1978, 374–382

1982   Uchida T., Kobayashi K.: "Micro-optic circuitry for fiber communication", in Japan Annual Review in Electronics, Computers, & Telecommunications, Vol. 1, Optical Fibers & Devices, ed. by Y. Suematsu (Ohm, Tokyo, Osaka, Kyoto; North-Holland, Amsterdam, New York, Oxford), pp. 172–189

1976   Ueno Y., Shimizu M.: Optical fiber fault location method. Appl. Opt. **15**, 1385–1388

1984   Uenoya T.: SM optical fiber splicing technique. Japan Telecommunic. Review **26**, 245–249

1985   Uenoya T., Murakami Y., Yamanouchi I.: A new field-use single-mode fiber splicer using a direct core observation technique and its application. Proc. SPIE **574**, 152–157

1981   Uesugi N., Ikeda M, Sasaki Y.: Maximum single frequency input power in a long optical fiber determined by stimulated Brillouin scattering. Electron. Lett. **17**, 379–380

1983   Uesugi N., Murakami Y., Tanaka C., Ishida Y., Mitsunaga Y., Negishi Y., Uchida N.: Infra-red optical loss increase for silica fiber cable filled with water. Electron. Lett. **19**, 762–764

1986   Uesugi N., Horiguchi T., Nakazawa M., Murakami Y.: Optical fiber cable measurement in the field. IEEE J. Select. Areas Commun. SAC-4, 732–736

1985   Uji T., Hayashi J.: High-power single-mode optical-fiber coupling to InGaAsP 1.3 μm mesa-structure surface-emitting LED's. Electron. Lett. **21**, 418–419

1985   Uji T., Shikada M., Fujita S., Hayashi J., Isoda Y.: 565 Mb/s-5 km and 140 Mb/s-25 km single-mode fiber transmission using 1.3 μm mesa-structure surface-emitting LEDs. 11th Europ. Conf. Opt. Commun. 1985, post-deadline papers, 57–60

1985a   Ulbricht L.W., Teare M.J., Olshansky R., Lauer R.B.: 140 Mb/s transmission over 30 km of single-mode fiber using an LED source. 11th Europ. Conf. Opt. Commun. 1985, 819–822

1985b   Ulbricht L.W., Teare M.J., Olshansky R., Lauer R.B.: Loss-limited transmission at 140 Mb/s over 30 km of single-mode fiber using a 1.3 μm LED. Electron. Lett. **21**, 860–861

1970   Ulrich R.: Theory of the prism-film coupler by plane-wave analysis. J. Opt. Soc. Am. **60**, 1337–1350

1971 Ulrich R.: Optimum excitation of optical surface waves. J. Opt. Soc. Am. **61**, 1467–1477

1977 Ulrich R.: Representation of codirectional coupled waves. Opt. Lett. **1**, 109–111

1980 Ulrich R., Rashleigh S.C.: Beam-to-fiber coupling with low standing wave ratio. Appl. Opt. **19**, 2453–2456

1976 Unger H.-G.: Optische Nachrichtentechnik. (Elitera, Berlin)

1977a Unger H.-G.: Planar Optical Waveguides and Fibres. (Clarendon Press, Oxford)

1977b Unger H.-G.: Optical pulse distortion in glass fibres at the wavelength of minimum dispersion. AEÜ, Arch. für Elektron. und Übertragungstech.: Electron. and Commun. **31**, 518–519

1984 Unger H.-G.: Optische Nachrichtentechnik, Teil I: Optische Wellenleiter. (Hüthig, Heidelberg)

1985 Uttam D., Culshaw B.: Precision time domain reflectometry in optical fiber systems using a frequency modulated continuous wave ranging technique. J. Lightwave Tech. **LT-3**, 971–977

1987 van Blitterswijk W., Smit K.M.: Index profile measurement of asymmetrical elliptical preforms or fibers. Fiber and Integrated Optics **6**, 55–62

1985 van Bochove A.C., Jacobs J.F., Nijnuis H.T.: Improved Raman-fiber pulse delay measurement system. Electron. Lett. **21**, 282–283

1970 Vanclooster R., Phariseau P.: The coupling of two parallel dielectric fibers. I. Basic equations. Physica **47**, 485–500

1983 van der Veeken, R.W.C.: Analysis of the power-coupling efficiency of lens connectors for single-mode fibers; derivation of design rules. Technische Hogeschool Delft, Department of Electrical Engineering, Laborat. Electromagnetic Research, Report No. 1983–02, 1–108

1984 van Leeuwen K.A.H., Nijnuis H.T.: Measurement of higher-order mode attenuation in single-mode fibers: effective cutoff wavelength. Opt. Lett. **9**, 252–254

1984 Varnham M.P., Payne D.N., Love J.D.: Fundamental limits to the transmission of linearly polarized light by birefringent optical fibers. Electron. Lett. **20**, 55–56

1977 Vassallo C.: Linear power responses of an optical fiber. IEEE Trans. **MTT-25**, 572–576

1981 Vassallo C.: Orthogonality and amplitude spectrum of radiation modes along open-boundary waveguides: comment. J. Opt. Soc. Am. **71**, 1282.

1983 Vassallo C.: Extension de la representation scalaire du champ aux fibres optiques à gaine finie. Opt. Quantum Electron. **15**, 349–353

1985 Vassallo C.: Scalar-field theory and 2-D ray theory for bent single-mode weakly guiding optical fibers J. Lightwave Tech. **LT-3**, 416–423

1983 Vatoux S., Combemale Y., Enard A., Arnoux J.M., Papuchon M.: Le couplage de l'énergie lumineuse entre guides optiques monomodes. Revue Techn. Thomson-CSF **15**, 663–710

1984a Vella P.J., Garel-Jones P.M, Lowe R.S.: Measuring the chromatic dispersion of single-mode fibers: a compliance test method. Conf. Opt. Fiber Commun. (OFC'84), 82–83

1984b Vella P.J., Garel-Jones P.M., Lowe R.S: Measurement of chromatic dispersion of long spans of single-mode fiber: a factory and field test method. Electron. Lett. **20**, 167–168

1981 Villarruel C.A., Moeller R.P.: Fused single-mode fiber access couplers. Electron. Lett. **17**, 243–244

1982 Villarruel C.A., Moeller R.P.: Optimized single-mode tapped tee data bus. Conf. Opt. Fiber Commun. (OFC'82), 24–25

1983 Villarruel C.A., Abebe M., Burns W.K.: Polarization preserving single-mode-fiber coupler. Electron. Lett. **19**, 17–18

1984 Vodhanel R.S., Ko J.S.: Reflection induced frequency shifts in single-mode laser diodes coupled to optical fibers. Electron. Lett. **20**, 973–974

1977 Vucins V.: Adjustable single-fiber connector with monitor output. 3rd Europ. Conf. Opt. Commun. 1977, 100–102

1982 Wagner R.E., Sandahl C.R.: Interference effects in optical fiber connectors. Appl. Opt. **21**, 1381–1385

1982 Wagner R.E., Tomlinson W.J.: Coupling efficiency of optics in single-mode fiber components. Appl. Opt. **21**, 2671–2688

1986 Walker S.S.: Rapid modeling and estimation of total spectral loss in optical fibers. J. Lightwave Tech. LT-**4**, 1125–1131

1982a Wang C.C., Villarruel C.A., Burns W.K.: Comparison of cutoff wavelength measurements by transmission and refraction for single-mode waveguides. Conf. Opt. Fiber Commun. (OFC'82), 64–65

1982b Wang C.C., Villarruel C.A., Burns W.K.: Comparison of cutoff wavelength measurements for single-mode waveguides. Techn. Dig. Sympos. Opt. Fiber Measur. 1982, NBS Publ. **641**, 97–100

1983 Wang C.C., Villarruel C.A., Burns W.K.: Comparison of cutoff-wavelength measurements for single-mode waveguides. Appl. Opt. **22**, 985–990

1985 Wang C.C., Burns W.K., Villarruel C.A.: 9 × 9 single-mode fiber-optic star couplers. Opt. Lett. **10**, 49–51

1986 Wang S.J.: Impact of reflection on the performance of single-mode fiber optic systems. Techn. Dig. Sympos. Opt. Fiber Measur. 1986, NBS spec. publ. **720**, 141–144

1981 Watanabe R., Fujii Y., Nosu K., Minowa J.: Optical multi/demultiplexers for single-mode fiber transmission. IEEE J. QE-**17**, 974–981

1983 Watanabe R.: "Optical multiplexer and demultiplexer", in Japan Annual Review in Electronics, Computers, & Telecommunications, Vol. 5, Optical Fibers & Devices, ed. by Y. Suematsu (Ohm, Tokyo, Osaka, Kyoto; North-Holland, Amsterdam, New York, Oxford), pp. 96–113

1979 Watkins L.S.: Laser beam refraction traversely through a graded-index preform to determine refractive index ratio and gradient profile. Appl. Opt. **18**, 2214–2222

1986 Wei L., Saravanos C., Lowe R.S.: Practical upper limits to cutoff wavelength for different single-mode fiber designs. Techn. Dig. Sympos. Opt. Fiber Measur. 1986, NBS spec. publ. **720**, 121–124

1987 Wei L., Lowe R.S., Saravanos C.: Measured dependence of modal noise on cutoff wavelength for single-mode fibers in a 1.2 Gb/s system. Conf. Opt. Fiber Commun. (OFC'87), THD5

1974 Weidel E.: Light coupling from a junction laser into a monomode fiber with a glass cylindrical lens on the fiber end. Opt. Commun. **12**, 93–97

1975 Weidel E.: New coupling method for GaAs laser fiber coupling. Electron. Lett. **11**, 436–437

1976 Weidel E.: Light coupling problems for GaAs laser – multimode fiber coupling. Opt. Quantum Electron. **8**, 301–307

1979 Weidel E., Gruchmann D.: Tee-coupler for single-mode fibers. Electron. Lett. **15**, 737–738

1978 Wells W.H.: Crosstalk in a bidirectional optical fiber. Fiber a. Integr. Optics **1**, 243–286

1979 Wemple S.H.: Material dispersion in optical fibers. Appl. Opt. **18**, 31–35

1981 Wenke G., Elze G.: Investigation of optical feedback effects on laser diodes in broadband optical transmission systems. J. Opt. Commun. **2**, 128–133

1983 Wenke G., Zhu Y.: Comparison of efficiency and feedback characteristics of techniques for coupling semiconductor lasers to single-mode fiber. Appl. Opt. **22**, 3837–3844

1985 Whalen M.S., Wood T.H.: Effectively nonreciprocal evanescent-wave optical-fiber directional coupler. Electron. Lett. **21**, 175–176

1985 Whalen M.S., Walker K.L.: In-line optical-fiber filter for wavelength multiplexing. Electron. Lett. **21**, 724–725

1986 Whalen M.S., Tennant D.M., Alferness R.C., Koren U., Bosworth R.: Wavelength-tunable single-mode fiber grating reflector. Electron. Lett. **22**, 1307–1308

1977 White I.A., Snyder A.W.: Radiation from dielectric optical waveguides: a comparison of techniques. Appl. Opt. **13**, 1470–1472

1979 White I.A.: Radiation from bends in optical waveguides: the volume-current method. Microwaves, Optics a. Acoustics, **3**, 186–188

1986 White I.A., Ludington P.D.: A novel single-mode fiber splice alignment and loss measurement using local detection. Techn. Dig. Sympos. Opt. Fiber Measur. 1986, NBS spec. publ. **720**, 61–64

1976 White K.I.: A calorimetric method for the measurement of low optical absorption losses in optical communication fibers. Opt. Quantum Electron. **8**, 73–75

1978 White K.I.: The measurement of the refractive index profiles of optical fibers by the refracted near-field technique. 4th Europ. Conf. Opt. Commun. 1978, 146–155

1979 White K.I.: Practical application of the refracted near-field technique for the measurement of optical fiber refractive-index profiles. Opt. Quantum Electron. **11**, 185–196

1981 White K.I., Hornung S., Wright J.V., Nelson B.P., Brierley M.C.: Characterization of single-mode optical fibers. Radio Electron. Eng. **51**, 385–391

1985 White K.I.: Methods of measurements of optical fiber properties. J. Phys. E.: Sci. Instrum. **18**, 813–821

1973 Wijngaard W.: Guided normal modes of two parallel circular dielectric rods. J. Opt. Soc. Am. **63**, 944–950

1985 Wilczewski Fr.: Analyse der Modeneigenschaften in Grundmodenfasern mit beliebigem radialen Brechzahlprofilverlauf. Diplomarbeit Universität Wuppertal 1985

1973 Williams C.S.: Gaussian beam formulas from diffraction theory. Appl. Opt. **12**, 872–876

1984 Williams J.C., May R.G., Althouse E.: High-performance bidirectional optical rotary joint. Conf. Opt. Fiber Commun. (OFC'84), 98–99

1981 Winzer G., Mahlein H.F., Reichelt A.: Single-mode and multimode all-fiber directional couplers for WDM. Appl. Opt. **20**, 3128–3135

1982 Winzer G.: Wavelength division multiplexing – status and trends. 8th Europ. Conf. Opt. Commun. 1982, 305–314

1984 Winzer G.: Wavelength multiplexing components – a review of single-mode devices and their applications. J. Lightwave Tech. LT-**2**, 369–378

1985 Wismeyer A.: Long-distance single-mode fiber transmission. Philips Telecommun. Review **43**(4), 43–49

1984 Wittmann J., Keil R., Auracher F.: Innovative method for laser-fiber alignment. 10th Europ. Conf. Opt. Commun. 1984, 188–189

1987 Wlodarczyk M.T., Seshadri S.R.: Radiation losses in a periodically deformed single-mode fiber. J. Lightwave Tech. LT-**5**, 163–173

1985 Wolffer N.: A method for the evaluation of optical fiber ends using spatial frequencies analysis. Proc. SPIE **584**, 115–119

1984a Woods J.G.: Single-mode fiber optic connectors and splices. Proc. SPIE **479**, 42–47

1984b Woods J.G.: Fiber optic splices. Proc. SPIE **512**, 44–56

1971 Worthington R.: An apparatus for the measurement of mode cut-off wavelengths of optical fiber waveguides. J. Phys. E: Sci. Instr. **4**, 1052–1054

1983a Wright J.V.: Microbending loss in monomode fibers: solution of Petermann's auxiliary function. Electron. Lett. **19**, 1067–1068

1985 Wright J.V.: Variational analysis of fused tapered couplers. Electron. Lett. **21**, 1064–1065

1986 Wright J.V.: Wavelength dependence of fused couplers. Electron. Lett. **22**, 320–321

1983a Wright S., Dalgoutte D.G., Salt S.: Characterization and fault location for single-mode fibers using long wavelength optical time domain reflectometry (OTDR). Proc. SPIE, **425**, 63–67

1983b Wright S., Richards K., Salt S.K., Wallbank E.: High dynamic range coherent reflectometer for fault location in monomode and multimode fibers. 9th Europ. Conf. Opt. Commun. 1983, 177–180

1985 Wright S., Epworth R.E., Smith D.F.: Performance of a coherent optical time domain reflectometer with semiconductor source. Conf. Opt. Fiber Commun. (OFC'85), 34–35

1983 Yabuta T., Yoshizawa N., Ishihara K.: Excess loss of single-mode jacketed optical fiber at low temperature. Appl. Opt. **22**, 2356–2362

1983 Yamabayashi Y., Saruwatari M.: Temporal coherence characteristics of InGaAsP laser diode with single-longitudinal-mode oscillation. Opt. Lett. **8**, 506–508

1980a Yamada J., Machida S., Mukai T., Tsuchiya H., Kimura T.: Long-span single-mode fiber transmission characteristics in long wavelength regions. IEEE J. QE-**16**, 874–884

1980b Yamada J., Murakami Y., Sakai J., Kimura T.: Characteristics of a hemispherical microlens for coupling between a semiconductor laser and single-mode fiber. IEEE J. QE-**16**, 1067–1072

1982 Yamada J.: Simple dispersion measurement of a long-span single-mode fiber using the longitudinal mode spacing of a semiconductor laser. Opt. Quantum Electron. **14**, 183–187

1979 Yamada M., Suematsu Y.: A condition of single longitudinal mode operation in injection lasers with index-guiding structure. IEEE J. QE-**15**, 743–749

1986 Yamada T., Ohsato Y., Yoshinuma M., Tanaka T., Itoh K.: Arc fusion splicer with profile alignment system for high-strength low-loss optical submarine cable. J. Lightwave Tech. LT-4, 1204–1210

1983 Yamagishi T., Fujii K., Kitano I.: Gradient-index rod lens with high N.A.. Appl. Opt. 22, 400–403

1981 Yamamoto A., Kimura T.: Coherent optical fiber transmission systems. IEEE J. QE-17, 919–935

1982 Yamamoto S., Sakaguchi H., Seki N.: Repeater spacing of 280 Mb/s single-mode fiber-optic transmission system using 1.55 μm laser diode source. IEEE J. QE-18, 264–273

1986 Yamamoto S., Sakaguchi H., Iwamoto Y.:Transmission performance impairment due to degradation of optical spectrum in 1.55 μm DFB laser. Electron. Lett. 22, 404–405

1985 Yamanouchi I.: High-performance optical fiber cable coverage throughout Japan. Proc. Intern. Wire & Cable Symp. 1985, 379–384

1985 Yamashita J., Suganuma R., Miyake Y., Takei T.: High-efficiency LD coupler with a truncated Si spherical lens. Conf. Opt. Fiber Commun. (OFC'85), 48–49

1982 Yamauchi R., Miyamoto M., Inada K.: Practical determination of equivalent step-index profiles for single-mode fibers. Electron. Lett. 18, 550–552

1985 Yang R., Unger H.-G.: Microbending loss in single-mode triple-clad fibers. AEÜ, Arch. für Elektron. und Übertragungstech.: Electron. and Commun. 39, 145–147

1971 Yariv A.: Introduction to Optical Electronics. (Holt, Rinehart and Winston, New York)

1973 Yariv A.: Coupled-mode theory for guided-wave optics. IEEE J. QE-9, 919–933

1981 Yata A., Ikuno H.: Cutoff frequencies of W-type fiber with polynomial profile core. Electron. Lett. 17, 9–11

1985 Yataki M.S., Varnham M.P., Payne D.N.: Fabrication and properties of very-long fused-taper couplers. Conf. Opt. Fiber Commun. (OFC'85), 108–110

1985 Yataki M.S., Payne D.N., Varnham M.P.: All-fiber wavelength filters using concatenated fused-taper couplers. Electron. Lett. 248–249

1978 Yeh C., Manshadi F., Casey K.F., Johnston A.: Accuracy of directional coupler theory in fiber or integrated optics applications. J. Opt. Soc. Am. 68, 1079–1083

1979 Yeh C.: Optical waveguide theory. IEEE Trans. Circuits a. Systems CAS-26, 1011–1019

1979a Yeh C., Brown W.P., Szejn R.: Multimode inhomogeneous fiber couplers. Appl. Opt. 18, 489–495

1979b Yeh C., Casperson L., Brown W.P.: Scalar-wave approach for single-mode inhomogeneous fiber problems. Appl. Phys. Lett. 34(7), 460–462

1984 Yoldas B.E., Partlow D.P.: Wide spectrum antireflective coating for fused silica and other glasses. Appl. Opt. 23, 1418–1424

1981 Yoshino T.: Fiber Fabry-Perot interferometers, Conf. Opt. Fiber Commun. (OFC'81), 128–129

1982 Yoshizawa T., Kurokawa T., Nara S.: High-precision moulded connector for single-mode optical fibers. Electron. Lett. 18, 598–599

1980 Young M.: Calibration technique for refracted near-field scanning of optical fibers. Appl. Opt. 19, 2479–2480

1981 Young M.: Optical fiber index profiles by the refracted-ray method (refracted near-field scanning). Appl. Opt. 20, 3415–3422

1982 Young M.: Refracted-ray scanning (refracted near-field scanning) for measuring index profiles of optical fibers. NBS spec. publ. 637, Vol. 1, 91–139

1980 Young W.C., Kaiser P., Cheung N.K., Curtis L., Wagner R.E., Folkes D.M.: A transfer-molded biconic connector with insertion losses below 0.3 dB without index match. 6th Europ. Conf. Opt. Commun. 1980, 310–313

1983 Young W.C., Curtis L.: Low-loss field-installable biconic connectors for single-mode fibers. Conf. Opt. Fiber Commun. (OFC'83), 14–15

1984 Young W.C.: Design considerations for single-mode fiber connectors. Proc. SPIE 479, 30–32

1986a Young W.C., Shah V., Curtis L.: Stress-induced increases in interference effects at fiber joints and a method to eliminate interference effects in optical fiber measurements. Techn. Dig. Sympos. Opt. Fiber Measur. 1986, NBS spec. publ. 720, 69–72

1986b Young W.C., Shah V., Curtis L.: Non-interferometric air-gap connectors for single-mode fiber cable loss measurements. Internat. Wire & Cable Sympos. Proceed. 1986, 224–230

1983   Youngquist R.C., Stoke L.F., Shaw H.J.: Effects of normal mode loss in dielectric
       waveguide directional couplers and interferometers. IEEE J. QE-19, 1888–1896
1984   Youngquist R.C., Brooks J.L., Shaw H.J.: Two-mode fiber modal coupler. Opt. Lett.
       9, 177–179

1974   Zaganiaris A., Bouvy G.: L'absorption dans les matériaux pour fibres optiques. Annales
       de Telecommun. 29, 189–194
1985   Zeidler G.H.: Stand und Weiterentwicklung der Einmoden-LWL-Technik. NTG-Fachbe-
       richte 89, 78–81
1976   Zemon S., Fellows D.: Tunneling leaky modes in a parabolic index fiber. Appl. Opt.
       15, 1936–1941
1985   Zengerle R., Leminger O.G.: Wavelength selective directional coupler made of non-
       identical single-mode fibers. 11th Europ. Conf. Opt. Commun. 1985, 563–566
1986   Zengerle R., Leminger O.G.: Wavelength-selective directional coupler made of non-
       identical single-mode fibers. J. Lightwave Tech. LT-4, 823–827
1987   Zengerle R., Leminger O.G.: Narrowband wavelength-selective directional couplers
       made of dissimilar single-mode fibers. Conf. Opt. Fiber Commun. (OFC'87), THB2
1982   Zolotov E.M., Kazanskii P.G., Prokhorov A.M.: Investigation of a contact between
       a channel waveguide in LiNbO$_3$ and a single-mode fiber. Sov. J. Quantum Electron.
       12(1), 107–110

Subject Index

Springer Series in Optical Sciences

Editorial Board: D.L. MacAdam A.L. Schawlow K. Shimoda A.E. Siegman T. Tamir